Springer Protocols Handbooks

For further volumes:
http://www.springer.com/series/8623

Patch Clamp Techniques

From Beginning to Advanced Protocols

Edited by

Yasunobu Okada

National Institute for Physiological Sciences, National Institutes of Natural Sciences, Okazaki, Japan

Springer

Editor
Yasunobu Okada
National Institute for Physiological Sciences
National Institutes of Natural Sciences
Okazaki, Aichi, Japan
okada@nips.ac.jp

ISSN 1949-2448 ISSN 1949-2456 (eBook)
ISBN 978-4-431-56105-7 ISBN 978-4-431-53993-3 (eBook)
DOI 10.1007/978-4-431-53993-3
Springer Tokyo Dordrecht Heidelberg London New York

Printed on acid-free paper

Springer is part of Springer Science+Business Media (www.springer.com)

Preface

Cell membranes are embedded with numerous proteins. Many of these membrane proteins are channels or transporters that mediate the movement of substances and transmission of signals between intracellular and extracellular compartments. Channels and transporters have recently been demonstrated to serve not only as transporting and transmitting pathways but also as dual- to multifunctional proteins simultaneously serving as regulators for other proteins and/or as sensors for environmental materials. Research on channels and transporters has become increasingly important for postgenomic life sciences. These studies are also of importance for medicine because a disorder or deficiency of some channels or transporters, called channelopathy or transporter disease, has recently been elucidated to be a main cause of a variety of human diseases.

Patch-clamping is an epochal, unique technique that allows researchers to carry out, in a quantitative manner, real-time measurements of ion channel functions at the single-molecule level and of ion transporter functions at the level of a cluster of molecules. Patch-clamp techniques currently comprise the most powerful approach for studying these electrogenic membrane proteins. Many variations of this technique have been developed. They not only can record the activities of ion channels and transporters but also can monitor the processes of exocytosis and endocytosis, detect bioactive environmental substances, and analyze single-cell gene expression. Thus, patch-clamp techniques, which were previously employed only in physiological, biophysical, or pharmacological studies, are now essential and widely used in all the life science fields and related industrial areas.

Each year, novices employ patch-clamp techniques with little guidance from expert electrophysiologists – not only at university and institutional laboratories but also in pharmaceutical, agricultural/agrichemical, and cosmetic companies and in companies manufacturing the patch-clamp apparatus itself. To date, there have been no textbooks that are comprehensible to beginners and practical enough for them to follow. Thus, the first purpose of this book is to give plain, practical descriptions as to what patch-clamping is and how to perform it, especially for beginners.

Since the establishment of conventional patch-clamp techniques – including cell-attached, excised patch, and whole-cell recordings – there have been many technical developments. Newly developed variations include the perforated patch-, slice patch-, blind patch-, in vivo animal patch-, in situ organelle patch-, imaging patch-, and smart patch-clamp. Much effort has gone into developing innovative applications of this technique, such as membrane capacitance measurements, single-cell gene analysis, and biosensor methods. Furthermore, the automated patch-clamp device, which is especially useful for drug screening, has recently become available. Here too, however, there is little written about such new variations and applications developed during the past few years. Hence, the second purpose of this book is to introduce nonbeginners, including expert electrophysiologists, to the wide and exquisite application of patch-clamp techniques and to guide them to expand the repertoire of the techniques in which they are already expert.

It is my hope that this book will allow patch-clamping to become more popular and less craftsman-based, to be employed by many more laboratory and industrial researchers, to have more variations and applications, and to lead to greater proliferation of its unique capabilities among the life sciences and their related fields.

I thank Ms. Tomomi Okayasu for her secretarial and editorial help.

Yasunobu Okada

Contents

Contributors

Tenpei Akita (Chapter 2) • *Department of Cell Physiology, National Institute for Physiological Sciences, National Institutes of Natural Sciences, Okazaki, Aichi, Japan*

Dirk Becker (Chapter 19) • *Molecular Plant Physiology and Biophysics, Julius-von-Sachs Institute for Biosciences, Biocenter, Würzburg University, Würzburg, Germany*

Hidemasa Furue (Chapter 11) • *Department of Information Physiology, National Institute for Physiological Sciences, National Institutes of Natural Sciences, Okazaki, Aichi, Japan; School of Life Science, The Graduate University for Advanced Studies, SOKENDAI, Okazaki, Aichi, Japan*

Kazuharu Furutani (Chapter 23) • *Department of Pharmacology, Graduate School of Medicine, Osaka University, Suita, Osaka, Japan*

Dietmar Geiger (Chapter 19) • *Molecular Plant Physiology and Biophysics, Julius-von-Sachs Institute for Biosciences, Biocenter, Würzburg University, Würzburg, Germany*

Seiji Hayashi (Chapter 21) • *Discovery Research Laboratory, Nippon Shinyaku Co., Ltd, Kyoto, Japan*

Akihiro Hazama (Chapters 17 and 18) • *Fukushima Medical University, Fukushima, Japan*

Rainer Hedrich (Chapter 19) • *Molecular Plant Physiology and Biophysics, Julius-von-Sachs Institute for Biosciences, Biocenter, Würzburg University, Würzburg, Germany; King Saud University KSU, Riyadh, Saudi Arabia*

Tetsuya Hori (Chapter 8) • *Okinawa Institute of Science and Technology, Kunigami, Okinawa, Japan; Department of Neurophysiology, Faculty of Life and Medical Sciences, Doshisha University, Kyotanabe, Kyoto, Japan*

Minoru Horie (Chapter 15) • *Department of Cardiovascular and Respiratory Medicine, Shiga University of Medical Sciences, Otsu, Shiga, Japan*

Keiji Imoto (Chapter 7) • *Department of Information Physiology, National Institute for Physiological Sciences, National Institutes of Natural Sciences, Okazaki, Aichi, Japan*

Tadashi Isa (Chapter 7) • *Department of Developmental Physiology, National Institute for Physiological Sciences, National Institutes of Natural Sciences, Okazaki, Aichi, Japan*

Hitoshi Ishibashi (Chapter 4) • *Department of Developmental Physiology, National Institute for Physiological Sciences, National Institutes of Natural Sciences, Okazaki, Aichi, Japan*

Andrew F. James (Chapter 27) • *School of Physiology and Pharmacology, Medical Sciences Building, University of Bristol, Bristol, UK*

Yasuo Kawaguchi (Chapter 7) • *Department of Cerebral Research, National Institute for Physiological Sciences, National Institutes of Natural Sciences, Okazaki, Aichi, Japan*
Kazuo Kitamura (Chapter 12) • *Department of Neurophysiology, Graduate School of Medicine, The University of Tokyo, Tokyo, Japan*
Yuri E. Korchev (Chapter 25) • *Division of Medicine, Imperial College London, Medical Research Council Clinical Science Centre, Faculty of Medicine, London, UK*
Fumio Kukita (Chapter 1) • *Department of Cell Physiology, National Institute for Physiological Sciences, National Institutes of Natural Sciences, Okazaki, Aichi, Japan*
Yoshihisa Kurachi (Chapter 23) • *Department of Pharmacology, Graduate School of Medicine, Osaka University, Suita, Osaka, Japan*
Toshiya Manabe (Chapter 9) • *Division of Neuronal Network, Department of Basic Medical Sciences, Institute of Medical Science, The University of Tokyo, Tokyo, Japan*
Irene Marten (Chapter 19) • *Molecular Plant Physiology and Biophysics, Julius-von-Sachs Institute for Biosciences, Biocenter, Würzburg University, Würzburg, Germany*
Yoshio Maruyama (Chapters 17 and 18) • *Department of Physiology, Tohoku University Graduate School of Medicine, Sendai, Miyagi, Japan*
Satoshi Matsuoka (Chapter 14) • *Department of Physiology and Biophysics, Graduate School of Medicine, Kyoto University, Kyoto, Japan; Center for Innovation in Immunoregulative Technology and Therapeutics, Graduate School of Medicine, Kyoto University, Kyoto, Japan*
Andrew J. Moorhouse (Chapter 4) • *Department of Physiology, School of Medical Sciences, University of New South Wales, Sydney, Australia*
Shigeru Morishima (Chapters 27 and 28) • *Division of Pharmacology, Department of Biochemistry and Bioinformative Sciences, School of Medicine, University of Fukui, Faculty of Medical Sciences, Fukui, Japan*
Junichi Nabekura (Chapter 4) • *Department of Developmental Physiology, National Institute for Physiological Sciences, National Institutes of Natural Sciences, Okazaki, Aichi, Japan*
Yukihiro Nakamura (Chapter 8) • *Okinawa Institute of Science and Technology, Kunigami, Okinawa, Japan; Department of Neurophysiology, Faculty of Life and Medical Sciences, Doshisha University, Kyotanabe, Kyoto, Japan*
Mami Noda (Chapter 13) • *Laboratory of Pathophysiology, Graduate School of Pharmaceutical Sciences, Kyushu University, Fukuoka, Japan*
Masahiro Ohara (Chapter 2) • *Section of Electron Microscopy, National Institute for Physiological Sciences, National Institutes of Natural Sciences, Okazaki, Aichi, Japan*
Harunori Ohmori (Chapter 6) • *Department of Physiology and Neurobiology, Faculty of Medicine, Kyoto University, Kyoto, Japan*
Shigetoshi Oiki (Chapters 1 and 16) • *Department of Molecular Physiology and Biophysics, School of Medicine, University of Fukui Faculty of Medical Sciences, Fukui, Japan*

Yasunobu Okada (Chapters 2, 25, and 26) • *National Institute for Physiological Sciences, National Institutes of Natural Sciences, Okazaki, Aichi, Japan*
M. Rob G. Roelfsema (Chapter 19) • *Molecular Plant Physiology and Biophysics, Julius-von-Sachs Institute for Biosciences, Biocenter, Würzburg University, Würzburg, Germany*
Ravshan Z. Sabirov (Chapters 25, 26, and 28) • *Laboratory of Molecular Physiology, Institute of Physiology and Biophysics, Uzbekistan Academy of Sciences, Tashkent, Uzbekistan; Department of Biophysics, National University, Tashkent, Uzbekistan*
Takeshi Sakaba (Chapter 17) • *Graduate School of Brain Science, Doshisha University, Kizugawa, Kyoto, Japan*
Kohei Sawada (Chapter 20) • *Global CV Assessment, BA-CFU, Eisai Co. Ltd, Tsukuba, Ibaraki, Japan*
Masahiro Sokabe (Chapter 5) • *Department of Physiology, Nagoya University Graduate School of Medicine, Nagoya, Aichi, Japan*
Hiroto Takahashi (Chapter 10) • *Howard Hughes Medical Institute, Department of Biological Sciences, Columbia University, New York, USA*
Tomoyuki Takahashi (Chapter 8) • *Okinawa Institute of Science and Technology, Kunigami, Okinawa, Japan; Department of Neurophysiology, Faculty of Life and Medical Sciences, Doshisha University, Kyotanabe, Kyoto, Japan*
Ayako Takeuchi (Chapter 14) • *Department of Physiology and Biophysics, Graduate School of Medicine, Kyoto University, Kyoto, Japan*
Makoto Tominaga (Chapters 21 and 22) • *Division of Cell Signaling, Okazaki Institute for Integrative Bioscience (National Institute for Physiological Sciences), National Institutes of Natural Sciences, Okazaki, Aichi, Japan; Department of Physiological Sciences, The Graduate University for Advanced Studies, Okazaki, Aichi, Japan*
Hiroshi Tsubokawa (Chapter 10) • *Faculty of Health Science, Tohoku Fukushi University, Sendai, Miyagi, Japan*
Kunitoshi Uchida (Chapter 22) • *Division of Cell Signaling, Okazaki Institute for Integrative Bioscience (National Institute for Physiological Sciences), National Institutes of Natural Sciences, Okazaki, Aichi, Japan*
Akihiro Yamanaka (Chapter 24) • *Division of Cell Signaling, Okazaki Institute for Integrative Bioscience (National Institute for Physiological Sciences), National Institutes of Natural Sciences, Okazaki, Aichi, Japan*
Takayuki Yamashita (Chapter 8) • *Okinawa Institute of Science and Technology, Kunigami, Okinawa, Japan; Laboratory of Sensory Processing, Brain Mind Institute, Faculty of Life Science, Ecole Polytechnique Federale de Lausanne (EPFL), Lausanne, Switzerland*
Hiromu Yawo (Chapter 3) • *Department of Developmental Biology and Neuroscience, Tohoku University Graduate School of Life Sciences, Sendai, Miyagi, Japan*
Takashi Yoshinaga (Chapter 20) • *Global CV Assessment, BA-CFU, Eisai Co. Ltd, Tsukuba, Ibaraki, Japan*

Chapter 1

Prologue: The Ion Channel

Fumio Kukita and Shigetoshi Oiki

Abstract

A historical overview describes how classic electrophysiological techniques, such as membrane potential measurement, seal formation, and the voltage clamp, have evolved into modern patch-clamp techniques. We show that old ideas from seminal papers on ion channels (i.e., gating, inactivation, ion permeation, ion selectivity) have remained as valid concepts for understanding the molecular properties of ion channels. With the currently available three-dimensional crystal structures of channel proteins, in combination with patch-clamping, novel experimental approaches that focus on the dynamic behavior of channel molecules are undergoing.

1.1. Introduction

The ion channel plays a fundamental role in cell signaling (1, 2). Various types of stimuli (e.g., ligand binding, membrane stretching, changes in the membrane potential) open the channel gate, and ions permeate through the pore (the ion permeation pathway). Unlike other signal transduction molecules, the output of the channel is ionic or electrical current across the membrane. The current spreads locally from the source of the open channels. By what means is this current utilized to generate messages? This question is fundamental because there is no current-sensing molecule. In the quest for insight into this fundamental issue, the ion channel mechanisms underlying signal transduction and the proper electrophysiological method for examining channel function have been brought to light.

We start with a historical overview of the method for detecting electrical signals from cells and describe how important advances have led to recent methods such as patch-clamping. In the following

Yasunobu Okada (ed.), *Patch Clamp Techniques: From Beginning to Advanced Protocols*, Springer Protocols Handbooks, DOI 10.1007/978-4-431-53993-3_1, © Springer 2012

sections, the molecular features of ion channels revealed by the electrophysiological studies are described. These characteristics are then linked to the structure of channel proteins.

1.2. Evolution of Electrophysiological Methods from Classic Voltage-Clamping to Patch-Clamping

The ion channel performs its function on the cell membrane by permeating ions across the membrane. This ionic flow can be detected as electrical current in electrophysiological measurements. The degree of activated channels and the number of active channels reflect the amplitude of the channel current. Thus, the behavior of channel molecules can be traced by continuously recording the channel current. There are technical tricks to measure the channel current, and an overview of the classic methods that forerunners in the field have developed can help to familiarize the reader with the present techniques and suggest insights into developing new techniques for future trials.

1.2.1. Measurement of the Membrane Potential: From Giant Cells to Small Cells

The cell membrane is an integrated continuous sheet wrapping the entire cell so its contents do not leak. An electrical potential is generated across the membrane that is stable in the steady state. The membrane potential is subject to rapid change, however, upon stimulation. Electrophysiological methods have been used to understand the origin of the membrane potential and its changes. For measuring membrane potential, two electrodes must be placed across the membrane. One electrode is placed outside the cell and one inside. In this regard, a key question was how to place the electrode inside the cell without breaking the membrane.

The membrane potential was first measured successfully in giant cells such as the squid giant axon, which has a diameter of 400–800 μm (3, 4). Early pioneers in the field found a way to make the intracellular space accessible. Squid giant axons were cut at the end, and the cut end was dried (air gap) (Fig. 1.1a). A wire electrode was inserted into the cytoplasm of the axon from the cut end in the axial direction without damaging the cell membrane. Placing another electrode in the extracellular space, the potential difference between the extracellular electrode and the intracellular axial electrode was measured as the membrane potential across the membrane. The cytoplasmic and extracellular spaces were electrically insulated by the dried cut end, so there was no leakage between these spaces. The potential difference between the two electrodes provides an accurate measurement of the membrane potential.

In the case of myelinated nerve fibers, because of the difficulty of inserting the electrode into an axon of small diameter, gap methods (sucrose gap, oil gap) were used to insulate the intact part of the nerve and the cut end (Fig. 1.1b). The cut end is bathed in

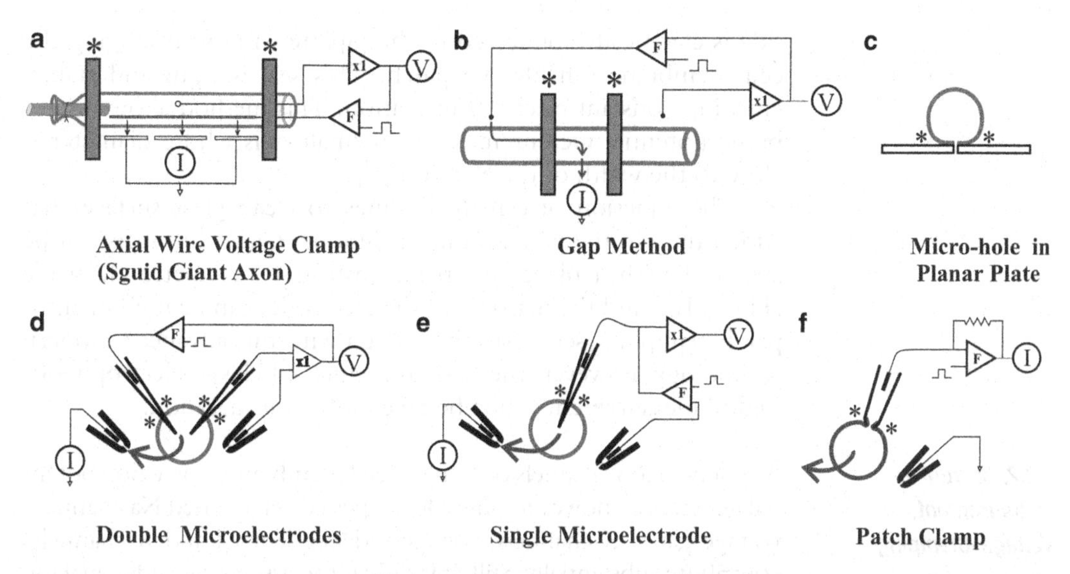

Fig. 1.1. Evolution of methods for the seal formation and the voltage clamp. Voltage-clamping of the membrane potential with electrodes sealed from the outside of the cell. The circuit diagram consists of a voltage measuring pre-amplifier (×1), a current measuring amplifier (I) and a feed-back amplifier (F). Upon disconnecting the current electrode from the output of a feed-back amplifier, the circuit is equivalent to a simple voltage-measuring circuit. In this case, the current stimulus is applied from a standard stimulator. *Asterisks* show the sealed regions.

electrolyte solution, to which the cytoplasm is electrically linked through the cut end. Thus, the membrane potential was measured between two compartments.

An alternative method has been developed in which an electrode with a fine tip impales the cell membrane (microelectrode; Fig. 1.1d,e). This method definitely breaks the membrane, but three modifications were applied to mitigate the damage. First, an insulated electrode was used. If an electrode made of conductive material is used, penetration immediately leads to an electrical shunt between extracellular and intracellular spaces. Second, an electrode with a very fine tip was used, which minimizes cell damage and leakage. Third, glass was used as the insulating material, which fits well with the edge of the broken membrane and thus minimizes leakage. These three modifications minimized the leakage of current at the impaled position. The "seal," which electrically separates the intracellular side from the outside of the cell, has been a special concern of electrophysiologists.

Neher and Sakmann (5) took a great leap forward in developing a method for the practical realization of the seal. Rather than making the pipette tip as fine as possible, they made a large-bore pipette and a soft-landing for the tip on the surface of the membrane (Fig. 1.1f). Neher found a method for attaining high seal resistance by applying negative pressure inside the pipette (6, 7). After forming the giga-ohm seal, the membrane patch at the pipette tip is broken by applying high negative pressure; thus, the cytoplasmic side is electrically connected to the pipette without damaging the cell membrane outside the patch. This seal is tight and stable, ensuring constant electrical insulation. This method allows membrane potential measurement from small cells whose diameter is close to the width of the electrode tip.

The adhesion of cell membranes to clean glass surfaces has wide utility in the patch-clamp technique. Giga-ohm seals can be achieved with a planar electrode instead of a pipette electrode (Fig. 1.1c), and the matrices of the electrodes can be used in auto-patch-clamping (see Chap. 24). Even with this advanced modern technology, however, the basic principles of classic electrophysiological measurements must be taken into consideration.

1.2.2. Current Measurement: Voltage-Clamping

A variety of ion channels exist in the cell membrane. For action potential generation, however, only a few types (voltage-gated Na channels, voltage-gated K channels, voltage-independent Cl and K channels) contribute substantially. Still, it is difficult to assess how individual ion channels function during generation of the action potential. Analysis of transient membrane current during action potential generation is especially complex (i.e., a derivative form of the potential) (Fig. 1.2a). To understand the mechanism underlying the action potential, an equivalent circuit for a membrane is depicted (Fig. 1.2c,d). There are multiple independent conductive channel pathways, which are represented as resistance (the inverse of channel conductance) (Fig. 1.2d). Because a specific ion permeates the conductive pathway driven by the difference in the electrochemical potential across the membrane ($V - E_X$, where V is the membrane potential and E_X the electrochemical potential for an X ion), a battery is placed in series with the resistance. In parallel to the resistance, electrical capacitance, representing the membrane of a thin insulator, is placed.

In this representative circuit, the membrane current is expressed as the sum of the current flowing through each instance of resistance and capacitance. The resistive current follows Ohm's law, but the capacitative current is related to the derivative of the membrane potential. Thus, if the membrane potential is kept constant, the membrane current represents the sum of the ionic currents without contribution from the capacitative current. In addition, the electromotive force of the battery is constant.

A constant voltage is attained by means of negative feedback circuitry (voltage clamp) (Fig. 1.1), in which the deviation of the membrane potential from a preset value is automatically corrected, and the current for correcting the voltage deviation is measured (9). For example, if a small current flows through the channels under the voltage clamp, the membrane potential is perturbed, and correcting the current to recover the membrane potential to the preset value is achieved with the operational amplifier. The membrane potential is a fixed constant even in the presence of the channel currents. As the channel current changes with time, the generated

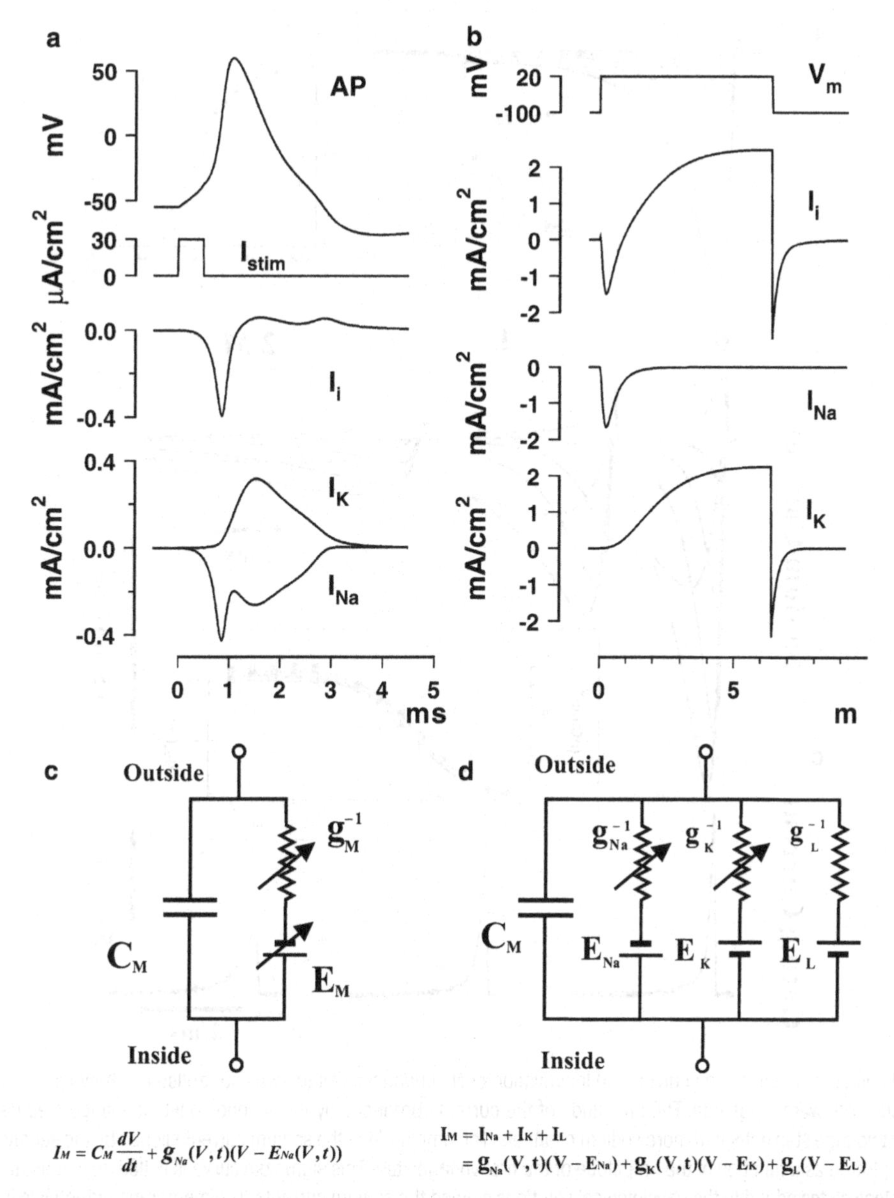

Fig. 1.2. Action potentials and membrane currents. (**a**) An action potential (AP) is generated with a current stimulus (I_{stim}). In this case, the current (I_i) is composed of a Na⁺ current (I_{Na}) and K⁺ current (I_K). (**b**) Membrane currents under the voltage clamp (I_i) are measured separately as a Na⁺ current (I_{Na}) and K⁺ current (I_K). (**c**) An equivalent circuit for the membrane potential is composed of a time-dependent conductance and time-dependent electromotive force in parallel to the membrane capacitance. (**d**) An equivalent circuit by Hodgkin and Huxley is composed of three fixed value electromotive forces with time-dependent internal conductances in series with them (adapted from Fig. 4 in Chap. 1 "Mechanism of Electrical Signal Generation") (8).

current follows such that the amplitude of the generated current is the same as that of the ionic current through the channels.

Ionic currents were first measured by voltage-clamping the squid giant axon (Fig. 1.1a) (9). In this process, when the membrane

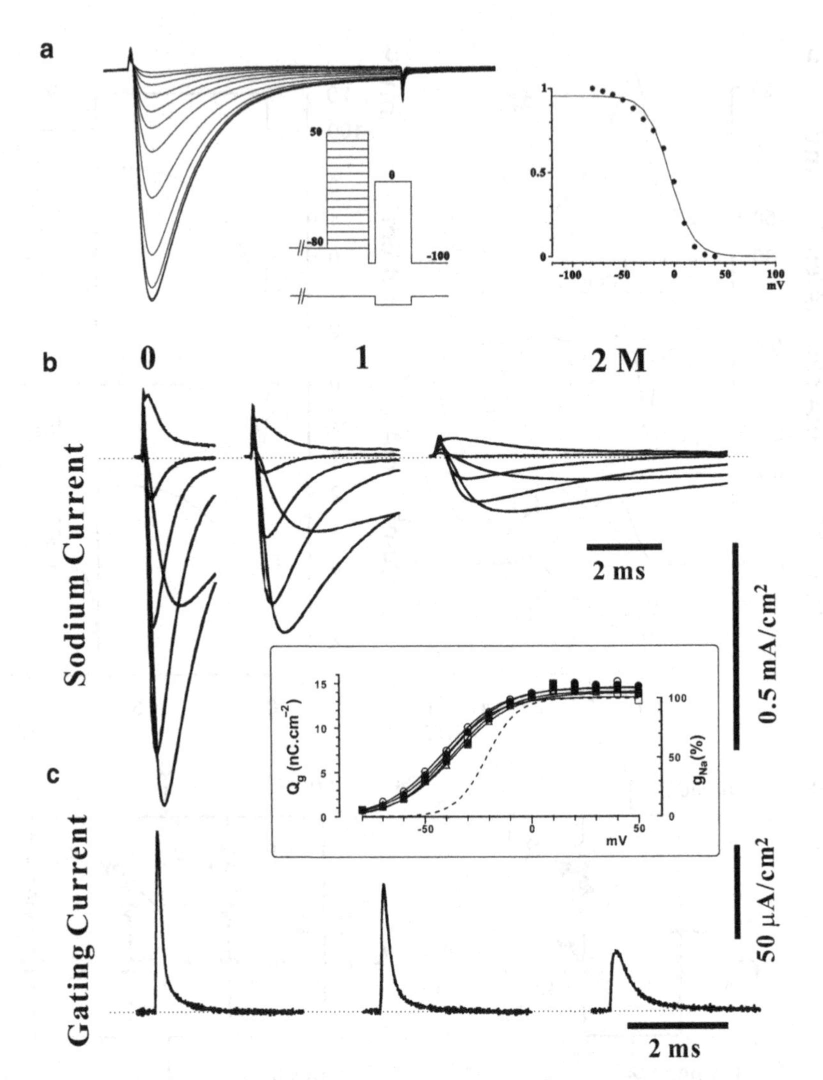

Fig. 1.3. Ionic current and gating current. (**a**) Inactivation of Na channels. Voltage was controlled as shown in the inset and sodium currents were measured. The amplitude of the currents produced by the second constant voltage step decreased as the first voltage step activated more sodium channels. The amplitude of the sodium current elicited by the second voltage step was plotted as a function of the amplitude of the first voltage step. This sigmoidal curve is called an inactivation curve. (**b**) Non-electrolytes added to the physiological solutions slowed the sodium currents. The membrane potential is in a range from −30 to 70 mV in 20 mV steps. (**c**) The gating currents in the sodium-free external solution containing tetrodotoxin were slowed with higher levels of non-electrolyte concentrations (modified from Kukita (10) Fig. 1.1).

potential is changed in the positive direction above the threshold level (depolarization), the inward Na^+ currents followed by the outward K^+ currents are recorded (Fig. 1.2, I_i). When the Na^+ ions are removed from the external solution, only the K^+ currents (I_K) are measured. When the K^+ ions are replaced with impermeable ions, only Na^+ currents are measured (Fig. 1.2b, I_{Na}; Fig. 1.3b). Hodgkin and Huxley (6, 7, 9) used the terminology the "independent pathways of Na^+ ions and K^+ ions," but later Hille and others (1) proposed the simpler term, the "ion channel."

Hodgkin and Huxley reproduced the action potential using an equivalent circuit (Fig. 1.2d). They assumed that the resistance representing the channel is not constant but, rather, varies as a function of the membrane potential over time. These properties were implemented into a set of differential equations that are integrated numerically to reproduce the action potentials and their conduction velocity (9, 11–14). These equations are commonly viewed as a heroic achievement in biological science of the twentieth century. Even now, the Hodgkin and Huxley differential equations are absolutely essential for describing neuronal circuits (8, 15).

The currents passing through the Na^+ channels (Na^+ current) decline time-dependently after having attained the maximum current level (Fig. 1.3a,b), a phenomenon called "inactivation" (14). The degree of inactivation is augmented in relation to increased channel activation. This fast inactivation (N-type inactivation) has been observed in voltage-gated Na^+ channels and voltage-gated K^+ channels. It is thought that a positively charged moiety of the channel protein (the inactivation ball), located on the cytoplasmic side, plugs the pore from the inside (the ball-and-chain model of N-type inactivation; see Sect. 1.3.2) (1).

1.3. Characteristics of Ion Channels

The molecular properties of ion channels revealed by electrophysiological methods are described in this section.

1.3.1. Gating Current and Voltage Sensor

Ions permeate through the pore when the gate opens. In voltage-gated channels, the opening and closing of gates are dependent on changes in the membrane potential. Huge numbers of ions permeate through the open pore and carry ionic current. This current either charges or discharges the membrane capacitance, leading to changes in the membrane potential.

Voltage sensitivity is the essential concept of the Hodgkin–Huxley model, and a mechanism for the voltage sensitivity has been proposed. For the channel to be sensitive, they assumed that there are electrical charges in the channel. In the present terminology, charged residues exist in the transmembrane domain of the voltage-gated channel proteins, and they are referred to as the voltage sensor. The voltage sensor moves along the electrical field in the membrane when the membrane potential is altered. The sensor completes the movement within a few milliseconds, and this transient charge movement is detected electrophysiologically as the gating current (Fig. 1.3c). This movement forces other parts of the channel molecule to change conformation, leading to opening of the gate. The gating current serves as a measure of the changes in the voltage sensor rather than the gate.

Gating currents themselves were predicted by Hodgkin and Huxley's early work (11); but the gating currents, which are two orders of magnitude smaller than the ionic current, were not detectable at that time. The first measurements were performed in giant cells, such as squid giant axons and muscle cells (16). More recently, huge numbers of exogenous channels have been expressed on *Xenopus* oocytes, by which the gating current was readily measured (17, 18).

The gating current precedes the ionic currents because it is elicited by the movement of the sensor, whereas the ionic current appears only after the gate opens (Fig. 1.3b,c). The gating current is not affected by the ion concentration or species of ion in the environment as it is an intrinsic property of channel proteins. The gating current–voltage relation is exhibited as a sigmoidal curve (Fig. 1.3c inset). The time course of the gating current is affected by temperature, pressure, and the physical properties of the environment (e.g., viscosity) (Fig. 1.3c) (10, 19).

The molecular mechanism underlying the gating current has been related to the molecular structure of the Na channel. The primary structure of the Na channel was shown to have four homologous repeats in the long stretch of the sequence, each of which has six transmembrane segments (S1–S6 × 4). In each repeat there is an unprecedented segment (S4) that positively charged residues appear every three residues (20). The hydrophobicity of the S4 segment is high, and S4 is considered to be transmembrane. Introducing mutations into S4 alters the voltage sensitivity, and S4 was thus assigned as the voltage sensor (21). Many hypotheses have been proposed regarding the orientation of S4 in the membrane (22). A recent model assumes that the full S4 region is not embedded in the hydrophobic membrane environment. Rather, most of the charged residues in S4 face the aqueous environment of either side of the membrane (19, 23). This discussion is still ongoing, even after publication of the crystal structure of voltage-gated channels.

1.3.2. Gating Behavior in Single-Channel Currents

Thousands of channels are expressed on the cell membrane, and ionic currents measured in a cell represent the average behavior of these channels. An electrophysiological technique was developed to allow detection of the behavior of single-channel molecules (the planar lipid bilayer method and the patch-clamp method; see Chap. 16) (5). This is a pioneering achievement, and was followed by dramatic advances in single-molecule studies of a variety of target molecules using various techniques.

Compared to measurements of cellular macroscopic currents (Figs. 1.2a, 1.3a,b and 1.4a), observing a single molecule from a fraction of the membrane area provides considerably more detailed information. Researchers need different vantage point to interpret the observed events (24, 25).

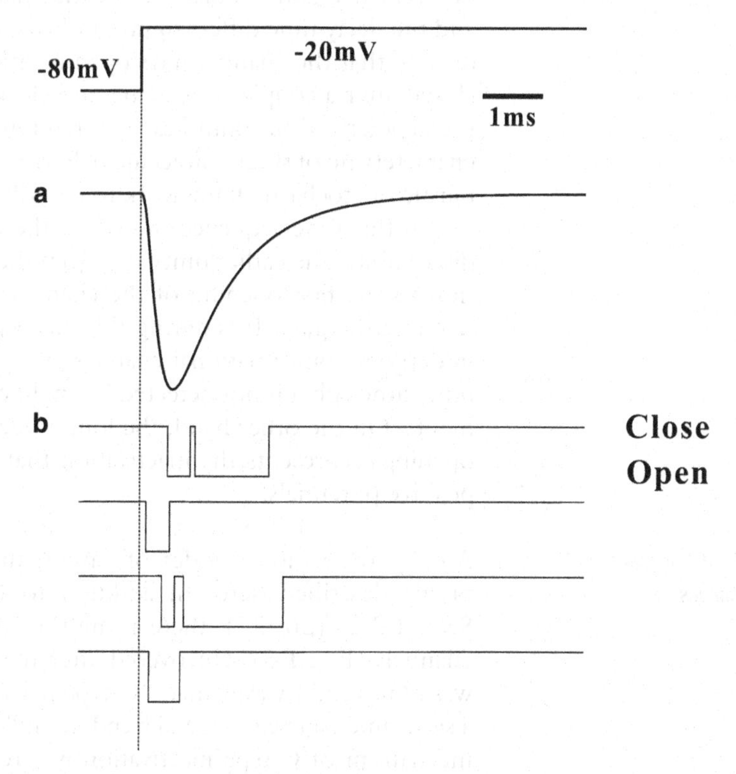

Fig. 1.4. Na channel currents. Time course of the current upon depolarization. (**a**) Macroscopic currents. (**b**) Single-channel currents. The channel opens after a delay from the time of depolarization. The downward deflections indicate the opening of the channel.

Single-channel currents and the macroscopic Na current are shown in Fig. 1.4. Single-channel currents appear as a series of discrete jumps between defined current levels: the zero current of the nonconductive closed conformation and the steady current amplitude of an open conformation.

During the "on" state, ion flux through the open pore is sufficiently high that it can be observed electrophysiologically as currents on the order of picoamperes (pA) in amplitude. Single-channel conductance data are a prerequisite for understanding the ion permeation mechanisms. On the other hand, the "off" or nonconductive state in the current measurements, called the "closed" state, is simply a silent state. One may gain the impression that these data on the closed states do not implicate important information of the gating, and this is true insofar as a short stretch of the current trace is concerned. However, collecting multiple traces under the same experimental conditions enables a statistical inference for gating kinetics. For example, when these traces are recorded from the same molecule repeatedly, each trace appears differently. Upon depolarization, the channel opens after a delay, and the delay time differs significantly for each trace. The duration of time that the channel stays open is different. The channel stays closed after a couple of opening and closing transitions as long as the depolarization continues. This random nature is a fundamental characteristic of single-molecule behavior and necessitates a probabilistic or stochastic framework to describe it appropriately (25).

In this time sequence of events, the underlying processes are discernible. The earlier time (i.e., from the onset of the depolarization to the first opening of the channel) represents the time for opening the gate. It is during this interval that the voltage sensor undergoes conformational changes and generates the gating current, although it is not detected in single-channel current measurements. On the other hand, the long nonconductive state after the openings represents the inactivation that becomes more stable at positive potentials.

1.3.2.1. Inactivation Kinetics

Among the various modes of gating, the inactivation process is briefly described here. In addition to N-type inactivation (see Sect. 1.2.2) (26, 27), there is another type of inactivation in K channels (Fig. 1.5). Uncovered after the fast N-type inactivation was removed, it was named "C-type inactivation" (28). It exhibits a slow time course of several hundred milliseconds. The underlying mechanism of C-type inactivation was revealed after finding that lowering extracellular K^+ facilitated the inactivation, suggesting that the selectivity filter may contribute to the inactivation. A recent crystal study unequivocally demonstrated that the selectivity filter is collapsed in the inactivated state (29).

1.3.3. Ion Permeation Mechanisms

Ion permeation is essential to the physiological functioning of ion channels. Generally, ions permeate through the channel at a rate on the order of 10^8/s, and certain specific ion species is selected out of high concentrations of various other ion species. People may have an impression that the mechanisms underlying the fast transfer rate and the high selectivity are in conflict with each other. Understanding this fundamental feature was of great interest for years during the history of channel investigation (1). Among the membrane transporting proteins, including transporters and pumps, we address which features are common and which are different in terms of the underlying mechanism? This question led us to conduct studies that would help to establish the conceptual identity of ion channels.

To understand the ion transfer mechanism, facilitated transporters offer a useful comparison with the ion channel (Fig. 1.6) (30). Both ion channels and facilitated transporters transfer passive movement of ions down the electrochemical gradient. One opinion holds that "channels transfer ions much faster than do transporters." This statement is roughly intuitive and does not discriminate fast transporters from low-conductance channels. Is there a qualitative

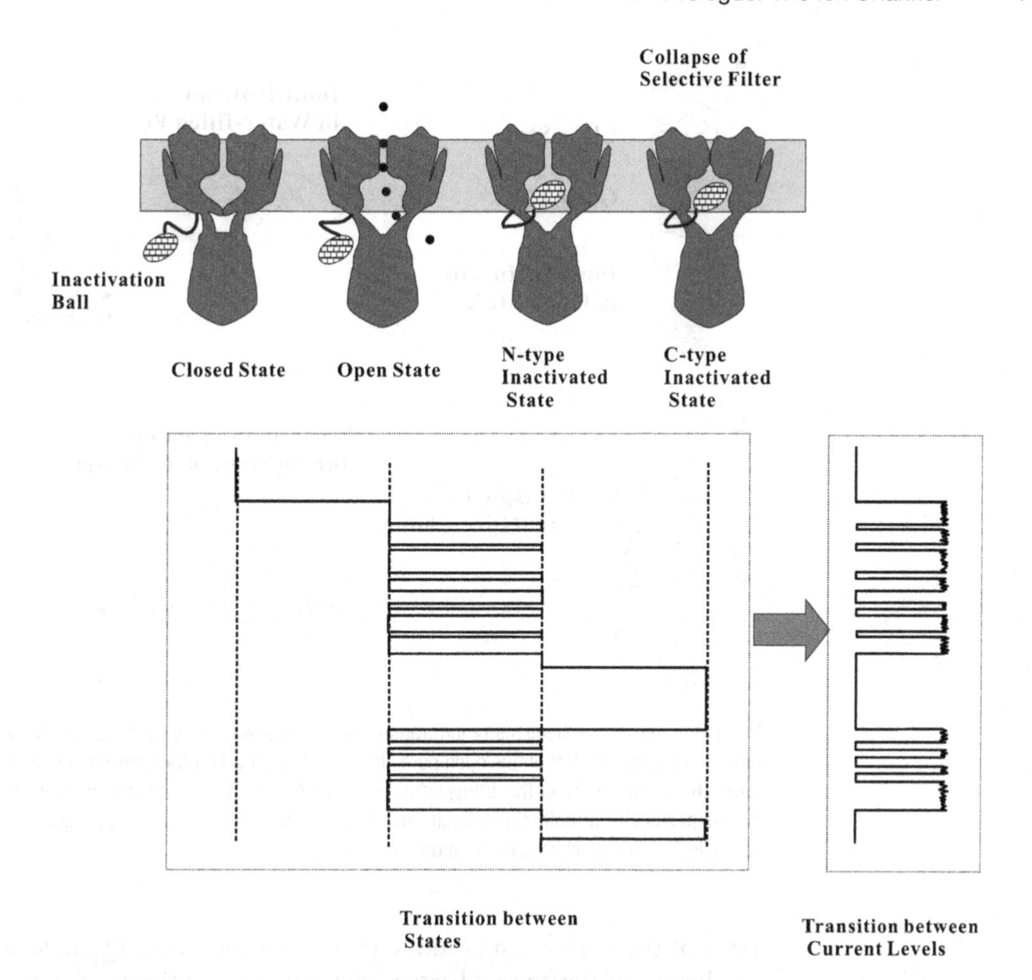

Fig. 1.5. Single-channel currents and conformational changes in the channel molecule. (**a**) A schematic representation of the conformational states: Closed, open, N-type inactivated and C-type inactivated. The channel becomes permeable when the activation gate located near the cytoplasmic side opens. Once the activation gate opens, the N-inactivation ball can access to the central cavity and the current is blocked (N-type inactivation). When the activation gate stays open, the selectivity filter tends to collapse and enters a long term inactivated state (C-type inactivation). (**b**) Single-channel current recordings and the underlying conformation transitions of a channel molecule. *Left*: Transitions between states. *Right*: Conductance changes.

difference between channels and transporters? For transporters, ion transfer across the membrane is accompanied by conformational changes in the membrane protein; whereas ions permeate through the open pore structure, to which ions have access from both sides of the membrane. The question then arises, how are these characteristics differentiated experimentally?

1.3.3.1. Q_{10} Value for Permeation Is Low for Channels

The Q_{10} value is a thermodynamic concept that expresses the changes in the rate of reaction at temperature differences of 10°C. In ion channels, the Q_{10} value for permeation is evaluated as the

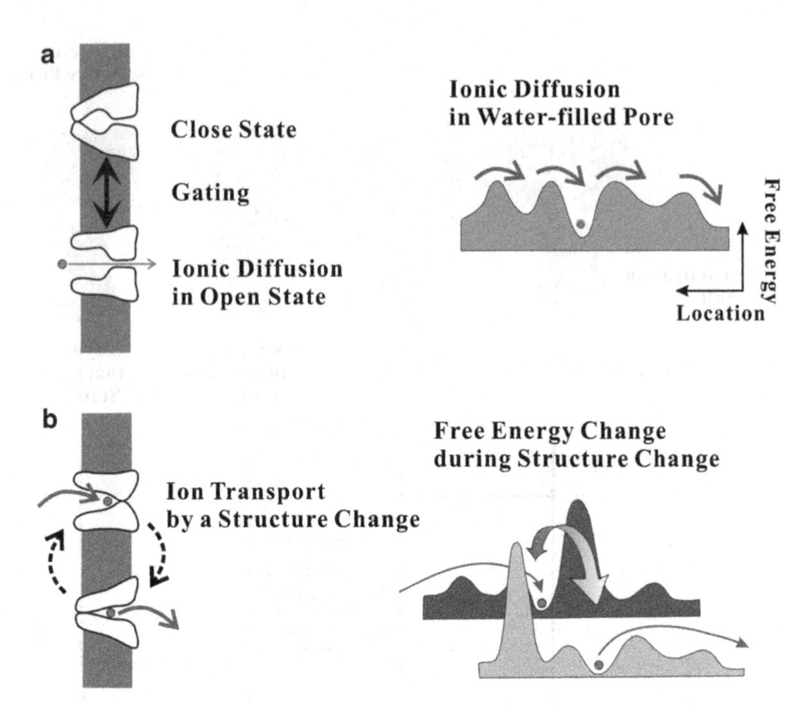

Fig. 1.6. Different mechanisms of ion transfer in channels and transporters. (**a**) In channels, ions permeate through the open pore structure that allows permeation in huge numbers. (**b**) In transporters, the transporting ion cannot gain access to the binding site simultaneously from both sides. Each time the ion binds to the site, the transporter is subjected to change in its conformation.

ratio of the current amplitudes at two temperatures. Q_{10} reflects the height of the energy barrier (the activation enthalpy) for ions to overcome during permeation (Fig. 1.5) (1). Ions must permeate through a narrow pore, or they must tear off the hydrating water molecules. Overall, these processes incur an energy cost of less than 20–30 kJ/mol. On the other hand, a transporter changes its conformation such that one conformation binds ions accessing from one side of the membrane and another conformation releases the ion to the other side. The rate of ion transfer in the transporter is determined by the rate of the conformational change, which is determined by the activation free energy (31) for converting different conformations. Generally, the activation enthalpy for the conformational change is much higher than that for ion permeation. The use of Q_{10} values as criteria for discriminating channel and transporter depends on the quantitative differences of the enthalpy for the ion transfer mechanism. Most of the ion channels display a Q_{10} value of <1.5, suggesting that ions are permeating through a water-filled pore (32).

1.3.3.2. Water Accompanies Ion Flux Through the Channels

Both channel and transporters allow water to flow in accompaniment with ion flow. In the transporter, water molecules around a bound ion are transported along with the ion upon conformational changes. In ion channels, hydrated ions in the bulk solution are partially dehydrated upon entering a water-filled pore, and the water molecules that remained on the ions accompany the ion flux. When the diameter of the pore is as narrow as the size of the permeating ions, the ions and water molecules cannot pass each other in the narrow pore, which leads to single-file permeation. In this case, the ion and water fluxes are not independent from one another (1). When ion permeates down the electrochemical gradient through the narrow pore, water molecules accompany the ions and are transferred uphill even without a driving force for water (osmotic pressure). Water flux driven by ion flux is referred to as electroosmosis (30). Conversely, differences in the osmotic pressure across the membrane drives water flux through the channel, and this water flux flushes out ions when they stay lodged in the narrow pore (Fig. 1.7). This accompanying ion flux, even without an electrochemical gradient, generates a potential difference across the membrane, known as the "streaming potential" (31, 33).

Compared to ion channels, the osmotic difference across the transporter does not generate any water flux because conformational changes are not affected by the osmotic gradient. Thus, transporters do not generate the streaming potential. On the other

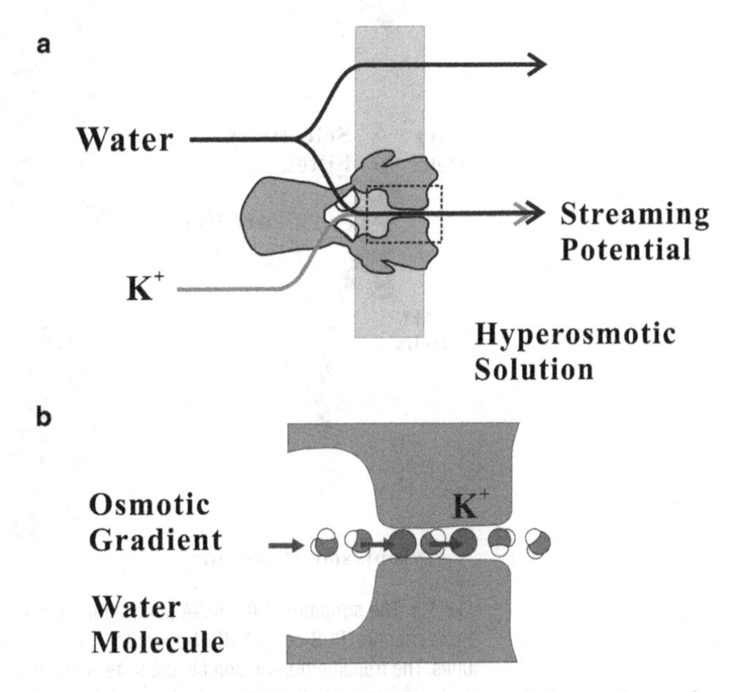

Fig. 1.7. The streaming potential. When osmotic pressure is applied across the membrane, a net flux of water occurs through the membrane and the channel pore. If ions are present in the narrow pore, the water flux flushes out the ions in the pore and generates the membrane potential, even in the absence of the electrochemical potential. From the value of the streaming potential, the number of water molecules carried with the ion can be evaluated.

hand, the channel allows water flux so long as the pore stays open and hence ion flux, thus generating the streaming potential. This is a clear distinction between channels and transporters.

The streaming potential reveals the length of the water-filled pore when permeation is occurring in a single-file manner (31). It also shows the underlying permeation process (34), which had not been predicted by other methods.

1.4. From Channel Structure to Channel Dynamics

In 1998, MacKinnon's group solved the crystal structure of the potassium channel (KcsA) (Fig. 1.8) (35, 36). It was a landmark study, dramatically advancing the field beyond its previous state. Before determining the structure of this channel, structure–function relations had focused on how a given sequence of amino acids takes a three-dimensional structure, which contributes to the channel's

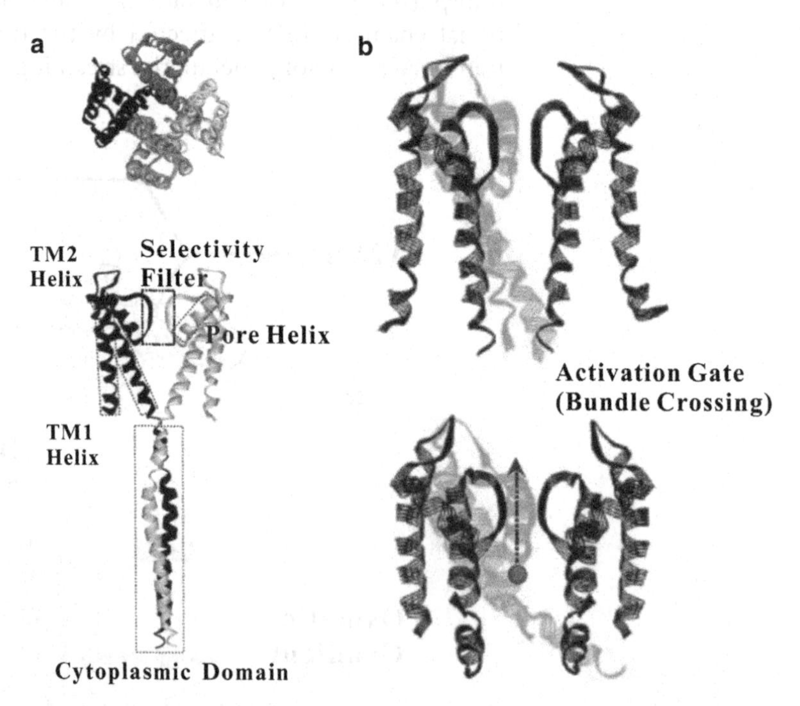

Fig. 1.8. The structure of the KcsA potassium channel. (**a**) The full-length structure of the KcsA channel (PDB code: 3eff). *Upper*: Top view; *Lower*: Side view of two diagonal subunits. The transmembrane domain consists of two transmembrane helices, TM1 and TM2, and a short pore helix. In the cytoplasmic domain, the long α-helices form a tight bundle. (**b**) The open and closed structure of the transmembrane domain. *Upper*: The closed conformation (PDB code: 1k4c). *Lower*: The open inactivated conformation (PDB code: 3f5w). In the closed conformation, four TM2 helices form a bundle crossing at the cytoplasmic end and prevent the permeation.

function. After the determination, attempts were made to decipher messages implicated in the structure, such as how ions are selected and how the gate is operated.

The KcsA channel, a member of the potassium channel family, has two transmembrane segments without the voltage sensor and forms a functional channel as a homotetramer. This family includes the inwardly rectifying potassium channels. As expected, the two transmembrane segments (TM1 and TM2) comprise the α-helical structure. Four TM2 helices from the four subunits are bundled to make a crossing at the cytoplasmic end (the bundle crossing), which prevents ion permeation. Unexpectedly, a short α-helix (pore helix) was found between two transmembrane helices located in the extracellular half of the transmembrane domain. In the amino acid sequence, the signature sequence for potassium selectivity (TVGYG) (37) follows immediately after the pore helix. This short stretch assumes a straight strand structure. Four of them are aligned in parallel, within which a narrow pore, or the selectivity filter, is formed. The pore diameter is 3 Å, and it is 12 Å long. Several types of potassium channel have been crystallized, and it was demonstrated that the architecture of the pore domain of the KcsA channel is shared by all members of the potassium channel family (38–41).

1.4.1. Ion Permeation and Selectivity

The crystal structure has provided insights into underlying mechanisms of ion permeation and selectivity. In the narrow selectivity filter of K channels, multiple K^+ ions are axially-aligned. In place of the hydrating water molecules around the ion, eight carbonyl oxygens from the backbone of the selectivity filter surround the dehydrated ions and stabilize them (42, 43). There are four binding sites in the short selectivity filter. In earlier studies, the presence of multiple ions in the pore (or multi-ion pore) has been proposed (1); this proposal, based on functional measurements, was proven with the crystal structure. Multiple ions in the short selectivity filter interact each other with electrostatic repulsion. This destabilized condition is further exacerbated when an additional ion comes in, and the ion closest to the exit is thus ejected from the opposite end by chain reaction collision of multiple ions in the filter (the knock-on mechanism) (44, 45). This mechanism accounts for the fast permeation rate.

For potassium channels, permeable ions have an ionic radius in the range of 1.3–1.7 Å. This selectivity is not simply governed by the cutoff size (46); it has been accounted for by the snug fit model, in which ions of a defined size can permeate. The underlying mechanism of the snug-fit has been proposed based on the crystal structure (47, 48).

1.4.2. Gating Dynamics

In respect to the gating property, there are two categories of potassium channel: voltage-independent and voltage-dependent channels. They differ by the presence or absence of the voltage-sensor domain. In the three-dimensional structure of the voltage-gated channel, the voltage-sensor domain forms an independent structural entity, and four of them surround the pore domain (41). Whichever sensor a channel bears (i.e., voltage or chemical), the message sensed by the sensor is transferred to the pore domain to regulate the opening and closing of the gate. Accordingly, the physical or chemical stimuli are converted to an electrical signal. Determination of the crystal structures for the closed and open conformations has helped to elucidate the mechanism of message transfer through the channel structure (49). How does the conformational change in the pore domain lead to opening the gate? These dynamic characteristics of the functioning channel cannot be predicted solely from a static picture of the crystal structure.

Electrophysiological methods have helped provide insight into the dynamic behavior of channel molecules. In particular the gating current measures changes in the charge distribution in the channel molecules, which is directly related to the conformational changes. Investigation has revealed that even though the basic idea of the Hodgkin–Huxley model is correct for the voltage-gated potassium channel, some refinements in the gating model are necessary (50). In the Hodgkin–Huxley model, the channel becomes conductive when the fourth, or last, voltage sensor turns on. In the refined model, the pore does not open even if the four voltage sensors enter the “on” state. Rather, the four “on” sensors send a message to the pore domain, and concerted conformational changes take place in the tetrameric pore domain in a voltage-independent manner.

Study of the gating mechanism was further advanced with the introduction of a method to detect signals related to conformational changes in the channel protein. A fluorescence probe labeled on the channel reports the environmental change (hydrophilic or hydrophobic) that the probe is surrounded (51). Among the single-molecule measurement techniques, finding a way to observe the conformational change in a more straightforward manner was urgent. X-ray diffraction methods provide pictures of molecules with high spatial resolution, but crystallization is required. In contrast, a recently developed method known as the diffracted X-ray tracking method (DXT) (52), detects changes in the conformation of single molecules by labeling the molecule with a gold nanocrystal (53). Diffraction from the nanocrystal continuously reports the conformational changes. Consequently, it has been revealed that the KcsA channel undergoes twisting conformational changes upon opening and closing of the gate (54).

1.5. Electrophysiological Methods in the Field of Channel Studies

The ion channel generates ionic currents upon activation, and this message is not directly recognized by any proteins. Instead, current across the membrane, by charging and discharging the membrane capacitance, yields changes in the membrane potential, which is sensed by voltage-gated channels. Furthermore, changes in the membrane potential lead to changes in the driving forces for passive movements of ions, which affects the rate of permeation and transport in both channels and transporters. Thus, electrophysiological methods preemptively detect signals from channels prior to any proteins can recognize. Together with the voltage measurements, electrophysiological method serves as the gold standard for studying channel-related phenomena. From neuronal connectivity to the molecular mechanism underlying selective ion permeation, electrophysiological methods cover a broad spectrum of physiological events.

Channel studies continue to expand our understanding of channel mechanisms. Researchers with expertise in both experimental and theoretical disciplines have joined the field to explore novel applications, which in turn has led to insights into the molecular mechanism of channel function. Among the evolving techniques, electrophysiological measurements, such as those of patch-clamping and the planar lipid bilayer, are dynamic in their nature and therefore best suited to capturing the dynamic characteristics of ion channels. With these and still emerging technical advantages, electrophysiological studies will continue to lead the study of ion channels onward by integrating the results from a variety of fields.

References

1. Hille B (2001) Ion Channels of Excitable Membranes, 3rd edn. Sinauer Associated, Inc., MA
2. Kukita F (2005) Progress in a study of ion channels for 50 years. Biophysics 45:10–15 (in Japanese)
3. Hodgkin AL, Huxley AF (1939) Action potentials recorded from inside a nerve fibre. Nature 144:710–711
4. Hodgkin AL, Katz B (1949) The effect of sodium ions on the electrical activity of giant axon of the squid. J Physiol 108(1):37–77
5. Sakmann B, Neher E (1995) Single-channel recording, 2nd edn. Plenum, New York
6. Sigworth FJ (2003) Molecular switches for "animal electricity". A century of nature: twenty-one discoveries that changed science and the world. University of Chicago Press, Chicago
7. Hamill OP, Marty A, Neher E, Sakmann B, Sigworth FJ (1981) Improved patch-clamp techniques for high-resolution current recording from cells and cell-free membrane patches. Pflugers Arch 391(2):85–100
8. Usui S (1997) Mathematical models in brain and nerve. New biophysics, vol 8. Kyoritsu Shuppan, Tokyo (in Japanese)
9. Hodgkin AL, Huxley AF, Katz B (1952) Measurement of current-voltage relations in the membrane of the giant axon of Loligo. J Physiol 116(4):424–448
10. Kukita F (2000) Effect of water on gating of voltage-gated ion channels. Biophysics 40:185–190 (in Japanese)

11. Hodgkin AL, Huxley AF (1952) A quantitative description of membrane current and its application to conduction and excitation in nerve. J Physiol 117(4):500–544
12. Hodgkin AL, Huxley AF (1952) The components of membrane conductance in the giant axon of Loligo. J Physiol 116(4):473–496
13. Hodgkin AL, Huxley AF (1952) Currents carried by sodium and potassium ions through the membrane of the giant axon of Loligo. J Physiol 116(4):449–472
14. Hodgkin AL, Huxley AF (1952) The dual effect of membrane potential on sodium conductance in the giant axon of Loligo. J Physiol 116(4):497–506
15. Koch C (2004) Biophysics of computation: information processing in single neurons. Computational neuroscience New Ed. Oxford University Press, Oxford
16. Armstrong CM, Bezanilla F (1974) Charge movement associated with the opening and closing of the activation gates of the Na channels. J Gen Physiol 63(5):533–552
17. Conti F, Stuhmer W (1989) Quantal charge redistributions accompanying the structural transitions of sodium channels. Eur Biophys J 17(2):53–59
18. Stefani E, Toro L, Perozo E, Bezanilla F (1994) Gating of Shaker K^+ channels: I. Ionic and gating currents. Biophys J 66(4):996–1010
19. Kukita F (2000) Solvent effects on squid sodium channels are attributable to movements of a flexible protein structure in gating currents and to hydration in a pore. J Physiol 522(Pt 3):357–373
20. Noda M, Shimizu S, Tanabe T, Takai T, Kayano T, Ikeda T, Takahashi H, Nakayama H, Kanaoka Y, Minamino N et al (1984) Primary structure of *Electrophorus electricus* sodium channel deduced from cDNA sequence. Nature 312(5990):121–127
21. Papazian DM, Timpe LC, Jan YN, Jan LY (1991) Alteration of voltage-dependence of Shaker potassium channel by mutations in the S4 sequence. Nature 349(6307):305–310
22. Yang N, George AL Jr, Horn R (1996) Molecular basis of charge movement in voltage-gated sodium channels. Neuron 16(1):113–122
23. Bezanilla F (2002) Voltage sensor movements. J Gen Physiol 120(4):465–473
24. Shelley C, Magleby KL (2008) Linking exponential components to kinetic states in Markov models for single-channel gating. J Gen Physiol 132(2):295–312
25. Colquhoun D, Hawkes AG (1995) The principles of the stochastic interpretation of ion-channel mechanisms. In: Sakmann B, Neher E (eds) Single-channel recording, 2nd edn. Plenum, New York, pp 397–482
26. Armstrong CM, Bezanilla F, Rojas E (1973) Destruction of sodium conductance inactivation in squid axons perfused with pronase. J Gen Physiol 62(4):375–391
27. Hoshi T, Zagotta WN, Aldrich RW (1990) Biophysical and molecular mechanisms of Shaker potassium channel inactivation. Science 250(4980):533–538
28. Lopez-Barneo J, Hoshi T, Heinemann SH, Aldrich RW (1993) Effects of external cations and mutations in the pore region on C-type inactivation of Shaker potassium channels. Recept Channel 1(1):61–71
29. Cuello LG, Jogini V, Cortes DM, Perozo E (2010) Structural mechanism of C-type inactivation in K(+) channels. Nature 466(7303): 203–208
30. Schulz SG (1980) Basic principles of membrane transport. Cambridge University Press, Cambridge
31. Finkelstein A (1987) Water movement through lipid bilayers, pores, and plasma membranes. Theory and reality. Wiley-Interscience, New York
32. Kuno M, Ando H, Morihata H, Sakai H, Mori H, Sawada M, Oiki S (2009) Temperature dependence of proton permeation through a voltage-gated proton channel. J Gen Physiol 134(3):191–205
33. Ando H, Kuno M, Shimizu H, Muramatsu I, Oiki S (2005) Coupled K^+-water flux through the HERG potassium channel measured by an osmotic pulse method. J Gen Physiol 126(5):529–538
34. Iwamoto M, Oiki S (2011) Counting ion and water molecules in a streaming file through the open-filter structure of the K channel. J Neuroscience 31(34):12180–12188
35. MacKinnon R, Cohen SL, Kuo A, Lee A, Chait BT (1998) Structural conservation in prokaryotic and eukaryotic potassium channels. Science 280(5360):106–109
36. Doyle DA, Morais Cabral J, Pfuetzner RA, Kuo A, Gulbis JM, Cohen SL, Chait BT, MacKinnon R (1998) The structure of the potassium channel: molecular basis of K^+ conduction and selectivity. Science 280(5360):69–77
37. Heginbotham L, Lu Z, Abramson T, MacKinnon R (1994) Mutations in the K^+ channel signature sequence. Biophys J 66(4):1061–1067
38. Jiang Y, Lee A, Chen J, Cadene M, Chait BT, MacKinnon R (2002) The open pore conformation of potassium channels. Nature 417(6888):523–526

39. Jiang Y, Lee A, Chen J, Ruta V, Cadene M, Chait BT, MacKinnon R (2003) X-ray structure of a voltage-dependent K+ channel. Nature 423(6935):33–41
40. Kuo A, Gulbis JM, Antcliff JF, Rahman T, Lowe ED, Zimmer J, Cuthbertson J, Ashcroft FM, Ezaki T, Doyle DA (2003) Crystal structure of the potassium channel KirBac1.1 in the closed state. Science 300(5627):1922–1926
41. Long SB, Campbell EB, MacKinnon R (2005) Crystal structure of a mammalian voltage-dependent Shaker family K+ channel. Science 309(5736):897–903
42. Morais-Cabral JH, Zhou Y, MacKinnon R (2001) Energetic optimization of ion conduction rate by the K+ selectivity filter. Nature 414(6859):37–42
43. Zhou Y, Morais-Cabral JH, Kaufman A, MacKinnon R (2001) Chemistry of ion coordination and hydration revealed by a K+ channel-Fab complex at 2.0 A resolution. Nature 414(6859):43–48
44. Hodgkin AL, Keynes RD (1955) The potassium permeability of a giant nerve fibre. J Physiol 128(1):61–88
45. Armstrong CM (1975) Potassium pores of nerve and muscle membranes. Membranes 3:325–358
46. Eisenman G, Horn R (1983) Ionic selectivity revisited: the role of kinetic and equilibrium processes in ion permeation through channels. J Membr Biol 76(3):197–225
47. Lockless SW, Zhou M, MacKinnon R (2007) Structural and thermodynamic properties of selective ion binding in a K+ channel. PLoS Biol 5(5):e121
48. Thompson AN, Kim I, Panosian TD, Iverson TM, Allen TW, Nimigean CM (2009) Mechanism of potassium-channel selectivity revealed by Na+ and Li+ binding sites within the KcsA pore. Nat Struct Mol Biol 16(12):1317–1324
49. Cuello LG, Jogini V, Cortes DM, Pan AC, Gagnon DG, Dalmas O, Cordero-Morales JF, Chakrapani S, Roux B, Perozo E (2010) Structural basis for the coupling between activation and inactivation gates in K+ channels. Nature 466(7303):272–275
50. Schoppa NE, Sigworth FJ (1998) Activation of Shaker potassium channels. III. An activation gating model for wild-type and V2 mutant channels. J Gen Physiol 111(2):313–342
51. Blunck R, McGuire H, Hyde HC, Bezanilla F (2008) Fluorescence detection of the movement of single KcsA subunits reveals cooperativity. Proc Natl Acad Sci USA 105(51): 20263–20268
52. Sasaki YC, Suzuki Y, Yagi N, Adachi S, Ishibashi M, Suda H, Toyota K, Yanagihara M (2000) Tracking of individual nanocrystals using diffracted x rays. Phys Rev E Stat Phys Plasmas Fluids Relat Interdiscip Topics 62((3 Pt B)):3843–3847
53. Oiki S, Shimizu H, Iwamoto M, Konno T (2012) Single molecular gating dynamics for the KcsA potassium channel. Adv Chem Phys 146: 147–193
54. Shimizu H, Iwamoto M, Konno T, Nihei A, Sasaki YC, Oiki S (2008) Global twisting motion of single molecular KcsA potassium channel upon gating. Cell 132(1):67–78

Chapter 2

Patch-Clamp Techniques: General Remarks

Tenpei Akita, Masahiro Ohara, and Yasunobu Okada

Abstract

The patch-clamp technique, invented by Erwin Neher and Bert Sakmann in 1976, was originally designed to detect the activity of single-ion channel proteins in the cell membrane. The technique is now widely used in the field of ion channel research for many purposes, such as to monitor changes in the total membrane current, intracellular voltage or cell membrane capacitance etc. in a living cell at high time resolution. We explain the basic principle of, and protocols for, the patch-clamp technique, including maneuvers for patch-clamp amplifier operation – basic knowledge required for applications explained in the following chapters.

2.1. Introduction

The patch-clamp technique, invented by Erwin Neher and Bert Sakmann in 1976 (1), was originally designed to detect the activity of one or several ion channel proteins in the cell membrane as ionic current fluxes through the proteins. With the establishment of the giga-seal technique and other variations (2, 3), this method has been widely applied to many types of cell since 1980 (3–5). The technique has revolutionized the field of physiological research at the cellular and molecular levels; and its effects, along with the advent of the new technology of molecular cloning, have propagated throughout the realm of life science research. Based on the invention of this technique, Neher and Sakmann won the Nobel Prize in 1991 (Fig. 2.1).

In this chapter, we explain the principle of, and protocols for, the patch-clamp technique, including maneuvers for patch-clamp amplifier operation.

Yasunobu Okada (ed.), *Patch Clamp Techniques: From Beginning to Advanced Protocols*, Springer Protocols Handbooks, DOI 10.1007/978-4-431-53993-3_2, © Springer 2012

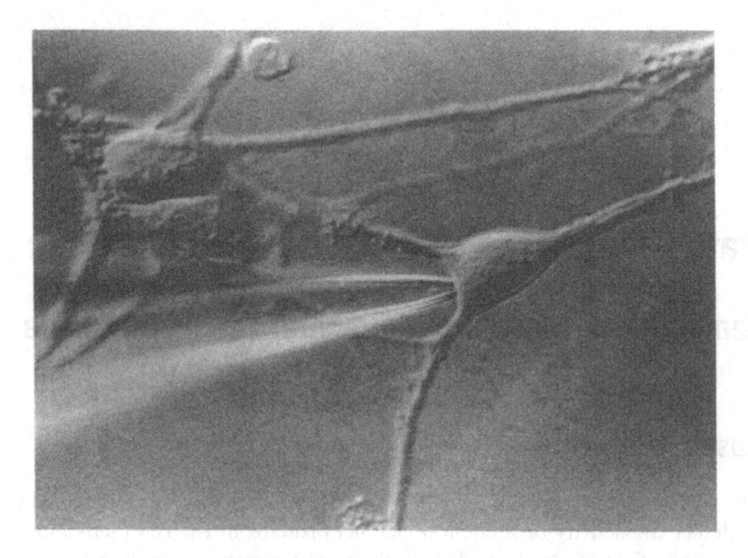

It was a great experience to receive messages from all around the globe, which unanimously expressed joy and approval. Many stated, that the decision of the Nobel Assembly will be beneficial to all research on ion channels. I would like to add, that it was the use of the patch clamp technique in so many laboratories which ultimately caused this decision.

Thank You,

E. Neher

With the very best wishes for a happy 1992!

Fig. 2.1. A letter sent by Prof. Erwin Neher in response to the congratulations from his friends all over the world after his receipt of the Nobel Prize.

2.2. What Is the "Patch-Clamp?"

2.2.1. Principle of the Patch-Clamp Technique

The patch-clamp technique uses a glass microcapillary tube as an electrode (called a "patch electrode" or "patch pipette"). The tip of the electrode is attached to the cell membrane so tightly that the electrical resistance between the inside and outside of the electrode becomes more than 1 GΩ (10^9 Ω, "giga-seal"). The minute region of the membrane in the inner tip of the electrode ("patch membrane") is then voltage-clamped ("patch-clamped"), and picoampere (10^{-12} A) levels of ionic currents through the ion channels in the patch membrane are detected through an amplifier (Fig. 2.2).

The basic circuitry of the amplifier for patch-clamp is an *I–V* converter composed of several operational amplifiers (OP amps) (Fig. 2.2). The inverting and noninverting inputs of an OP amp are clamped at the same voltage level through a feedback resistor;

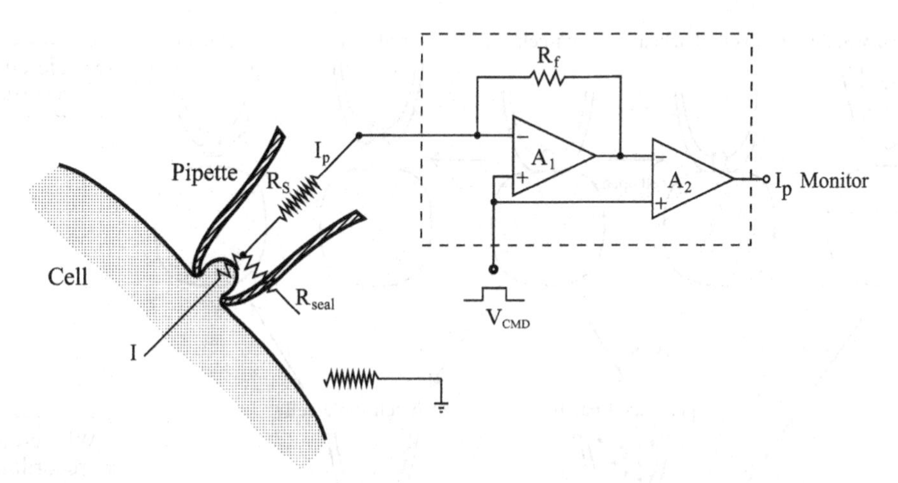

Fig. 2.2. Principle of patch-clamp technique. R_S represents the series resistance (or access resistance) linked serially to the patch membrane resistance. R_{seal} means the seal resistance. Usually, R_S is 1–5 MΩ. When R_{seal} is more than 10 GΩ, $I_p/I = R_{seal}/(R_S + R_{seal}) \sim 1$. This I_p is detected as a voltage drop across the high-resistance feedback resistor (R_f) in the *I–V* converter circuitry (surrounded by *dotted lines*). Since the output of the OP amp A_1 actually contains the command voltage (V_{CMD}), this is to be subtracted through the next OP amp A_2.

and by application of a command voltage (V_{CMD}) to the noninverting input, the voltage level of the patch membrane linked to the inverting input is clamped to V_{CMD} (virtual short).

When the tip of the electrode is sealed with the patch membrane at more than 10 GΩ (10^{10} Ω), the shunt current through the seal becomes negligible and the recorded current (I_p) purely reflects the current through the patch membrane (Fig. 2.2, I).

2.2.2. Variations in the Mode of Patch-Clamp

Figure 2.3 indicates some variations on the patch-clamp technique. The cell-attached mode was designed first, to record single-channel currents (1); the inside-out mode and outside-out mode were then added (2). Subsequently, the open cell-attached inside-out mode (6) and the perforated vesicle outside-out mode (7) were invented. The conventional whole-cell mode (2) and the perforated patch mode (8) were designed for recording the total current across the whole membrane of a single cell.

2.2.2.1. Cell-Attached Mode (On-Cell Mode)

The cell-attached mode records single-channel currents under giga-seal conditions (called "On Cell" mode in the amplifiers made by HEKA Elektronik). In this mode, channel activity is observed with minimum disturbance to the intracellular environment. However, intracellular conditions cannot be controlled directly, and the actual transmembrane voltage remains unknown because there is no information on the intracellular membrane potential. Furthermore, membrane impermeable drugs or active substances added to the bathing solution cannot reach ion channels in the patch membrane. Nevertheless, if addition of these substances to the bath elicits some responses in the recorded currents, we can

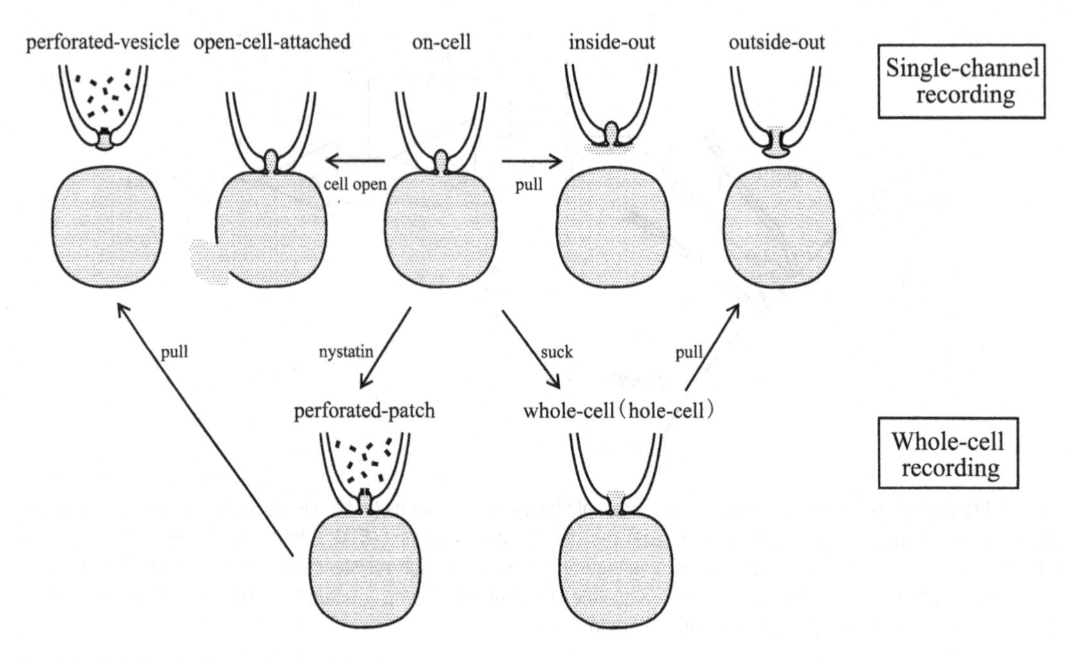

Fig. 2.3. Variations in the mode of patch-clamp.

conclude that their effects on the ion channels must be mediated by some intracellular second messengers.

2.2.2.2. Inside-Out Mode

When the electrode in the cell-attached mode is lifted or slid, the patch membrane can be excised from the cell and the inside-out mode established ("In Out" mode in HEKA amplifiers). In this mode, the conditions on the intracellular side of the patch membrane can be controlled via bath perfusion, and single-ion channel activity can be observed independently of cellular activity. However, it should be noted that some cytosolic regulatory factors are washed out under this condition. In other words, if some properties of the channel activity disappear ("run-down") or emerge ("run-up") after excision, we can surmise that the channel activity must be regulated by some cytosolic factors under normal conditions.

2.2.2.3. Outside-Out Mode

When the electrode in the whole-cell mode (see below) is detached from the cell, the intracellular side of the excised membrane may face the inner solution of the patch pipette. This is called the "outside-out" mode ("Out Out" mode in HEKA amplifiers). In this mode, the conditions on the extracellular side of the patch membrane can be varied via bath perfusion during recordings of single-channel currents. Note that a similar run-down or run-up problem may also occur in this mode.

2.2.2.4. Open Cell-Attached Inside-Out Mode

In this mode, a part of the cell membrane distant from the electrode in the cell-attached mode is mechanically ruptured to make a hole. The intracellular environment is then controlled by perfusion through the hole during single-channel recordings. The speed of washout of cytosolic factors becomes slower when a smaller hole is made in a large cell.

2.2.2.5. Perforated Vesicle Outside-Out Mode

When an electrode in the perforated patch mode (see below) is carefully detached from the cell, it is possible to make a membrane vesicle containing some intracellular organelles such as mitochondria at the tip of the electrode. In this mode, the outside-out recording of single-channel currents under nearly physiological (signaling and metabolic) conditions becomes possible.

2.2.2.6. Conventional Whole-Cell Mode (Hole-Cell Mode)

When the patch membrane is ruptured after making the cell-attached configuration, the total current across the whole membrane of the cell can be recorded. This is called the "whole-cell" mode. (To distinguish the perforated mode explained next, this mode may be recognized as a "conventional" one or called the "hole-cell" mode (9).) In this mode, the intracellular milieu is dialyzed with the pipette solution, so both extracellular and intracellular conditions can be controlled. Furthermore, the intracellular membrane potential can be controlled or monitored reliably by voltage-clamp or current-clamp, respectively. However, one must note that the run-down or run-up problem may also occur if some unknown cytosolic regulatory factors are washed out into the patch pipette.

2.2.2.7. Perforated Patch Mode (Slow Whole-Cell Mode)

To minimize the washout problem under the whole-cell mode, Horn and Marty (8) added pore-forming ionophores, nystatin (or amphotericin B), to the pipette solution. The ionophores are permeable only to univalent ions. They are incorporated into patch membranes containing cholesterol under the cell-attached configuration and make electrically conductive pores in the membrane. The whole-cell current becomes measurable in this condition. In this mode, the washout problem is almost negligible. However, the series resistance (R_s in Fig. 2.2) is inevitably higher than in the conventional whole-cell mode, and the speed of voltage-clamp is necessarily slower; hence, this mode may be called the "slow whole-cell" mode. Instead of nystatin and amphotericin B, use of β-escin, a saponin derivative that is water-soluble and forms highly permeable holes, has been reported (10). To maintain in vivo intracellular anion concentrations, gramicidin, an ionophore permeable only to univalent cations, can be used (11).

2.2.3. Advantages and Disadvantages of the Patch-Clamp Technique

The first advantage of the patch-clamp technique is that it enables reliable voltage-clamp because of the giga-seal that is formed. However, giga-seal formation necessitates invagination of the patch membrane into the tip of the electrode, resulting in mechanical stress on the membrane.

The second advantage is very low background noise. Because the variance of thermal (Johnson) noise is inversely proportional to the resistance value, the noise generated around the patch membrane is extremely low under giga-seal conditions. However, in addition to noise around the patch membrane, noise may originate from the resistance and capacitance along the patch electrode, as explained later, and from the *I–V* converter circuitry in the headstage of the amplifier. Because the noise from these sources is significant, especially in the high-frequency range, the actual observable amplitude and open time of single-channel currents are limited by these noise sources (2, 3).

Classic voltage-clamp methods use two microelectrodes inserted into a single cell or double insulating gaps made with Vaseline or sucrose on a cell; either of these methods can be applied only to a very large cell. The patch-clamp technique, by contrast, can be applied to a wide variety of cells, which is a major advantage. An additional advantage is that the technique enables direct control of the intracellular environment through the pipette solution. However, this may be a drawback in terms of the problems associated with washout.

2.3. Protocols for Patch-Clamp Techniques

2.3.1. Fabrication of Patch Electrodes

2.3.1.1. Pulling

Patch electrodes are fabricated from glass microcapillary tubes with a device called a "puller." Common soft soda glass may be acceptable for whole-cell recording. However, for low noise recording, as for single-channel recordings, hard borosilicate (Pyrex) or aluminosilicate glass, which has a much lower dielectric constant, is desirable.

To ensure that the cell is not impaled by the patch electrode, the tip of the electrode must be blunted; and a shorter length of the tapered part of the tip is required for reducing R_S. Such electrode shapes can be obtained using a microprocessor-driven puller, which melts and pulls the glass tube at multiple steps (Fig. 2.4).

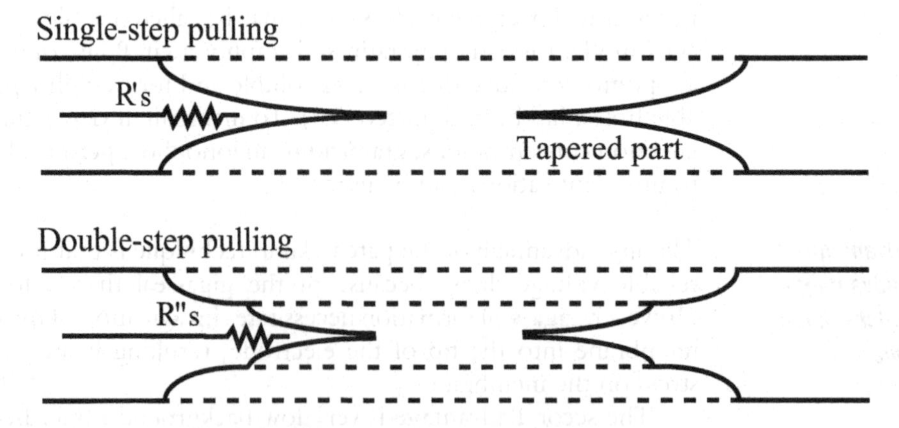

Fig. 2.4. Difference in tip shape between pipettes made by single-step pulling and by double-step pulling (*R*′s> *R*″s is evident).

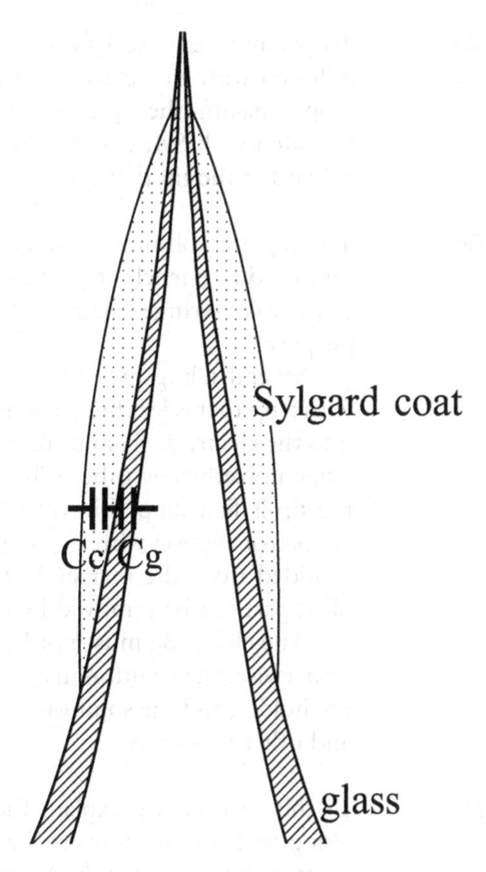

Fig. 2.5. Reduction of pipette stray capacitance by Sylgard coating. $C_S = C_c C_g/(C_c + C_g)$. Thus, if $C_g \gg C_c$, $C_S \sim C_c$.

Usually, electrodes with a tip diameter of 1–5 μm and resistance of 1–10 MΩ when filled with saline solution are used. For single-channel recordings a finer tip (with high pipette resistance) is preferred because of high seal resistance, and for whole-cell recordings a blunter one (with low pipette resistance) is preferred because of reduced R_S effects. Because dust at the tip interferes with giga-seal formation, it is recommended that pipettes be fabricated on the day of use.

2.3.1.2. Sylgard Coating

During recordings of single-channel events, in which opening and closing at very high frequencies is observed, it is crucial to avoid noise from the capacitance of the glass electrode (stray capacitance, C_S) (2). To reduce this C_S noise, the outer shank of the electrode can be coated with a hydrophobic insulating material, such as Sylgard (Dow Corning 184). Sylgard is applied to the electrode, except at its very tip, and dried and cured in a Nichrome coil heater. This makes the surface of the electrode hydrophobic and prevents the bathing solution from creeping up the electrode wall, thereby reducing C_S. The Sylgard layer also reduces greatly the intrinsic capacitance of the glass because the low capacitance coat is connected serially to it (Fig. 2.5).

2.3.1.3. Heat Polishing

To promote giga-seal formation, the tip of the electrode can be polished with an electrically heated platinum wire under a microscope, making the tip broader and smoother. When the electrode is coated with Sylgard, this step is necessary to remove any excess Sylgard at the tip that may prevent giga-seal formation.

2.3.1.4. Solution Filling

The pipette solution must be filtered before electrode filling to remove dust particles that may inhibit giga-seal formation. A small disposable syringe filter (0.2 μm pore size) is useful for this purpose.

Several filling procedures are used. When using an electrode with a tip that is less fine, it is sufficient simply to inject the solution into the electrode from its rear end using a long injection needle or a fine polyethylene tube. When the tip diameter is relatively small, the tip is first dipped into the solution to fill it through capillary action (or suction with a syringe), and then the rest of the solution is added from the rear end. Any small bubbles remaining in the filled part can be removed by gently tapping the electrode.

The electrode must not be filled completely with the solution. Otherwise, the solution may spill when inserting the electrode into the holder, and the solution inside the holder may cause large noise and other problems.

2.3.2. Giga-Seal Formation

In this section, we explain the procedure of giga-seal formation along with the maneuvers for amplifier operation. Currently, the most widely used patch-clamp amplifiers are the Axopatch 200B made by Molecular Devices and the EPC-10 (or -9) by HEKA Elektronik. The HEKA amplifiers are fully controlled digitally via software (PATCHMASTER or PULSE) on a personal computer (PC). Here we explain the procedure with reference to the panel indicators on these amplifiers or the software. The indicator labels will be identified with quotation marks. References will be made in the order (Axopatch)/(PATCHMASTER or PULSE) unless otherwise stated. Operation of the amplifiers provided by other manufacturers is more or less the same.

Before amplifier operation, we should know about the liquid junction potential that is generated at the interface between different solutions. The junction potential is generated so differences in the mobility of ions are canceled out by the potential and the net charge movement across the interface becomes zero (see also Chap. 28 in this book). For example, when the principal ions included in the bathing and pipette solutions are relatively small (e.g., Na^+, K^+, Cl^-, Cs^+), the junction potential generated would be several millivolts. This may not cause a serious problem. However, when large anions (e.g., aspartate, gluconate) are used as the principal anions in the pipette solution, the low mobility of these anions is compensated for by generation of a junction potential >10 mV higher on the bathing solution side of the interface; this nullifies the net charge movement across the interface. Therefore, this potential must be taken into account even when no current is generated in the circuitry of the amplifier after immersing the pipette tip into the bathing solution. The junction potential can easily be measured in advance. First, the mounted pipette tip containing the pipette solution is submerged in a bath containing the same pipette solution, and voltage is offset (zeroing done) through the amplifier using a saturated (or 3 M) KCl electrode as a bath reference. Then the solution in the bath is exchanged with the bathing solution that will be used in the experiments. The voltage value indicated by the amplifier at this moment is the junction potential. The junction potential can also be estimated by calculations using software (e.g., JPCalc in Clampex provided by Molecular Devices). In the case of the HEKA amplifier, the value of the junction potential (with respect to the potential inside the pipette) should be entered into the "LJ" box in the Amplifier window.

For experiments, before immersing the pipette tip in the bath, positive pressure should be applied from the back of the pipette through the pipette holder. (Weak pressure is already applied through hydrostatic pressure resulting from the height difference between the pipette solution and the bath. Additional pressure may be applied by the use of a syringe or expiration through the experimenter's mouth.) This pressure can blow away dust particles floating on the surface of the bath solution and in the bath so the particles do not stick to the tip and interfere with giga-seal formation. After the tip is submerged, the gain of the amplifier is set to a relatively lower level (1–5 mV/pA) and the recording mode is set to "TRACK" (which keeps the pipette current zero on average through a slow feedback loop including an integrator circuit)/set to "On Cell" and the "Track" button is to be clicked. This may prevent the monitored current from going off-scale during movement of the pipette tip. (Small voltage changes generated around the tip may accumulate during movement, causing a large current change.)

While advancing the pipette tip toward the cell, current responses to repetitive small voltage pulses (~5 mV amplitude, 5–10 ms duration) should be monitored. The pulses can be applied by switching the "EXT. COMMAND" to "SEAL TEST"/by turning on the "Test Pulse." The amplitude of the response is inversely proportional to the pipette resistance. When the tip is moved within several tens of micrometers from the cell, the "SEAL TEST" should be switched off in the Axopatch 200B amplifier; in addition, the value in the "METER" under the "V_{TRACK}" mode should be adjusted to indicate the junction potential by turning the "PIPETTE OFFSET" (in the case of the Axopatch amplifier, the polarity of the potential must be *inverted* – i.e., the potential with respect to that outside the pipette). Another method is to switch the recording mode to "V-CLAMP" at this moment, set the "HOLDING COMMAND" to the junction potential, and then adjust the value in the "METER" under the "I" mode to 0 with the "PIPETTE OFFSET." (The reason for the *inverse* setting of the potential is that the Axopatch amplifier always indicates the voltage with respect to the reference electrode. In the HEKA amplifiers, by contrast, the polarity of the voltage is automatically changed so the potential on the extracellular side becomes the voltage reference. Therefore, in the "On Cell" and "In Out" modes the reference is switched to the intrapipette side, whereas in the "Whole Cell" and "Out Out" modes it is switched to the bath electrode side.) When using the HEKA amplifier, adjustment is not necessary at this moment because the amplifier in the "Track" mode periodically adjusts the offset voltage in "V_0" so the basal current becomes zero on the assumption that the junction potential entered into "LJ" is generated. (Indeed, in the "V-membrane" box in the Amplifier window, the same value as in "LJ" must be indicated under this condition.)

When the pipette tip is moved nearly to the surface of the cell, the flow of pipette solution released from the tip should produce a small "dimple" on the surface if the condition of the cell is good. Even if this is not visible, small reductions in the current responses are seen because of a slight increase in the resistance at the tip (Fig. 2.6). How close the tip must be located to the cell for good giga-seal formation depends on the cell type, the solution composition, or the strength of the positive pressure applied to the pipette. Trial and error is unavoidable. (As a rule of thumb, one may use as a reference an ~1.5 MΩ increase in resistance or some degree of fluctuation in current responses.) When the tip is located appropriately, the recording mode is switched to "V-CLAMP", / the "Track" button is re-clicked to be switched off, and the positive pressure is released. The giga-seal may be formed by a simple release of pressure in some cases; if not, weak negative pressure (how weak depends on the case) must be applied by suction with a pump, a syringe, or inspiration through the experimenter's mouth. When the current response becomes almost flat, except for the small "spikes" generated at the beginning and end of the pulse, and the current noise level also becomes very small, the giga-seal has been attained. (The time required for giga-seal may also vary on a case-by-case basis.) After giga-seal formation, the negative pressure must be released. If a giga-seal is not attained for a long period, the pipette should be replaced with a new one. If the tip has touched a cell once, the pipette should not be reused because of adherence of cell debris or scratches on the tip.

In some cases, a giga-seal is attained only after brief treatment of the cell with some proteases to smooth its surface.

2.3.3. Single-Channel Recording

The "spikes" observed at the beginning and the end of the current response under the giga-sealed condition are called fast capacitive transients and originate from the stray capacitance of the pipette.

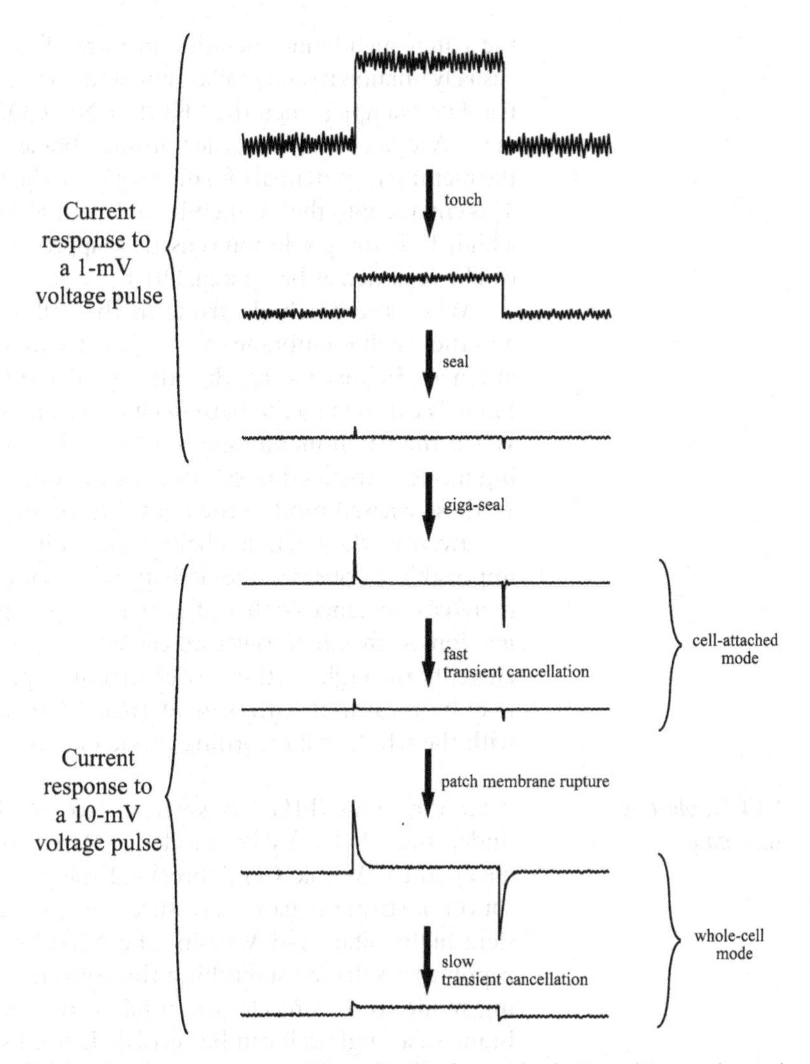

Fig. 2.6. Current responses used to monitor the changes in pipette resistance, giga-seal formation and fast and slow transient cancellation.

The transients can be electronically canceled out by adjusting "FAST" and "SLOW" in "PIPETTE CAPACITANCE COMPENSATION" / by clicking "Auto" in the "C-fast" box (Fig. 2.6) (the principle of this compensation is described later in detail). Because the stray capacitance is proportional to the surface area of the pipette submerged in the bath, a shallower depth of the bath chamber is preferable.

The mode of recording single-channel currents under this condition is called the "cell-attached" or "on-cell" recording mode. In this mode, the gain of the amplifier should be switched into the higher range (>50 mV/pA). (In addition, in the Axopatch amplifier, the "CONFIG." should be set to "PATCH (β=1).") Note that in this recording mode the transmembrane voltage across the patch membrane includes an intracellular membrane potential (usually unknown) and that a command voltage is added to the potential. For example, when the "HOLDING COMMAND" is set to V_p in the Axopatch amplifier, the transmembrane potential (V_m) becomes the membrane potential (V_i) minus V_p. (In the HEKA amplifier, when V_p is entered into the "V-membrane" box, V_m becomes $V_i + V_p$.) When a high-K^+ bathing solution is used, V_i approaches zero. Therefore, V_m can be regarded as being equal to V_p.

When the patch electrode in the cell-attached mode is lifted and the patch membrane excised, the inside-out recording mode is attained. In this mode, the intracellular side of the patch membrane is exposed to the bath, so its environment can be controlled. When the patch membrane is excised from the whole-cell recording mode (explained next), the outside-out mode may be attained. In these excised modes, the contribution of V_i should disappear.

Even in these single-channel recording modes, however, it is impossible to observe the activity of a channel with a sub-picosiemens conductance or that of a carrier or pump protein, which transfers ions with much lower efficiency. Nevertheless, the sum of the currents through all the carrier or pump proteins in a membrane may be measured with a giant (macro) membrane patch (12) or with the whole-cell recording mode (see also Chaps. 13 and 14).

2.3.4. Whole-Cell Recording

After the "CONFIG." is switched to "WHOLE CELL (β=1)" under the "V-CLAMP" mode, / After the recording mode is changed to "Whole Cell," breaking the patch membrane by application of strong negative pressure (−30 to −200 cm·H_2O) or transient high voltage (~1 V) using the "ZAP"/"Zap" function (called zapping) results in establishing the conventional whole-cell recording mode. In this mode, the total current across the whole membrane of a single cell can be recorded. It must be noted that the V_i is immediately clamped to V_p after breaking the patch. Thus, it is good practice to enter a presumed value of the resting potential of the cell into the "HOLDING COMMAND" / "V-membrane" box before breaking, especially for excitable cells. When ionophores such as nystatin are included in the pipette solution, simply waiting under the cell-attached mode may produce a perforated patch membrane through spontaneous incorporation of the ionophores into the membrane, which results in another slow whole-cell mode (explained in Sect. 2.3.5).

In either whole-cell mode, because the currents charging the cell capacitance must be generated additionally, a voltage pulse should elicit much larger and wider capacitive surge currents at the beginning and the end of the response (Fig. 2.6). These are called slow capacitive transients. These big transients may also be canceled out partially (the principle is explained later in detail) by adjusting "WHOLE CELL CAP" and "SERIES RESISTANCE" in "WHOLE-CELL PARAMETERS" / by adjusting the "Range," "C-slow," and

"R-series" values (it is recommended that "Auto" not be used initially in this case). After the adjustment, we can get approximate values of the cell capacitance and R_S from the panel indicators.

To measure nonstationary current responses accurately, it is necessary to clamp the membrane potential to V_{CMD} as quickly as possible. Furthermore, when a large membrane current is generated, error due to the voltage drop across the R_S is no longer negligible and should be corrected to some extent. For these purposes, the circuitry for series resistance compensation (explained later in detail) is operated by adjusting "% COMPENSATION" and "LAG" in "SERIES RESISTANCE COMP" / by adjusting the two values (2–100 ms and 0–95%) in the "Rs Comp" box. For this operation, the cancellation of fast and slow capacitive transients and a slight readjustment of their parameters become necessary.

When the amplifier mode is switched to "I-CLAMP NORMAL" or "I-CLAMP FAST"/"C-Clamp" under this condition, the voltage recording mode under current-clamp is attained. The *I–V* converter circuitry in the amplifier is intrinsically not suitable for voltage recording. However, as a result of the manufacturer's inclusion of ingenious compensating circuitries, the amplifiers currently produced can track membrane voltage changes almost as accurately as amplifiers based on voltage-follower circuits. Thus, the generation and recording of neuronal action potentials, for example, can be done with sufficient reliability using these amplifiers.

In the whole-cell configuration, the bathing solution can be freely exchanged with different solutions during recordings. However, it should be noted that the voltage difference that is generated initially between the bathing solution and the reference electrode (which has been canceled out by voltage offsets before giga-seal formation) may change, and in some cases (especially in the case of replacing most of the principal small ions such as Na^+ or Cl^- with much larger ions) this change may cause serious voltage errors. In such cases, possible errors may be reduced substantially by measuring in advance the voltage shift due to the solution exchange or by using a salt bridge made with saturated (or 3 M) KCl as the reference electrode. Exchanging the pipette solution is also possible with some devices (13) (see also Chap. 15). However, a similar problem (due to the voltage difference between the solution and the Ag–AgCl electrode wire) may also occur.

When the single-channel conductance is so small that individual channel activity cannot be resolved in the on-cell or excised patch mode (e.g., (14)), the conductance can be roughly estimated by analyzing the noise contained in the whole-cell current (see Chap. 6 for details). When the density of the channel is very low (and the probability of capturing the channel in the patch membrane is therefore small), the whole-cell recording mode may be suitable for detecting single-channel events if the channel conductance itself is sufficiently large (e.g., (15)).

2.3.5. Perforated Patch Recording with Nystatin

Perforated patch (slow whole-cell) recording with nystatin (or amphotericin B) was epoch making because it could overcome the washout problem encountered frequently under the conventional whole-cell recording mode. However, this mode is accompanied by several technical difficulties. Nystatin may be dissolved in dimethylsulfoxide (DMSO) at a high concentration (~50 mg/ml) for a stock, but the stock becomes inert within several days even when frozen at −20°C. On the day of use it is usually prepared at a working concentration of 50–100 μg/ml, but it cannot be dissolved completely even with sonication. The remaining undissolved aggregates may interfere with giga-seal formation. Once dissolved, it becomes inert within 1–2 h. Moreover, high concentrations of DMSO may damage the patch membrane. Horn and Marty (8) resolved these problems by applying the nystatin solution afterward via intrapipette perfusion with a fine polyethylene tube. However, pipette perfusion is not easy, and it takes time for nystatin to reach the patch membrane. Therefore, many researchers usually fill the pipette tip first with a nystatin-free solution and then add the nystatin-containing solution into the pipette from its backside. Controlling the time that nystatin reaches the membrane is still difficult, however, and the damaging effect of high concentrations of DMSO is not yet resolved. Yawo and Chuhma (16) found a way to alleviate these problems: The stock solution in a freezer can be preserved for a much longer time when nystatin is dissolved in methanol with fluorescein-Na at a ratio of 1:10 and protected from light. They also found that the amphiphilic dye fluorescein greatly facilitates nystatin dissolution in the pipette solution when it is dissolved after the methanol is evaporated with nitrogen gas just before use.

When the patch membrane is excised after establishing the perforated mode, it is possible that a vesicle containing part of the cytosol is formed if the excised edges of the membrane neatly fuse. In this condition, outside-out single-channel recording under a condition of nearly intact cytosolic metabolic and signaling cascades may become possible (7).

2.3.6. Methods of Cell Capacitance Measurement

An electrical model of a cell membrane can be divided into a resistance component and a capacitance component (Fig. 2.7). The resistance component reflects the ion permeability of the cell membrane, whereas the capacitance component is due to the formation of an electrical "capacitor" by the lipid bilayer of the cell membrane; its magnitude is proportional to the surface area of the membrane (the unit capacitance of the membrane is about 1 μF/cm^2). Using methods to measure cell capacitance, the cell surface area and the increases and decreases in the area caused by exocytosis and endocytosis, respectively, can be tracked in real time. Two representative methods are as follows.

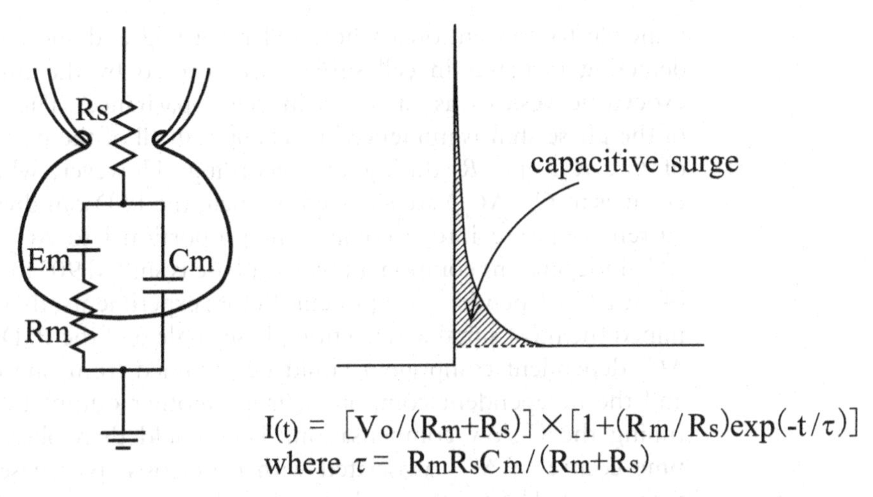

Fig. 2.7. Equivalent circuit for the whole-cell recording mode and a capacitive surge current induced by a square voltage pulse in the circuit.

2.3.6.1. Capacitive Surge Measurement (Time-Domain Technique)

The time course of a slow capacitive transient generated by a square voltage pulse (V_0) under whole-cell patch-clamp conditions (the equivalent circuit is shown in Fig. 2.7 in the left panel) can be described by the equation shown in Fig. 2.7 in the right panel. Thus, by fitting the equation to the actual transient trace, we can determine the cell capacitance (C_m). This equation holds unless the parameters C_m, R_m, and R_s do not vary during the pulse. To determine the parameters reliably, the record of the transient must be of sufficient length (at least five times longer than the decay time constant of the transient). The time resolution of the real-time C_m measurement is therefore limited by this factor.

If $R_m \gg R_s$ holds during the C_m measurement, the time integral of the exponential term in the equation (i.e., the total charge q accumulated by the cell capacitor due to the voltage pulse) becomes nearly V_0C_m. Thus, the C_m value can be obtained more simply by measuring the surge area q (shaded region in Fig. 2.7 in the right panel). As another alternative, when a linear ramp voltage pulse ($V=\beta t$) is applied to the cell, the current response should consist of the resistance component increasing linearly with a slope of β/R_m and the capacitance component with a constant value of βC_m. Thus, the changes in C_m can be evaluated from jumps in the capacitance component.

2.3.6.2. Phase-Sensitive Detection Method

When a high-frequency AC voltage signal (500 Hz to 1.5 kHz) is applied instead of a square voltage pulse, the current response must also be an AC signal. However, the phase of the current signal is shifted because of the presence of the cell capacitance. This phase shift can be detected and tracked using a lock-in amplifier (phase-sensitive detector; PSD). Neher and Marty (17) applied this principle to conventional whole-cell recording and succeeded in detecting increases in cell surface area caused by the fusion of exocytotic vesicles as increases in cell capacitance. The degree of the phase shift is influenced by changes in all of the parameters of C_m, R_m, and R_s during the recording. However, when the changes in C_m (ΔC_m) are sufficiently small, the PSD can divide the current response into a component proportional to ΔC_m and a ΔC_m-independent component whose phase is shifted 90° from that of the ΔC_m-dependent component. Before experiments, they determined the most suitable detection phase angle ϕ of the PSD so the ΔC_m-dependent component could be obtained from one output and the independent component from another output by hand-tuning the C-slow compensation. They could then observe the time course of ΔC_m associated with exocytosis (piecewise-linear technique). The ϕ may vary because of changes that accumulate in C_m, R_m, and R_s during the measurement; but a compensation procedure in which ϕ is automatically adjusted by estimating the ϕ periodically with an artificially varied R_S through computer calculation can be used (18).

The PSD gives two outputs for the amplitudes of the current components, which are perpendicular (normal) to each other in phase. If the equilibrium potential of the cell membrane, E_m, is known, all three parameters of C_m, R_m, and R_s must be determined by calculation (Fig. 2.7). Lindau and Neher (19) proposed a method for calculating these three parameters continuously through a computer with reference to the two PSD outputs and the response to a DC voltage signal on the assumption of a constant E_m (sine + dc method). Because errors in E_m little affect the calculated C_m value under the condition $R_m \gg R_s$ (20), the C_m can be tracked with sufficient reliability at high time resolution using the E_m value in conditions in which R_m undergoes a large decrease during measurement, if the value is known. To deal with a situation in which E_m is unknown, we previously proposed a procedure in which low-frequency alternating square voltage pulses are applied in addition to the AC signal to obtain all the parameters continuously regardless of unexpected E_m changes during measurement (21). With this procedure, we succeeded in recording several exocytotic events (21, 22).

The HEKA amplifiers (EPC-10 and EPC-9) include the PSD function in their operating software. These phase-tracking methods can therefore be easily used with these amplifiers.

2.3.7. Methods of Capacitance and Series Resistance Compensation

2.3.7.1. Pipette Capacitance Compensation

The fast capacitive transients generated under giga-seal conditions mainly reflect the currents charging the stray capacitance (C_p) of the pipette electrode. This current is proportional to the amplitude of the command voltage pulse (V_{CMD}), and a large current amplitude may cause saturation of the amplifier. Even if it is not as large as the saturation level, it may cause distortion of the fast current response generated at the beginning of V_{CMD}. As described above,

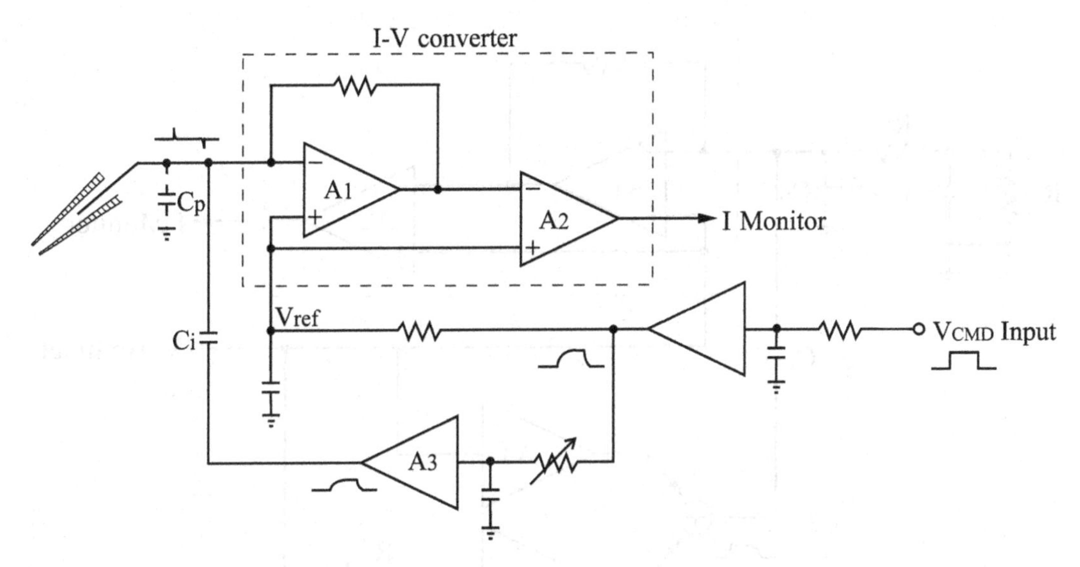

Fig. 2.8. Circuitry for pipette capacitance compensation.

because the C_p is proportional to the surface area of the pipette submerged in the bath, the submerged part should be as small as possible. Also, because the capacitance is inversely proportional to the thickness of the glass and proportional to its dielectric constant, thick and hard glass capillaries are more suitable for reducing the amplitude of the transient. The C_p can also be greatly reduced by coating the pipette with Sylgard.

The remaining C_p can be canceled out electronically through the specialized circuitry in the amplifier (Fig. 2.8). First, the rising and falling phases of V_{CMD} are slightly blunted by the first integrator circuit just after the V_{CMD} input. By sending this blunted V_{CMD} to the noninverting input of the *I–V* converter, the amplitudes of fast transients are slightly reduced. In the HEKA amplifier, the bluntness can be varied at two levels (time constant of 2 μs or 20 μs) through the "Stim" box in the Amplifier window. Second, to provide the C_p charging currents directly to the pipette through a circuit different from the *I–V* converter, the same blunted V_{CMD} is sent to another integrator circuit just before the OP amp A_3 in Fig. 2.8, further modulating the bluntness. The time constant of the bluntness can be varied through the potentiometer of the second integrator (the "τ" in "PIPETTE CAPACITANCE COMPENSATION" / the lower box in "C-fast"), and the amplitude of the blunted V_{CMD} can be varied by varying the gain of A_3 (the "MAG" in "PIPETTE CAPACITANCE COMPENSATION"/ the upper box in "C-fast"). This modulated V_{CMD} is then differentiated by a capacitor C_i, and the resulting currents are sent directly to the pipette simultaneously with V_{CMD} from the *I–V* converter. In this way, pipette charging is accomplished without detection of the charging currents by the *I–V* converter (3).

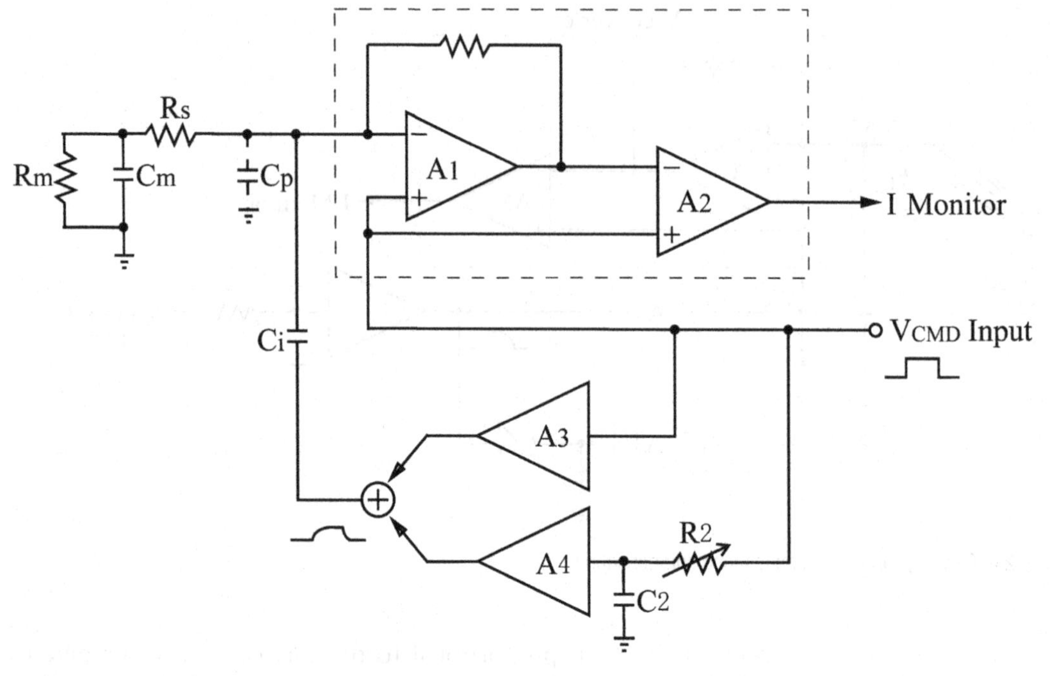

Fig. 2.9. Circuitry for whole-cell capacitance compensation.

2.3.7.2. Whole-Cell Capacitance Compensation

Under the whole-cell recording mode, in addition to C_p, the cell capacitance C_m is charged through the series resistance R_S. Slow capacitive transients are therefore additionally generated. When $R_S = 5$ MΩ and $C_m = 20$ pF, for example, the time constant of the slow transient becomes 100 μs (in the equation in Fig. 2.7, $\tau \approx R_S C_m$ when $R_m >> R_S$). This slow transient may be canceled out partially through different circuitry in the amplifier (Fig. 2.9), which uses a procedure similar to that for pipette capacitance compensation. V_{CMD} is sent to an integrator circuit specially designed for cell capacitance compensation, and its rising and falling time constants are modulated through the potentiometer R_2 (the "SERIES RESISTANCE" in "WHOLE CELL PARAMETERS" / "R-series"; the time constant can be further modulated in the HEKA amplifier by choosing C_2 with different capacities through "Range"), and its amplitude is modulated through the gain of A_4 (the "WHOLE CELL CAP." in "WHOLE CELL PARAMETERS" / "C-slow"). The output of A_4 is added to the output of A_3 derived from the pipette capacitance compensation circuitry, the total signal is differentiated by the capacitor C_i, and the resulting currents are sent directly to the pipette (3). In this way, the input into the *I–V* converter of a large current signal that may be generated by a large V_{CMD} or result from series resistance compensation (explained below in detail) is avoided, and saturation of the amplifier is prevented.

This circuitry, however, can only cancel out transients described by a single time constant (as in the equation in Fig. 2.7). Thus, the circuitry cannot perfectly cancel out transients generated in cells having a complex morphology of abundant fine cellular processes. (In the HEKA amplifier especially, if one tries to adjust R_2 and C_2 in such cases at one time with the "Auto" button in "C-slow" it frequently induces overcompensation and thus cannot be done.) When substantial cancellation is needed even in such cases, one may use a voltage pulse much smaller in amplitude than V_{CMD} to extract the linearly varying current components due to uncompensated transients and leak conductance. The expanded version of these components fitted to the amplitude of V_{CMD} can then be used to subtract the uncompensated transients (with leak currents) from the records.

2.3.7.3. Series Resistance Compensation

The series resistance (R_S) originates from the resistance other than that of the cell membrane and consists mainly of a component from the pipette shank and a component from the pipette tip that contains some broken patch membranes. The presence of R_S prevents accurate voltage-clamp owing to three effects.

1. The time constant of the slow capacitive surge ($\tau \approx R_S C_m$) indicates the time required for charging the cell capacitance (i.e., the time for the membrane potential to reach V_{CMD}). Therefore, the larger that R_S and C_m are, the longer is the time required for reaching V_{CMD}. As a result of this longer time, a fast current response cannot be detected accurately.
2. When a large current is generated in the whole-cell mode, a voltage drop across R_S yields an error in V_{CMD}. For example, when $R_S = 5$ MΩ and the current amplitude is 2 nA, the error becomes 10 mV.
3. When a nonstationary current response occurs, the actual membrane potential should vary according to the product of the varying current amplitude and R_S even when a constant V_{CMD} is applied. The resulting varying membrane potential should induce a continuously varying capacitive surge that distorts the current response.

The capacitance compensation procedures described above provide charging currents independent of the I–V converter. Thus, even when the capacitive transients seem to disappear in the oscilloscope monitor, the actual speed of voltage-clamp is unchanged. To accelerate the charging of the cell capacitance, a large transient voltage pulse (~10 μs) is concomitantly applied at the beginning of V_{CMD}. This results in increased charging speed (supercharging). Furthermore, to alleviate the effects of the voltage drop at R_S, some fraction of the voltage drop is brought back through a feedback loop to V_{CMD} (3). The fraction can be determined by tuning

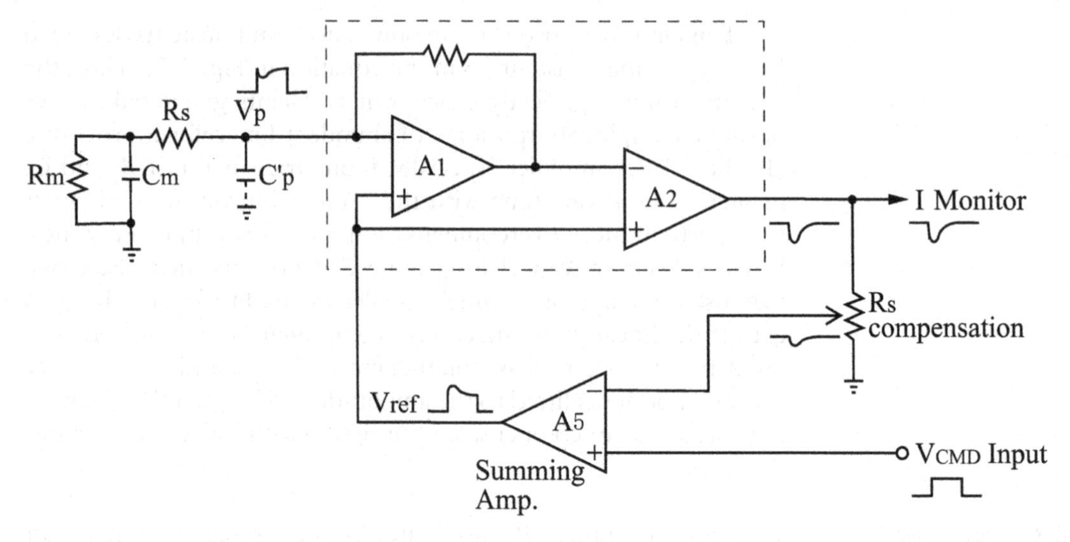

Fig. 2.10. Circuitry for series resistance compensation.

the potentiometer linked to the current output (indicated with "R_S compensation" in Fig. 2.10). With the Axopatch amplifier, the degree of supercharging (as the percent decrease in the time constant of the slow capacitive transient) can be determined through the "PREDICTION" potentiometer of the "%COMPENSATION" in the "SERIES RESISTANCE COMP." panel, and the fraction of R_S feedback can be determined through the "CORRECTION" potentiometer; the speed of feedback is set with "LAG (μs)." (With the HEKA amplifier, the left box in "Rs Comp" determines the speed of feedback, and the right box determines the degree of supercharging and R_S feedback with the common percent setting.) It should be noted that the degree of supercharging may be limited by the maximum voltage that can be applied through the amplifier and that R_S feedback may be disturbed by intrinsic errors and instability in the feedback circuitry; thus, 100% compensation cannot be attained for both types of compensation.

References

1. Neher E, Sakmann B (1976) Single-channel currents recorded from membrane of denervated frog muscle fibres. Nature 260:799–802
2. Hamill OP, Marty A, Neher E, Sakmann B, Sigworth FJ (1981) Improved patch-clamp techniques for high-resolution current recording from cells and cell-free membrane patches. Pflugers Arch 391:85–100
3. Sakmann B, Neher E (1995) Single-channel recording, 2nd edn. Plenum, New York
4. Neher E (1992) Ion channels for communication between and within cells. Science 256: 498–502
5. Sakmann B (1992) Elementary steps in synaptic transmission revealed by currents through single ion channels. Science 256:503–512
6. Kakei M, Noma A, Shibasaki T (1985) Properties of adenosine-triphosphate-regulated potassium channels in guinea-pig ventricular cells. J Physiol 363:441–462
7. Levitan ES, Kramer RH (1990) Neuropeptide modulation of single calcium and potassium channels detected with a new patch clamp configuration. Nature 348:545–547
8. Horn R, Marty A (1988) Muscarinic activation of ionic currents measured by a new

whole-cell recording method. J Gen Physiol 92:145–159

9. Armstrong DL, White RE (1992) An enzymatic mechanism for potassium channel stimulation through pertussis-toxin-sensitive G proteins. Trends Neurosci 15:403–408
10. Fan JS, Palade P (1998) Perforated patch recording with beta-escin. Pflugers Arch 436:1021–1023
11. Abe Y, Furukawa K, Itoyama Y, Akaike N (1994) Glycine response in acutely dissociated ventromedial hypothalamic neuron of the rat: new approach with gramicidin perforated patch-clamp technique. J Neurophysiol 72: 1530–1537
12. Hilgemann DW (1989) Giant excised cardiac sarcolemmal membrane patches: sodium and sodium-calcium exchange currents. Pflugers Arch 415:247–249
13. Soejima M, Noma A (1984) Mode of regulation of the ACh-sensitive K-channel by the muscarinic receptor in rabbit atrial cells. Pflugers Arch 400:424–431
14. Sakai H, Okada Y, Morii M, Takeguchi N (1992) Arachidonic acid and prostaglandin E2 activate small-conductance Cl^- channels in the basolateral membrane of rabbit parietal cells. J Physiol 448:293–306
15. Loirand G, Pacaud P, Baron A, Mironneau C, Mironneau J (1991) Large conductance calcium-activated non-selective cation channel in smooth muscle cells isolated from rat portal vein. J Physiol 437:461–475
16. Yawo H, Chuhma N (1993) An improved method for perforated patch recordings using nystatin-fluorescein mixture. Jpn J Physiol 43:267–273
17. Neher E, Marty A (1982) Discrete changes of cell membrane capacitance observed under conditions of enhanced secretion in bovine adrenal chromaffin cells. Proc Natl Acad Sci USA 79:6712–6716
18. Fidler N, Fernandez JM (1989) Phase tracking: an improved phase detection technique for cell membrane capacitance measurements. Biophys J 56:1153–1162
19. Lindau M, Neher E (1988) Patch-clamp techniques for time-resolved capacitance measurements in single cells. Pflugers Arch 411:137–146
20. Gillis KD (1995) Techniques for membrane capacitance measurements. In: Sakmann B, Neher E (eds) Single-channel recording, 2nd edn. Plenum, New York
21. Okada Y, Hazama A, Hashimoto A, Maruyama Y, Kubo M (1992) Exocytosis upon osmotic swelling in human epithelial cells. Biochim Biophys Acta 1107:201–205
22. Hashimoto A, Hazama A, Kotera T, Ueda S, Okada Y (1992) Membrane capacitance increases induced by histamine and cyclic AMP in single gastric acid-secreting cells of the guinea pig. Pflugers Arch 422:84–86

Chapter 3

Whole-Cell Patch Method

Hiromu Yawo

Abstract

With the whole-cell mode of patch-clamp the membrane current is recorded from a cell under the voltage-clamp. The membrane potential response of a cell is also measured and manipulated under a current-clamp. The method enables one to correlate the macroscopic aspects of ion channels with their microscopic properties. It also demonstrates how these ion channels regulate the membrane potential. This chapter summarizes the basic concepts of whole-cell patch-clamp recording and the practical procedures and protocols used to examine currents through channels/transporters across entire cells.

3.1. Introduction

The improved patch-clamp technique reported by Hamill et al. in 1981 (1) was remarkable and attracted the attention of many physiology researchers. In the improved technique, a patch electrode was pressed onto a cell surface to form a seal with the cell membrane, and negative pressure was applied to rupture the membrane directly under the electrode, thereby allowing communication between the electrolyte solution contained in the electrode and the intracellular fluid. The implication of this whole-cell recording technique was that patch-clamp techniques could be used not only in biophysical research of channels but also in research on cell physiology.

Hamill et al. noted that the improved technique made it possible to measure the membrane potential and the membrane current of small cells, a task that had been impossible with the previous glass microelectrode method. This new measurement capability enabled elucidation of the function/mechanism of ion channels in the cell. The physiological significance of single-channel (microscopic) recordings, described in Chap. 5, can only be understood in comparison with macroscopic current recordings.

Yasunobu Okada (ed.), *Patch Clamp Techniques: From Beginning to Advanced Protocols*,
Springer Protocols Handbooks, DOI 10.1007/978-4-431-53993-3_3, © Springer 2012

Another advantage of the new technique is that it became possible to perfuse intracellular fluid with an artificial fluid (see Chap. 15). Although intracellular perfusion had been tried on giant cells, such as snail ganglion neurons, it requires highly sophisticated techniques (2). Experimental alteration of the intracellular fluid enabled analysis of the intracellular signal transduction mechanism. In fact, during the three decades following the report of Hamill et al., whole-cell recording in cell physiological research has made dramatic progress. A close relation between cell function and membrane potential has been revealed in many types of cell (3, 4).

Yet another advantage offered by the whole-cell patch-clamp method was the ability to measure membrane capacitance, which was discussed in a 1982 paper by Neher and Marty (5). This ability enabled the measurement of important cell functions such as exocytosis and endocytosis (see Chap. 17).

The range of applications with whole-cell patch-clamp techniques is still expanding. Attempts to use a cell monitored under whole-cell recording as a biosensor may make it possible to apply the whole-cell recording method to the field of engineering (see Chap. 21). Moreover, the method of obtaining mRNA from a single cell using whole-cell patch-clamp techniques enables analysis of the spatial and temporal diversity of living cells (see Chap. 24).

3.2. Basic Experimental Techniques

Whole-cell patch-clamp techniques are described in detail elsewhere (6, 7). *The Axon Guide* published by Axon Instruments (Molecular Devices) contains a detailed description of the basic principles and techniques of patch-clamp methods (8). In this chapter, only practical information extracted from the above-mentioned literature is discussed.

3.2.1. Specimen Preparation

In the early stages of its development, the patch electrode technique was limited to dissociated cells or cultured cells (primary culture and cultured cell lines). There were two reasons for this limitation: (1) both the cell and the tip of the electrode must be visible under a microscope when pressing the patch electrode onto the cell surface and (2) in normal tissues, connective tissues, including collagen fibers, are present in intercellular spaces. These connective tissues must be removed to form a tight (high-resistance) seal between the patch electrode and the cell membrane (Box 3.1). In other words, the patch electrode method can be applied only if

Box 3.1
Collagenase Treatment

Dorsal root ganglion, autonomic ganglion, and retinal ganglion cell layers, for example, are coated by a membranous tissue. The patch electrode method can be used on these neurons after removing the coating tissue. Thin-slice preparations of brain tissues contain connective tissues in intercellular spaces, which prevent the patch electrode from reaching the cells located deep inside the tissues. Treating the tissues with collagenase can solve these problems (11, 12).

Method 1

1. Dissolve collagenase (2,000 U/ml) (type II; Sigma) and thermolysin (20 U/ml) (Sigma) in the solution from which the debris has been previously removed (e.g., using a syringe filter). It is a good idea to divide this solution into small samples at each session of the experiments and store them in a freezer.
2. Prepare a glass pipette with a tip diameter of 20–30 μm in the same way a patch electrode is prepared.
3. Fill a glass pipette with the collagenase/thermolysin mixture and connect the pipette to the patch electrode holder.
4. Gently place the tip on the target under a microscope, add a small positive pressure, and hold the pipette steady to administer the enzyme mixture from the tip of the glass pipette. Ideally, adjust the positive pressure as needed and administer the enzyme mixture contained in the pipette slowly over 10 min. You should see the membranous tissue being removed under the microscope.
5. Wash out the enzyme by incubating the specimens in the perfusate for about 10 min.

Method 2

1. Dissolve collagenase (200 U/ml) of (type II, Sigma) and thermolysin (2 U/ml) (Sigma) in the perfusate. It is a good idea to use a tenfold diluted solution of the stock solution prepared in Method 1.
2. Incubate the specimen in the enzyme mixture at 35–38°C for 0.5–1.0 h.

the specimens meet these two conditions. Therefore, the patch electrode method is now used on thin-slice preparations of tissues (9) (see Chaps. 7–10), autonomic ganglion cells (10–12), and so on, while preserving their tissue structure. In some cases, condition (1) is not necessary. For example, in tissues in which homogeneous cells form a cluster (e.g., hippocampus in the brain), inserting an electrode into the tissue while monitoring electrode resistance can create a high-resistance seal. Therefore, in the above procedure, the individual cells do not need to be visible (blind patch method) (see Chaps. 9 and 11) (13–15). In other words, with a good experimental design, the patch electrode method can be used on all kinds of cells and tissues.

3.2.2. Microscopes

Different microscopes are used depending on the specimen. A dissecting stereomicroscope is suitable for the blind patch method. Phase-contrast microscopes can be used for single-cell-layer specimens. The surface of a cell can be clearly observed using a Nomarski differential interference contrast (DIC) microscope. Upright Nomarski DIC microscopes are usually used for thin-slice preparations. For autonomic ganglia, individual cells can be identified under a usual upright microscope using oblique or reflected illumination (10, 16). It is reasonable to use an upright microscope for specimens consisting of layers of cells because the cell on which the patch electrode is pressed is usually in the top layer of the tissue sample. An infrared camera microscope can capture the internal structure of a thick tissue specimen, and the patch electrode can be attached to the interior of the specimen (17).

As a general precaution, line-frequency noise (hum) from the power lines must be eliminated. The most effective way to accomplish this is to connect the main body of the microscope with a ground. If the hum is originating from the microscope's light, the simplest solution is to turn off the light while carrying out the electrophysiological measurement. It is also a good idea to use a DC power supply for the microscope lamp. An expensive regulated DC power supply is not necessary, however. Rectifying the output current with diodes and a high-capacity condenser to DC and connecting either the positive or negative pole to a ground can also remove the hum. When a water-immersion objective lens is used with an upright microscope, the lens must be isolated to prevent hum, which has been done in recent models of water-immersion objective lenses. In previous measurements done at our laboratory, an insulator was placed between the water-immersion objective lens and the main body of the microscope.

3.2.3. Electrode Preparation

Electrodes used for whole-cell recordings are nearly the same as those used for single-channel recordings. However, the current measured in whole-cell recordings is usually larger than that in single-channel recordings; and the membrane area subjected to measurement is usually larger, which increases the noise level. Because of this, the noise from stray capacitances between the patch electrode and the solution is often relatively permissible. Thus, even for a thin electrode, it is not necessary to provide Sylgard coating all the way up to the tip. Stray capacitances need to be reduced to a certain degree, however, because otherwise the capacitance compensation described earlier does not work.

Sometimes the whole-cell capacitive current of a small cell is as same as or even less than the capacitive current derived from the stray capacitance. We use a thin electrode that has Sylgard coating on the area in contact with the solution, up to 1 mm from the tip. Alternatively, a thick-walled electrode is used without coating. However, the same precaution that is taken during single-channel recordings should be applied when measuring a small current. In other words, the signal-to-noise ratio should always be taken into consideration if experiments are to be conducted under ideal conditions.

Figure 3.1a shows an equivalent circuit of the whole-cell recording. Feedback for voltage clamping is effective up to the patch electrode solution. In the single-compartment model where the membrane potential is held constant under a voltage-clamp, the following relation holds under a steady state:

$$Vc = \frac{R_m V_p}{R_m + R_s} \tag{3.1}$$

where V_p denotes the electrode potential, V_c is the intracellular potential, R_m is the membrane resistance, and R_s is the access (series) resistance between the electrode and the cell. When the specific membrane resistance is constant, R_m is inversely proportional to the cell surface area. In other words, the membrane potential of a small cell can be controlled with ease; however, to control the membrane potential of a large cell, R_s should not be too large. Moreover, the specific membrane resistance changes if the membrane potential is changed. For example, when the membrane

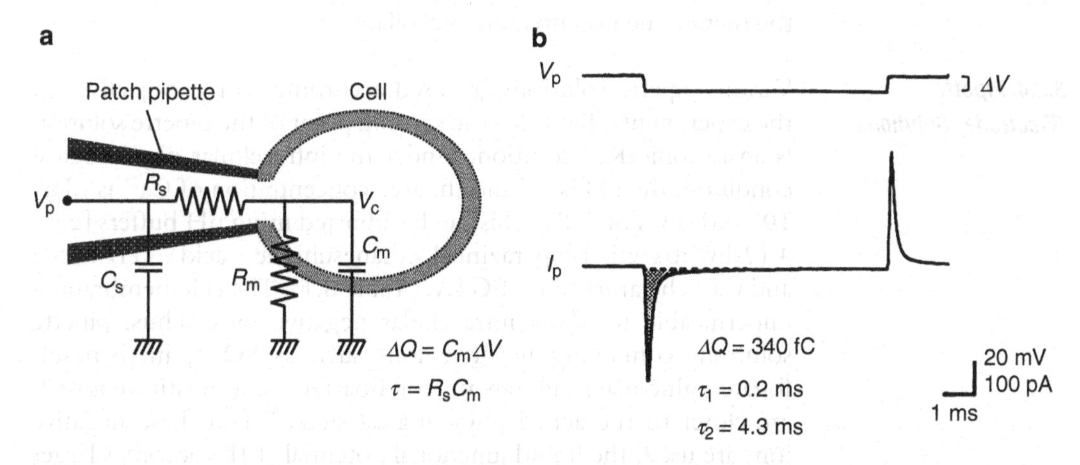

Fig. 3.1. Whole-cell patch recording. (**a**) Equivalent circuit in a whole-cell state. R_s: access/series resistance of patch electrode; R_m: cell membrane resistance; C_m: cell membrane capacitance; C_s: stray capacitance. (**b**) Whole-cell capacitive current (*bottom trace*) recorded in response to a hyperpolarizing pulse (top trace, $\Delta V = 10$ mV). Holding potential: −60 mV. The whole-cell membrane capacitance $C_m = 34$ pF is calculated from the patch area ΔQ. In this case, the falling phase of the capacitive current was best fitted to the sum of two exponential functions. Based on the two-compartment model (26), $C_1 = 16$ pF and $C_2 = 14$ pF were estimated from the time constants $\tau_1 = 0.2$ ms and $\tau_2 = 4.3$ ms. In this case, $R_s = 13$ MΩ.

potential changes from near zero to a positive potential, ion channels such as K^+ channels are activated, which reduces the specific membrane resistance. Under the above condition, R_s needs to be small enough to control the membrane potential. When the current to be measured is large, it is better to reevaluate whether the membrane potential is adequately controlled during voltage-clamp experiments (18).

The access resistance R_s is mostly dependent on the shape of the patch electrode (pipette). For electrodes with a conical tip, the patch electrode resistance R_p can be expressed using the following equation:

$$R_p = \frac{\rho l}{\pi r_s^2} + \frac{\rho \cot(\phi/2)}{\pi}\left(\frac{1}{r_t} - \frac{1}{r_s}\right) \tag{3.2}$$

Here, r_s and r_t are the internal radius at the base of the glass tube (the length l) and that at the tip, respectively; ϕ is the angle of the tip; and ρ is the specific resistance of the internal fluid: ρ is 50 Ωcm for 150 mM KCl (at 25°C). In other words, R_p becomes smaller as the tip diameter or tip angle increases. In general, R_s is two to five times larger than R_p. This is because in a whole-cell state the tip of the electrode contains a cell membrane and intracellular components.

With respect to the polishing at the tip of the electrode, it is not absolutely necessary to form a high-resistance seal. However, in our experience, even a small amount of heat polishing often helps hold the whole-cell state longer. Electrodes without heat polishing tend to allow current leakage from the membrane during the measurement; this leakage gradually increases and often makes the membrane potential uncontrollable.

3.2.4. Pipette (Electrode) Solutions

Various pipette solutions are used according to the objectives of the experiments. Basically, the starting point of the pipette solution is an isotonic KCl solution. Under the intracellular physiological condition, the pH is ~7 and the free concentration of Ca^{2+} is ~10–100 nM (pCa of 7–8). This can be adjusted using pH buffers [e.g., 4-(2-hydroxyethyl)piperazine-1-ethanesulfonic acid (HEPES)) and Ca^{2+} chelators (e.g., EGTA) (Table 3.1). The cell membrane is impermeable to many intracellular negative ions. Thus, pipette solutions containing negative ions such as SO_4^{2-}, methanesulfonate$^-$, gluconate$^-$, glutamate$^-$, or aspartate$^-$ as a substitute for Cl^- are closer to the actual physiological state. When these negative ions are used, the liquid junctional potential (LJP) becomes larger than the potential of the KCl solution. The LJP can be corrected on the basis of actual measurements or with calculations (19, 20) (see Chap. 28). When the negative ions are organic acids, the dissociation state is dependent on both the pH and temperature; thus, if the osmotic pressure is the problem, it should be corrected on the basis of actual measurements.

The susceptibility to microbial contamination should be considered. It is a good idea to divide the pipette solution into small samples and store them at −20°C. On the other hand, positive ions are often replaced with organic bases such as tetraethylammonium (TEA) and *N*-methyl-D-glucamine. The use of organic bases may shift the membrane potential dependence of ion channels to the negative potential side (21). One of the reasons for this is a change in the surface potential on the intracellular side of the plasma membrane.

The concentration of the Ca^{2+} chelator should be minimized when measuring the current of Ca^{2+}-dependent ion channels, studying the cell function regulated by intracellular Ca^{2+} (e.g., exocytosis) or measuring the intracellular concentration of Ca^{2+}. When the concentration of the Ca^{2+} chelator is high, the free Ca^{2+} concentration is fixed at a low level; this is not appropriate for observing the Ca^{2+}-dependent phenomena. Table 3.1 shows the compositions of the pipette solution commonly used to measure the intracellular concentration of Ca^{2+}. On the other hand, in some cases, Ca^{2+}-dependent phenomena (e.g., Ca^{2+}-dependent inactivation of Ca^{2+} channels) must be eliminated as much as possible. In such cases, chelators [e.g., 1,2-bis(2-aminophenoxy)ethane-*N*,*N*,*N'*,*N'*-tetraacetic acid

Table 3.1
Compositions of pipette solutions used in whole-cell recordings

Purpose	Substances	Ref.
Basic composition	KCl 140, $MgCl_2$ 1, $CaCl_2$ 1, EGTA 10, Mg-ATP 2, NaOH-HEPES 10; pH 7.3	(9)
Measure intracellular Ca^{2+}	Cs-glutamate 125, TEA-glutamate 20, $MgSO_4$ 2, K_5fura-2 0.1, Na_4ATP 2, HEPES 10; pH 7.0	(86)
Measure Ca^{2+} current	CsCl 140, NaOH 10, $MgCl_2$ 1, BAPTA 5, Mg-ATP 5, GTP 0.3, CsOH-HEPES 20; pH 7.4	(12)
ATP-regenerating system	Creatine phosphokinase 50 U/ml, Na_2 creatine-phosphate 20, Mg-ATP 5, GTP 0.04 added to an appropriate pipette solution	(22)

Numbers following each substance are the concentration in millimoles
BAPTA: 1,2-bis(2-aminophenoxy)ethane-*N*,*N*,*N'*,*N'*-tetraacetic acid, *EGTA*: ethyleneglycol-bis-(2-aminoethylether)-*N*,*N*,*N'*,*N'*-tetraacetic acid, *GTP*: guanosine triphosphate, *HEPES*: 4-(2-hydroxyethyl)piperazine-1-ethanesulfonic acid, *TEA*: tetraethylammonium

(BAPTA)] with a high binding speed are used instead of EGTA. Table 3.1 shows the composition of the pipette solution commonly used for measuring Ca^{2+} current. In these solutions, K^{+} is occasionally substituted by Cs^{+}. The purpose of this is to block the K^{+} channels with the intracellular Cs^{+} to isolate the Ca^{2+} current that is measured. This substitution can also reduce the whole-cell noise originating from various K^{+} channels.

One of the problems associated with whole-cell recordings is the run-down of the cell functions. Run-down occurs because the intracellular fluid is replaced with the pipette solution, and the soluble components necessary to maintain cell function are washed out. In previous attempts to prevent run-down, the pipette solution was supplemented with a soluble component. For example, supplying Mg-adenosine triphosphate (ATP) to the pipette solution can reduce the run-down speed of the Ca^{2+} current. This can be explained because Ca^{2+} channels have sites for phosphorylation, and maintenance of their function requires the phosphorylated state. With the reduction of ATP concentration, the equilibrium should shift to the side of dephosphorylation. However, supplementing with ATP increases the adenosine 5′-diphosphate (ADP) concentration near the channels, thereby promoting dephosphorylation. To prevent this, an ATP-regenerating system is used (22) (Table 3.1). In some systems, G-proteins are involved in receptor-mediated signaling. To maintain the function of these receptors, the pipette solution must be supplemented with Mg-guanosine triphosphate (GTP) 0.1–0.3 mM. Presynaptic terminals may use glutamic acid as a neurotransmitter, and the run-down of this neurotransmitter can be prevented by using glutamate as the negative ion for the pipette solution (23). ATPs and GTPs used for metabolism are usually chelating Mg^{2+}; thus, the pipette solution needs to contain at least an equimolar amount of Mg^{2+}.

To analyze intracellular signaling pathways, various endogenous or exogenous substances are experimentally included in the pipette solutions. For example, enzymes (e.g., A-kinase, C-kinase, proteases), phosphatase inhibitors, intracellular factors (e.g., calmodulin), second messengers [e.g., cyclic adenosine monophosphate (cAMP), inositol 1,4,5-triphosphate (IP_3)], agents that inhibit these substances, or selective toxins can be administered to the cell. These agents include a group of compounds called caged substances (or caged compounds) that are biologically inactive but activated after being degraded by exposure to intense light. In a previous attempt, the molecules that were thought to be involved in cell function were identified by introducing antibodies or competing peptide fragments in the cytoplasm. Cell function can be studied by introducing Ca^{2+}-sensitive or pH-sensitive dyes; in addition, cell morphology can be studied by introducing biocytine or fluorescent dyes such as Lucifer yellow. The pipette solution can be replaced while maintaining a whole-cell state (intracellular perfusion method), which enables qualitative and quantitative analysis of intracellular factors. The intracellular perfusion method is explained in Chap. 15.

3.2.5. Pipette Pressure Control

Various methods are used to control the positive or negative pressure at the tips of the electrodes. One of the better known methods is to apply pressure or suction by mouth from the end of the tube connected to the pipette fluid. Here, we introduce the method we use that requires a syringe and a simple manometer (Fig. 3.2). The syringe is connected between a *U*-shaped tube filled with colored water and a patch electrode holder via a three-way stopcock. Another three-way stopcock is connected at the end of the *U*-shaped tube. This system is kept closed except when refilling water as it evaporates. A positive or negative pressure is applied to the patch electrode as follows:

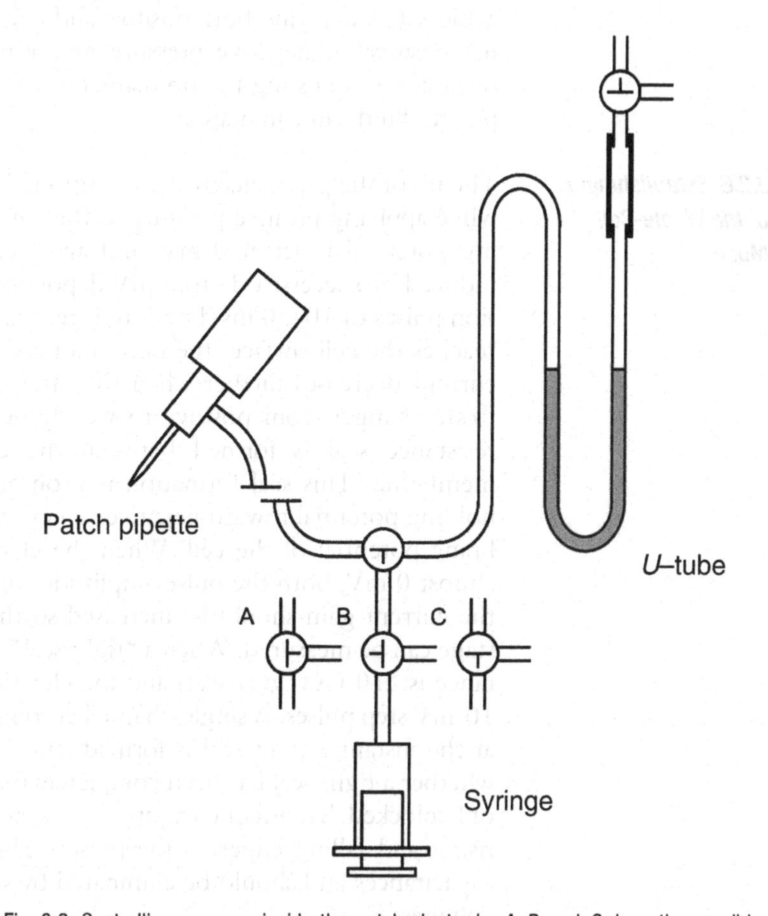

Fig. 3.2. Controlling pressure inside the patch electrode. *A*, *B*, and *C* show the possible positions of a three-way stopcock.

1. Pull the syringe to draw air with the three-way stopcock in the *A* position.
2. Place the three-way stopcock in the *B* position and apply a positive pressure by pressing the syringe.
3. Retain the positive pressure by turning the three-way stopcock to the *C* position.
4. Release the positive pressure by turning the stopcock to the *A* position again.
5. Apply negative pressure by pulling the syringe after turning the stopcock to the *B* position. Retain this negative pressure by turning the stopcock to the *C* position.
6. Release the negative pressure by turning the stopcock to the *A* position again.

It is difficult to read the precise value of the applied pressure from the level of the colored water in the *U*-shaped tube. However, the pressure required to form or rupture a patch can be familiar based on one's experience, and good reproducibility can be achieved by applying both positive and negative pressures. A residual positive or negative pressure may remain after the pressure release, but adjusting the tip diameter, angle, or the height of the pipette fluid can eliminate it.

3.2.6. Establishment of the Whole-Cell Mode

The tip of the patch electrode is positioned close to the cell surface while applying positive pressure to the patch electrode. The holding potential is set at 0 mV, and an electrical current, which is induced by successive 1- to 5-mV depolarization or hyperpolarization pulses of 10–20 ms duration, is recorded. When the electrode reaches the cell surface, the pulse-induced current decreases. The current decreases further when the internal pressure of the electrode changes from positive to weakly negative because a high-resistance seal is formed between the electrode and the cell membrane. This seal formation is promoted by a change in the holding potential toward negative values around the resting membrane potential of the cell. When the electrical current becomes almost 0 mV, both the pulse amplitude (up to about 10 mV) and the current gain should be increased so the seal resistance at this stage can be measured. When a "tight seal" is formed, the seal resistance is ≥10 GΩ (giga-seal) and the electrical current is ≤1 pA for 10-mV step pulses. A single-channel current is sometimes observed at the instant a tight seal is formed, and it is a good indicator of whether a tight seal has been completely formed or the electrode is only blocked. Transient currents are observed synchronously at the rising and falling edges of the pulses. They originate from stray capacitances and should be eliminated by supplying compensation voltages.

Next, a relatively large negative pressure is applied to the patch electrode to rupture the membrane patch. There are also other methods to rupture the membrane, such as giving a single, short, high-intensity pulse (±0.5–5.0 V for 2–20 ms) or applying a high-frequency voltage (Zap method). Large transient currents are observed at the onset and offset of the 10-mV step pulse, just after rupture of the membrane patch (Fig. 3.1b). They are capacitive currents (capacitative currents, capacitance currents, capacity currents) originating from the cell membrane. We can determine the membrane capacitance of the cell (C_m), the membrane resistance (R_m), and the access resistance between the electrode and the cell (R_s) by analyzing the capacitive currents (Fig. 3.1b). The capacitive currents are eliminated as much as possible by providing compensation voltages. When the circuit equivalent to a whole-cell patch can be expressed as the model in Fig. 3.1a, the capacitive currents could be, in principle, compensated completely. C_m and R_s can also be determined from the current required for the capacitive current compensation. However, this compensation is difficult in practice; it is only possible with almost globular, small cells. Compensation in asymmetrical cells (e.g., nerve cells with dendrites) is limited to the area around the patch electrode because of the high resistance of the cytoplasm. In this case, the compensation value corresponds to the membrane capacitance of the restricted area where the voltage is clamped. The C_m, R_m, and R_s of such cells can be determined by an equation deduced from a circuit electrically equivalent to such cells (7, 24–27).

Whole-cell recordings involve current-clamp and voltage-clamp modes. In the former, we can determine changes in the membrane potential by providing a constant or modulated current. For example, the native response of cells (e.g., resting membrane potential, spontaneous oscillations in the membrane potential, spontaneous impulses) can be obtained when the current is fixed at zero. Modulated currents generated according to various protocols, such as a square pulse, a ramp wave, a sinusoidal wave, and a Zap-function wave (28–30), have been applied to manipulate the cell (Fig. 3.3a). Current-clamp experiments are important for elucidating the contribution of each ion current (which can be identified and measured under a voltage-clamp mode) to the formation of the membrane potential. The classification of neurons is now underway based on an analysis of their response characteristics under a current-clamp mode.

The success of the voltage-clamp depends on the current amplitude and the rate of current change. To keep the membrane potential constant when the current flows across the membrane, the feedback system of the instrument must respond rapidly enough to compensate for the potential change. For example, in nerve cells, the Na^+ current flows rapidly in response to a depolarization potential. However, when the feedback response is slow, positive

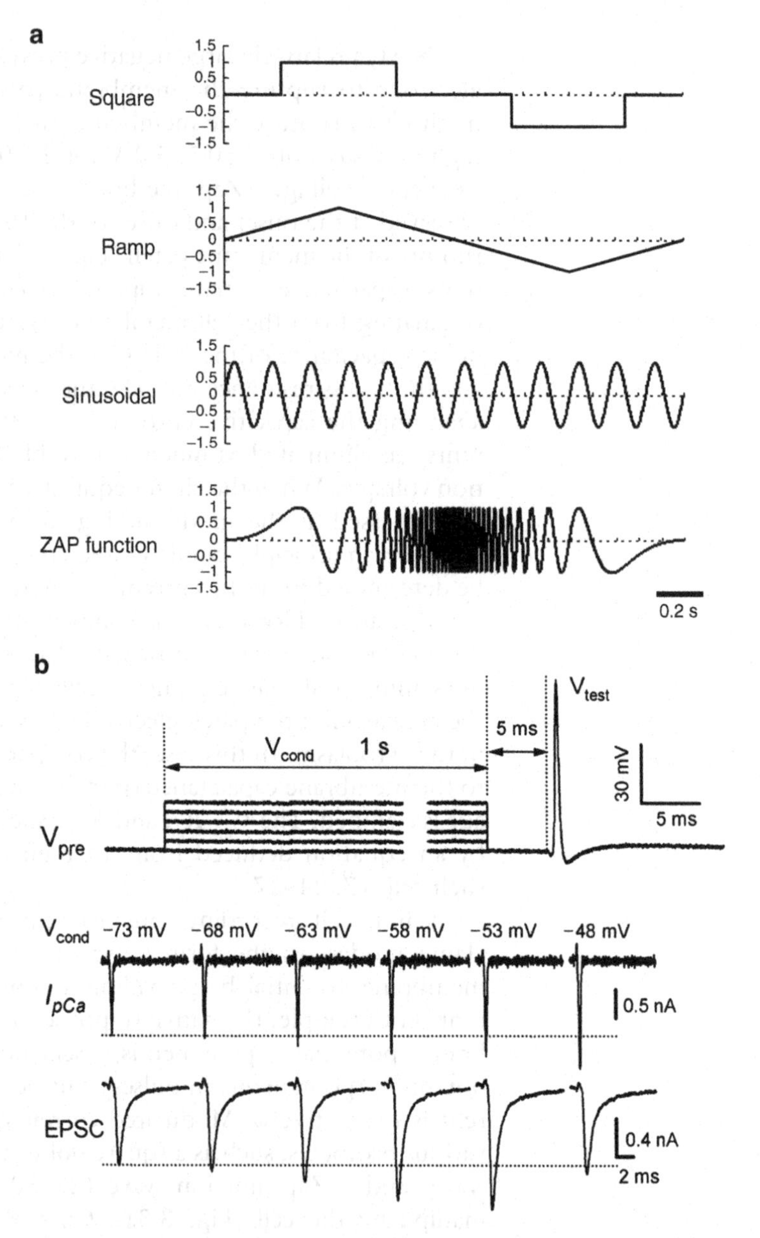

Fig. 3.3. Waveforms used in current-clamp and voltage-clamp experiments. (**a**) From *top* to *bottom*: square pulses, a ramp wave, a sine wave (3 Hz), and a Zap-function wave (0.3 → 30 → 0.3 Hz). Maximum amplitudes of these waves are normalized to 1. (**b**) Voltage-clamp experiments in which voltage is changed according to the waveform of the action potential. *Top panel*: One-second prepulses (−73, −68, −63, −58, −53, and −48 mV) were applied to the presynaptic terminal of the calyx of Held in mice under voltage-clamp conditions, and the waveforms of the action potential were determined. Normal cerebrospinal fluid, but with NaCl replaced by TEA-Cl, was used as extracellular fluid to minimize Na^+ and K^+ currents. Tetrodotoxin (TTX) was also included. In addition, cations in the inner-patch solution were replaced with Cs^+ and TEA^+ so only Ca^{2+} currents were detected, with negligible Na^+ or K^+ currents. *Middle* and *lower panels*: Ca^{2+} currents from the presynaptic terminal and excitatory postsynaptic currents (EPSCs) from the postsynaptic cell, respectively, after the prepulses. (**b**, from (32), with permission).

charges accumulate in the cell, and the membrane potential shifts farther to the depolarization side. This results in greater enhancement of Na^+ flow and eventually generates the escape response of an action potential. The response time of the instrument's feedback system is given by the following equation:

$$\tau = \left[R_s' R_m / \left(R_s' + R_m \right) \right] C_m \quad (3.3)$$

where R_s' is $R_s'(1-f)$ (f is the feedback compensation value); τ is nearly equal to $R_s' C_m$ because $R_m >> R_s'$ usually holds. This means that the voltage-clamp conditions can be easily established by decreasing both the access resistance and the area of the cell surface. In addition, a decrease in the current amplitude has a favorable effect on voltage-clamp performance. For example, in Na^+ current measurements, compensating for a change in the membrane potential may be possible by decreasing the Na^+ concentration in the extracellular fluid or by decreasing the number of active channels by using tetrodotoxin (TTX).

Modulated voltages generated according to various protocols, such as a square pulse, a ramp wave, a sinusoidal wave, and a Zap-function wave, have also been considered in the voltage-clamp experiments. For example, when voltage is changed following to the waveform of the action potential (Fig. 3.3b), the size of each ion current accompanied by action potentials can be measured as a function of time (31, 32). The voltage-clamp with the waveform of ramp pulses is a convenient way to determine the current–voltage relation (*I–V* relation) quickly (as noted later), but cautions are necessary when the current is changed in a time-dependent manner.

3.2.7. Problems of Whole-Cell Recording

As soon as the whole-cell state is established, the intracellular fluid is perfused with a pipette solution (washout). Often this greatly affects the small cells. In other words, the whole-cell state is an artificial environment for the cell. The whole-cell patch method allows us experimentally to control the intracellular fluid, but there is a risk of losing normal cell function. For example, soluble functional molecules contained in the cell may be lost. Some of them are involved in the channel activities and the intracellular signaling cascades. In the case of Ca^{2+} channels, the run-down that is often observed during whole-cell recording can be explained by the above-described mechanism (33–36). Moreover, in whole-cell recordings, the mechanism of receptor-mediated signal transduction is often lost (37). Another problem related to whole-cell recording is that there is a change in the buffering ability of intracellular Ca^{2+} (38–40).

This problem can be solved to a certain degree by supplying substances, which can be identical to those washed out, to the pipette solution. Current run-down can be reduced by supplying

Mg^{2+}-ATP, active subunits of cAMP-dependent protein kinases, protease inhibitors such as leupeptin, or strong Ca^{2+} buffers to the pipette solutions. However, most of the substances lost by washout are difficult to identify. In addition, normal cell function often depends on dynamic quantitative/qualitative changes in the functional molecules. When these molecules are supplied to the pipette solution, the fixed concentrations of the functional molecules make it difficult to observe the above-mentioned dynamic changes. Perforated patch-clamp recordings or gramicidin perforated patch recordings can be used to solve these problems (see Chap. 4).

3.3. Applications of Whole-Cell Recording

3.3.1. Macroscopic Characterization of Ion Channels

In general, individual ion channels are either "open" or "closed." The shift from one state to another is very quick (on the order of microseconds), and the subconductance state is rarely observed. However, some ion channels are known to have two or more open states (41–47). From this point of view, the properties of ion channels are stochastically dependent on time and space. In other words, the number of ions (i.e., current) flowing through a certain type of ion channel is determined by the time for a single channel dwelling in the open state during a certain observation time period and the number of ion channels in the open state. The latter cannot be determined from the activity of a single channel but can be found by macroscopically measuring the current with a whole-cell clamp. Assuming that individual channels open or closed independently (Markov process), the ratio of the channel's dwelling time in the open state to the observation time period (open probability p, a microscopic term) should be equal to the ratio of the number of functional channels dwelling in the open state to the total number of functional channels (N) in the cells during a very short time around a given time instant (a macroscopic term). In other words, the single channel current i and the whole-cell current I have the following relation:

$$I = Npi \tag{3.4}$$

3.3.1.1. Membrane-Potential Dependence of Channel Activity

One of the important factors that define the state of a channel is the membrane potential. Many ion channels show dependence on the membrane potential. Figure 3.4 shows the membrane-potential dependence of various ion channels in whole-cell current measurements. In Fig. 3.4 (bottom panels), the Y axis represents the current, and the X axis denotes the membrane potential (the current–voltage curve, I–V plot). The membrane potential at zero current is called the reversal potential (E_{rev}), or the zero-current potential, and this potential is dependent on the ion selectivity of

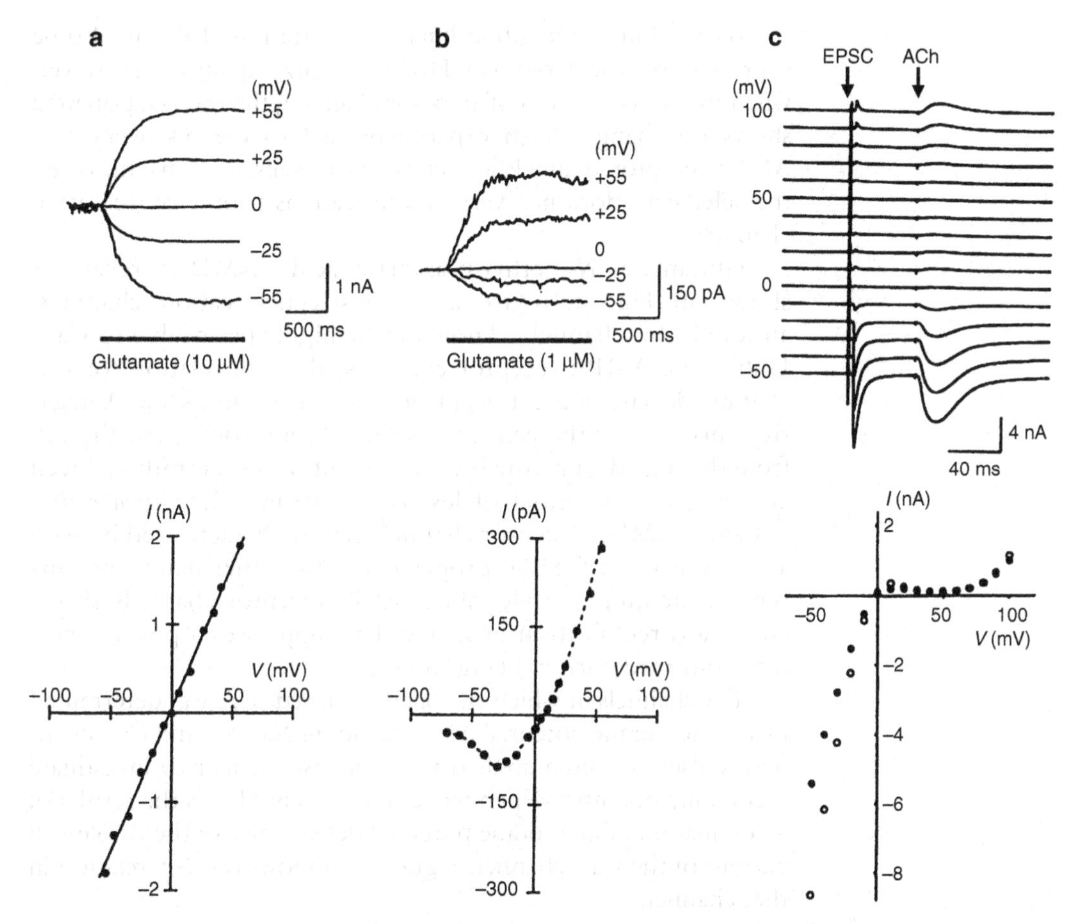

Fig. 3.4. Membrane potential dependence of receptor channel currents. In each part of the figure, the sample current recordings are presented at the top and the current–voltage (*I–V*) curves for the peak response are at the bottom. (**a**) Glutamate AMPA response in rat cerebellar Purkinje cells. (**b**) Outward rectifier response of glutamate NMDA response in rat cerebellar granule cells. (**c**) Inward rectifier response of nicotinic acetylcholine receptor in mouse submandibular ganglion neurons. Open circles: response to excitatory postsynaptic current (EPSC) by preganglionic fiber stimulation; filled circles: response to iontophoretically applied acetylcholine (ACh). (**a**, **b**, from (87), with permission. **c**, from (10)).

the channel and the ion concentrations (e.g., $[K]_o$, $[Cl]_i$) on both sides of the membrane. For example, the reversal potential for glutamate α-amino-3-hydroxy-5-methyl-4-isoxazolepropionic acid (AMPA) receptor channels is about 0 mV in Fig. 3.4a, but it is dependent on the extracellular concentration of Na^+ ($[Na]_o$). Their relation can be almost precisely expressed by the following Goldman–Hodgkin–Katz equation with the appropriate permeability coefficient of each ion (e.g., P_K, P_{Na}, P_{Cl}).

$$E_{rev} = \frac{RT}{F} ln \frac{P_K[K]_o + P_{Na}[Na]_o + P_{Cl}[Cl]_i}{P_K[K]_i + P_{Na}[Na]_i + P_{Cl}[Cl]_o} \qquad (3.5)$$

Its relation to the intracellular concentration of K^+ can also be expressed by the Goldman–Hodgkin–Katz equation. However, when the concentration of anions is changed, the reversal potential shows no change. Such experiments led to the discovery that AMPA receptor channels are selective to alkaline cations; however, the selectivity does not vary among cations (nonselective cation channels).

Glutamate *N*-methyl-D-aspartic acid (NMDA) receptors shown in Fig. 3.4b are also nonselective cation channels, although this channel is known to be highly permeable to Ca^{2+}. Unlike the AMPA receptor channels, the *I–V* relation for this channel deviates a great deal from the linear Ohm's law. A negative current (for the cell, this is the current going into the cell from the outside, i.e., the inward current) is significantly reduced at a negative potential of less than −50 mV (hyperpolarizing voltage). NMDA receptor channels can be characterized by such an outward rectification property. On the other hand, the current of nicotinic acetylcholine (ACh) receptor channels shows an inward rectification property that suppresses a positive current (outward current) (Fig. 3.4c).

For channels in which the open or closed state was determined by the membrane potential itself, including Na^+, K^+, and Ca^{2+} channels, pulses are often applied in a stepwise manner to investigate membrane potential dependence. Figure 3.5a shows the results of an evaluation of membrane potential dependence of the whole-cell current of the Ca^{2+} channel. Figure 3.5b shows the *I–V* relation in that channel.

3.3.1.2. Tail Current Analysis

In voltage-dependent channels, transient currents (tail currents) can be observed during stepwise changes in the membrane potential. Figure 3.5c shows an example of tail currents that originated from Ca^{2+} channels. For example, when the voltage was changed from the holding potential of −70 mV to +20 mV in a stepwise manner, a relatively small whole-cell current flowed. The reason for this is that because the voltage was close to the reversal potential of Ca^{2+} channels the driving force was relatively small. However, when the membrane potential was returned to −70 mV, a large transient current flowed. The reason for this is that because the driving force was relatively large a relatively large current flowed through the individual channels. However, each of the channels starts to close in a stochastic manner. Let us assume that the channel has *n* closed states and one open state, i.e.,

$$C_n \underset{\alpha_n}{\overset{\beta_n}{\rightleftharpoons}} \rightleftharpoons C_2 \underset{\alpha_2}{\overset{\beta_2}{\rightleftharpoons}} C_1 \underset{\alpha_1}{\overset{\beta_1}{\rightleftharpoons}} O \tag{3.6}$$

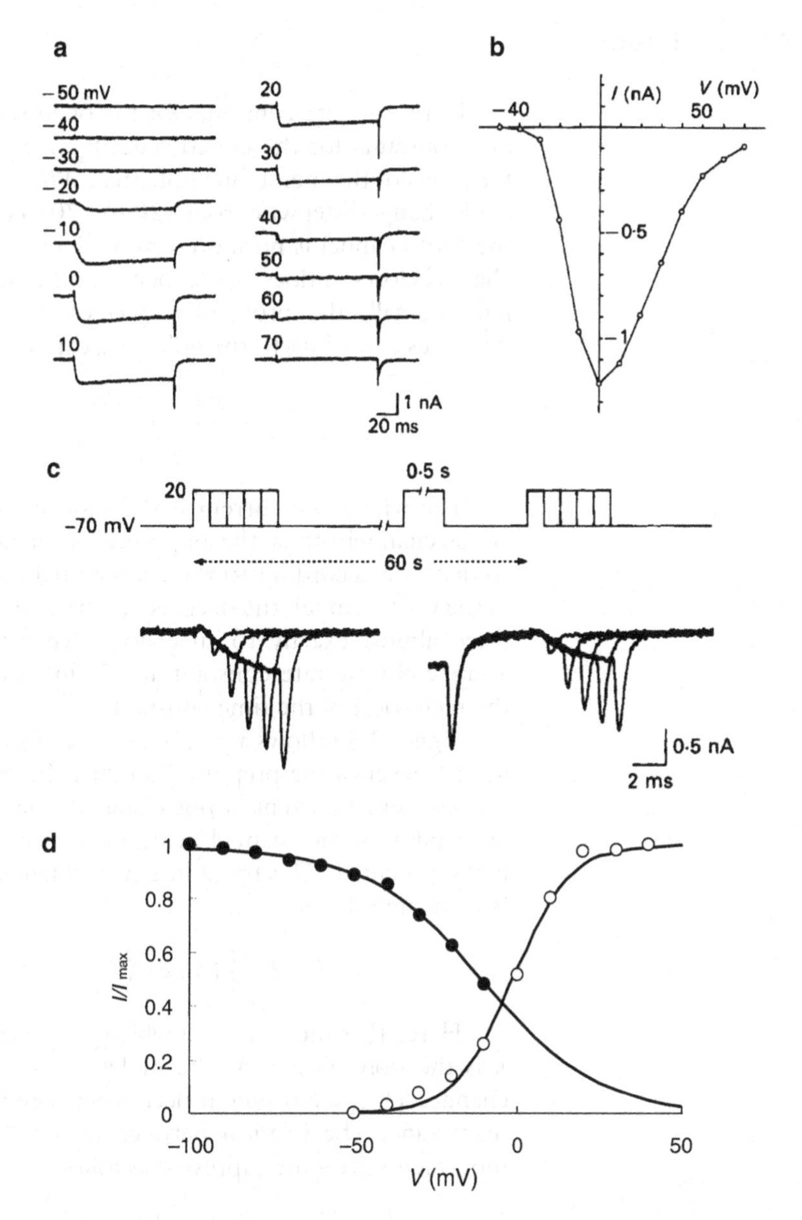

Fig. 3.5. Membrane potential dependence of Ca^{2+} channel activity. (**a**) Sample recordings of Ca^{2+} current in a presynaptic terminal of a chick ciliary ganglion: response to 100-ms depolarizing pulses at the holding potential of –80 mV. (**b**) *I–V* curve for the peak Ca^{2+} currents in the same sample. (**c**) Ca^{2+} tail currents in a postsynaptic neuron of a chick ciliary ganglion. The voltage was changed from the holding potential of –70 mV to +20 mV for a variable duration and then changed to –70 mV in a stepwise manner (*top*). The capacitive and leak currents were digitally subtracted by adding 10 successive responses of 1/10 voltage command with the opposite polarity. The Ca^{2+} current responses are presented in the *bottom superimposed traces*. With the prepulse to +20 mV for 0.5 s each Ca^{2+} current was partially inactivated (*right*). (**d**) Membrane potential dependence of activation (*open circles*) and inactivation (*filled circles*) of high-voltage-activated Ca^{2+} currents in the ciliary neurons. The membrane potential dependence of the activation was determined by plotting the relative amplitudes of tail currents (relative to the maximum value) to the prepulse level. With respect to the membrane potential dependence of the inactivation, the Ca^{2+} currents were recorded with various holding potential levels. The relative current amplitudes are plotted for different prepulse voltages. Each fitted line was obtained on the basis of the Boltzmann relation (see text). (**a**, **b**, from (12); **c**, from (88)).

Here, the rate constants for the open state, $\alpha_1 \ldots \alpha_n$, and the rate constants for the closed state, $\beta_1 \ldots \beta_n$, can be expressed as functions of the membrane potential. When the membrane potential is changed stepwise from +20 to −70 mV starting at time $t=0$, the Ca^{2+} channel is no more activated at −70 mV; thus, once the channel closes, it does not reopen. In other words, when β_1 is negligibly small, the probability $p(t)$ of the channel being open decreases according to the following exponential function:

$$p(t) = p(0)e^{-t/\tau} \tag{3.7}$$

$$\tau = \alpha_1^{-1} \tag{3.8}$$

The whole-cell current is the sum of currents through individual channels; thus, the amplitude of the tail current is expected to decrease according to the exponential function. For example, in the Ca^{2+} channel, the decreasing phase of the tail current can be fitted almost exactly to an exponential function (48), and the average closing rate constant at −70 mV can be calculated from the reciprocal of the time constant.

Figure 3.5d shows a plot of tail currents recorded while changing the level of the prepulse potential. In the above, because the single-channel current is not changed, the relation describes the probability of the channel being open as a function of the membrane potential. The fitted line was obtained on the basis of the Boltzmann relation:

$$I = I_{\max}\left[1 + exp\left\{(V_h - V)/\kappa\right\}\right]^{-1} \tag{3.9}$$

Here, V_h is the voltage at which half of the channels open, and κ is the slope factor. As the molecular structure of the channel changes, charge movement occurs between inside and outside the membrane. The relation between κ and the number of charge movements Z can be expressed as follows (49):

$$Z = \frac{kT}{e\kappa} \approx \frac{24}{\kappa} \tag{3.10}$$

In the Ca^{2+} channel shown in Fig. 3.5d, $V_h = -1.5$ and $\kappa = 8.5$ mV, thus, the number of charge movements is estimated to be three.

3.3.1.3. Inactivation of Channels

Some potential-dependent channels are known to be inactivated depending on the membrane potential. Figure 3.6a shows the inactivation of N-type Ca^{2+} channels at a single-channel level. The opening and closing of individual channels occur stochastically; thus, the time-dependent activity of the channel can be elucidated from averaging repeated events (ensemble average) (Fig. 3.6b). On the other hand, whole-cell currents represent spatial averaging of the channel activity. The time dependence is nearly the same for both the ensemble and the whole-cell current (Fig. 3.6b). This is consistent with the idea that individual channels open and close independently.

In general, a channel's inactivation rate is dependent on the membrane potential, as is the degree of inactivation in the steady state. Figure 3.5d displays an evaluation of the membrane potential dependence of the steady-state inactivation of Ca^{2+} channels. The Ca^{2+} channels are inactivated by changing the holding potential prior to the pulse; the corresponding current is then measured. Taking I_c as the current component that remains non-inactivated, this relation can be expressed using the Boltzmann relation,

$$I = I_c + I_{max}\left[1 + exp\left\{(V - V_h)/\kappa\right\}\right]^{-1} \quad (3.11)$$

In this case, $V_h = -11$ and $\kappa = 18$ mV, and the number of charge movements was estimated to be one or two.

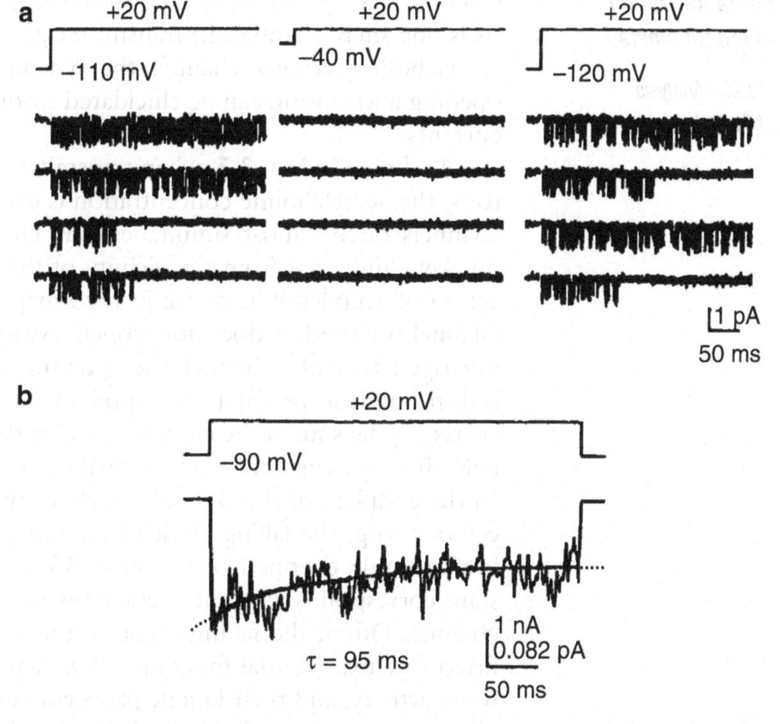

Fig. 3.6. Inactivation of Ca^{2+} channels. (**a**) Single-channel recording of an N-type Ca^{2+} channel. *Top row*: Each part shows a voltage command protocol. Two traces in the top row on the left, no change was observed in the channel activity during the depolarization; whereas in the two traces in the two rows from the bottom, the channels in the open state are found only immediately after depolarization. N-type Ca^{2+} channels characteristically move back and forth between these two to three modes (89, 90). Almost no channels were open while holding the depolarizing potential. (**b**) Ensemble-averaged single-channel current and the whole-cell current of N-type Ca^{2+} channels were overlayed after adjusting amplitudes. The result of best fitting to the exponential function and the constants is indicated by a *dotted line*. (From (91), with permission).

3.3.1.4. Gating Currents

The charge movements during the activation/inactivation of channels are directly measured as a small current under whole-cell recording (gating current) (50). A gating current associated with activation is observed before ion currents. To extract the gating current alone, the ion current is measured under voltage-clamp with channel blockers or with the elimination of permeable ions. After subtracting linear ion current components and capacitive current components, asymmetrical capacitive current components can be extracted as the gating current. An example of the gating current analysis using the whole-cell patch method has previously been reported for Ca^{2+} channels in the myocardial cell (51). During excitation and contraction of cultured skeletal muscle cells, charge movements similar to gating currents were observed. It was suggested that these charge movements were caused by a structural change in the dihydropyridine receptor located in the T-tubule (52).

3.3.2. Microscopic Characterization of Ion Channels

3.3.2.1. Analysis of Synaptic Currents

Ion channels can be characterized microscopically through an analysis of whole-cell currents. The aforementioned tail current analysis is one such example: In transmitter-gated channels such as the acetylcholine receptor channel, the dynamic property of a channel opening and closing can be elucidated through analysis of synaptic currents.

As shown in Fig. 3.7, when transmitters are released by exocytosis, the acetylcholine concentration is extremely high, and many channels open almost simultaneously. However, acetylcholine is quickly eliminated from the vicinity of the channel mainly by the action of acetylcholinesterase in the synaptic cleft; thus, once the channel is closed, it does not reopen. Assuming that there is only one open state of a channel, the open time for individual channels is distributed according to an exponential function. The whole-cell current is the sum of the currents flowing through individual channels; thus, the current decays according to an exponential function. In the endplate of the skeletal muscle or the sympathetic ganglion cell of a frog, the falling phase of the synaptic currents almost follows a single exponential function (53–55). Here, the time constant corresponds with the average open time of the acetylcholine channel. Often, the falling phase of the synaptic current is better fitted to biexponential function. Such channels show a burst-and-wane activity, and their kinetic properties are expected to be complex (e.g., presence of a blocked state in addition to the closed and open states). The larger time constant corresponds to the average bursting period; the smaller one corresponds to the average open time during the burst (56).

When the membrane potential is changed in a stepwise manner in the presence of the receptor ligands, a transient current similar to a tail current can be observed. In the acetylcholine channel of the skeletal muscle cell, this transient current follows an exponential function; in the above, the time constant is the same as the

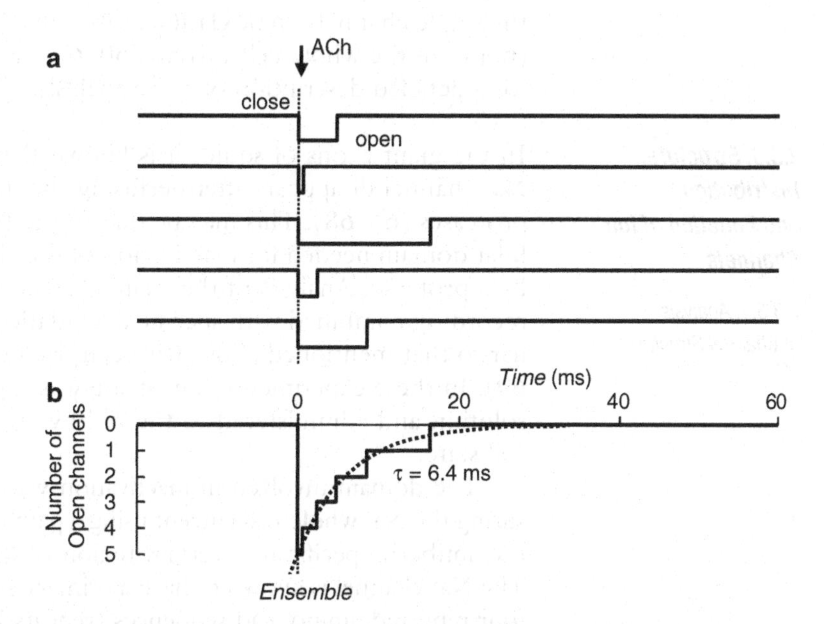

Fig. 3.7. Relation of the falling phase of the synaptic current to the average open time in a single channel. (**a**) Acetylcholine (*ACh*) released in the synaptic cleft is rapidly eliminated; thus, once the channel is closed, it does not reopen. The open time of individual simulated channels stochastically varies. (**b**) The falling phase of ensemble follows an exponential function. Because the ensemble is assumed to be equivalent to the whole-cell current, the falling phase of the synaptic current is presumed to follow an exponential function (*dotted line*).

average open time at that membrane potential (voltage-jump analysis) (57–61). With regard to the acetylcholine channel in the autonomic ganglion cell, the transient current that is recorded while changing the membrane potential in a stepwise manner is in good agreement with biexponential function (62). This again indicates that the channel exhibits a burst-and-wane activity. Moreover, with regard to the time constants, the larger one corresponds to the average bursting period; the smaller one corresponds to the average open time during the burst.

3.3.2.2. Measurement of Single-Channel Activity

Whole-cell recording is often performed from a small cell with small membrane capacitance. As the background noise is low in such a case, stepwise activity of the single channel can be ascertained (1, 19, 47).

3.3.2.3. Analysis of Whole-Cell Noise

As the number of open channels at a given time varies stochastically, the amplitude of whole-cell current fluctuates around a certain average. Assuming that there are only two states (open/closed) in individual channels, and that the shift from open to closed or from closed to open is instantaneous, the property of the single channel can be clarified through analysis of fluctuations (noise) in the whole-cell current (60, 63–65). Please see Chap. 6 for a detailed description of noise analysis.

3.3.3. Structure, Distribution, and Function of Ion Channels

3.3.3.1. Analysis of Channel Structure

In the giant axons of squid, it is known that inactivation of the Na^+ channel disappears after perfusing the inside of the cell with proteases (66–68). This may be due to the fact that the intracellular domain needed for inactivation of the channel is hydrolyzed by a protease. Analysis of the channel structure using whole-cell recording combined with specific degradation by an enzyme similar to that mentioned above has been tried using small cells (34, 69). In these experiments, a protease was supplied to the pipette solution and administered to the cell by establishing the whole-cell state.

The domain involved in inactivation was determined by measuring the Na^+ whole-cell current using a patch electrode containing the antibody specific to a certain region of the Na^+ channel (70). The Na^+ channel is known to have a primary structure consisting of four repeated amino acid sequences (repeats I–IV), each of which has six transmembrane domains (3, 4, 71). The antibody against the peptide consisting of 20 amino acids as an intracellular segment between repeats III and IV is known to delay the onset of Na^+ current inactivation, and the delay is positively related to the time after the whole-cell state is established. However, when both the antibody and antigen peptide were supplied to the pipette solution, the onset of inactivation was not delayed. This can be interpreted as demonstrating that inactivation is inhibited as the antibody binds to the intracellular segment of the Na^+ channel. In other words, this experiment suggests that the intracellular segment between repeats III and IV is involved in inactivation of the Na^+ channel.

3.3.3.2. Studies on Channel Distribution

To understand the cell function mediated by ion channels, knowledge about the channel distribution and density is indispensable. For example, in the adrenal medulla, three Ca^{2+} channel subtypes (facilitation-type, P-type and N-type) have been identified, and these subtypes can be pharmacologically isolated. When the amount of increase per Ca^{2+} ion in the membrane capacitance was compared among these subtypes, the efficiency of the facilitation-type Ca^{2+} channel was found to be about five times greater than that of the P- or N-type Ca^{2+} channels. Based on whole-cell recordings, an increase in the membrane capacitance corresponds to exocytosis; thus, the facilitation-type Ca^{2+} channels were suggested to be localized in the vicinity of the exocytosis sites (72).

In general, ion channels are localized, and this is closely related to cell function. The Ca^{2+}-dependent K^+ current caused by the inward flow of Ca^{2+} in the hair cell is not inhibited by intracellular administration of the Ca^{2+} chelator ethyleneglycol-bis-(2-aminoethylether)-*N*,*N*,*N'*,*N'*-tetraacetic acid (EGTA) through a patch electrode but by the administration of BAPTA (73). The binding potential to Ca^{2+} is nearly the same between EGTA and BAPTA; however, the Ca^{2+}-binding speed of BAPTA is greater than that of EGTA, indicating that the Ca^{2+}-dependent K^+ channels are localized near the Ca^{2+} channels (74–77).

3.3.3.3. Studies on Subcellular Components

Neurons are typical cells that have polarity. Membrane potential changes are conducted from the dendrites to the soma where they are integrated. Once the membrane potential at the initial segment of the axon exceeds the threshold value, an action potential is generated and is conducted along the axon. When the action potential reaches the presynaptic terminal, the transmitter is released, which changes the membrane potential of the postsynaptic membrane. In neurons, whole-cell recordings have been made from the soma, dendrite, and presynaptic terminal (see Chaps. 8 and 10). For example, in the cerebral cortical neuron, whole-cell current-clamp recordings were made from the soma and the dendrite at the same time (17). Although action potentials were recorded from the soma and the dendrite, as the Na^+ channel blocker QX314 diffused from the patch electrode at the dendrite only the action potential from the dendrite diminished. In other words, Na^+ channels also exist in the dendrite, and the action potential retrogradely spreads to the dendrite (78). Exact control of the membrane potential can be established by directly measuring the ion channel activity from a fragment (dendrosome) of the dendrite (79). In general, the presynaptic terminal is very small; however, whole-cell recording can be applied for exceptionally large presynaptic terminals. Whole-cell currents through Ca^{2+} (Fig. 3.5a,b), K^+, or Na^+ channels have been recorded previously from various large presynaptic terminals (12, 23, 32, 80, 81). In a developing neuron, growth cones are formed at the tip of the neurites. The whole-cell recording has previously been attempted using a large growth cone (82, 83). The ion current in a growth cone was also measured by a method in which perfusion with the electrolyte solution is applied topically while carrying out whole-cell recordings from the soma in a solution without electrolytes (84, 85).

References

1. Hamill OP, Marty A, Neher E, Sakmann B, Sigworth FJ (1981) Improved patch-clamp techniques for high-resolution current recording from cells and cell-free membrane patches. Pflugers Arch 391:85–100
2. Brown AM, Wilson DL, Tsuda Y (1985) Voltage clamp and internal perfusion with suction-pipette method. In: Smith TG Jr, Lecar H, Redman SJ, Gage PW (eds) Voltage and patch clamping with microelectrodes. American Physiological Society, Bethesda, pp 151–169
3. Hille B (1992) Ionic channels of excitable membranes, 2nd edn. Sinauer, Sunderland
4. Nicholls JG, Martin AR, Wallace BG (1992) From neuron to brain, 3rd edn. Sinauer, Sunderland
5. Neher E, Marty A (1982) Discrete changes of cell membrane capacitance observed under conditions of enhanced secretion in bovine adrenal chromaffin cells. Proc Natl Acad Sci USA 79:6712–6716
6. Marty A, Neher E (1983) Tight-seal whole-cell recording. In: Sakmann B, Neher E (eds) Single-channel recording. Plenum, New York, pp 107–122
7. Marty A, Neher E (1995) Tight-seal whole-cell recording. In: Sakmann B, Neher E (eds) Single-channel recording, 2nd edn. Plenum press, New York, pp 31–52
8. Yamane D (ed) (2008) The axon guide, 3rd edn. MDS analytical technologies, Sunnyvale http://www.moleculardevices.com/pdfs/Axon_Guide.pdf
9. Edwards FA, Konnerth A, Sakmann B, Takahashi T (1989) A thin slice preparation for patch clamp recordings from neurones of the mammalian central nervous system. Pflugers Arch 414:600–612
10. Yawo H (1989) Rectification of synaptic and acetylcholine currents in the mouse submandibular ganglion cells. J Physiol Lond 417:307–322
11. Yawo H (1999) Involvement of cGMP-dependent protein kinase in adrenergic potentiation of transmitter release from the calyx-type presynaptic terminal. J Neurosci 19:5293–5300
12. Yawo H, Momiyama A (1993) Re-evaluation of calcium currents in pre-and postsynaptic neurones of the chick ciliary ganglion. J Physiol Lond 460:153–172
13. Blanton MG, Lo Turco JJ, Kriegstein AR (1989) Whole cell recording from neurons in slices of reptilian and mammalian cerebral cortex. J Neurosci Methods 30:203–210
14. Coleman PA, Miller RF (1989) Measurement of passive membrane parameters with whole cell recording from neurons in the intact amphibian retina. J Neurophysiol 61:218–230
15. Manabe T, Renner P, Nicoll RA (1992) Postsynaptic contribution to long-term potentiation revealed by the analysis of miniature synaptic currents. Nature 355:50–55
16. Purves D, Hadley RD, Voyvodic JT (1986) Dynamic changes in the dendritic geometry of individual neurons visualized over periods of up to three months in the superior cervical ganglion of living mice. J Neurosci 6:1051–1060
17. Stuart GJ, Dodt H-U, Sakmann B (1993) Patch-clamp recordings from the soma and dendrites of neurons in brain slices using infrared video microscopy. Pflugers Arch 423:511–518
18. Armstrong CM, Gilly WF (1992) Access resistance and space clamp problems associated with whole-cell patch clamping. Methods Enzymol 207:100–122
19. Fenwick EM, Marty A, Neher E (1982) A patch-clamp study of bovine chromaffin cells and of their sensitivity to acetylcholine. J Physiol Lond 331:577–597
20. Neher E (1992) Correction for liquid junction potentials in patch clamp experiments. Methods Enzymol 207:123–131
21. Malecot CO, Feindt P, Trautwein W (1988) Intracellular n-methyl-D-glucamine modifies the kinetics and voltage-dependence of the calcium current in guinea pig ventricular heart cells. Pflugers Arch 411:235–242
22. Forscher P, Oxford GS (1985) Modulation of calcium channels by norepinephrine in internally dialyzed avian sensory neurons. J Gen Physiol 85:743–763
23. Borst JGG, Helmehen F, Sakmann B (1995) Pre- and postsynaptic whole-cell recordings in the medial nucleus of the trapezoid body of the rat. J Physiol Lond 489:825–840
24. Jackson MB (1992) Cable analysis with the whole-cell patch clamp, theory and experiment. Biophys J 61:756–766
25. Jackson MB (1993) Passive current flow and morphology in the terminal arborizations of the posterior pituitary. J Neurophysiol 69: 692–702
26. Llano I, Marty A, Armstrong CM, Konnerth A (1991) Synaptic- and agonist-induced excitatory currents of purkinje cells in rat cerebellar slices. J Physiol Lond 434:183–213
27. Major G, Evans JD, Jack JJ (1993) Solutions for transients in arbitrary branching cables: I. voltage recording with a somatic shunt. Biophys J 65:423–449

28. Gutfreund Y, Yarom Y, Segev I (1995) Subthreshold oscillations and resonant frequency in guinea-pig cortical neurons: physiology and modeling. J Physiol Lond 483: 621–640
29. Hutcheon B, Yarom Y (2000) Resonance, oscillation and the intrinsic frequency preferences of neurons. Trends Neurosci 23:216–222
30. Puil E, Gimbarzevsky B, Miura RM (1986) Quantification of membrane properties of trigeminal root ganglion neurons in guinea pigs. J Neurophysiol 55:995–1016
31. Llinas R, Sugimori M, Simon SM (1982) Transmission by presynaptic spike-like depolarization in the squid giant synapse. Proc Natl Acad Sci USA 79:2415–2419
32. Hori T, Takahashi T (2009) Mechanisms underlying short-term modulation of transmitter release by presynaptic depolarization. J Physiol Lond 587:2987–3000
33. Belles B, Malecot CO, Hescheler J, Trautwein W (1988) "Run-down" of the Ca current during long whole-cell recordings in guinea pig heart cells, role of phosphorylation and intracellular calcium. Pflugers Arch 411:353–360
34. Hescheler J, Trautwein W (1988) Modification of L-type calcium current by intracellularly applied trypsin in guinea-pig ventricular myocytes. J Physiol Lond 404:259–274
35. Kameyama M, Hescheler J, Hofmann F, Trautwein W (1986) Modulation of Ca current during the phosphorylation cycle in the guinea pig heart. Pflugers Arch 407:123–128
36. Shuba YM, Hesslinger B, Trautwein W, McDonald TF, Pelzer D (1990) Whole-cell calcium current in guinea-pig ventricular myocytes dialysed with guanine nucleotides. J Physiol Lond 424:205–228
37. Horn R, Marty A (1988) Muscarinic activation of ionic currents measured by a new whole-cell recording method. J Gen Physiol 92:145–159
38. Beech DJ, Bernheim L, Mathie A, Hille B (1991) Intracelluar Ca^{2+} buffers disrupt muscarinic suppression of Ca^{2+} current and M current in rat sympathetic neurons. Proc Nat Acad Sci USA 88:652–656
39. Neher E (1988) The influence of intracellular calcium concentration on degranulation of dialysed mast cells from rat peritoneum. J Physiol Lond 395:193–214
40. Neher E, Augustine GJ (1992) Calcium gradients and buffers in bovine chromaffin cells. J Physiol Lond 450:273–301
41. Bormann J, Hamill OP, Sakman B (1987) Mechanism of anion permeation through channels gated by glycine and gamma-aminobutyric acid in mouse cultured spinal neurones. J Physiol Lond 385:243–286
42. Cull-Candy SG, Usowicz MM (1987) Multiple-conductance channels activated by excitatory amino acids in cerebellar neurons. Nature 325:525–528
43. Hamill OP, Sakmann B (1981) Multiple conductance states of single acetylcholine receptor channels in embryonic muscle cells. Nature 294:462–464
44. Hamill OP, Bormann J, Sakmann B (1983) Activation of multiple-conductance state chloride channels in spinal neurones by glycine and GABA. Nature 305:805–808
45. Jahr CE, Stevens CF (1987) Glutamate activates multiple single channel conductances in hippocampal neurons. Nature 325:522–525
46. Smith SM, Zorec R, McBurney RN (1989) Conductance states activated by glycine and GABA in rat cultured spinal neurones. J Membr Biol 108:45–52
47. Takahashi T, Momiyama A (1991) Single-channel currents underlying glycinergic inhibitory postsynaptic responses in spinal neurons. Neuron 7:965–969
48. Swandulla D, Armstrong CM (1988) Fast-deactivating calcium channels in chick sensory neurons. J Gen Physiol 92:197–218
49. Hille B (1984) Ionic channels of excitable membranes. Sinauer, Sunderland
50. Armstrong CM (1992) Voltage-dependent ion channels and their gating. Physiol Rev 72:S5–S13
51. Field AC, Hill C, Lamb GD (1988) Asymmetric charge movement and calcium currents in ventricular myocytes of neonatal rat. J Physiol Lond 406:277–297
52. Adams BA, Tanabe T, Mikami A, Numa S, Beam KG (1990) Intramembrane charge movement restored in dysgenic skeletal muscle by injection of dihydropyridine receptor cDNAs. Nature 34:569–572
53. Kuba K, Nishi S (1979) Characteristics of fast excitatory postsynaptic current in bullfrog sympathetic ganglion cell. Effects of membrane potential, temperature and Ca ions. Pflugers Arch 378:205–212
54. Lipscombe D, Rang HP (1988) Nicotinic receptors of frog ganglia resemble pharmacologically those of skeletal muscle. J Neurosci 8:3258–3265
55. Magleby KL, Stevens CF (1972) The effect of voltage on the time course of end-plate currents. J Physiol Lond 223:151–171
56. Colquhoun D, Hawkes AG (1983) The principles of the stochastic interpretation of ion-channel mechanisms. In: Sakmann B, Neher E

(eds) Single-channel recording. Plenum, New York, pp 135–175

57. Adams PR (1977) Relaxation experiments using bath-applied suberyldicholine. J Physiol Lond 268:271–289
58. Adams PR (1977) Voltage jump analysis of procaine action at frog end-plate. J Physiol Lond 268:291–318
59. Ascher P, Large WA, Rang HP (1979) Studies on the mechanism of action of acetylcholine antagonists on rat parasympathetic ganglion cells. J Physiol Lond 295:139–170
60. Colquhoun D, Dreyer F, Sheridan RE (1979) The actions of tubocurarine at the frog neuromuscular junction. J Physiol Lond 293: 247–284
61. Neher E, Sakmann B (1975) Voltage-dependence of drug-induced conductance in frog neuromuscular junction. Proc Nat Acad Sci USA 72:2140–2144
62. Rang HP (1981) The characteristics of synaptic currents and responses to acetylcholine of rat submandibular ganglion cells. J Physiol Lond 311:23–55
63. Anderson CR, Stevens CF (1973) Voltage clamp analysis of acetylcholine produced end-plate current fluctuations at frog neuromuscular junction. J Physiol Lond 235:655–691
64. Colquhoun D, Hawkes AG (1977) Relaxation and fluctuations of membrane currents that flow through drug-operated channels. Proc Roy Soc Lond B 199:231–262
65. Howe JR, Colquhoun D, Cull-Candy SG (1988) On the kinetics of large-conductance glutamate-receptor ion channels in rat cerebellar granule neurons. Proc Roy Soc Lond B 233:407–422
66. Armstrong CM, Bezanilla FM, Rojas E (1973) Destruction of sodium conductance inactivation in squid axons perfused with pronase. J Gen Physiol 62:375–391
67. Oxford GS, Wu CH, Narahashi T (1978) Removal of sodium channel inactivation in squid axons by n-bromoacetamide. J Gen Physiol 71:227–247
68. Wang GK, Brodwick MS, Eaton DC (1985) Removal of Na channel inactivation in squid axon by an oxidant chloramine-T. J Gen Physiol 86:289–302
69. Gonoi T, Hille B (1987) Gating of Na channels. Inactivation modifiers discriminate among models. J Gen Physiol 89:253–274
70. Vassilev PM, Scheuer T, Catterall WA (1988) Identification of an intracellular peptide segment involved in sodium channel inactivation. Science 241:1658–1661
71. Catterall WA (2000) From ionic currents to molecular mechanisms: the structure and function of voltage-gated sodium channels. Neuron 26:13–25
72. Artalejo CR, Adams ME, Fox AP (1994) Three types of Ca^{2+} channel trigger secretion with different efficacies in chromaffin cells. Nature 367:72–76
73. Roberts WM (1993) Spatial calcium buffering in saccular hair cells. Nature 363:74–76
74. Gola M, Crest M (1993) Colocalization of active K_{Ca} channels and Ca^{2+} channels within Ca^{2+} domains in Helix neurons. Neuron 10:689–699
75. Lancaster B, Nicoll RA (1978) Properties of two calcium-activated hyperpolarizations in rat hippocampal neurones. J Physiol Lond 389:187–203
76. Roberts WM, Jacobs RA, Hudspeth AJ (1990) Colocalization of ion channels involved in frequency selectivity and synaptic transmission at presynaptic active zones of hair cells. J Neurosci 10:3664–3684
77. Robitaille R, Garcia ML, Kaczorowski GJ, Charlton MP (1993) Functional colocalization of calcium and calcium-gated potassium channels in control of transmitter release. Neuron 11:645–655
78. Stuart GJ, Sakmann B (1994) Active propagation of somatic action potentials into neocortical pyramidal cell dendrites. Nature 367:69–72
79. Kavalali ET, Zhuo M, Bito H, Tsien RW (1997) Dendritic Ca^{2+} channels characterized by recordings from isolated hippocampal dendritic segments. Neuron 18:651–663
80. Meir A, Ginsborg S, Butkevich A, Kachalsky SB, Kaiserman I, Ahdut R, Demirgoren S, Rahamimoff R (1999) Ion channels in presynaptic nerve terminals and control of transmitter release. Physiol Rev 79:1019–1088
81. Geiger JRP, Jonas P (2000) Dynamic control of presynaptic Ca^{2+} inflow by fast-inactivating K^{+} channels in hippocampal mossy fiber boutons. Neuron 28:927–939
82. Haydon PG, Man-Son-Hing H (1988) Low- and high-voltage-activated calcium currents, their relationship to the site of neurotransmitter release in an identified neuron of Helisoma. Neuron 1:919–927
83. Man-Son-Hing H, Haydon PG (1992) Modulation of growth cone calcium current is mediated by a PTX-sensitive G protein. Neurosci Lett 137:133–136
84. Gottmann K, Roher H, Lux HD (1991) Distribution of Ca^{2+} and Na^{+} conductances during neuronal differentiation of chick DRG precursor cells. J Neurosci 11:3371–3378
85. Streit J, Lux HD (1989) Distribution of calcium currents in sprouting PC-12 cells. J Neurosci 9:4190–4199

86. Thomas P, Surprenant A, Almers W (1990) Cytosolic Ca^{2+}, exocytosis, and endocytosis in single melanotrophs of the rat pituitary. Neuron 5:723–733
87. Kataoka Y, Ohmori H (1994) Activation of glutamate receptors in response to membrane depolarization of hair cells isolated from chick cochlea. J Physiol Lond 477:403–414
88. Yawo H (1990) Voltage-activated calcium currents in presynaptic nerve terminals of the chicken ciliary ganglion. J Physiol Lond 428:199–213
89. Delcour AH, Tsien RW (1993) Altered prevalence of gating modes in neurotransmitter inhibition of N-type calcium channels. Science 259:980–984
90. Plummer MR, Hess P (1991) Reversible uncoupling of inactivation in N-type calcium channels. Nature 351:657–659
91. Plummer MR, Logothetis DE, Hess P (1989) Elementary properties and pharmacological sensitivities of calcium channels in mammalian peripheral neurons. Neuron 2:1453–1463

Chapter 4

Perforated Whole-Cell Patch-Clamp Technique: A User's Guide

Hitoshi Ishibashi, Andrew J. Moorhouse, and Junichi Nabekura

Abstract

The patch-clamp technique has revolutionized the study of membrane physiology, enabling unprecedented resolution in recording cellular electrical responses and underlying mechanisms. The perforated-patch variant of whole-cell patch-clamp recording was developed to overcome the dialysis of cytoplasmic constituents that occurs with traditional whole-cell recording. With perforated-patch recordings, perforants, such as the antibiotics nystatin and gramicidin, are included in the pipette solution and form small pores in the membrane attached to the patch pipette. These pores allow certain monovalent ions to permeate, enabling electrical access to the cell interior, but prevent the dialysis of larger molecules and other ions. In this review we give a brief overview of the key features of some of the perforants, present some practical approaches to the use of the perforated patch-clamp mode of whole-cell (PPWC) recordings, and give some typical examples of neuronal responses obtained with the PPWC recording that highlight its utility as compared to the traditional whole-cell patch recording configuration.

4.1. Introduction: Rationale for the Perforated Whole-Cell Patch-Clamp Technique

The patch-clamp recording technique represents a significant advance for cellular physiology and neuroscience in that it has enabled direct measurement of the activity of individual ion channels that provide the molecular basis of cellular excitability (1). The utility of the technique is illustrated by the further various configurations that can be achieved to record electrical activity in excised patches or for the whole cell (2). One of the features of this "whole-cell" recording technique is that the relatively vast solution inside the recording electrode (~10 μl) can completely replace the much smaller volume of intracellular solution (~10 pl). Such control over the internal ionic concentration can be advantageous for certain experiments, such as isolating particular ionic currents,

Yasunobu Okada (ed.), *Patch Clamp Techniques: From Beginning to Advanced Protocols*, Springer Protocols Handbooks, DOI 10.1007/978-4-431-53993-3_4, © Springer 2012

loading specific compounds into cells, or careful quantification of ion channel selectivity profiles. However, this intracellular dialysis is a double-edged sword: Soluble components that influence cellular excitability and contribute to signaling pathways are dialyzed from the cell. One frequently encountered consequence is that receptor-mediated responses and/or ionic currents that require soluble second messengers are absent or rapidly "run down" during conventional whole-cell recordings (3, 4). Furthermore, intracellular ion concentrations are disrupted, hampering efforts to determine physiologically relevant responses or properties of channels and receptors (5, 6).

The problems of cell dialysis have been effectively overcome with the perforated patch-clamp whole-cell (PPWC) technique. With this approach, a membrane-perforating agent is used to provide access to the cell interior without mechanical rupture of the membrane attached to the patch pipette. The idea was initiated by Lindau and Fernandez (3), who used ATP (~0.5 mM) to permeabilize the patch membrane and record the passive electrical properties of mast cells during antigen-induced degranulation, a response that was absent in cells dialyzed during conventional whole-cell recordings. The access resistances (0.1–5.0 GΩ) were very high, however, and the use of ATP as a perforant is limited to membranes containing ionotropic P2X receptors. An improvement was the use of the antibiotic nystatin, which provided lower (18–50 MΩ) and stable access resistance (4). Much wider applicability of the technique has resulted from use of this and other perforants (described below) that can spontaneously form channels in numerous membranes and result in access resistances that under the most optimal conditions may approach values comparable to those obtained with conventional whole-cell recordings.

4.2. Comparison of Perforants Commonly Employed

Table 4.1 compares a number of commonly used perforants (see also Fig. 4.1a). Nystatin (4) and amphotericin B (11) are structurally similar antibiotics derived from streptomycetes (*Streptomyces*) bacteria. They contain a large, amphiphatic polyhydrophylic lactone ring that allows them to form aqueous pores in the cell membrane. The properties of these polyene antibiotics have been well characterized in lipid bilayers. Their effect on membrane conductance (reflecting their stability as membrane pores) is highly dependent on the antibiotic concentration, membrane composition, and temperature (7, 22). For example, sterols in the membrane enhance membrane conductance, and increasing temperature reduces it. The ion permeation properties of the two agents are similar (Table 4.1), both being modestly selective for monovalent cations

Table 4.1
Characteristics of perforants used in perforated patch-clamp recordings

Perforant	Nystatin	Amphotericin B	Gramicidin	β-Escin
Class	Polyene antibiotic, *Streptomyces nousei*	Polyene antibiotic, *Streptomyces nodosus*	Linear polypeptide antibiotic, *Bacillus brevis*[a]	Saponin derivative
Pore structure	Lactone ring, interacts with sterols to forms a barrel-like pore	Lactone ring, interacts with sterols to forms a barrel-like pore	15 Alternating DL-amino acids in β-sheet as end-on-end dimer	?
Pore radius	~0.4–0.5 nm[b]	~0.4–0.5 nm[b]	≤0.35 nm[c]	>10 kDa
Selectivity	Weakly cationic[d] (P_K:P_{Cl} = 30:1) (P_K:P_{Cl} = 40:1)	Weakly cationic[d] (P_K:P_{Cl} = 30:1) $K^+ \geq Na^+ \geq Li^+$	Highly cationic,[d] P_{Cl} ~0 $Cs^+ \geq K^+ \geq Na^+ \geq Li^+ > Tris^+$	Nonselective,[d] Ca^{2+} permeable
Conductance	~5 pS[e]	~6 pS[e]	~50 pS[f] $H^+ > Cs^+ > K^+ > Na^+ > Li^+$	?
Stock conc.[g]	Methanol (5–10 mg/ml) DMSO (25–50 mg/ml)	DMSO (25–100 mg/ml)	Methanol (10 mg/ml) DMSO (2–50 mg/ml)	H_2O (25–50 mM)
Final conc.[g]	20–400 μg/ml	25–250 μg/ml	5–100 μg/ml	25–50 μM
Key refs.	(4, 7–10)	(7, 10–12)	(5, 12, 13)	(14, 15)

[a]Gramicidin used in most laboratories is commonly gramicidin D, a combination of gramicidin A, B, and C.
[b]Inferred from lipid bilayer experiments with nystatin or amphotericin applied to both sides of the lipid bilayers (and different from the channel from one-sided application as used for PPWC recordings).
[c]Myers and Haydon (16) reported that tetramethylammonium (diameter 0.7 nm) was impermeant.
[d]The selectivity of the polyene antibiotics has not been quantified, although a value of 10:1 appears in the literature. The values cited here are calculated using the Goldman–Hodgkin–Katz equation and the ~50 mV potential across nystatin-doped bilayers in a tenfold KCl activity gradient (17, 18), and from the 15- to 16-mV shift in V_{rev} shown in Fig. 4.3 of (4) in a twofold KCl gradient, with concentrations converted to activities. β-Escin is likely nonselective. Permeability sequences are from (17) for amphotericin B and from (16, 19) for gramicidin.
[e]Conductance values from lipid bilayers in symmetrical 2 M KCl (nystatin (17); amphotericin (20)). Lower (<1 pS) values would be predicted in physiological ionic strength (20).
[f]Conductance values are from lipid bilayers in symmetrical 2 M KCl (21). In approximate physiological ionic strengths (0.1 M), conductances were about 10 pS (KCl) and 5 pS (NaCl).
[g]A range of values are given in the literature (including those cited under key references) and are based on personal experience. For nystatin and gramicidin, DMSO and methanol can be interchanged. Little information is published about concentration dependence. Tajima et al. (19) noted less frequent perforation (and higher R_a) as gramicidin was lowered from 100 μg/ml, although Kyrizos and Reichling (13) reported more frequent spontaneous membrane rupture at >20 μg/ml.

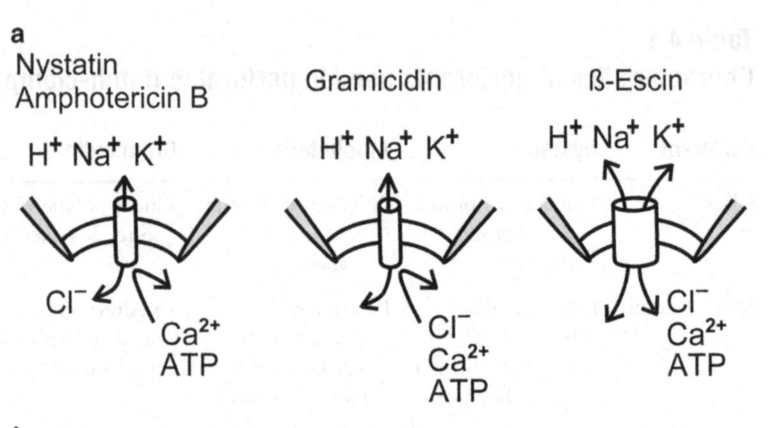

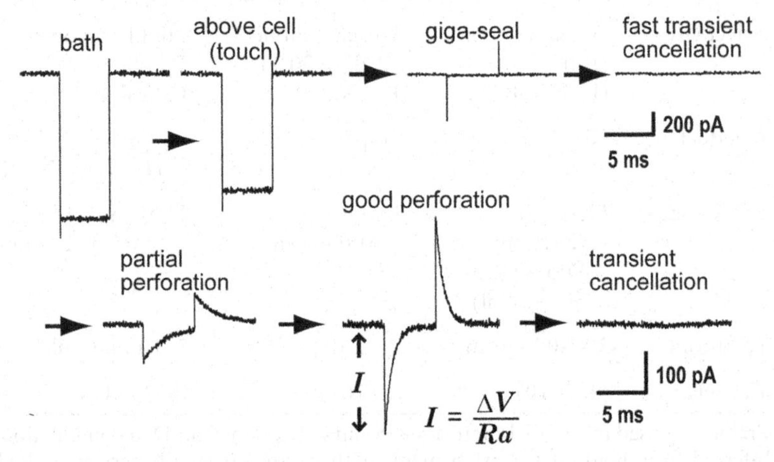

Fig. 4.1. Overview of cell perforation. (**a**) Relative size and selectivity of commonly used perforants. (**b**) Current responses to a −5 mV voltage pulse (ΔV) during the stages of sealing a patch pipette to a cell to form a GΩ seal (*upper drawing*) followed by subsequent perforation of the patch membrane and cancelation of the current transients (*lower traces*). The recordings were obtained using an acutely isolated hippocampal neuron.

and impermeant to divalent ions (17); they also have low single channel conductance and a pore diameter small enough to exclude molecules larger than glucose (~0.8 nm) or sucrose (~1 nm). Remarkably, the permeation properties of the amphotericin B and nystatin pores change to predominantly anion-selective when added to both sides of the membrane (18). Fortunately, however, their lateral diffusion is limited by the seal (23), and access from the interior membrane side does not occur in PPWC recordings (unless the membrane ruptures).

The modest ion selectivity of these antibiotics results in eventual equilibration of *all* intracellular monovalent ions with the pipette concentrations. In contrast, the use of the highly cation-selective linear polypeptide antibiotic gramicidin as the perforant results in preservation of the physiological intracellular Cl^- concentration (5, 12). Gramicidin has negligible permeability to multivalent ions and monovalent anions under physiological conditions in both bilayers and cells (Table 4.1) (16, 19). Although it has a minimal pore diameter similar to those of amphotericin B and nystatin, it has much larger single-channel conductance. Gramicidin PPWC recordings have resulted in a greater understanding of Cl^- homeostasis in cells under various conditions and of the physiological responses of Cl^- permeant ionotropic γ-aminobutyric acid (GABA) and glycine receptors (5, 12, 24). Intracellular Cl^- may also modify aspects of cell signaling or ion channel gating (25, 26). Although amphotericin B has been reported to produce slightly better and faster perforation than either nystatin or gramicidin (11, 13), all three perforants can produce access resistances ~10–20 MΩ or lower under optimal conditions. Factors beyond the perforant may be more important for low access recordings, such as appropriate tip-filling times, a steeply tapering pipette shank with low resistance, and a freshly made perforant–pipette solution (11).

The need to solubilize the above ionophores in nonaqueous solvents and dissolve them in pipette solution led to an evaluation of alternate simpler (and less expensive) perforants such as the saponin derivative β-escin. β-Escin can produce greater and more frequent perforation than either amphotericin B or nystatin (14, 15), it is more stable in aqueous solution, and it is less costly. However, β-escin pores are significantly larger, with divalent ions and even high-molecular-weight compounds (up to 10–15 kDa) permeating; hence, it is not suitable when native concentrations or small signaling molecules need to be maintained. The larger pore size has been utilized to fill cells with fluorophores such as fluo-2 (14).

4.3. Practical Guide to Perforated Patch-Clamp Recordings

The following is a step-by-step description of our typical method for using amphotericin B or nystatin followed by that for gramicidin.

4.3.1. Amphotericin B or Nystatin

1. A stock solution of amphotericin B or nystatin is prepared at a concentration of 50–100 mg/ml in dimethylsulfoxide (DMSO) on the day of the recording. This stock solution is kept away from light and at room temperature.
2. The stock solution (2 μl) is added to 1 ml of internal pipette solution and mixed using ultrasonication, yielding a final concentration of 100–200 μg/ml. The DMSO concentration is 0.2% (vol/vol). We avoid filtering these solutions, although Rae et al. (11) reported that amphotericin B may be filtered through 0.2-μm filters without loss of activity. Fresh nystatin or amphotericin should be added to the pipette solution every 2–3 h.

3. Both polyene antibiotics and gramicidin at the tip of the patch pipette may impair the initial GΩ seal formation; consequently, prefilling the tip of the pipette by brief immersion into an antibiotic-free solution may be necessary. The optimal time for pipette tip immersion depends on multiple factors, including pipette shape and diameter [larger tips require less filling time, smaller-tip pipettes (~>5 MΩ) may not need filling], the perforant concentration used (a lower concentration requires little or no filling), and the time required between filling and sealing (with slower procedures requiring more tip filling). The time can be determined empirically, aided by microscopic inspection of tip geometries and filling heights. In our experience with filament-free borosilicate glass capillaries of 3–7 MΩ and with short and blunt tapers, filling for about 2 s is optimal. The remainder of the pipette is back-filled with the internal pipette solution containing the perforant. Small air bubbles at the tip are rapidly and easily removed by tapping the pipette shaft with the index finger.
4. The patch pipette is secured in the pipette holder and dipped into the external solution in the recording chamber and guided toward the cell as rapidly as possible, before the antibiotic diffuses to the tip. No positive pressure is applied during the approach. After gently pushing the pipette tip onto the cell membrane, causing a slight increase of the pipette resistance, gentle negative pressure/suction is applied to obtain the GΩ seal (Fig. 4.1b). Canceling the fast capacitive transients helps with subsequent monitoring of perforation.
5. The pipette potential is subsequently typically held between -40 and -70 mV, gradually approaching the cell's resting membrane potential as perforation proceeds. The progress of perforation is monitored by visualizing at high temporal resolution the current response to repetitive hyperpolarizing voltage pulses of about 5–10 mV (ΔV) (Fig. 4.1b). As the access resistance (R_a) decreases and electrical contact with the cell improves, a current transient due to the cell membrane capacitance (C_m) is observed. The amplitude of this current transient is given by $\Delta V/R_a$ (hence, the current response increases as perforation proceeds), and the decay time constant is given by $C_m \times R_a$ (hence, the current decay gets sharper/faster as perforation proceeds). The values of C_m and R_a can be quantified by analyzing the current transient or read from the capacitance and series resistance cancelation dials off the patch-clamp amplifier. Recordings can commence once the R_a has stabilized at a suitable value and then is compensated for. We routinely obtain stable $R_a < 30$ MΩ by 30 min. R_a should be monitored and adjusted during the experiment, and data should be used only during periods when this value is relatively stable.

4.3.2. Gramicidin Perforated Patch

The general procedure for the gramicidin-perforated patch is the same as for amphotericin or nystatin, with some additional points as follows. The stock solution is prepared by dissolving gramicidin D in DMSO at a concentration of 2–50 mg/ml and may be stored for 1–2 days at -20°C. The patch pipette solution contains gramicidin at 5–100 μg/ml, with the most appropriate concentration determined in preliminary experiments. Higher concentrations accelerate perforation speed but may result in spontaneous rupturing of the patch membrane (this seems to depend on the preparation) and requires prefilling of the pipette tip with gramicidin-free solution. A fresh pipette-filling solution needs to be prepared every 1–2 h to obtain low-access resistance recordings.

4.3.3. Some Alternatives

The lipophilic nature of the polyene antibiotics and gramicidin results in difficulties in using them in physiological solutions, including time-consuming preparation, large solvent concentrations, rapid loss of potency in pipette solutions, and interference with GΩ seal formation. Some of these problems can be overcome by using β-escin, but this large pore is not suitable for many applications. Horn and Marty (4) perfused the pipette with nystatin after forming the GΩ seal, but it requires a pipette perfusion system.

An alternative method (27, 28) used the bipolar moleculues fluorescein or N-methyl-D-glucamine (NMDG) to help disperse and solubilize nystatin or amphotericin B. A stock solution is made from 5 mg nystatin and 20 mg fluorescein in 1 ml methanol, or from a 0.1 M solution of NMDG dissolved in methanol (pH adjusted to about 7 with methanesulfonic acid in the presence of 0.01 M phenol red) to which nystatin (5 mg/ml) was added. Immediately before use, 50 μl of the stock solution is placed in a polyethylene test tube and dried completely with a stream of N_2 gas. Pipette solution (1 ml) is added to the tube and briefly vortexed. The pipette solution can be filtered through a 0.22-μm syringe filter; tip filling is not required (hence perforation is achieved more rapidly). Positive pressure can be applied during the approach to the cell, which makes this particularly useful for blind patch-clamping in tissues slices. As with other perforants, flouroscein is light sensitive. In fact, whole-cell access can be reversibly closed by the microscope light (27).

4.3.4. Perforated Vesicles Preparation

The activity of some ion channels requires cytoplasmic constituents that are lost with the formation of cell-free patches, such as outside-out patch configuration. Withdrawal of the pipette following formation of the PPWC configuration can result in formation of a perforated excised vesicle, analogous to an outside-out excised membrane patch but in which larger intracellular molecules and hence signaling pathways are maintained (8). The perforated vesicle allows recording of single channels without marked run-down. It also allows investigation of local signal transduction pathways that modulate single-channel activity.

4.4. Examples of the Utility of Perforated Patch-Clamp Recordings

We illustrate here some of the benefits of PPWC recordings with a few brief examples of responses from acutely isolated and primary cultured neurons.

4.4.1. Intracellular Ca^{2+} Homeostasis

Dialysis of cytoplasm and/or the presence of exogenous Ca^{2+} chelators (e.g., EGTA) results in a gradual loss of Ca^{2+} from internal stores, particularly when these stores are evoked to release Ca^{2+}. This is illustrated in Fig. 4.2, which shows both conventional and PPWC recordings of K^+ currents in response to repetitive bath application of caffeine, which releases Ca^{2+} from intracellular stores and activates Ca^{2+}-activated K^+ channels on the cytoplasmic membrane (29). The response to caffeine gradually decreases in amplitude on conventional whole-cell recordings but is constant for about ≥1 h on the amphotericin B PPWC recordings.

4.4.2. Run-Down of Voltage-Activated Ca^{2+} Currents

A peculiar property of high-voltage-activated Ca^{2+} currents is that they show a time-dependent decrease in amplitude during conventional whole-cell recordings, a process termed "run-down." Various

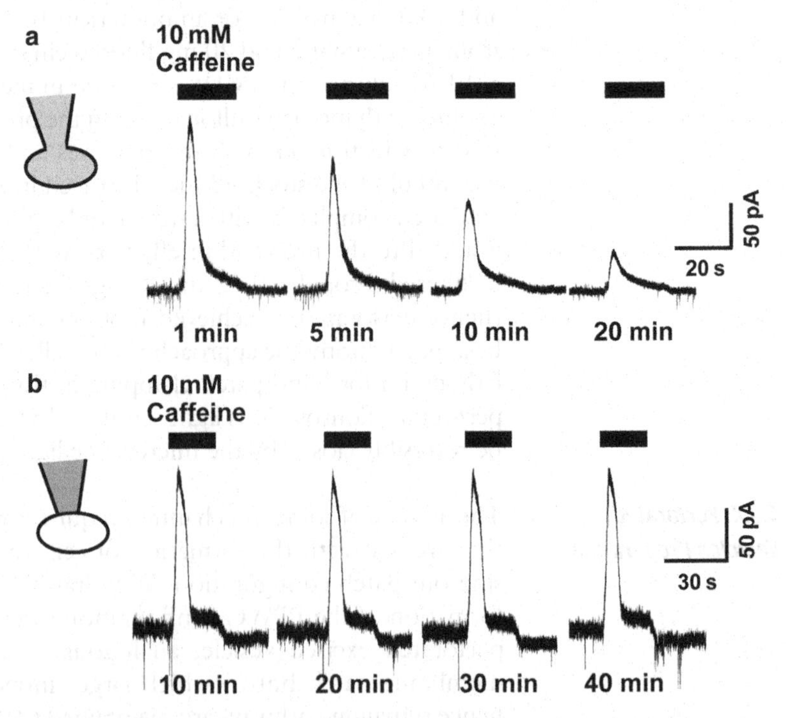

Fig. 4.2. Perforated patch-clamp whole-cell (PPWC) mode maintains Ca^{2+} stores. Current responses to 10 mM caffeine in acutely isolated rat CA1 hippocampal neurons recorded under conventional whole-cell recordings (**a**) or amphotericin B PPWC recordings (**b**) Note that the response amplitude runs down in (**a**) but is constant in (**b**). Caffeine was applied each ~5 min. Recordings were performed at a holding potential (V_H) of −50 mV.

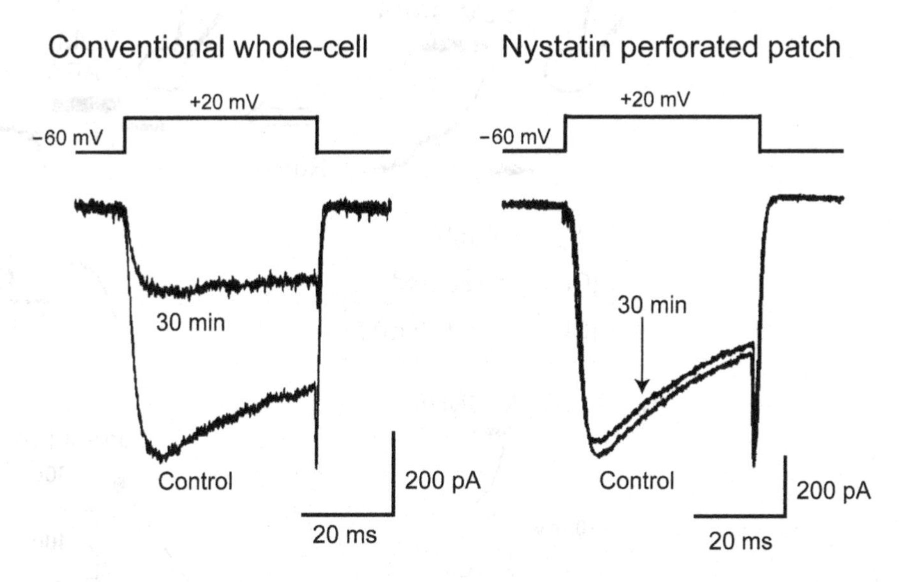

Fig. 4.3. Nystatin PPWC mode prevents the run-down of Ca^{2+} channel currents. Typical traces of high-voltage activated Ca^{2+} currents recorded from acutely isolated rat intracardiac ganglion cells using conventional (*left panel*) or PPWC (*right panel*) recordings. Current traces obtained at the beginning of the experiment (control) and 30 min later (*30 min*) are superimposed.

attempts have been made to prevent run-down, including application of substances to maintain channel phosphorylation and/or protect against proteolysis, but none of these efforts has been satisfactory. In contrast, run-down of voltage-activated Ca^{2+} channels is markedly reduced/delayed using nystatin PPWC recordings (Fig. 4.3). The stable recording of Ca^{2+} channel currents has allowed, for example, detailed pharmacological dissection of the contribution of different voltage-activated Ca^{2+} channels in different preparations (30, 31).

4.4.3. GABA and Glycine Responses Recorded by the Gramicidin-Perforated Patch Recording

On conventional whole-cell recordings and PPWC recordings using the polyene antibiotics or β-escin, the intracellular Cl^- concentration ($[Cl^-]$) equilibrates with that in the pipette $[Cl^-]$. In contrast, gramicidin PPWC preserves the "normal" intracellular $[Cl^-]$ and enables one to record the physiological response of Cl^--permeant channels; it also allows us to investigate the modulation of intracellular Cl^- homeostasis. Figure 4.4a compares the response of ionotropic hippocampal GABA receptors when recorded using a conventional whole-cell or gramicidin PPWC technique at the same holding potential (V_H) of –50 mV. The direction of the current response is completely different, reflecting the different intracellular $[Cl^-]$ and hence driving force. The physiological range of intracellular $[Cl^-]$ (~5–30 mM) is less than often used in pipette solutions in conventional whole-cell recordings (~150 mM). The gramicidin PPWC technique can be used to measure the intracellular $[Cl^-]$ as shown in Fig. 4.4b. Currents through the anion-selective ionotropic GABA or glycine receptors are measured at different holding potentials, and a current–voltage curve can be

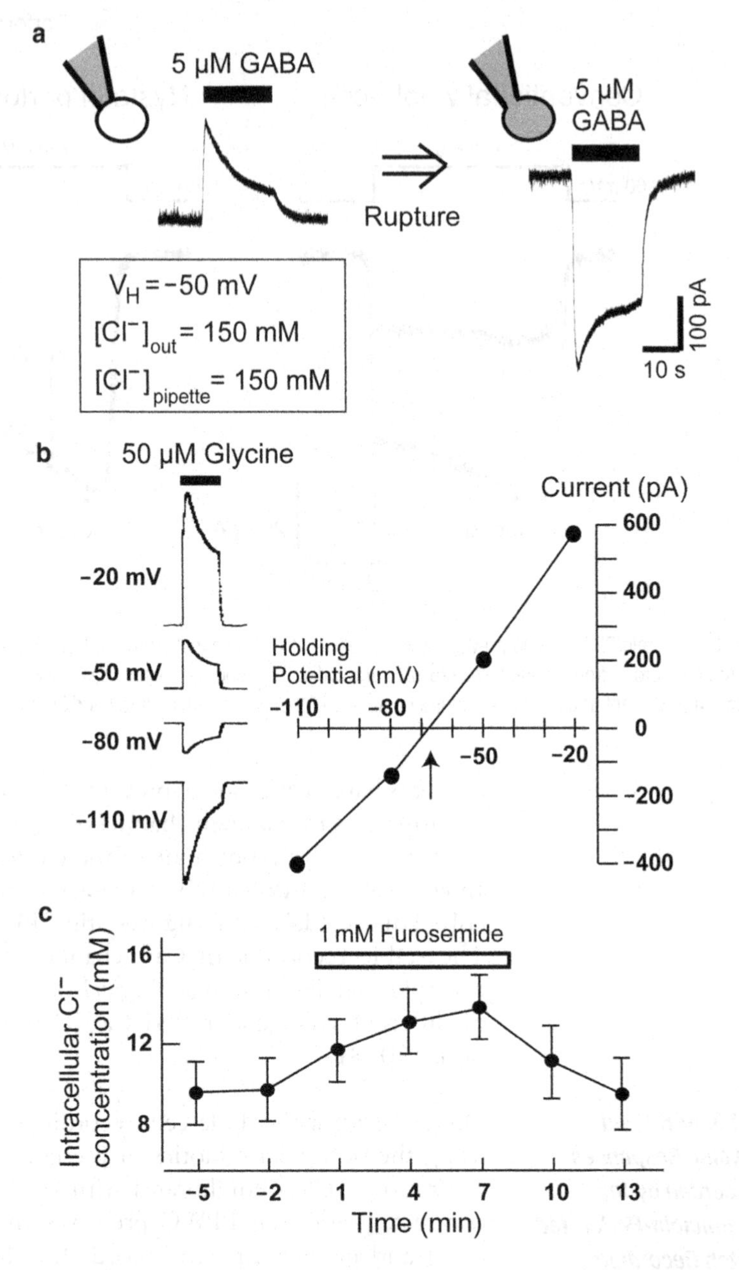

Fig. 4.4. Physiological γ-aminobutyric acid (GABA) and glycine responses and intracellular [Cl⁻] measurements using the gramicidin PPWC technique. (**a**) GABA-induced outward and inward currents recorded at a V_H of −50 mV before (*left trace*) and after (*right trace*) rupture of the membrane during a gramicidin PPWC recording. Membrane rupture results in a much higher intracellular [Cl⁻] as the cell equilibrates with the pipette [Cl⁻] (~150 mM). Hence, the GABA response changes from an outward current (representing Cl⁻ efflux) to an inward current (Cl⁻ influx). (**b**) Glycine-induced currents recorded using the gramicidin PPWC recording at various V_Hs values in cultured spinal cord neurons are measured to construct a corresponding current–voltage curve from which the V_{rev} (*arrow*) is determined. The intracellular [Cl⁻] is estimated from a derivation of the Nernst equation: $[Cl^-]_{in} = [Cl^-]_{out}\ \exp(V_{rev}F/RT)$, where F is Faraday's constant (96,485 Cmol⁻¹), R is the gas constant (8.3145 VCmol⁻¹ K⁻¹), and T is absolute temperature (293.15 K at 20°C). (**c**) Such measurements are used to demonstrate that furosemide causes a reversible increase in the intracellular [Cl⁻]. This is due to direct inhibition of K⁺-coupled Cl⁻ efflux via the neuronal KCC2 transporter. The graph plots the mean ± SEM from five experiments.

plotted. If using ramp voltage responses, control currents in the absence of GABA or glycine should be subtracted from those in the presence of GABA or glycine. The X axis intercept of this curve (the reversal potential, V_{rev}) gives the equilibrium potential where the driving force due to the membrane voltage cancels out that from the concentration gradient. This is expressed mathematically by the Nernst equation, from which the intracellular $[Cl^-]$ can be estimated (Fig. 4.4b). Precise quantification of V_{rev} and intracellular $[Cl^-]$ requires consideration of all permeant ion species, activity coefficients, and liquid junction potentials (32); the latter can be up to ~10 mV or more if using larger ions in the pipette solution. These recordings should use a K^+-based solution in the pipette, as the intracellular $[Cl^-]$ is sensitive to transmembrane K^+ (and, to a lesser extent, Na^+) gradients, and some other cations (e.g., Cs^+) can significantly inhibit some of the key Cl^- transporters (33). Figure 4.4c shows the effect of furosemide, a Cl^- transporter inhibitor, on the intracellular $[Cl^-]$ in isolated hippocampal neurons. The ability to measure the intracellular $[Cl^-]$ by the gramicidin PPWC technique has facilitated revealing developmental- and injury-induced changes in intracellular Cl^- homeostasis (34, 35).

4.5. Drawbacks of the Perforated Patch-Clamp Technique

This chapter has highlighted some of the benefits of the PPWC technique, but there are also some drawbacks potential users need to consider. First, PPWC recordings are much more time-consuming than conventional whole-cell recordings: fresh perforant-containing pipette solutions are required, and there is a wait (often ≥30 min) for perforation and stabilization to occur. Second, only under the most optimal conditions can access/series resistances similar to those under conventional whole-cell recordings be obtained. Usually they are much higher and one must (as with conventional whole-cell recordings) be aware of the voltage errors and filtering effects of this access resistance. A voltage offset (=pipette current × R_a) must be subtracted from V_H, and the −3 dB of the filtering effect = $1/[2\pi \times R_a \times C_m]$. Series resistance compensation can reduce these errors. Finally, one must also be aware of a Donnan potential between the V_H and the real membrane potential that arises as the large anions in the cell cannot equilibrate with the patch pipette solution. As discussed by Horn and Marty (4), this may be as high as ~10 mV with a KCl pipette solution. This potential can be reduced by including large impermeant ions in the pipette, but it is difficult to measure or estimate it precisely. Hence in PPWC recordings, one must be aware of potential inaccuracies in citing absolute voltages associated with current responses (e.g., K_d for activation or inactivation).

4.6. Conclusion

The patch-clamp technique has revolutionized the study of membrane proteins and their contribution to physiology, pharmacology, and neuroscience. This chapter briefly reviewed the procedures and applications of the various configurations of the perforated patch-clamp configuration. It is hoped the information included will enable electrophysiologists to make more informed decisions about adding the perforated patch-clamp technique to their repertoire of means to study cellular excitability.

References

1. Neher E, Sakmann B (1976) Single-channel currents recorded from membrane of denervated frog muscle fibres. Nature 260:799–802
2. Hamill OP, Marty A, Neher E, Sakmann B, Sigworth FJ (1981) Improved patch-clamp techniques for high-resolution current recordings from cells and cell-free membrane patches. Pflugers Arch 391:85–100
3. Lindau M, Fernandez M (1986) IgE-mediated degranulation of mast cells does not require opening of ion channels. Nature 319:150–153
4. Horn R, Marty A (1988) Muscarinic activation of ionic currents measured by a new whole-cell recording method. J Gen Physiol 92:145–159
5. Rhee JS, Ebihara S, Akaike N (1994) Gramicidin perforated patch-clamp technique reveals glycine-gated outward chloride current in dissociated nucleus solitarii neurons of rat. J Neurophysiol 72:1103–1108
6. Ebihara S, Shirato K, Harata N, Akaike N (1995) Gramicidin-perforated patch recording: GABA response in mammalian neurones with intact intracellular chloride. J Physiol 484:77–86
7. Akaike N, Harata N (1994) Nystatin perforated patch recording and its applications to analyses of intracellular mechanisms. Jpn J Physiol 44:433–473
8. Levitan ES, Kramer RH (1990) Neuropeptide modulation of single calcium and potassium channels detected with a new patch clamp configuration. Nature 348:545–547
9. Korn SJ, Horn R (1989) Influence of sodium-calcium exchange on calcium current rundown and the duration of calcium-dependent chloride currents in pituitary cells, studied with whole cell and perforated patch recording. J Gen Physiol 94:789–812
10. Marsh SJ, Trouslard J, Leaney JL, Brown DA (1995) Synergistic regulation of a neuronal chloride current by intracellular calcium and muscarinic receptor activation: a role for protein kinase C. Neuron 15:729–737
11. Rae J, Cooper K, Gates P, Watsky M (1991) Low access resistance perforated patch recordings using amphotericin B. J Neurosci Methods 37:15–26
12. Reichling DB, Kyrozis A, Wang J, MacDermott AB (1994) Mechanisms of GABA and glycine depolarization-induced calcium transients in rat dorsal horn neurons. J Physiol 476:411–421
13. Kyrozis A, Reichling DB (1995) Perforated-patch recording with gramicidin avoids artifactual changes in intracellular chloride. J Neurosci Methods 57:27–35
14. Fan JS, Palade P (1998) Perforated patch recording with β-escin. Pflugers Arch 436:1021–1023
15. Sarantopoulos C, McCallum JB, Kwok WM, Hogan Q (2004) β-escin diminishes voltage-gated calcium current rundown in perforated patch-clamp recordings from rat primary afferent neurons. J Neurosci Methods 139:61–68
16. Myers VB, Haydon DA (1972) Ion transfer across lipid membranes in the presence of gramicidin A. II. The ion selectivity. Biochim Biophys Acta 274:313–322
17. Kleinberg ME, Finkelstein A (1984) Single-length and double length channels formed by nystatin in lipid bilayer membranes. J Membr Biol 80:257–269

18. Marty A, Finkelstein A (1975) Pores formed in lipid bilayer membranes by nystatin; Differences in its one-sided and two-sided action. J Gen Physiol 65:515–526
19. Tajima Y, Ono K, Akaike N (1996) Perforated patch-clamp recording in cardiac myocytes using cation-selective ionophore gramicidin. Am J Physiol 271:C524–C532
20. Ermishkin LN, Kasumov KM, Potzeluyev VM (1976) Single ionic channels induced in lipid bilayers by polyene antibiotics amphotericin B and nystatine. Nature 262:698–699
21. Hladky SB, Haydon DA (1972) Ion transfer across lipid membranes in the presence of gramicidin A. I. Studies of the unit conductance channel. Biochim Biophys Acta 274:294–312
22. Holz R, Finkelstein A (1970) The water and nonelectrolyte permeability induced in thin lipid membranes by the polyene antibiotics nystatin and amphotericin B. J Gen Physiol 56:125–145
23. Horn R (1991) Diffusion of nystatin in plasma membrane is inhibited by a glass-membrane seal. Biophys J 60:329–333
24. Farrant M, Kaila K (2007) The cellular, molecular and ionic basis of $GABA_A$ receptor signaling. Prog Brain Res 160:59–87
25. Lenz RA, Pitler TA, Alger B (1997) High intracellular Cl^- concentrations depress G-protein-modulated ionic conductances. J Neurosci 17:6133–6141
26. Beato M (2008) (2008) The time course of transmitter at glycinergic synapses onto motoneurons. J Neurosci 28:7412–7425
27. Yawo H, Chuhma N (1993) An improved method for perforated patch recordings using nystatin-fluorescein mixture. Jpn J Physiol 43: 267–273
28. Endo K, Yawo H (2000) μ-Opioid receptor inhibits N-type Ca^{2+} channels in the calyx presynaptic terminal of the embryonic chick ciliary ganglion. J Physiol 524:769–781
29. Uneyama H, Munakata M, Akaike N (1993) Caffeine response in pyramidal neurons freshly dissociated from rat hippocampus. Brain Res 604:24–31
30. Ishibashi H, Akaike N (1995) Somatostatin modulates high-voltage-activated Ca^{2+} channels in freshly dissociated hippocampal neurons. J Neurophysiol 74:1028–1036
31. Ito Y, Murai Y, Ishibashi H, Onoue H, Akaike N (2000) The prostaglandin E series modulate high-voltage-activated calcium channels probably through the EP3 receptor in rat paratracheal ganglia. Neuropharmacology 39: 181–190
32. Barry PH, Lynch JW (1991) Liquid junction potentials and small cell effects in patch clamp analysis. J Membr Biol 121:101–117
33. Kakazu Y, Uchida S, Nakagawa T, Akaike N, Nabekura J (2000) Reversibility and cation selectivity of the K^+-Cl^- cotransport in rat central neurons. J Neurophysiol 84:281–288
34. Kakazu Y, Akaike N, Komiyama S, Nabekura J (1999) Regulation of intracellular chloride by cotransporters in developing lateral superior olive neurons. J Neurosci 19:2843–2851
35. Nabekura J, Ueno T, Okabe A, Furuta A, Iwaki T, Shimizu-Okabe C, Fukuda A, Akaike N (2002) Reduction of KCC2 expression and $GABA_A$ receptor-mediated excitation after in vivo axonal injury. J Neurosci 22: 4412–4417

Chapter 5

Methods for Processing and Analyzing Single-Channel Data

Masahiro Sokabe

Abstract

Analysis of data from single-channel studies can provide us with insights into the detailed physicochemical features of channel proteins, such as ion permeation rate, ion selectivity, and gating. However, single-channel analysis, particularly gating analysis, is sometimes time-consuming, unconvincing, and unproductive. Notwithstanding, if it is combined with structure information and site-directed mutagenesis of channel proteins, it can prove to be a powerful tool that moves us toward the ultimate understanding of structure–function of channel proteins. This chapter is composed of roughly two parts. The first part describes the technical aspect how to set up the hardware for correct acquisition of the tiny, long-lasting single-channel currents against background noises and drifts. The second part deals with the statistical estimation of the current amplitude and dwell times of the open and closed states and how to fit the estimated data to an appropriate reaction model. Each section is described in a step-by-step fashion so even beginners can understand and experience the course of single-channel study, from data acquisition to model fitting.

5.1. Introduction: What Is Single-Channel Current?

Capturing the movement of single molecules in function mode is an ultimate technique in scientific measurements. Except in certain cases, most measurements are performed on a mass of molecules. The behavior of single molecules is estimated afterward with the help of statistical mechanics. Having this in mind, the technique for real-time observation of the behavior of single ion channel molecules is astonishing, even though the observation may be indirect.

"Indirect" here means that what we observe is not the conformational change of a single-channel molecule but the resultant change in its current. We are able to observe the current because the exceptionally high enzymatic activity of ion channels converts

Yasunobu Okada (ed.), *Patch Clamp Techniques: From Beginning to Advanced Protocols*, Springer Protocols Handbooks, DOI 10.1007/978-4-431-53993-3_5, © Springer 2012

the activation (gating) of a single-channel molecule into the flow (current) of an enormous number of ions. For example, the opening of an ion channel with a conductance of 10 pS (picosiemens: 10^{-12} A/V) at 100 mV membrane potential generates a current of 1 pA (10^{-12} A). Conversion to the number of monovalent ions by elementary electric charge (1.6×10^{-19} coulomb) indicates that the effective transfer occurs at a rate of 6.25×10^{6} (i.e., about six million) monovalentions per second. Ion channels are indeed a current converter with a super-high gain. Simple estimation from this value suggests that the transfer rate of ions is about 160 nanoseconds (ns)/channel. In other words, a single ion passes through the channel every 160 ns on average. If there were a super-high-speed electrical charge counter, the current through an ion channel should be able to be detected as discrete time series of each ion transfer, and more detailed information on the mechanism of ion passage would be available. However, the reality is that what we observe as a single-channel current is merely an integral of the charge during 100 μs at least due to the limitation of the response speed of current amplifiers. For the same reason, the progress of channel gating and high-speed gating itself cannot be fully time-resolved. Moreover, various background noises prevent us from detecting minute and fast changes in the current.

Although this ultimate technique has the limitations mentioned here, most of the major ion channels known so far have fortunately been correlated with physiological (cellular) responses within these limitations. As for the data processing explained below, the focus is on techniques practicable in average laboratories while keeping these limitations in mind.

5.2. Experimental Techniques: Measures Against Drift

The methods for measuring single-channel currents include a planar lipid bilayer method and a patch-clamp method, whose details can be found elsewhere (1, 2). Here, the patch-clamp method is described concisely.

The most basic mode is the "cell-attached (on-cell) mode," followed by the "excised (inside-out, outside-out) mode." New techniques of "open cell-attached inside-out" and "perforated vesicle outside-out" modes have also been developed. As for these techniques and their advantages and disadvantages, refer to Chap. 2. In either method, measurements of tiny single-channel currents demand (1) a reduction of background noise and (2) capacitance compensation. In addition, because the time required is markedly longer for single-channel recordings than for whole-cell recordings, (3) measures against drift are also important. Because items (1) and (2) were mentioned in Chap. 2, only points that should be considered with respect to item (3) are discussed here. Note, however, that drifts of the amplifier and electrical system are out of the question.

A severe problem in the cell-attached mode is the mechanical drift from manipulators. This can result in changes in seal resistance and fluctuations in the leak current and noise level. One of the sources of drifts is the mechanical shear between microscope stage and manipulator. This shear can be eliminated by fixing the manipulator firmly to the stage or the microscope body and by making the distance between the fixing point of the amplifier head stage and the electrode tip as short as possible. A more difficult problem is the drift that derives from thermal expansion when manipulators driven by oil or water pressure are used. The manipulators driven by water pressure, which has a lower thermal expansion coefficient, are naturally advantageous; and the current versions (Narishige, Newport) are improved in terms of practical performance although they had problems at the time of the first release. Mechanical manipulators are more stable. However, the manual manipulator has the disadvantage that a large drift can occur upon touching it with a hand. Therefore, remote control systems driven by stepping motors (Narishige), servomotors (Newport, Eppendorf), and Piezo actuators (Sutter, Burleigh) are favored, but they are expensive. If the budget allows, these remote control systems are recommended.

Another problem is that the solution may evaporate when a small-volume chamber is used. Although this causes little problem during a single recording session, repeated experiments result in changes in the solution concentration and a mismatch in compensated capacitance due to changes in the level of the water surface. It is not advisable to add distilled water at a rough estimate. Depending on the experiments, the evaporation problem can be address by perfusion or by adding small amounts of silicone oil to cover the water surface.

5.3. Data Recording and Signal Preprocessing

5.3.1. Data Recording[1]

A typical setup for channel current recording is shown in Fig. 5.1. In the case of whole-cell recordings, data are recorded directly onto the hard disk of a computer simultaneously with the stimulation signals. For single-channel recordings, because the single-channel current measurement generates an enormous amount of

[1]The description on the data recording in this section seems to be rather out of date given the huge capacity of recent hard disks. However, this bygone description is left for the sake of researchers that use old computers due to A/D converters and software.

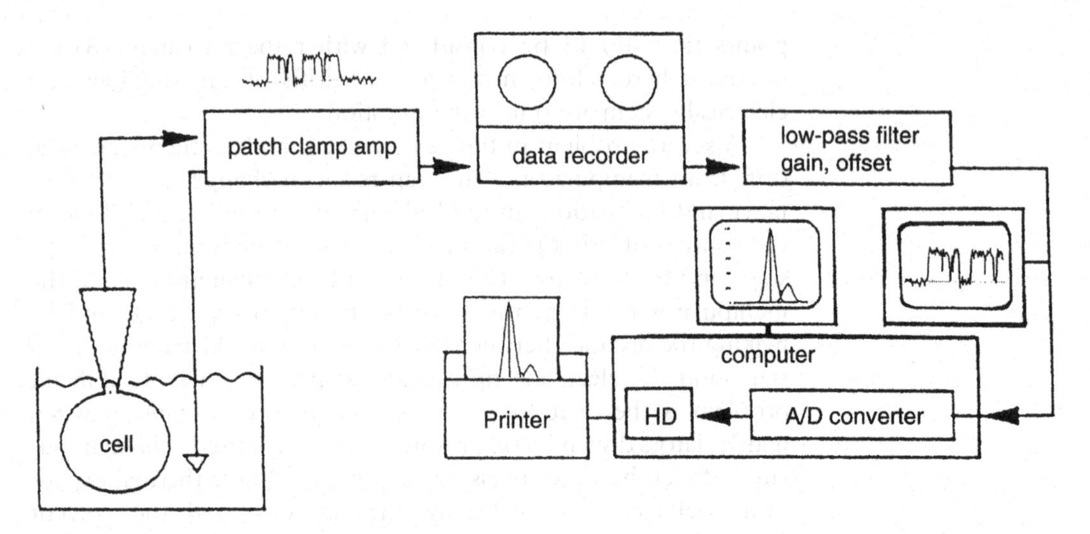

Fig. 5.1. General instruments for recording and analyzing single-channel current. The unprocessed data from a patch-clamp amplifier are fed to a data recorder and recorded simultaneously on the hard disk of computer from the monitor output of the recorder by way of a filter and an A/D converter. On offline analysis, recorded tapes are played back with the same setup.

data, they are stored on a data recorder followed by offline analyses. For example, when the kinetics of channel opening and closing are analyzed, more than 1,000 events are generally required: If opening events that last for 1 ms on average occur once every 1 s on average, a recording as long as 1,000 s is needed at a 10-kHz sampling rate. If a 12-bit A/D converter, which requires 2 bytes per sample, is used, a total of $2 \times 1{,}000 \times 10{,}000 = 2 \times 10^7 = 20$ MB of memory is consumed. Although the situation is not as extreme as this, it is not practical to store all data on a hard disk because at least several megabytes are required for each experiment. Nevertheless, data need to be recorded simultaneously on a hard disk to analyze the data promptly. Therefore, data are recorded in a parallel manner on a data recorder and a hard disk. The former is usually sampled at a frequency that is as high as possible; the latter is stored after subjecting it to an appropriate filter because it is also used for monitoring the waveform and is deleted from the disk after a preliminary analysis. When recording on a data recorder, the gain of the amplifier needs to be adjusted so the input dynamic range (typically ±10 V) of the data recorder is covered as effectively as possible.

Whereas FM, PCM, and DAT recorders are used as data recorders, the PCM type (Instrutech VR series, for example), which shows a good cost performance ratio, seems to be used most commonly. This type has a characteristic frequency at 44 kHz and is practically without problems, although it is somewhat troublesome to search target positions with the tape recorder while playing back. This is because recapturing into the computer is often needed. Because PCM recorders use a home-use VTR as the recorder, the portion to be played back is generally sought with the help of a time counter. For the sake of exactness of the playing position, it is best to stamp a time code on the tape and search for it; however, analysis software that supports this function may be difficult to find. The second best way may be to use the index marker of the VTR and PCM converter or to record trigger signals in the auxiliary channel, which are used as marks for initiating acquisition into the computer. It is also possible to select and analyze the required part from the played data with software after the data are acquired roughly. Note that recent dramatic progress in the capacity of memory devices allows us to record all data on the hard disk, move the desired data to various types of external memory media (DVD, BD), and use them for the analysis. It goes without saying that this procedure is advantageous over recording on tape media in terms of efficiency.

5.3.2. Preprocessing of Signals: Gain Adjustment and Filter Setting

Several points should be taken into consideration in cases when a computer acquires the played-back signals from data recorders through the A/D converter. As already mentioned, the output voltage range should be adjusted to the input range (typically ±5–10 V) of the A/D converter as much as possible. Once data are converted from analog to digital, the precision cannot be recovered by any subsequent amplification. If the voltage range of the data recorded on tapes does not match the input range of the A/D converter, a booster amplifier should be inserted between them or the input range of a programmable A/D converter should be adjusted on software. The offset voltage sometimes requires adjustment.

Nevertheless, the most important point is setting the low pass filter, which is used for two purposes. First, the filter removes high-frequency noise components present in signals. The performance required in the filter here is the faithful reproduction of waveform and steep cutoff characteristics, with the former having priority in the biological signals. In this sense, the Bessel filter, which has flat group delay characteristics with small phase shift (phase shift causes distortion of waveforms), is the best choice. The disadvantage of this filter is its mild cutoff characteristics, but this can be overcome by using fourth to eighth order (−24 to −48 dB/oct) filter. As discussed later, determination of the cutoff frequency (the frequency at which signals decay by −3 dB) of the filter is a baffling problem that should be judged under the compromise between faithful reproduction of signals and noise removal. Some software is equipped with a digital Gaussian filter. If the processing speed is without problems, this filter is useful because data obtained at high frequency can be processed multiple times. The characteristics of the Gaussian filter are almost the same as those of the high-order Bessel filter.

The second reason for using a filter is to avoid aliasing due to sampling with the A/D converter. This is because false signals (aliasing) may be generated if signal components include higher-frequency components than the sampling frequency. According to the sampling theorem, the original signal can be faithfully reproduced if the sampling frequency is twice the highest frequency component of the signal to be acquired. Conversely, all that is needed is the relation that the sampling frequency is twice the cutoff frequency of the filter. However, because signals are not actually completely cut off at the cutoff frequency, sampling frequency needs to be at least 5–10 times the frequency of the cutoff frequency to ensure faithful reproduction of the signals that have passed the filter. Because the highest sampling frequencies of A/D converters differ among products, the final cutoff frequency should be determined taking this into consideration. The commercially available Bessel filter equipped with gain and offset adjustments (Warner, Frequency Device, and Dagan) conveniently allows us to perform the signal preprocessing mentioned here with a single instrument.

5.4. Computer, Analysis Software, and A/D Converter

The choice of computer depends on the analysis software to be used. After all, the software (e.g., pCLAMP (Axon), PAT (Dagan], ISO2 (MKF)] that runs on IBM PC/AT compatible machines (PC machines) is overwhelmingly dominant. Although troubles arose due to incompatibility with software and A/D converters (e.g., the timing shift with ISA bus) in previous PC/AT-compatible machines, there seems to be no problem these days if a well-known motherboard is used. TAC (HEKA), which runs on Mac computers (Apple), has been released also as a Windows version, and most researchers seem to have shifted to PC-compatible machines running Windows. If you want to conduct secondary data processing on the same computer, you may need to prepare a PC at least with Core2Duo·CPU, 4 GB RAM, >500 GB HD, and DVD or BD drive, on which Windows runs smoothly. Of course, an A/D converter that is supported by your software should be used. Representative converters are Digidata (Axon Instruments) or LabmasterDMA (Scientific Solution) if pCLAMP is used, and LabmasterDMA/DT2801A(DataTranslation)orLabPC+(National Instruments) if PAT is used. The price may vary mainly depending on the difference in the highest sampling frequency, and the converter that fits the purpose should be selected. As already mentioned, a measure is the sampling frequency at least five times as high as the frequency (cutoff frequency) you want to analyze. Because A/D converters generally monitor input waveforms and, at the same time, write signals onto the hard disk while generating stimulus by the auxiliary D/A converter, it is advisable to use a converter compatible with DMA (direct memory access), which allows high-speed data transfer. Note that purchasing bundles of software and hardware are becoming more common these days. For example, the bundle of pCLAMP and Digidata (A/D and D/A converters) by Axon Instruments and the bundle of TAC and EPC series (patch-clamp amplifier and A/D and D/A converters are integrated in most of this series) by HEKA are representative. Although I have used pCLAMP and PAT for a long time, the following examples of analysis are mostly based on the latter.

5.5. Amplitude Analysis

Single-channel current analysis is generally composed of two steps: amplitude analysis and gating kinetics analysis. The starting point is the all-point histogram of amplitude. The graph, on which current and frequency are plotted on the abscissa and ordinate, respectively, provides three types of information. The first is obviously the channel conductance, the second is the open probability of ion channels, and the last type is the lower limit of the number of ion channels in the patch membrane.

5.5.1. Estimation of Conductance

For instant estimation of conductance, all you have to do is to read the peak values of open currents with a cursor by taking the averaged closed currents as zero. For more quantitative analysis, the peak values are estimated from a Gaussian distribution curve. The Gaussian approximation can generally be applied simultaneously to multiple peaks; but if this does not work, approximation is best applied sequentially from the clearest peaks. This technique works effectively when visual distinction of the peaks is difficult due to high noise levels (Fig. 5.2a,b). The problem is that the frequency limitation of the measuring system results in the incorporation of unresolved events in which high-speed openings and closings appear as incomplete openings and closings. In this case, the distribution of peaks is skewed asymmetrically from side to side, and estimation of peaks by the Gaussian curve deviates from the true value (Fig. 5.3a). In this case, it is advisable to fit only to the data on the right shoulder, which does not contain unresolved events (Fig. 5.3b).

When a large noise level causes a considerable overlap of the gating peaks, precise estimation of the peak values by the Gaussian approximation is sometimes difficult. In such a case, the noise must be reduced by some means. Although the simplest means is removal of high-frequency noise by a low-pass filter, it is not ideal for the subsequent gating analysis because information about high-speed

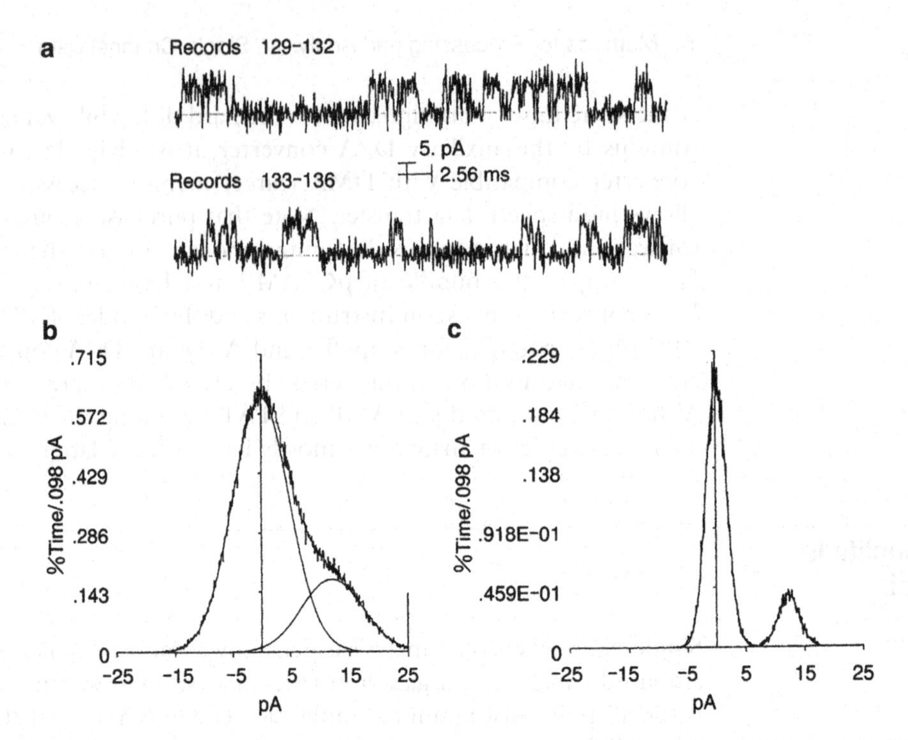

Fig. 5.2. Estimation methods for single-channel current. (**a**) Example of single-channel current (average 12.49 pA). (**b**) Amplitude histogram and fitting with the Gaussian distribution (*smooth curve*) (estimated peak value 12.09 pA). (**c**) Amplitude histogram after Patlak's moving average processing (eight points) (estimated peak value: 12.3 pA). Note that this processing results in clear separation of peaks but that area ratio of the peaks is far from the true value.

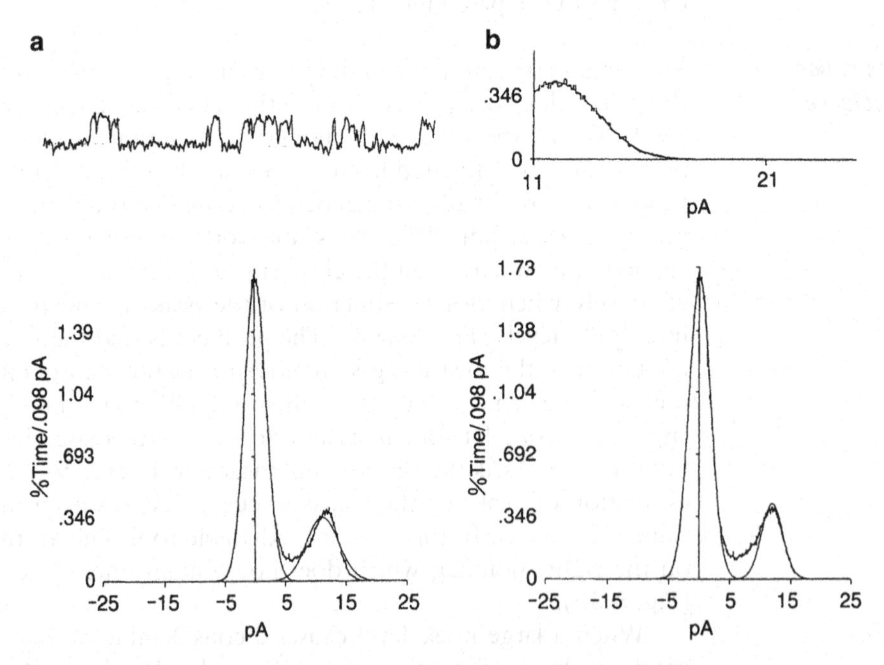

Fig. 5.3. Fitting to amplitude histograms whose peaks do not show normal (Gaussian) distribution. (**a**) Example of a recording (*top*) and the amplitude histogram (*bottom*) in which high-speed openings and closings are not fully time-resolved because of an overly low cutoff frequency of the filter. The values estimated by the Gaussian distribution greatly deviate from the true value. (**b**) Example of fitting by eliminating the processed region to the right shoulder of the second peak (*top*) and subsequent refitting of the entire data by fixing this estimate (*bottom*).

events would be lost by the filtering (note that this is not the case for a software-based digital filter). The most effective means is Patlak's moving average (3). The moving average with appropriate width of window (the number of data points) gives us a histogram with clearer peaks than the simple all-points histogram (Fig. 5.2c). If we make use of the feature of this method that the moving average, including transition points of openings and closings, have a large deviation from the average of clean openings and closings only, such specific moving average values can be removed automatically by software. If the separation of peaks is still difficult, the only way to calculate the averaged values for openings and closings is to repeat the procedure and assign the range (including undoubted openings and closings) by cursor while displaying the original trace and to measure individual values by cursor.

The channel conductance obtained by the amplitude analysis is used exclusively for analysis of the ion permeation mechanisms of ion channels. The analysis informs us, for example, about the (1) voltage dependence, (2) dependence on permeable ion concentration, (3) dependence on blocker concentration, and (4) voltage dependence of blocking. The Eyring's absolute reaction rate theory is adopted for analysis of ion permeation mechanisms on the basis of these relations. That is, an ion channel is expressed as several energy barriers against permeable ions and blockers, and the number, relative position, and energy levels of the peaks and wells are estimated from the above relations. The details can be seen in Hille (4). Progress in the determination of amino acid sequences and configurations in the ion permeation pathway enables us to simulate the process of ion permeation by molecular dynamics.

5.5.2. Estimation of Open Probability and the Number of Channels

If the patch includes only a single-ion channel, the channel open probability (P_o) can easily be calculated from the area ratio of each peak approximated by the Gaussian distribution. Alternatively, the totals of open time and closed time, which the software reports on gating analysis, are also usable. If the patch contains multiple ion channels, the open probability can be estimated from $P_o = 1 - P_c^{1/N}$ provided that the open probability is small and the number of ion channels (N) is known, where P_c is the proportion of the closed state during the total time of recording. This is because the equation $P_c = P_{c_i}^N = (1 - P_o)^N$ holds true with P_c, which is the probability that all channels are closed simultaneously, where P_{c_i} represents the closing probability of a single channel. If the open probability is large, the number of ion channels (N) coincides with the maximum number of channels that open simultaneously. If the open probability is low and simultaneous opening events are not or rarely observed, the number of channels should be tested statistically by estimating the mean open and closed times through gating analysis as described in the next section. The mean open time can be calculated directly in this case, but the true mean closed time can be assumed to be about N times the apparent mean closed time if the number of channels is N. Accordingly, the expected values for the probability of simultaneous opening of N channels can be calculated, and the hypothesis that N channels are present can be tested. Obviously, a large number of events are required to offset the low probability of simultaneous opening. For instance, if the open probability is 0.01, the probability of simultaneous opening of two channels is 0.0001, which demands data with at least several thousands of events for a reliable test (5). On the contrary, if the mean open and closed times of a specific channel under a given condition are available, the number of channels can be estimated from the observed apparent mean of the closed time. Alternatively, estimation is also possible by assuming that the frequency of simultaneous channel opening follows the binomial distribution (6).

5.6. Gating Analysis

The amplitude analysis accounts for the static characteristics of ion channels, but another important factor that determines the characteristics of ion channels is the transition rate constants between each open and closed state. They not only determine the macroscopic stationary response by way of channel open probability, they dominate the macroscopic response rate against the stimulus change. The rate constants can be estimated by analyzing the open and closed dwell times in single-channel currents. The analysis of dwell times is composed of three steps: (1) detection of opening and closing transitions, (2) construction and analysis of a dwell time histogram, and (3) decision of the reaction model.

5.6.1. Detection of Opening and Closing Transitions

5.6.1.1. Setting the Threshold Level

Before computers became popular, data on a chart recorder had been analyzed manually, which was time-consuming and often inaccurate. It should be noted that automation is also not easy unless the record is exceptionally accurate. This is because discrimination of states on records is often difficult even for trained researchers because of contamination with background noise, high-speed openings, high-speed closings during burst, presence of substates, and simultaneous openings of multiple channels.

Here, we begin a step-by-step description of a simple situation. The most common procedure is to set an appropriate threshold level and then judge the transition: The threshold is usually set at 50% of the single-channel conductance obtained by amplitude analysis. Setting the level at other values would cause errors in the estimation of dwell time. The reason is that the waveform of open current is not a complete square wave but is deformed to a trapezoid owing to the effect of the low-pass filter of the measuring system.

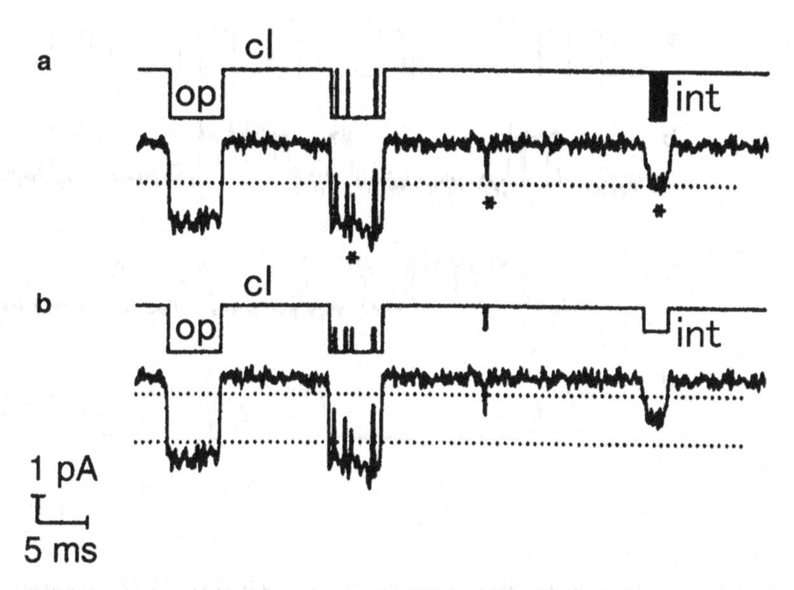

Fig. 5.4. An example of the improvement of transition detection by two-level thresholding. (**a**) Detection by common one-level thresholding (50%). Idealized trace after detection is shown on the *top*. *Asterisks* indicate unresolved opening, substate, and erroneous detection by noise. *op*: *open state*; *cl*: *closed state*; *int*: *intermediate state*. (**b**) Example of detection by two-level thresholding (25% and 75%). The erroneous detection in (**a**) is resolved into a substate level. (From (7)).

A threshold larger than 50% would result in underestimation of the open time, and a threshold smaller than 50% would result in overestimation. The opposite occurs on the closed time estimation. With this method, however, it is difficult to detect properly the substates and high-speed openings and closings that are routinely observed, resulting in a large error (Fig. 5.4a). An effective method to minimize the error is two-level thresholding (7). For example, two thresholds are set at two levels – 25% and 75% – from the closed level, and the states are divided into three states: closed, intermediate, and open (Fig. 5.4b). This method sometimes bears great merit over sacrificing some accuracy of the estimation of each dwell time.

5.6.1.2. Reduction of Background Noise

What is most troublesome in the detection of transition is the high-frequency background noise. Let us refer to practical data. Figure 5.5a shows computer-generated ideal openings and closings without noise. The single-channel current is set at 12.5 pA, and the mean open and closed times are set at 1.03 and 1.65 ms, respectively. Next, 2.5-pA (rms) noise is superimposed on these data (Fig. 5.5b). Although the effect is seemingly absent, the mean open and closed times as estimated by 50% thresholding are 0.607 and 1.04 ms, respectively, which are smaller than the true values by about 40%. Appropriate application of a Gaussian filter to the data generates new data, as shown in Fig. 5.5c. The mean open and

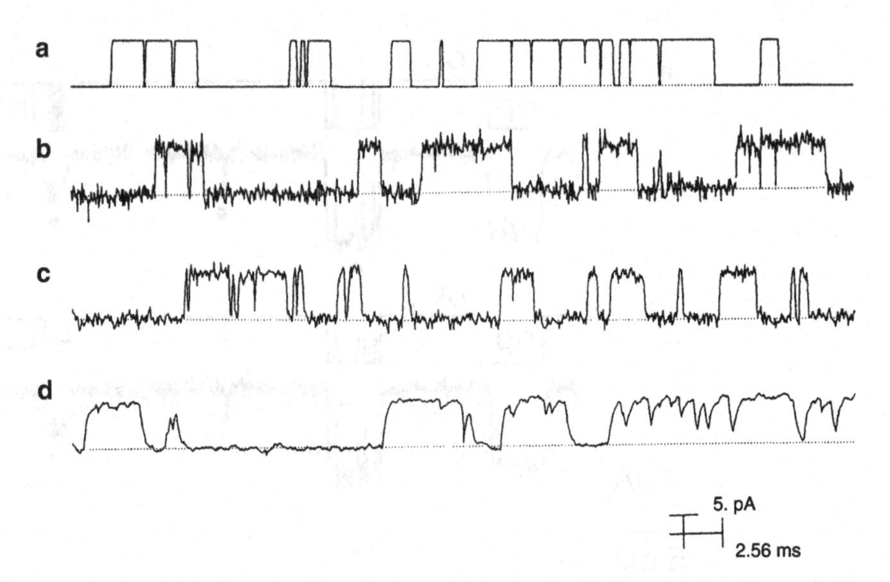

Fig. 5.5. Effect of noise and a filter on estimation of open and closed times. (**a**) Trace of ideal openings and closings generated by a computer (single channel current 12.5 pA, mean open time 1.03 ms, mean closed time 1.65 ms, open probability 0.38). (**b**) Noise (rms = 0.25 pA) is applied to trace A (mean open time 0.607 ms, mean closed time 1.04 ms, open probability 0.37). (**c**) Proper Gaussian filter (deviation coefficient 0.5) is applied to trace B (mean open time 1.08 ms, mean closed time 1.85 ms, open probability 0.37). (**d**) A strong filter (deviation coefficient 0.9) was applied to trace A (mean open time 1.95 ms, mean closed time 3.22 ms, open probability 0.38). The values in parentheses represent the estimated values.

closed times are now 1.08 and 1.85 ms, showing recovery nearly to the true values. However, if an excessively strong filter is applied, the original waveform is skewed considerably, and the mean open and closed times are increased to 1.95 and 3.22 ms, which are about two times the true values (Fig. 5.5d). This simulation shows how important the filter setting is for estimating the dwell time.

How can the cutoff frequency of the filter be determined properly? The decision is made as follows. The conductance level (−100%) is set in the direction opposite to the actual direction of openings against the baseline, and the 50% threshold level (−50%) is set relative to this level. When transitions are detected under this condition, the cutoff frequency of the filter is reduced until the apparent opening events due to noise are not detected. Detection of false events occurred in the data in Fig. 5.5b due to noise can thus be avoided. Needless to say, we should be aware that this procedure also causes high-speed openings and closings to be missed (Fig. 5.5d). Some compromise at the proper point is required depending on the purpose of the analysis. It should be noted that if false events are present the mean open and closed times are doubly underestimated. This underestimation derives from a large number of short-lived false open and closed events and the resulting interruption of long open and closed events. By contrast, the presence of missing events results in overestimation of the open and closed times, but the error is smaller than with false events because the missed short events are extrapolated by fitting (described below).

5.6.1.3. Baseline Correction

The drift of the baseline has not been taken into consideration so far, but in fact low-frequency fluctuation and drift that cannot be removed by filter are sometimes superimposed. The threshold level, which is set last, can become meaningless. A method for the automatic correction of this is the zero-crossing method (7). The minimum level is set first, and the level is increased gradually until the number of zero crossings becomes maximum, at which the baseline is set. That is to say, while searching for the average noise level, the baseline is set at this level. This method is not, however, always effective if the data contain long open events and frequent bursts. For the method used most commonly at present, after calculating the moving average of closed events with a given dwell time, the weighted average with the mean of the nearest previous closed event is calculated to represent a new baseline (8). For example, if the nearest previous baseline is 1 pA and the moving average of the previous closed level is 2 pA, the weight of 0.1 gives us a new baseline, $1 \times 0.9 + 2 \times 0.1 = 1.1$ pA. Most software is equipped with such a baseline correction tool, which can be used depending on the condition. Time-course fitting is also proposed as a more precise method but is not always practical. For more information, refer to Sigworth and Sine (9).

The above-described methods and warnings for event detection, however, may not be perfect. As the final finish, the experimenter should, after all, confirm visually that every event is correctly detected and classified; and artifacts should be removed. Most software is equipped with such an editing tool.

5.6.2. Analysis of Open and Closed Dwell Times

5.6.2.1. Construction of Dwell Time Histograms

The series of dwell times for each event is referred to as the "event list" and can be stored as an ASCII file. The next step is to assemble this list into a histogram with the dwell time as abscissa and the number of events as ordinate. The bin width of this histogram should be set properly depending on the number of events. It is known that the bin width is reasonably set at 10–20% of the smallest dwell time to be analyzed. In addition, the number of bins recommended is 50–100, taking into consideration the subsequent fitting to exponential functions. As a result, data with at least about 1,000 events are preferred. Note, however, that this method is reliable only if the histogram can be well fitted with two exponential functions at most and their time constants differ by one order of magnitude. If the contained exponential functions are more than three and the time constants differ by more than two orders of magnitude, reliable fitting to exponential functions cannot be expected unless fitting is applied progressively from the short

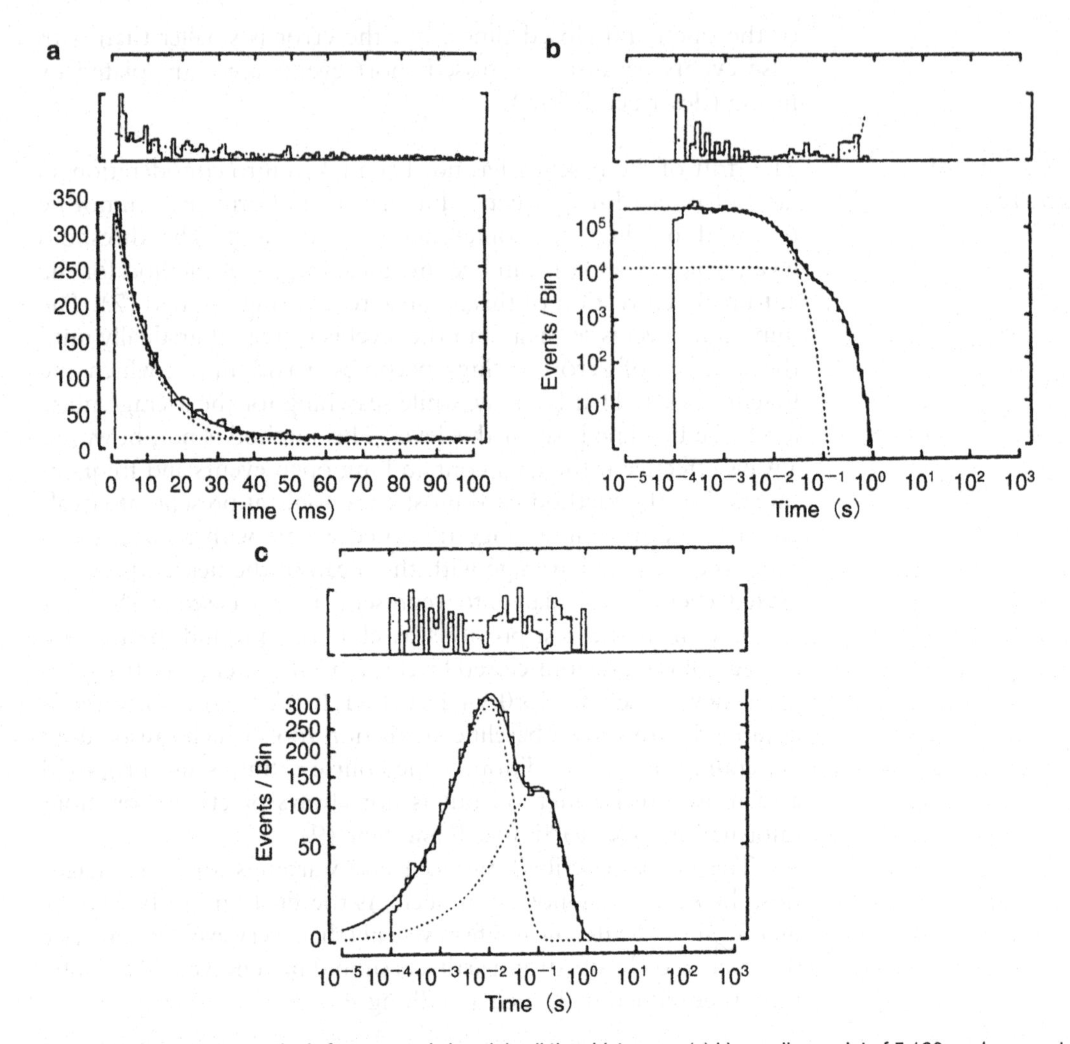

Fig. 5.6. Various plotting methods for open and closed dwell time histogram. (**a**) Linear–linear plot of 5,120 random events that show the distribution with time constants of 10 ms (frequency 70%) and 100 ms (frequency 30%), with 1 ms bin width. The deviation from the theoretical value (*broken line*) is shown on the *top*. The 100 ms component is underestimated with this plot. (**b**) Log–log plot. Bin width increases logarithmically. Note that the ordinate adopts the number of events divided by each bin width (**c**) Logarithms of time and the square root of the number of events are plotted as the abscissa and ordinate, respectively. Note that the deviation shown on the *top* is almost constant irrespective of time. (From (10)).

components, or the components with fast time constants are emphasized on log–log plot or log-square root plot (Fig. 5.6). Various limitations are present even in these cases; therefore, consult Jackson (10) for details.

5.6.2.2. Fitting to Exponential Functions

Because the opening and closing reactions of ion channels can be considered to follow Markov chain process, the distribution histogram of open and closed dwell times can be approximated to a probability density function, $f(t) = (1/\tau)\exp(-t/\tau)$, where $f(t)$ is the probability density function representing the probability that events with dwell time (t) occur, and τ represents the average dwell time (lifetime), which is referred to as the time constant. This can be considered as follows: In an opening and closing reaction

$$\text{closed} \underset{\beta}{\overset{\alpha}{\rightleftharpoons}} \text{open}$$

the dwell time (lifetime) of the open state, $F(t)$, is given by $dF(t)/dt = -\beta F(t)$. The integration of this equation gives $F(t) = \exp(-\beta t)$. In other words, $F(t)$ represents the probability that the lifetime of the open state is t or longer. Therefore, the probability that the lifetime is t or shorter is $1-\exp(-\beta t)$. Differentiation of this equation gives the probability density function, $f(t) = \beta\exp(-\beta t)$, which is the probability that the lifetime is t. Here, replacement of β with $1/\tau$ provides equation $f(t) = (1/\tau)\exp(-t/\tau)$ (see Appendix 1). The lifetime of the closed state can be considered in the completely same manner, but β should be replaced with α. As is evident in this example, the average lifetime of a given state is the inverse of the rate constant (typically, the sum of them) of the transition from the given state to a different state. Therefore, this reaction model allows us to directly obtain the rate constants and determine the characteristics of the reaction by calculating the time constants from the histogram.

If the reaction contains multiple closed (open) states whose conductance cannot be discriminated, the histogram for the closed (open) states is generally the superposition of multiple exponential functions that correspond to the number of states (n), and the theoretical curve is given by $f(t) = \Sigma(a_i/\tau_i)\exp(-t/\tau_i)$, where a_i is the relative frequency that is proportional to the total number of events of each component and $\Sigma a_i = 1$. In other words, a_i corresponds to the proportion of the area of each component. This can easily be confirmed, for example, by the fact that the integral of the component i $\{f_i(t)\}$ (probability of appearance of the state i) of the above function from 0 to ∞ is equal to a_i. As shown in this example, if the histogram is composed of multiple exponential functions, the rate constants for each reaction are generally given by considerably complex functions containing τ_i and a_i. A simple example is provided in the next section.

The Levenberg–Marquardt method (nonlinear least-square method) (11) is practical for fitting the dwell time histogram to exponential functions, but the error becomes larger if the size of the data is small and the histogram contains empty bins. In this case, better results can be obtained by the maximum likelihood estimation(12). Most software is bundled with the former, but either can be selected in some software (pCLAMP6 or later version). The number of exponential components for fitting can only be determined through trial and error while monitoring the results.

5.7. (*) Appendix 1

Let me explain more carefully the development of the solution and the meaning of this differential equation. Assuming that $F(t)$ represents the probability that the gate is open continuously from time 0 to t, let us obtain the probability $\{F(\Delta t + t)\}$ that the gate remains open for Δt. This is, in other words, the product of the probability that the gate is open for t seconds and the probability that the gate does not close for the subsequent Δt (because these are independent events). Now, because the probability that the gate closes during Δt is $\beta\Delta t$, the probability that the gate does not close during Δt is $(1-\beta\Delta t)$. Therefore,

$$F(t+\Delta t) = F(t)(1-\beta\Delta t)$$

is obtained. Transposing the first term on the right side of this equation and dividing both sides by Δt give

$$\{F(t+\Delta t) - F(t)\} / \Delta t = -\beta F(t)$$

If Δt approaches 0, this equation is expressed as:

$$\mathrm{d}F(t) / \mathrm{d}t = -\beta F(t)$$

which gives us the differential equation for the open lifetime $F(t)$. Integration of this equation gives us

$$F(t) = \exp(-\beta t)$$

because $F(0) = 1$. As is evident from the process of deviation, $F(t)$ is the cumulative density function (probability distribution function) representing the probability that the open dwell time is longer than t ($>t$). Next, to obtain the probability function $f(t)$ at which the dwell time is t [$f(t)$ is referred to as the probability density function and is obtained by differentiating the probability distribution function], we differentiate the probability that the open lifetime is t or shorter ($\leq t$) – i.e., the probability distribution function, $1-F(t) = 1-\exp(-\beta t)$ – and obtain the aimed equation.

$$f(t) = \beta\exp(-\beta t)$$

The above differential equation may more easily be understood if we obtain it directly by assuming that the probability $\beta\Delta t$ is equal to $-\{F(t+\Delta t) - F(t)\}$, taking into account the case that the gate is open for t seconds and is closed for the subsequent Δt. This equation has absolutely the same form as the differential equations that express the process of population decrease among living organisms without regeneration or the process of isotope decay.

5.7.1. Determination of Reaction Model

The final purpose of dwell time analysis is determination of the reaction model of channel gating and the rate constants between each state. The number of time constants in the respective open and closed states obtained from the above dwell time histogram is used as a starting point because it provides the minimum number of states involved in the respective open and closed states. However, it cannot readily be determined whether these states could be connected linearly or cyclically, or with a combination of the two. These can only be determined through trial and error based on the channel characteristics and experimental conditions. Semiempirical methods using computer simulations are often used here. That is, simulated current traces are generated by assuming a proper reaction model on the basis of the estimated number of states and rate constants obtained from the fitting of a dwell time histogram to exponential functions and the calculated a_i and τ_I, respectively. Dwell time histograms are then constructed from some simulated current traces, and the best-fit reaction model is determined through comparison with actual histograms (13). Axon Instruments offers a program (CSIM) that generates channel current data from a reaction model; it can perform incredibly complex simulations for up to 12 states. A general procedure for solving the relation between a_i and τ_i obtained from histograms and rate constants in a reaction model is given by Colquhoun and Hawkes (14). Because the exposition of this procedure is beyond the scope of this chapter, its mathematical essence is summarized below.

If we assume n closed states with identical conductance, the probability density function of the closed dwell time follows an nth-order linear differential equation. Rearranging this equation into an intuitively plain one provides simultaneous first-order differential equations on n closed states. Here, the coefficients of the simultaneous differential equations (consisting of transition rate constants between each state) compose nth-order square matrix (see Appendix 2). Now the general solution of nth-order linear differential equations is given by $f(t)=\Sigma w_i \exp(-\lambda_i t)$. Given that $\lambda_i = 1/\tau_i$, where τ_i corresponds to the time constant mentioned in the previous section, λ_i can be obtained as the root of the characteristic equation for nth-order differential equations [nth-order equations obtained by replacing $d_j f(t)/dt_j$ with λ_j), and w_i (here $\Sigma w_i/\lambda_i = 1$)] can be obtained from the initial value (in matrix expression, λ_i can be obtained as the eigenvalue of the matrices). These are the mathematical grounds on which a dwell time histogram can be approximated by the linear sum of exponential functions. If the number of states exceeds three, the relationship between λ_i and the rate constants generally becomes extremely complex, and the equation can be effectively solved by the matrix method, the details of which can be found in Colquhoun and Hawkes (15).

5.7.2. Example of Analysis of Open Channel Blocking

The model to be analyzed is:

$$\text{closed} \underset{\beta}{\overset{\alpha}{\rightleftharpoons}} \text{open} \underset{k^-}{\overset{C_B k^+}{\rightleftharpoons}} \text{blocked}$$

where C_B represents the blocker concentration. The single-channel currents generated by this reaction model display burst-like traces in which short blocks are inserted in the open state. Because the conductance of the closed state and the blocked state in this reaction represent the same level and are indistinguishable, the closed dwell time histogram should have a short time constant τ_B, which derives from the blocked state, and a long time constant τ_c, which derives from the true closed state between bursts. By contrast, the open dwell time histogram is given by a single exponential function with a single time constant (τ_o). These time constants can be readily obtained from $\tau_B = 1/k^-$, $\tau_c = 1/\alpha$, and $\tau_o = 1/(\beta + C_B k^+)$. From this equation, the probability density function $f(t)$ approximating the closed dwell time histogram can be expressed as $f(t) = a_1 k^- \exp(-k^- t) + a_2 \alpha \exp(-\alpha t)$ according to the previous section. Because a_1 and a_2 represent the existence probability (relative probability) of the blocked state and the closed state, respectively, they can be given by $a_1 = C_B k^+/(\beta + C_B k^+)$ and $a_2 = \beta/(\beta + C_B k^+)$ using the rate constants β and $C_B k^+$. Based on these relations, four rate constants (α, β, k^+, k^-) can be determined from the experimental values $\alpha = 1/\tau_c$, $\beta = a_2/\tau_o$, $k^+ = a_1/\tau_o C_B$, and $k^- = 1/\tau_B$.

Incidentally, k^-/k^+ corresponds to the dissociation constant (K_B) of the blocker. After a simple calculation, the mean total open time per burst is expressed as $1/\beta$, and the mean total closed time is $C_B/K_B \cdot \beta$. It follows that the mean burst duration is $(1 + C_B/K_B)/\beta)$ (15).

5.8. (*) Appendix 2: Solution by Q Matrix

Let O, B, and C be the existence probability in the open, blocked, and closed states, respectively, in the previous reaction model. The differential equation expressing the time course is then as follows:

$$dO/dt = -(\beta + C_B k^+)O + k^- B + \alpha C$$

$$dB/dt = C_B k^+ O - k^- B$$

$$dC/dt = \beta O - \alpha C$$

Matrix expression gives

$$\begin{bmatrix} O' \\ B' \\ C' \end{bmatrix} = \begin{bmatrix} -(\beta + C_B k^+) & k^- & \alpha \\ C_B k^+ & -k^- & 0 \\ \beta & 0 & -\alpha \end{bmatrix} \begin{bmatrix} O \\ B \\ C \end{bmatrix}$$

Moving the column vector on the right side to the front transposes the matrix as follows:

$$\begin{bmatrix} O' \\ B' \\ C' \end{bmatrix} = \begin{bmatrix} O \\ B \\ C \end{bmatrix} \begin{bmatrix} -(\beta + C_B k^+) & C_B k^+ & \beta \\ k^- & -k^- & 0 \\ \alpha & 0 & -\alpha \end{bmatrix}$$

This transposed matrix is expressly called the Q matrix. If the column vector on the right side is expressed as P, the above equation can be simplified to

$$dP(t)\,/\,dt = P(t)\cdot Q(t)$$

The general solution for this equation is $\mathbf{P}(t) = e^{Q(t)}$. After this calculation, the eigenvalues (λ_i) for Q matrix (actually, $-Q$) are obtained according to the common procedures in matrix calculation. Here, for the purpose of obtaining the correlation with the dwell time histograms, which are separated into open and closed states, the above Q matrices are separated (broken lines) into minor matrices ($\boldsymbol{A}$, $\boldsymbol{B}$) that correspond to the open (O) and closed (B, C) states, respectively; and each eigenvalue is obtained from the characteristic equations ($|-\boldsymbol{A}-\lambda \boldsymbol{I}| = 0$, $|-\boldsymbol{B}-\lambda \boldsymbol{I}| = 0$), where $\boldsymbol{I}$ is unit matrix. Because $\boldsymbol{A}$ is a first-order matrix and $\boldsymbol{B}$ is a diagonal matrix in this example, it goes without saying that each eigenvalue is $(\beta + C_B k^+)$, k^-, and α, respectively, which coincide with the results mentioned above. This Q matrix method is a common method with which complex reaction model can be analyzed just by generating the Q matrix. A simple procedure for the generation of the Q matrix is as follows:

If we carefully examine the matrix in which the on-diagonal elements of the above transposed matrix are set at 0, we see that

$$\begin{matrix} O \\ B \\ C \end{matrix} = \begin{matrix} \quad O \quad B \quad C \\ \begin{bmatrix} 0 & C_B k^+ & \beta \\ k^- & 0 & 0 \\ \alpha & 0 & 0 \end{bmatrix} \end{matrix}$$

is the transition probability matrix from i (row) to j (column). Moreover, the on-diagonal elements of the above transposed matrix are the negatives of the lifetime (time constant) of each state, meaning that the sum of each row is zero. That is, the on-diagonal elements on each row can be obtained simply by reversing the sign of the sum of the row elements of the transition matrix (this can easily be assembled from the reaction model). It is wise to commit the calculation of the eigenvalues of complex matrices to an appropriate program package.

5.9. Conclusion

I believe that although the analysis proceeds smoothly up to level of amplitude analysis, the dwell time analysis takes enormous time and one cannot come to any conviction until the end of the analysis. On the other hand, the higher-order structures of ion channel proteins have been revealed gradually, and introduction of structural modification has become possible through site-directed mutagenesis. Analysis of subtle changes in conductance and gating due to mutation enables in-depth research on the structure–function relation on the atomic level. In this sense, single-channel data analysis has become much more important. I hope that this chapter is helpful to researchers who aim to uncover the structure–function relations of ion channels.

References

1. Miller C (ed) (1986) Ion channel reconstitution. Plenum, New York
2. Sakmann B, Neher E (eds) (1995) Single-channel recording, 2nd edn. New York, Plenum
3. Patlak JB (1988) Sodium channel subconductance levels measured with a new variance-mean analysis. J Gen Physiol 92:413–430
4. Hille B (1992) Ionic channels of excitable membranes, 2nd edn. Sinauer, Sunderland
5. Colquhoun D, Hawkes AG (1983) The principles of the stochastic interpretation of ion-channel mechanisms. In: Sakmann B, Neher E (eds) Single-channel recording. Plenum, New York, pp 135–189
6. Sachs F, Neil J, Barkakati N (1983) The automated analysis of data from single ionic channels. Pflugers Arch 395:331–340
7. Dempster J (1993) Computer analysis of electrophysiological signals. Academic Press, London, p167
8. Colquhoun D (1987) Practical analysis of single channel records. In: Standen NB, Gray PTA, Whitaker MJ (eds) Microelectrode techniques. The Plymus workshop handbook. Company of Biologist Ltd, Cambridge, pp 83–104
9. Sigworth FJ, Sine SM (1987) Data transformations for improved display and fitting of single channel dwell time histograms. Biophys J 48:149–158
10. Jackson MB (1992) Stationary single-channel analysis. Method Enzymol 207:729–746
11. Nakagawa T, Koyanagi Y (1982) Analysis of experimental data by least square method (in Japanese). University of Tokyo Press, Tokyo
12. Awaya T (1991) Data analysis (2nd edition, in Japanese). Japan Scientific Societies Press, Tokyo
13. French RJ, Wonderli WF (1992) Software for acquision and analysis of ion channel data:choices, tasks, and strategies. Method Enzymol 207:711–728
14. Colquhoun D, Hawkes AG (1981) On the stochastic properties of single ion channels. Proc R Soc Lond B 211:205–235
15. Colquhoun D, Hawkes AG (1983) On the stochastic properties of bursts of single ion channel openings and clusters of bursts. Phil Trans R Soc Lond B 300:1–59

Chapter 6

Channel Noise

Harunori Ohmori

Abstract

Patch-clamp is the best method for analyzing single-channel current activities today. The method is not good, however, at measuring the total number of channels expressed on the cell membrane that are responsible for the current of concern. In contrast, noise analysis has some advantage over the single-channel recording as a method to estimate the total number of ion channels. Also, noise analysis can obtain some channel information regarding size and kinetics even when individual channel events are so small to be detected directly as single-channel events. In this chapter the logic of noise analysis for stationary currents and nonstationary currents follows a description of several classic experiments conducted during the 1970s–1980s. Several technical issues frequently encountered during experiments are also addressed.

6.1. Introduction

Cellular function is achieved in many respects by the activities of various ionic channels distributed in the membrane at a density of a few channels to thousands of channels per square micron. These ionic channels are activated by changes in membrane potential or by association with ligands. Ions passing through channels are recorded as currents.

Patch-clamp is the best method today for analyzing single-channel currents. With single-channel recordings, we can gain an explicit understanding of gating, ionic selectivity, and the unitary current. This method, however, is applicable only to a patch membrane that isolates a few channels in it. Thus, a disadvantage of the method, for example, lies in the difficulty of measuring the total number of channels expressed in the entire plasma membrane of a cell. Moreover, single-channel recording experiments must be conducted under the optimized condition of a high signal-to-noise (S/N) ratio, making application of this method limited. In contrast, noise analysis has some advantages over single-channel recording,

Yasunobu Okada (ed.), *Patch Clamp Techniques: From Beginning to Advanced Protocols*, Springer Protocols Handbooks, DOI 10.1007/978-4-431-53993-3_6, © Springer 2012

particularly as a method to estimate the total number of ion channels expressed on a cell membrane. Noise analysis can also obtain some channel information even when individual channel events are too small to be detected directly as single-channel currents. These are the reasons noise analysis is still used in many experiments.

As early as the 1950s, noise analysis was applied to neurobiological experiments (1) on the fluctuation of spike frequency (2). During the early 1970s or even in some preceding years, physiologists had an interest in channel noise, which we address in this chapter, including Katz and Miledi (3, 4), who applied the technique to voltage fluctuation at the endplate. Stevens (5) published a theoretical review of channel noise in 1972, and the following year Anderson and Stevens applied the theory to endplate current under a voltage-clamp (5, 6). These papers are classic, and the noise analysis treated in this chapter starts from the explanation of their treatment of current noise in the steady state and the more recent approach to nonstationary noise by Sigworth (7, 8).

6.2. Electrical Noise Generated in Biological Membranes

An equivalent circuit in the biological membrane is the parallel resistor R and capacitor C, which represent pathways of currents across the membrane, such as ion channels and the nonconductive lipid membrane structure, respectively. Both generate electrical noise, which becomes background noise to the channel noise.

Background noise consists of many sources, such as the thermal noise that originates most in the resistive part of the cell membrane, the shot noise that originates from the passage of ions across the membrane, and $1/f$ noise. *Shot noise* is a high-frequency noise with much higher frequency than channel noise, whereas *1/f noise* overlaps the frequency range and power of channel noise and becomes serious background noise. The $1/f$ noise, also called *pink noise*, is frequently observed in various natural phenomena (9, 10); how it is generated is still an enigma.

Thermal noise is affected greatly by the impedance of the cell membrane under conditions of recording and recording devices. It reflects the fluctuation of electrons in conducting materials.

When the cell membrane is expressed by an equivalent circuit of membrane resistance R and capacitance C, the impedance $Z(f)$ is expressed as a function of frequency (6.1); here $Z_R = R$, $Z_C = \frac{j}{\omega C}$, and $j^2 = -1$, $\omega = 2\pi f$

$$Z_m = \frac{1}{\frac{1}{Z_R} + \frac{1}{Z_C}} = \frac{R}{1+(\omega CR)^2}(1 + j\omega CR) \tag{6.1}$$

Thus, the power spectrum of voltage generated by this impedance is as follows, (6.2).

$$S_V(f) = 4kT \cdot \mathrm{Re}(Z(f)) \quad (6.2)$$

Here, k and T are the Boltzmann constant and the absolute temperature, respectively. Re($Z(f)$) is a real part of impedance $Z(f)$.

Equation (6.3) is the expression of the current noise power spectrum.

$$S_I(f) = 4kT \cdot \mathrm{Re}(1 / Z(f)) \quad (6.3)$$

In the equivalent membrane circuit, the corresponding power spectra are as follows:

$$S_v(f) = 4kT \cdot \frac{R}{1 + (2\pi f \cdot RC)^2} \quad (6.4)$$

$$S_I(f) = 4kT \cdot \frac{1}{R} \quad (6.5)$$

These formulations indicate that thermal noise is reduced when the membrane resistance is large (6.5), and its effect on the channel noise analysis is reduced. However, there is another source of background noise in real measurements: the noise that originates from the voltage noise of the head stage amplifier, σ_v^2. This noise, when coupled with a stray capacitance C_s around the electrode and feedback resister, becomes a source of capacitative current noise, σ_{Icap}^2 and increases the background current noise significantly (6.6).

$$\sqrt{\sigma_{\mathrm{Icap}}^2} = C_S \frac{\mathrm{d}\sqrt{\sigma_V^2}}{\mathrm{d}t} \quad (6.6)$$

6.3. Current Noise Analysis in the Steady State

Anderson and Stevens (6) analyzed the fluctuation of acetylcholine (ACh)-induced ionic current under a voltage-clamp at the endplate of frogs. Prior to their work, Katz and Miledi (3, 4) observed the fluctuation of the ACh-induced voltage response and interpreted the phenomena as due to the activity of ionic channels of 100 pS unitary conductance. Anderson and Stevens (6) analyzed the current noise by applying a low concentration of ACh to the endplates to open channels while simultaneously maintaining the channel open probability low. ACh generated a current of 120 nA. When observed at a larger gain, the current exhibited fluctuation much larger than that in the resting state (Fig. 6.1a). These authors calculated the variance of conductance fluctuation and plotted it against the mean conductance (6.7 and 6.8) as in Fig. 6.2. There was a linear relation between the variance and the mean conductance.

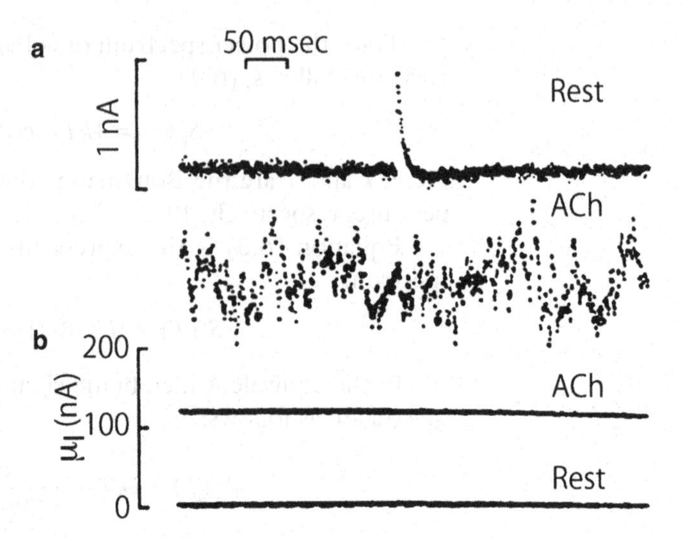

Fig. 6.1. Acetylcholine (ACh) channel currents under a voltage clamp. (**a**) Two traces show current fluctuation at rest and after ACh application. Miniature endplate current is included in the resting current. (**b**) Two traces show the mean current corresponding to the two traces in (**a**). (From Anderson and Stevens (6), with permission).

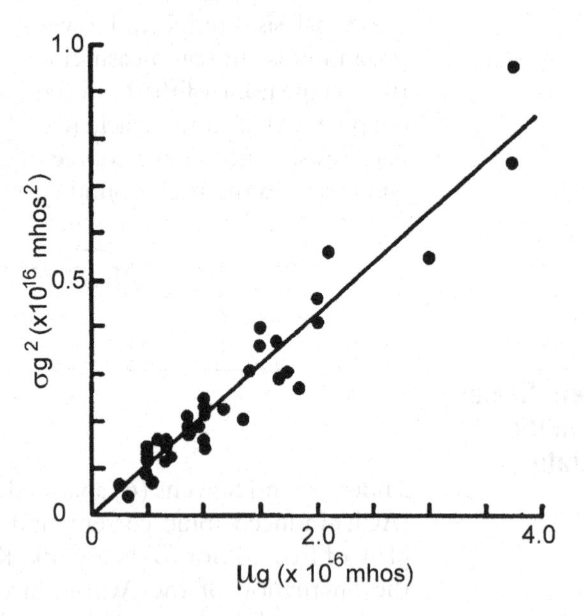

Fig. 6.2. Variance of conductance plotted against the mean current. Recorded from −140 and +60 mV. Single-channel conductance of the ACh receptor channel is 19 pS. (From Anderson and Stevens (6), with permission).

This is what is expected when the mean open probability of channels is low ($p \ll 1$). If each ACh channel opens with the probability of p, and closes with $1-p$, the mean open probability is $\mu = p$ and variance is $\sigma^2 = p(1-p) \approx p$, from the theory of binomial distribution. Conductance of channel and the current passing through the channel are related as $\mu_I = \mu_g\ (V - V_{eq})$ and variance of current

$\sigma_I^{\,2}$ and conductance $\sigma_g^{\,2}$ are as follows: $\sigma_I^{\,2} = \sigma_g^{\,2}(V - V_{\rm eq})^2$. Here V and $V_{\rm eq}$ represent the membrane potential and equilibrium potential, respectively. When the channel has a conductance γ, the variance and mean conductance are expressed as (6.7) and (6.8), respectively. Thus, the slope of the plot, the variance against the mean, gives the single-channel conductance γ (6.9) (Fig. 6.2).

$$\sigma_g^2 = \gamma^2 \cdot N \cdot p(1-p) \approx \gamma^2 \cdot N \cdot p \tag{6.7}$$

$$\mu_g = \gamma \cdot N \cdot p \tag{6.8}$$

$$\sigma_g^{\,2} = \gamma \cdot \mu_g \tag{6.9}$$

Anderson and Stevens (6) calculated further a power spectrum of ACh current noise using the fast Fourier transform (FFT) (Fig. 6.3). The power spectrum of Lorentzian type was obtained in which the power is constant at low frequency $S(0)$ and decreases with a relation of $1/f^2$ at higher frequencies; also, the transition frequency is called a corner frequency f_c and is determined by open–close kinetics of the channel. Anderson and Stevens (6) considered the open–close kinetics of the ACh channel as the following formulation (6.10) and interpreted f_c using the closing rate constant α (6.11).

$$nT + R \underset{}{\overset{K}{\rightleftarrows}} TnR \underset{\alpha}{\overset{\beta}{\rightleftarrows}} TnR^* \tag{6.10}$$

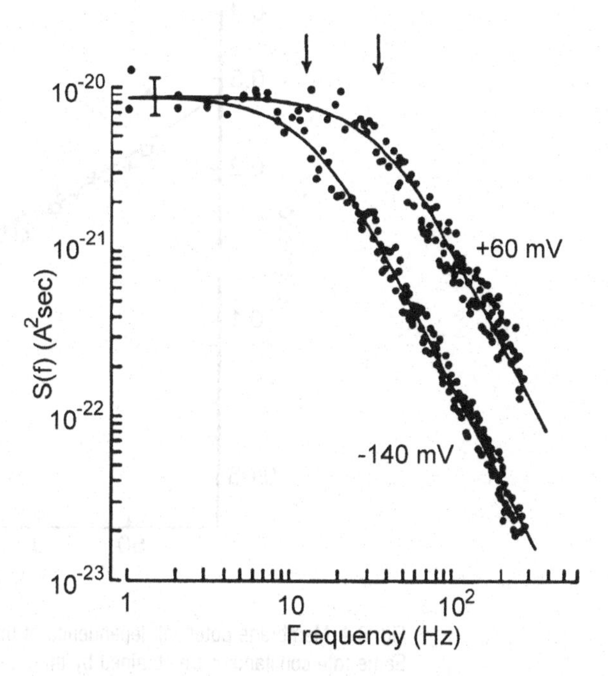

Fig. 6.3. Lorentzian-type power spectrum of ACh current noise. Corner frequency (f_c) depends on the membrane potential. (From Anderson and Stevens (6), with permission).

Here, K is the equilibrium constant of ACh binding with the receptor; n is the number of ACh molecules that bind to a receptor; T represents the ACh molecule; R is the ACh receptor; TnR and TnR^* represent the closed state and the open state of the ACh receptor channel, respectively; α and β are rate constants for the closing and opening transitions, respectively. If c is the concentration of ACh, $c = [T]$, then at the steady state TnR, the ACh receptor channel that is associated with n molecules but is still in a closed state does exist with a probability of Kc^n. Thus, the rate of transition from the closed state to the open state is β times Kc^n. If this opening rate βKc^n is much smaller than the closing rate α, the corner frequency f_c is simplified to (6.11), and the power spectrum function $S_I(f)$ (6.12) is also simplified.

$$f_C = \alpha / 2\pi \tag{6.11}$$

$$S_I(f) = \frac{S_I(0)}{1+(f/f_C)^2} = \frac{2\gamma\mu_I(V - V_{eq})/\alpha}{1+(2\pi f/\alpha)^2} \tag{6.12}$$

Anderson and Stevens (6) compared closing rate constants obtained by three independent methods, and all were consistent. Figure 6.4 plots the voltage dependence of α estimated from f_c, the corner frequency of the power spectrum (□), two α-s estimated from the falling phase of the endplate current (EPC, +), and miniature endplate current (mEPC, Δ). Thus, the falling phase of the endplate current corresponds to the closing transition of ACh channels.

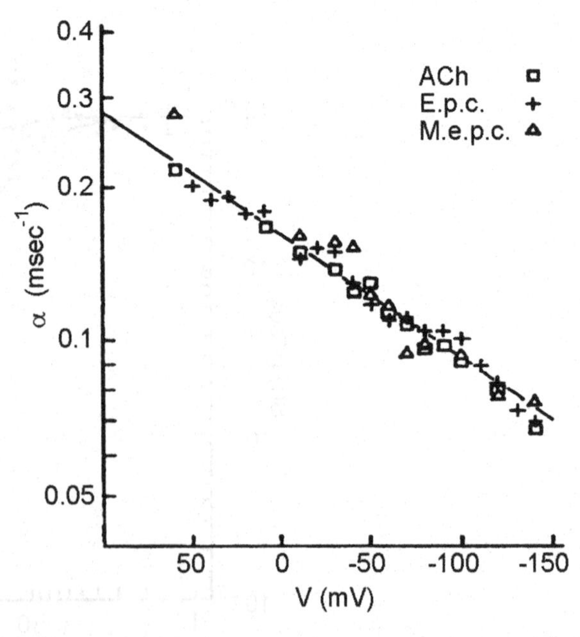

Fig. 6.4. Membrane potential dependence of the closing rate constant of ACh receptors. Same rate constants were obtained by three separate experiments. Corner frequency of ACh noise, end-plate current (E.p.c.), and miniature endplate current (M.e.p.c.). (From Anderson and Stevens (6), with permission).

Equation (6.12) is a most important one for analyzing current noise. The exact formulation needs precise modeling on the gating property of the channel, which means that the estimates of unitary conductance γ and rate constant α depend on a specific kinetic model. Therefore, it is important to prove the consistency of data with separate, independent approaches.

6.4. Details of Current Noise Analysis by Anderson and Stevens

If we assume (1) the concentration of ACh at the endplate synapse c (c=[T]), (2) that binding between the ACh molecule T and the receptor R is fast enough with an equilibrium constant K, and (3) the current fluctuation is generated by transition between two states – TnR (close) and TnR^* (open) – the stochastic process of channel opening and closing is defined as (6.13) by using a conditional probability $p\ (k|t)$, which is the probability of being open at time t when the channel is either open or closed at time zero (the condition is denoted by k). Here, the equation is simplified by assuming that the opening rate βKc^n is smaller than the closing rate α; k is either o or c, corresponding to being either open or closed at time zero, respectively.

$$\begin{aligned}\frac{\mathrm{d}p(k\,|\,t)}{\mathrm{d}t} &= \beta{\cdot}Kc^n - (\alpha + \beta{\cdot}Kc^n){\cdot}p(k\,|\,t)\\ &= \beta{\cdot}Kc^n - \alpha{\cdot}p(k\,|\,t)\end{aligned} \tag{6.13}$$

Open probability at the steady state p_∞ has a relation with the conditional probability $p(k|t)$, as (6.14).

$$p_\infty = p(o\,|\,\infty) = p(c\,|\,\infty) = \frac{\beta}{\alpha}Kc^n \tag{6.14}$$

Thus, $p(c|t)$ and $p(o|t)$ are expressed as follows in (6.15) and (6.16), with initial conditions of $p(c|0)=0$ and $p(o|0)=1$, respectively:

$$p(c\,|\,t) = p_\infty(1-\exp(-\alpha t)) = \frac{\beta}{\alpha}Kc^n[1-\exp(-\alpha t)] \tag{6.15}$$

$$\begin{aligned}p(o\,|\,t) &= p_\infty + (1-p_\infty)\exp(-\alpha t)\\ &= \frac{\beta}{\alpha}Kc^n + \exp(-\alpha t)\end{aligned} \tag{6.16}$$

We are now prepared to calculate the power spectrum function of current noise for the case a single ACh receptor channel to fluctuate between two conductance states of open and closed states. In the case of N channels, as long as individual channels operate independently, the power spectrum function becomes N times of the one calculated for a single channel. We also assume a steady state in which stochastic features are independent of time, called ergodicity; thus, we consider current fluctuation around time zero.

Current noise is sampled at regular time intervals of Δt; $\mu_1(t=j.\Delta t, j=1,2,\ldots,L)$, and the mean at each time point $<\mu_1(t)>$ is given by a sum over samples divided by L (6.17).

$$\begin{aligned}\langle \mu_1(t) \rangle &= \frac{\sum_{j=1}^{L} \mu_1(j \cdot \Delta t)}{L} \\ &= \mu_{1\infty} = p_\infty \gamma (V - V_{\text{eq}})\end{aligned} \quad (6.17)$$

Autocorrelation, $C_1(\tau)$, is defined as the mean of multiplication of samples at time t and time $t+\tau$, a sample displaced by τ. Current fluctuation generated in one channel model is expressed as $\Delta\mu_1(t)=\mu_1(t)-\mu_{1\infty}$, and the autocorrelation of the fluctuation, $C_1(\tau)$, then becomes:

$$\begin{aligned}C_1(\tau) &= \langle \Delta\mu_1(0) \cdot \Delta\mu_1(\tau) \rangle = \langle \mu_1(0) \cdot \mu_1(\tau) \rangle - {\mu_{1\infty}}^2 \\ &= \gamma^2 (V - V_{\text{eq}})^2 [p_\infty \cdot p(o \mid \tau) - {p_\infty}^2]\end{aligned} \quad (6.18)$$

Here, $\langle \mu_1(0) \cdot \mu_1(\tau) \rangle = \gamma^2 (V - V_{\text{eq}})^2 p_\infty \cdot p(o \mid \tau)$ and $\langle \mu_1(\tau) \rangle = \mu_{1\infty} = \gamma . P_\infty (V - V_{eq})$. Equation (6.18) is then further transformed to (6.19) using (6.14) and (6.16):

$$C_1(\tau) = \gamma^2 (V - V_{\text{eq}})^2 \frac{\beta}{\alpha} Kc^n \cdot \exp(-\alpha\tau) \quad (6.19)$$

Because noise that originates from activities of N channel is N times of the noise generated by the activity of one channel, as long as each channel opens and closes independently the autocorrelation of N channel model is given by N times the autocorrelation of one channel model. The power spectrum function is given by FFT of the autocorrelation (6.20); detailed discussions are found in Lee (11) and in Bendat and Piersol (12).

$$\begin{aligned}S_N(\omega) &= N \int_{-\infty}^{\infty} C_1(\tau) \cdot \exp(-j\omega\tau) d\tau \\ &= \gamma^2 (V - V_{\text{eq}})^2 \frac{\beta}{\alpha} Kc^n \cdot N \cdot \left[2 \int_0^{\infty} \exp(-\alpha\tau) \cdot \cos(\omega\tau) d\tau \right] \\ &= \frac{2\gamma^2 (V - V_{\text{eq}})^2 \cdot \beta Kc^n \cdot N}{\alpha^2 + \omega^2} \\ &= \frac{2\gamma (V - V_{\text{eq}}) \cdot \frac{\mu_N}{\alpha}}{1 + \left(\frac{\omega}{\alpha} \right)^2} \\ &= \frac{2\gamma (V - V_{\text{eq}}) \cdot \frac{\mu_N}{\alpha}}{1 + \left(\frac{2\pi f}{\alpha} \right)^2} = S_I(f)\end{aligned} \quad (6.20)$$

where μ_N is the mean current passing through N of ACh channels and is given by (6.21):

$$\mu_N = \gamma(V - V_{eq})p_\infty \cdot N = \gamma(V - V_{eq})\frac{\beta}{\alpha}Kc^n \cdot N \tag{6.21}$$

The variance of current passing through N of ACh channels is given by (6.22):

$$\sigma_N{}^2 = NC_1(0) = N\gamma^2(V - V_{eq})^2 \frac{\beta}{\alpha}Kc^n \tag{6.22}$$

Therefore, the relation between the variance and the mean of the current passing through N of ACh channels is as follows (6.23). Note that the equation is the same as (6.9):

$$\sigma_N^2 = \gamma(V - V_{eq})\mu_N \tag{6.23}$$

When the transition rate βKc^n, the rate from the closed state to the open state, is not negligible in regard to the closing rate α, p_∞ is not significantly less than 1 and remains in (6.14) as $p_\infty = \beta Kc^n / (\alpha + \beta Kc^n)$. Single-channel conductance is therefore affected by p_∞.

6.5. Stochastic Approach to Current Noise Analysis in the Steady State

Small open probability p_∞ is not always satisfied. In the following, we consider a more general gating property based on the one-gate Hodgkin–Huxley model. The analysis can be extended to a channel model with more gates by modifying the expression of p_∞ (6.25) and the conditional probability $p(o|t)$ (6.28).

The model we consider below is based on three fundamental assumptions of Hodgkin–Huxley model: (1) each channel has an open and a closed state; (2) transition is made between these two conductance states with rate constants for closing α and opening β; (3) interstate transitions of individual channels are independent.

We start discussion of one-channel behavior:

$$\text{Close state} \underset{\alpha}{\overset{\beta}{\rightleftarrows}} \text{Open state} \tag{6.24}$$

Channel open probability p_∞ and closed probability $1 - p_\infty$ in the steady state are expressed as follows, from (6.24):

$$p_\infty = \frac{\beta}{\alpha + \beta}, \quad 1 - p_\infty = \frac{\alpha}{\alpha + \beta} \tag{6.25}$$

When open probability changes as a function of time, it is expressed by the following stochastic equation.

$$\frac{dp}{dt} = \beta - (\alpha + \beta)p \tag{6.26}$$

Then, the conditional probability of channel being open is 1 at time 0 and p_∞ in the steady state when the channel is open at time 0 (6.27). The conditional probability of being open at time t is given by (6.28).

$$p(o|0) = 1 \tag{6.27}$$

$$p(o|\infty) = p_\infty$$

$$p(o\,|\,t) = p_\infty + (1 - p_\infty)\cdot\exp\{-(\alpha+\beta)t\} \tag{6.28}$$

An autocorrelation of one-channel fluctuation is defined in (6.29).

$$C_1(\tau) = \langle \Delta i(\tau)\cdot\Delta i(t+\tau)\rangle \tag{6.29}$$

In the steady state, the autocorrelation at time t is the same as the autocorrelation of time 0 (ergodicity). Because the mean current through a single channel is expressed as $\mu_1 = \gamma(V - V_{eq})\cdot p_\infty$, the autocorrelation of current fluctuation for one channel is given as follows:

$$\begin{aligned} C_1(\tau) &= \langle \Delta i(0)\cdot\Delta i(\tau)\rangle \\ &= \langle (i(0) - \mu_1)\cdot(i(\tau) - \mu_1)\rangle \\ &= \langle i(0)i(\tau)\rangle - \mu_1{}^2 \\ &= \gamma^2\left(V - V_{eq}\right)^2\{p_\infty\cdot p(o\,|\,\tau) - p_\infty{}^2\} \\ &= \gamma^2(V - V_{eq})^2\, p_\infty(1 - p_\infty)\cdot\exp\{-(\alpha+\beta)\tau\} \end{aligned} \tag{6.30}$$

For N channels, autocorrelation is expressed as N times $C_1(\tau)$, because of condition 3 (above) (6.31).

$$C_N(\tau) = N\cdot\gamma^2(V - V_{eq})^2\, p_\infty(1 - p_\infty)\cdot\exp\{-(\alpha+\beta)\tau\} \tag{6.31}$$

Thus, the power spectrum function could be formulated as (6.32) because the mean current passing through these N channels is $\mu_N = N.\mu_1$.

$$\begin{aligned} S(f) &= \int_{-\infty}^{\infty} C_N(\tau)\cdot\exp(-j\omega\tau)\mathrm{d}\tau \\ &= 2\int_0^{\infty} C_N(\tau)\cdot\cos(\omega\tau)\mathrm{d}\tau \\ &= 2N\gamma^2(V - V_{eq})^2\;\; p_\infty(1 - p_\infty) \\ &\quad \int_0^{\infty}\exp\{-(\alpha+\beta)\tau\}\cdot\cos(\omega\tau)\mathrm{d}\tau \end{aligned} \tag{6.32}$$

Because $\omega = 2\pi f$ (6.32) is further transformed to (6.33) using expression of the mean current μ_N.

$$S(f) = \frac{2\mu_N^{\ 2}(1-p_\infty)\tau_m}{Np_\infty\left[1+(2\pi f\cdot\tau_m)^2\right]} = \frac{2N\gamma p_\infty(V-V_{eq})\mu_N(1-p_\infty)\cdot\tau_m}{Np_\infty\left[1+(2\pi f\cdot\tau_m)^2\right]} = \frac{2\gamma(V-V_{eq})\mu_N(1-p_\infty)\cdot\tau_m}{1+(2\pi f\cdot\tau_m)^2} \tag{6.33}$$

Here $\tau_m = \dfrac{1}{\alpha+\beta}$ is the time constant that determines the kinetics of m gate of the Hodgkin–Huxley formulation. Then, (6.33) tells us that if we have knowledge of channel open probability p_∞ and the mean current μ_N we can obtain an estimate of single-channel conductance γ from the power spectrum. τ_m can be estimated separately from the corner frequency f_c of the power spectrum as $\tau_m = 1/(2\pi f_c)$.

6.6. Noise Analysis in the Nonstationary Condition

The open probability of channels usually changes with time, and even under this condition it is possible to estimate properties of a single channel from the relation between the mean and the variance. Sigworth (7) estimated single Na channel conductance at 7 pS and the density at 1,000 channels per square micron at the nodes of Ranvier of frog myelinated fibers. Na channels are activated by a depolarizing pulse and then inactivated. Because of this inactivation, we cannot apply the technique of steady-state noise to the Na current. Sigworth analyzed the Na channel current noise by focusing on the current fluctuation that was visible after subtracting an ensemble (averaged) trace from individual Na current traces. Figure 6.5a illustrates eight traces of Na current that was activated repeatedly ($M=8$). Currents were sampled at time intervals of Δt and are expressed as $I_{jk}(t)$ ($k=1,M$; $j=1$, L; $t=j\cdot\Delta t$). Figure 6.5b illustrates a component of fluctuation calculated by subtracting the ensemble (averaged) Na current trace $\mu_j(t)$ from an individual current record $\{I_{jk}(t)\}$ at each sample point. Figure 6.5c illustrates the variance calculated at each sample point.

The ensemble averaged current μ_j and the variance $\sigma_j^{\ 2}$ are calculated at time $t=j\cdot\Delta t$ from a set of current traces of M ($k=1$, M) and are expressed as follows as a function of current $I_{jk}(t)$:

$$\mu_j(t) = \frac{1}{M}\sum_{k=1}^{M} I_{jk}(t) \tag{6.34}$$

$$\sigma_j^{\ 2} = \frac{1}{M-1}\sum_{k=1}^{M}[I_{jk}(t)-\mu_j(t)]^2 \tag{6.35}$$

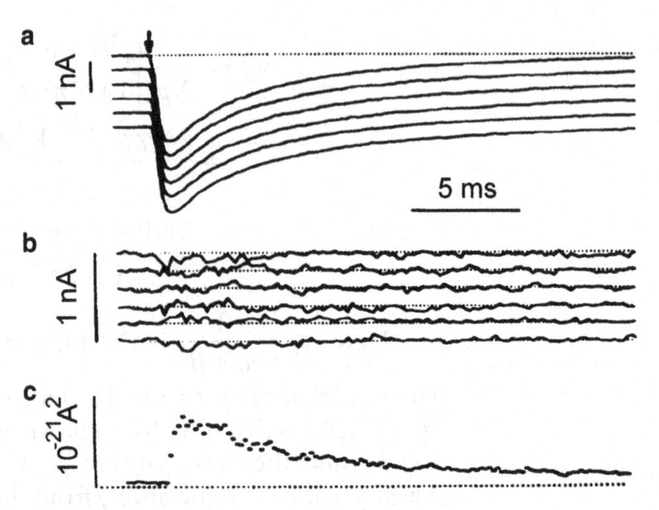

Fig. 6.5. Na channel current and ensemble noise. Eight traces were processed in a set, and the fluctuations of the Na channel currents were extracted as a component around the mean. (**a**) Na channel currents. (**b**) Difference between the means and individual currents. (**c**) Na channel current variance at each time point. (From Sigworth (7), with permission).

These functions are further transformed using single-channel conductance γ and the channel's open probability $p(t)$ as follows:

$$\mu_N(t) = N\,p(t)\gamma(V - V_{\text{eq}}) \tag{6.36}$$

$$\sigma_N(t)^2 = N\,p(t)[1 - p(t)]\gamma^2(V - V_{\text{eq}})^2 \tag{6.37}$$

The ratio between $\sigma_N(t)^2$ and $\mu_N(t)$ at the corresponding time t gives (6.38) and (6.39), which illustrate that the ratio changes with time as a function of $p(t)$ (6.38) or the ensemble mean current $\mu_N(t)$ (6.9).

$$\frac{\sigma_N(t)^2}{\mu_N(t)} = (1 - p(t))\gamma^2(V - V_{\text{eq}}) \tag{6.38}$$

$$\frac{\sigma_N(t)^2}{\mu_N(t)} = \gamma(V - V_{\text{eq}}) - \frac{\mu_N(t)}{N} \tag{6.39}$$

Thus, unitary current $i = \gamma(V - V_{\text{eq}})$ and/or unitary conductance γ can be estimated directly by plotting the ratio against the ensemble current mean at corresponding times. The number of channels N can be estimated from the slope of a fitting line (13). A somewhat tricky problem is the divergence of (6.39) that occurred when the ensemble mean crossed the current zero level, which sometimes occurs because of fluctuation of the real data.

By plotting $\sigma_N(t)^2$ against the ensemble current mean $\mu_N(t)$ gives a hyperbolic function:

$$\sigma_N(t)^2 = \gamma(V - V_{\text{eq}})\cdot\mu_N(t) - \frac{1}{N}\cdot\mu_N(t)^2 \tag{6.40}$$

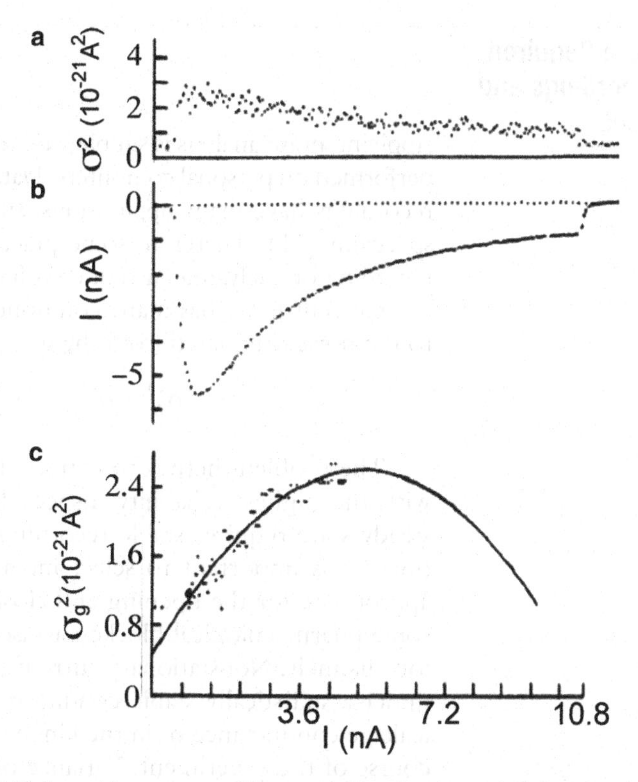

Fig. 6.6. Variance (**a**) mean ensemble current (**b**) and variance plotted against the mean (**c**). Convexed hyperbolic function was fitted to the plot. Estimated single Na channel current was 0.55 pA, and the channel number was 20,400. (From Sigworth (8), with permission).

The variance could be maximum at $p(t) = 0.5$, as is shown in Fig. 6.6c. Unitary conductance γ and number of channels N can be estimated from the hyperbolic fitting.

One advantage of analyzing nonstationary ensemble noise rather than stationary noise analysis is that the former does not need a specific gating model for interpretation of the data; it needs only the condition that fluctuation follows a binomial distribution, which reflects fluctuation of two states: open and closed. Thus, the condition is fulfilled in most cases and makes ensemble noise analysis highly applicable.

Ensemble noise analysis has been applied to estimate synaptic single-channel conductance (14, 15). Fluctuation of synaptic conductance is not only generated by the open–close transition of postsynaptic receptor channels, it is affected by the fluctuation of transmission from presynaptic terminals. Therefore, any analysis must test the constancy of transmitter release during a large number of repeated activations of synapses; otherwise, some compensation for the fluctuation in transmitter release is needed.

6.7. Care Required for Recordings and Analyses

Applying noise analysis is simple, and most experiments are currently performed on personal computers. Data acquisition systems for patch recordings have many applications, such as FFT (16) and a power spectrum (11). Therefore some practical matters that arise during recording or analyzing current or voltage noise data must be noted.

Current noise has many components of noise σ_j^2 in addition to the noise originated from the gating of channels.

$$\sigma_I^2 = \sigma_{\text{channel}}^2 + \sum_j \sigma_j^2 \quad (6.41)$$

The problem here is to extract the channel noise component with the highest reliability. Especially the noise recorded in the steady state requires stable recording conditions for quite a long time. It is important to select membrane potential ranges most appropriate for the opening and closing of target channels or use some pharmacological blockers to isolate activities of the targeted ion channels. Nonstationary current noise should also be recorded under a statistically stable condition, without large differences in leakage conductance or in the kinetics of current traces during the course of the experiment. Variance of current noise is likely to be increased for many reasons; therefore, if possible the variance that originates from sources other than channel activities should be subtracted appropriately.

The speed of analog to digital (A/D) conversion deserves some attention. If we sample data at a rate of h, the frequency components contained in data become lower than the Nyquist frequency:

$$f_{\text{Nyquist}} = \frac{1}{2 \cdot h} \quad (6.42)$$

Frequency components higher than f_{Nyquist} cannot be sampled correctly; moreover, they are distorted by aliasing and appear as fake low-frequency components that are lower than the Nyquist frequency.

It is essential to use a low pass filter, with a corner frequency set around the Nyquist frequency. A low pass filter with Butterworth character is appropriate for stationary noise analysis made on a frequency axis. A Butterworth filter produces smooth a cutoff of the power spectrum, although the filter generates a transient response on the time axis to a step change of input. A filter of Bessel character is then appropriate for the analysis of nonstationary transient current response on the time axis. A filter with Bessel characteristics does not generate any transient response at the onset or offset of step inputs. However, the Bessel filter has a less steep cutoff than the Butterworth filter near the corner frequency, and sometimes the filter is not appropriate for power spectrum analysis. The cutoff frequency of a low pass filter should be chosen after considering the speed of the gating phenomenon, and the sampling rate of the A/D conversion should be appropriately set.

Estimates of single-channel conductance are occasionally lower when attained by measuring noise than those measured by patch recording (17). The difference largely depends on the frequency responsiveness of the recording systems. Single-channel current has a square shape of a fast rise and fall, and current noise is formed as an ensemble of these single-channel currents (18). When step responses such as single-channel events are transformed to a spectrum, there should be repeated frequency components, although the power gradually decreases (11). The limited frequency responsibility of a recording system and the nature of the A/D conversion cause the higher-frequency power of noise to be lost, thereby making the variance smaller and the estimate of single-channel conductance lower.

It is advisable to evaluate the noise data basing on multiple criteria, such as by using pharmacological methods to block other channel activities, considering the membrane potential dependence of the targeted channel's activities, and comparing the noise records with macroscopic current records for both stationary and nonstationary noise data.

References

1. Hagiwara S (1954) Analysis of interval fluctuation of the sensory nerve impulses. Jpn J Physiol 4:234–240
2. Verveen AA, DeFelice LJ (1974) Membrane noise. Prog Biophys Mol Biol 28:189–265
3. Katz B, Miledi R (1970) Membrane noise produced by acetylcholine. Nature 226: 962–963
4. Katz B, Miledi R (1970) Further observations on acetylcholine noise. Nature New Biol 232:124–126
5. Stevens CF (1972) Inferences about membrane properties from electrical noise measurements. Biophys J 12:1028–1047
6. Anderson CR, Stevens CF (1973) Voltage clamp analysis of acetylcholine produced endplate current fluctuations at frog neuromuscular junction. J Physiol 235:655–691
7. Sigworth FJ (1977) Na channels in nerve apparently have two conductance states. Nature 270:265–267
8. Sigworth FJ (1980) The variance of Na current fluctuations at the node of Ranvier. J Physiol 307:97–129
9. Verveen AA, Derksen HE (1965) Fluctuations in membrane potential of axons and problem of coding. Kybernetik 2:152–160
10. Poussart D (1971) Membrane current noise in lobster axon under voltage clamp. Biophys J 11:211–234
11. Lee YW (1960) Statistical theory of communication. Wiley, New York
12. Bendat JS, Piersol AG (1971) Random data. Wiley-Interscience, New York
13. Ohmori H, Yoshida S, Hagiwara S (1981) Single K^+ channel currents of anomalous rectification in cultured rat myotubes. Proc Natl Acad Sci USA 78:4960–4964
14. Robinson HPC, Sahara Y, Kawai N (1991) Nonstationary fluctuation analysis and direct resolution of single channel currents at postsynaptic sites. Biophys J 59:295–304
15. Traynelis SF, Jaramillo F (1998) Getting the most out of noise in the central nervous system. Trends Neurosci 21:137–145
16. Brigham EO (1974) The fast Fourier transform. Prentice-Hall, Upper Saddle River
17. Hille B (1992) Ionic channels of excitable membranes. Sinauer Associates, Sunderland
18. Ohmori H (1981) Unitary current through sodium channel and anomalous rectifier channel estimated from transient current noise in the tunicate egg. J Physiol 311:289–305

Chapter 7

Slice Patch Clamp

Tadashi Isa, Keiji Imoto, and Yasuo Kawaguchi

Abstract

Since late 1980s, the whole-cell patch-clamp technique has been used as a powerful tool for analyzing local circuits of the central nervous system in a brain slice preparation in which the fundamental architecture of local circuits is mostly maintained. When combined with intracellular staining techniques, the technique allows high-resolution analysis of the membrane and the synaptic and morphological properties of the recorded cells. We offer a brief introduction to slice patch-clamp recording, including acute slice preparation, selection of intracellular and extracellular solutions, techniques for forming giga-ohm seals and whole-cell recordings, and staining of the recorded neurons. Much effort has gone into describing a number of small but important technical issues that can be useful for beginners attempting this technique.

7.1. Introduction

The patch-clamp method, developed by Neher and Sakmann in 1976 (1), was initially used to record tiny currents flowing through single channels on the cell surface. It led to the epoch-making achievement of proving the existence of the ion channel. In 1981, Hamill and others reported the whole-cell patch-clamp method (2), which enabled the measurement of total ionic currents flowing through the channels on the whole-cell surface as well as the ability to record from small cells. The patch-clamp method requires intimate contact between the tip of the electrode and the cell surface such that the electrode makes a tight seal with a resistance greater than a giga-ohm (10^9 Ω). Because of this requirement, application of the whole-cell patch-clamp method has been limited to cultured cells whose cell surface was exposed. In 1989, however, Edwards and others (3) developed the patch-clamp method for brain slice preparations, and it became possible to record from neurons in brain tissues where local neuronal circuits were preserved, analyze

Yasunobu Okada (ed.), *Patch Clamp Techniques: From Beginning to Advanced Protocols*, Springer Protocols Handbooks, DOI 10.1007/978-4-431-53993-3_7, © Springer 2012

synaptic transmissions in the central nervous system and their modification mechanisms, and investigate how neuronal signals are transmitted in local neuronal circuits. As slice patch-clamp recordings are usually performed under upright microscopic observation, we have to use thin slices from young animals. However, with the development of the blind patch-clamp method, which does not require an upright microscope, it became possible to record from thick slices from mature animals.

With the slice patch-clamp method, measurements are taken from neurons with widely extended dendrites. Because of the complicated shape of neurons, it is practically impossible to escape the space clamp problem,[1] and the accuracy of measurements from neurons is not as good as that from cultured cells. Despite this compromised accuracy, the fact that synaptic currents can be measured in the central nervous system is truly marvelous and rewarding. The recent main developments of the slice patch-clamp method include recordings from presynaptic terminals and dendrites and combining the method with various other techniques including the single-cell real-time polymerase chain reaction (PCR), imaging methods, application of caged compounds, the perforated patch-clamp method, and the intracellular perfusion method. Because the slice patch-clamp method is still being modified and developed, it continues to be a powerful method for analyzing brain functions. In this chapter, we describe the basic technique using an upright microscope and give examples of its application.

7.2. Basic Technique of the Slice Patch-Clamp Method

7.2.1. Overview

One of the advantages of the patch-clamp method is that the recording mode can be easily switched between the current-clamp recording (voltage recording) and the voltage-clamp recording (current recording). The merits and demerits of whole-cell patch-clamp recording, in comparison with the sharp electrode method, are summarized in Table 7.1.

Research targets suitable for the slice patch-clamp method are as follows:

1. Electrophysiological properties of the cell membrane

   Basic membrane properties (e.g., membrane resistance, membrane capacitance)

   Firing properties in response to current injection (e.g., firing frequency)

[1] It is often difficult to clamp the whole membrane potential of neurons with their complex shape through all of the soma and elongated dendrites. It is referred to as a "space clamp problem."

Table 7.1
Merits and demerits of intracellular recordings and whole-cell patch-clamp recordings

	Intracellular recordings with sharp electrodes	Whole-cell patch clamp recordings
Merits	Cells are not dialyzed with pipette solution Sharp electrodes can be used for relatively thick slices and in vivo experiments	Cytoplasmic composition can be changed by dialysis with pipette solution Intracellular staining is easy Voltage-clamp is well controlled Recordings are stable
Demerits	Cytoplasmic composition cannot be dialyzed Intracellular staining is difficult Voltage-clamp is poorly controlled Recordings are unstable	Cells are dialyzed with pipette solution It is difficult to use for thick slices and in vivo experiments (it is feasible with the blind patch-clamp method or two-photon laser scanning microscopy)

2. Postsynaptic potentials (EPSP, IPSP)

   PSPs evoked by electrical stimulation

3. Properties of voltage-dependent ion channels and pumps

   Mechanisms of activation and inactivation

   Ion permeability, ion selectivity, voltage dependence

4. Postsynaptic currents (EPSC, IPSC)

   Receptor currents evoked by agonist application (rectification, ion permeability)

   Spontaneous and miniature synaptic currents

   Synaptic currents evoked by electrical stimulation

5. Perforated patch-clamp method

   Synaptic transmission under conditions maintaining intracellular functional molecules

   Synaptic currents mediated by $GABA_A$ receptors (gramicidin-perforated patch-clamp)

6. Recordings from presynaptic terminals (limited to special preparations, e.g., calyx of Held)

   Properties of voltage-dependent ion channels and various kinds of receptors in presynaptic terminals

   Simultaneous recordings with postsynaptic neurons

7. Dendrite recordings

   Properties of voltage-gated ion channels and various receptors expressed on the dendrites

Analyses of synaptic transmission onto the dendrites and dendritic action potentials

8. Simultaneous recordings from multiple neurons

   Unitary synaptic currents and potentials

   Synaptic connectivity in local neuronal networks

9. Neuron staining with biocytin (optical microscope, electron microscope)

   Identification of recorded neurons and morphological observation (dendrites, axons, synapses)

10. Combination with immunohistochemistry

    Relation with specifically expressed molecules

11. Combination with imaging methods

    Calcium imaging: simultaneous recording with intracellular calcium dynamics

12. Combination with molecular biological techniques

    Single-cell PCR: relation with molecular markers in recorded neurons

    Gene-modified animals: functional studies of targeted molecules

7.2.2. Preparation of Brain Slices

Successful slice patch-clamp recording is entirely dependent on the quality of the brain slice preparations. The standard methods for brain slice preparations are described along with several important points regarding each step. Although the methods described are mainly for young mice, essentially similar methods are used for other animals. Some of the difficulties associated with brain slice preparations from mature animals are mentioned at the end of the chapter.

7.2.2.1. Preparation of Solutions; Tools; Other Considerations

- Normal extracellular Ringer's solution: artificial cerebrospinal fluid is used for maintaining brain slices and for recording. The composition of the Ringer's solution is as follows (in mM): 125 NaCl, 2.5 KCl, 2 $CaCl_2$, 1 $MgCl_2$, 26 $NaHCO_3$, 1.25 NaH_2PO_4, 11 glucose. The solution is used after bubbling with 95% O_2 and 5% CO_2.
  - As $CaCl_2$ and $MgCl_2$ react with $NaHCO_3$ to make precipitates, $CaCl_2$ and $MgCl_2$ should be added after $NaHCO_3$ is completely dissolved.
  - The composition of Ringer's solution has to be modified to match the aim of the experiments. Appropriate references are consulted when necessary.
  - To reduce cell death during brain slice preparation, lower the temperature of the brain tissue and reduce the oxygen demand. Ice blocks or sherbet made from Ringer's solution bubbled with mixed gas are used to prevent the solution in the preparation chamber from warming. To further

reduce cell death, modified Ringer's solutions are frequently used (see Note 1, below).

- A small beaker is used as an incubation (holding) chamber for brain slices. Ringer's solution is poured into the beaker and bubbled with mixed gas.
- A razor blade is used to trim and slice the brain. As oil and adhesive compounds attached to the blade are harmful for cells, they are thoroughly cleaned using a cotton swab soaked with an acetone-ethanol (1:1) mixture.
- A petri dish (100 mm in diameter) is filled with ice, and a lid is placed on the dish. A piece of filter paper is put on the lid and soaked with ice-cold Ringer's solution (the brain is trimmed on this paper).

7.2.2.2. Decapitation and Removal of the Brain

- Procedure for removing the brain (Fig. 7.1a): A scissor blade is inserted into the decapitated head at the spinal cord or the medulla, which is exposed, and the occipital and parietal bones are cut anteriorly along the midline to the olfactory bulb (Fig. 7.1a, ①). The skull is cut in the lateral direction at the anterior and posterior parts (Fig. 7.1a, ②, ③). The skull is opened laterally in both directions using forceps (Fig. 7.1a, ④), and the brain is removed using a spatula.
- Procedure for removing the cerebellum (Fig. 7.1b): A scissor blade is inserted laterally into the decapitated head and is advanced to the lateral part of the occipital and parietal bones

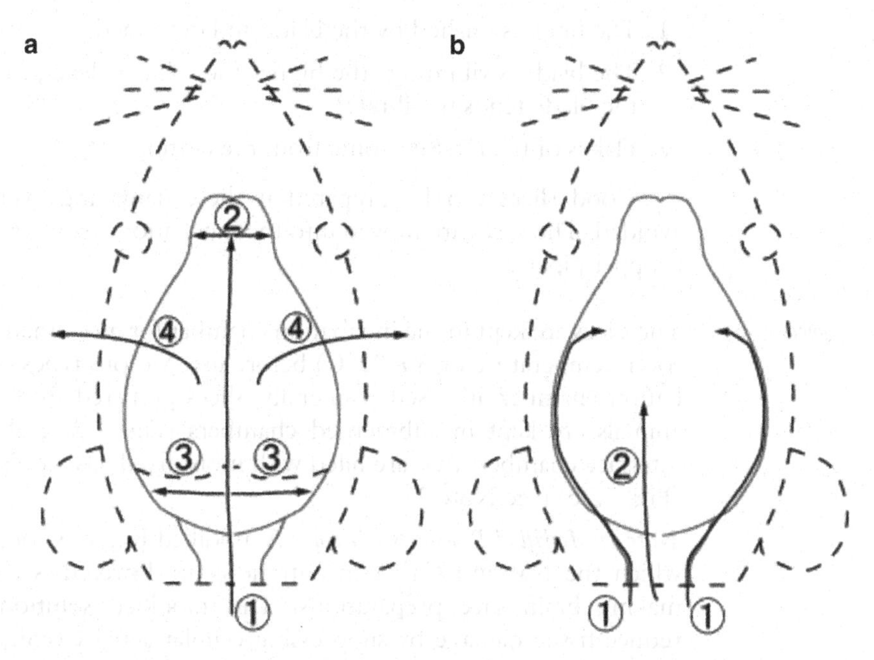

Fig. 7.1. Procedures for opening the skull. (**a**) Removing the cerebrum. (**b**) Removing the cerebellum. (Courtesy of Dr. Daisuke Kase).

(Fig. 7.1b, ①). A cut is made on both sides, and the occipital bone is elevated using forceps (Fig. 7.1b, ②), and the brain is removed using a spatula.

7.2.2.3. Trimming and Fixation to the Microslicer

The isolated brain is cooled in ice-cold Ringer's solution for 1–3 min. The brain is put on a petri dish and is trimmed with a razor blade to remove the unnecessary parts. It is important to pull the blade during trimming because pushing it against the brain damages the tissue. The trimmed block is cooled again in ice-cold Ringer's solution for 3 min. The brain block is very gently wiped to remove Ringer's solution and is then fixed to the slicing chamber (it should be cooled in advance) using instant adhesive. A good balance between the height and the bottom area of the block is crucial to maintain stability during the slicing procedure. A small block (e.g., spinal cord) or a tissue with an irregular shape that is difficult to trim may be covered with agar melted in Ringer's solution or embedded in agar.

7.2.2.4. Slicing

Wait until the instant adhesive agent becomes dry. If fixation of the block is insufficient, the block may be dislocated from the slicer or the tissue may be distorted. The slice chamber is filled with ice-cold Ringer's solution (always bubbled with 95% O_2/5% CO_2), and the chamber is fixed to the slicer. The used parts of the brain should be sliced slowly, whereas the unused parts are cut rapidly, so the slicing time is minimized.

The following are points of caution during the slicing procedure and should be watched for under a stereoscopic microscope.

1. The brain is pushed by the blade and deformed.
2. The blade is vibrating (the higher the solution level, the more the blade tends to vibrate).
3. Flakes of brain tissue come from the cutting site.

Good slices can be prepared if these damaging events are avoided. The slices are moved into the incubation chamber using a dropper pipette.

7.2.2.5. Maintaining the Slices

The slices are kept in the incubation chamber for more than 1 h (at room temperature or 30–32°C) before use. Various types of incubation chamber are used. Generally, slices prepared from young animals are kept in submerged chambers (Fig. 7.2a), although interface chambers that are filled with moist fixed gas are also used (Fig. 7.2b) (see Note 2).

Note 1: Modified Ringer's solution. A modified Ringer's solution in which the Na^+ and Ca^{2+} concentrations are lowered is used for making brain slice preparations. This modified solution helps reduce tissue damage by suppressing cellular activity (e.g., action potentials). NaCl is replaced with sucrose or choline chloride, and

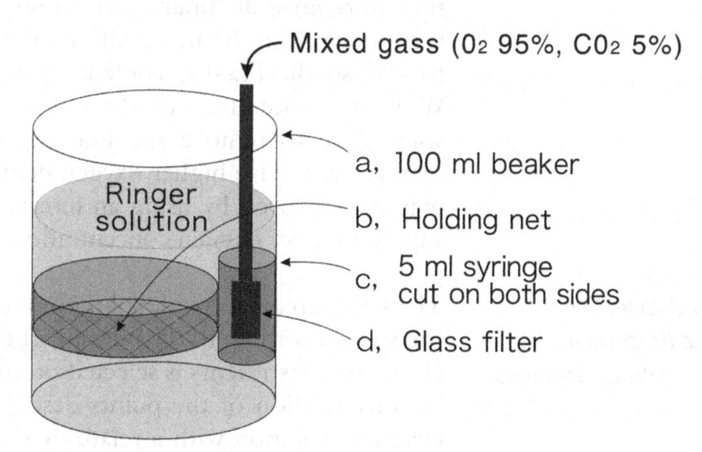

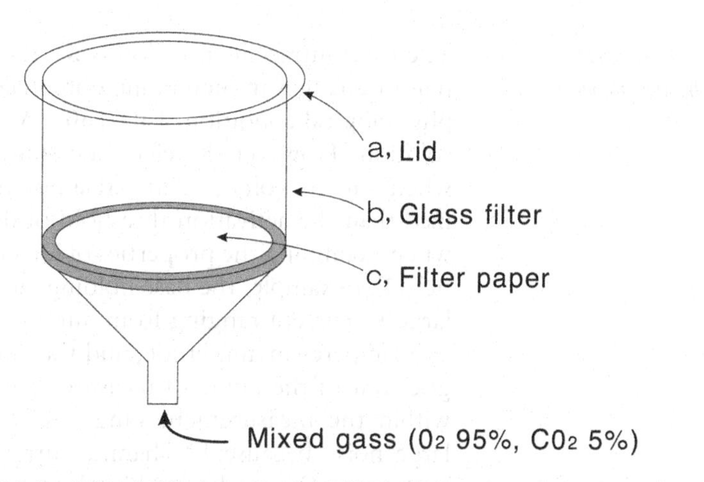

Fig. 7.2. Incubation (holding) chambers. (**a**) Simple submerged chamber. The holding net is made by putting a piece of cotton gauze between a 35-mm plastic dish whose lid and bottom have been removed. Slices are put on the holding net. (**b**) Interface chamber. Filter paper is put on the glass filter and soaked with Ringer's solution. Slices are placed on a piece of semipermeable cellophane sheet and put on the filter paper.

Ca^{2+} is replaced with Mg^{2+}. An example of the composition is shown below. The solution is bubbled with mixed gas (95% O_2/5% CO_2) and ice-cold chilled (or cooled until the solution enters the sherbet state) before use.

- Sucrose solution (in mM): 234 sucrose, 2.5 KCl, 1.25 NaH_2PO_4, 10 $MgSO_4$, 0.5 $CaCl_2$, 26 $NaHCO_3$, 11 glucose
- Choline chloride solution (in mM): 120 choline chloride, 3 KCl, 1.25 NaH_2PO_4, 28 $NaHCO_3$, 8 $MgCl_2$, 25 glucose

Note 2: Brain slices from mature animals. The largest problem in making brain slices from mature animals is that it requires a longer time to remove the brain. To circumvent this problem, the animal is perfused from the heart with modified Ringer's solution cooled to 4°C so the brain is cooled before decapitation and removal. With this procedure, cell survival is improved even if it requires some time to remove the brain. Furthermore, neurons in the mature brain have higher oxygen demands, and their survival rate may be improved by using an interface-type chamber, which can supply a higher oxygen concentration.

7.2.3. Selection and Preparation of Electrode Solutions

The selection of electrode solutions is essential, and they have to be prepared according to the purpose of the experiment. An appropriate set of constituents is selected on the basis of reference articles, in consideration of the points described later. An example of an electrode solution with a relatively simple composition is the following (in mM): 140 K-gluconate, 5 KCl, 0.2 EGTA, 2 $MgCl_2$, 2 Na_2ATP, 10 HEPES (pH is adjusted to 7.3 with KOH).

7.2.3.1. Selection of the Main Cation: K^+ or Cs^+?

The main intracellular cation is K^+; therefore, when action potentials or changes in membrane potentials are recorded in an almost physiological condition, 140–150 mM K^+ is used for the electrode solution. However, K^+ electrode solutions often cause problems when cells are voltage-clamped at potentials that are more depolarized than the activation threshold of delayed rectifier K^+ channels when examining the properties of ion channels other than K^+ channels. For example, the base holding current is contaminated with a large K^+ current ranging from hundreds of picoamperes to several nanoamperes in amplitude, and the current to be measured often goes out of the range of measurement. Even when the current is within the measurement range, K^+ currents make unacceptably large noise because K^+ channel currents are usually unstable. In such cases, Cs^+ can be used in place of K^+ because Cs^+ is as permeable as K^+ to glutamate receptors and other channels.

7.2.3.2. Selection of the Main Anion

The anion most frequently used for pipette solutions is Cl^-. Cl^- generates only a small liquid junctional potential (~4 mV), and is easily prepared by dissolving KCl or CsCl. On the other hand, high concentrations of Cl^- shift the Cl^- current reversal potential in the depolarizing direction. When a long-lasting recording is intended, or when hyperpolarizing inhibitory postsynaptic potentials (IPSPs) are observed around the resting membrane potential under the current-clamp recording, it is better to use gluconate or methanesulfonate as the main anion and to lower the Cl^- concentration to 5–6 mM.

7.2.3.3. Selection of Chelators (EGTA or BAPTA)

Chelators are used to control intracellular divalent ions, particularly Ca^{2+}. When intracellular Ca^{2+} is held constant during voltage-clamp mode recordings, a high concentration (10 mM) of ethyleneglycol-bis-(2-aminoethylether)-*N*,*N*,*N'*,*N'*-tetraacetic acid (EGTA) is included in the pipette solution. When the physiological consequences of intracellularly fluxed Ca^{2+} are allowed to occur, a lower concentration (~0.2 mM) of EGTA is used. For example, if 10 mM EGTA is used when the firing response to a constant current injection is examined, Ca^{2+} entering the neuron in association with action potentials is chelated, and Ca^{2+}-activated K^+ channels, which contribute to after-hyperpolarization, are gradually inhibited to alter the spike frequency adaptation of the neuron. When the firing property of the neuron is investigated, 0.2 mM EGTA is used. In contrast, when Ca^{2+} rise needs to be quickly chelated, 1,2-bis(2-aminophenoxy)ethane-*N*,*N*,*N'*,*N'*-tetraacetic acid (BAPTA), which is a faster chelator, is used at concentrations of 10–20 mM.

7.2.3.4. Osmolarity of the Pipette Solution

The osmolarity of the pipette solution is adjusted to 280–290 mOsm/L to match the osmolarity of the extracellular solution (usually 300–310 mOsm/L). A poor balance in osmolarity often causes unstable recordings (e.g., the cell becomes swollen). When the pipette solution is frozen and stored, tubes with an O-ring should be used.

7.2.4. Whole-Cell Recordings from Brain Slices

7.2.4.1. Observation of Cells

Brain slices are maintained in normal Ringer's solution bubbled with mixed gas (95% O_2–5% CO_2) at room temperature. After 1–1.5 h, slices are moved to the recording chamber and fixed with a "grid" (Fig. 7.3a). The grid is made of a platinum bar (1 mm diameter) hit by a hammer, flattened, and bent as a U-shape on which thin nylon threads are loosely attached.

Neurons appropriate for whole-cell recording are selected under an upright phase-contrast microscope. By enhancing the contrast of video images taken by the CCD camera, cells can be

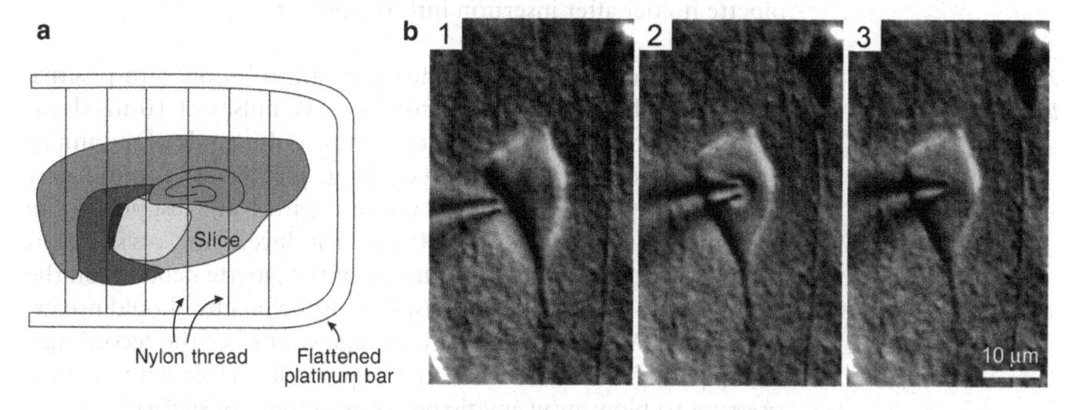

Fig. 7.3. Forming a giga-seal on neurons in brain slices. (**a**) Fixation of a brain slice with a "grid." (**b**) The method to form a giga-ohm seal (photomicrographs *1–3* are from a neocortical neuron in a mouse). *1.* Approach the cell with a pipette while blowing away the tissue covering the cell surface by applying positive pressure. *2.* Make a "dimple" on the cell surface. *3.* The cell surface automatically becomes attached to the pipette when the positive pressure is removed.

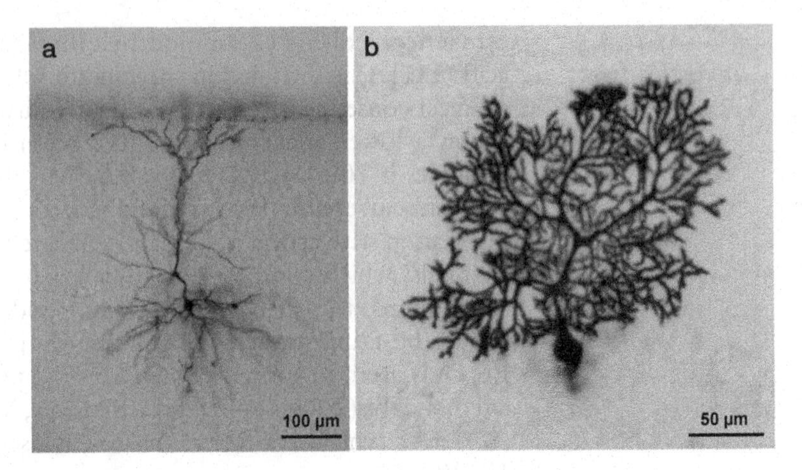

Fig. 7.4. Cells stained with biocytin. (**a**) Neocortical pyramidal cell. (**b**) Cerebellar Purkinje cell.

clearly visualized (Fig. 7.4b). By using the infra-red differential interference contrast filter (IR-DIC), cells can be more easily observed for recordings, even cells deep in the slices or in slices from mature animals, which are usually difficult to observe with ordinary devices.

7.2.4.2. Filling the Pipette with the Pipette Solution

First, the tip of the pipette is dipped in the intracellular solution to fill the tip using the capillary phenomenon (this procedure prevents an air bubble plugging the tip). Next, the pipette solution is injected into the pipette from the tail end by using a plastic pipette tip that is thinly extended or with a thin injection needle (or "Microfill"). By tapping the pipette, air-bubbles in the pipette should be removed (check carefully whether small air-bubbles remain at the tip of the pipette). Be careful not to inject too much solution into the pipette, to avoid overflow of the solution into the pipette holder after insertion into the holder.

7.2.4.3. Approaching the Cells

The recording pipette is put into the bath solution with positive pressure applied. While 1 mV square-wave pulses of 10 ms duration are repetitively applied under the test-pulse mode of the amplifier, the resistance of the recording pipette is continuously monitored (if 200 pA current is induced with 1 mV test pulses, the resistance is 5 MΩ, and if 500 pA is induced, the resistance is 2 MΩ). The appropriate tip diameter of the pipette depends on the cell size, but in general, the resistance of the pipette should preferably be 2–3 MΩ for appropriate voltage control of the recordings. The pipette is moved toward the target cell surface with positive pressure to blow away any tissue covering the cell surface.

7.2.4.4. Formation of a Giga-Ohm Seal

As the pipette tip comes close to the cell surface, a small "dimple" is formed on the surface, which indicates the close contact of the tip to the cell surface (Fig. 7.3b,7.2). In general, the patch pipette can be pushed onto the cell surface more strongly than when approaching a monolayer of cells in a culture dish because cells in the slices are tightly fixed in the tissue. We recommend pressing the pipette against the cell rather strongly. If the dimple is small and confined, it means there are no obstructions between the pipette tip and the cell surface. However, if it is a much larger dimple compared to the pipette tip, it means that some tissue material (e.g., membrane from broken cells) is obstructing the attachment of the pipette tip to the cell surface. If the pipette tip is properly placed on the cell surface, the cell surface automatically becomes attached to the pipette by removing the positive pressure applied to the pipette. The seal resistance increases markedly at this moment (sometimes a giga-ohm seal can be formed with this procedure alone). By applying some negative pressure to the pipette, a giga-ohm seal can then be formed successfully (Fig. 7.3b-3). When the seal resistance exceeds 1 GΩ, the voltage is set close to the resting membrane potential (e.g., –60 to –70 mV). A transient current is observed after formation of the giga-ohm seal. This is a capacitative current caused by charging the floating capacitance of the pipette. This capacitative current considerably affects the recording (e.g., saturation of the inputs to the amplifier or distortion of the electrical responses with a fast time constant); therefore, the electrical component of the floating capacitance is compensated for by adjusting the capacitance and the time constant.

7.2.4.5. Whole-Cell Recordings

After seal formation, the cell membrane attached to the pipette tip is broken by applying negative pressure to the pipette. As the cell membrane breaks, a transient current component is observed in response to the test pulses. This is a capacitative current that charges the membrane capacitance of the recorded cell through the series resistance (access resistance) between the pipette tip and the cell membrane. It is necessary to compensate for this current component because it considerably affects the nonstationary current recording under the voltage-clamp mode. This compensation is made by adjusting two parameters – capacitance and series resistance – in parallel, to eliminate the capacitative component on the screen as much as possible. If the series resistance is too large, the compensation ratio (percent) of the series resistance should be changed to reduce the error as much as possible because otherwise the drop of the membrane potential through the series resistance becomes non-negligible. If the whole-cell recording is of long duration, it is recommended that the series resistance be continuously monitored simultaneously with the recording. The series resistance sometimes changes considerably during the experiment and can seriously affect the amplitude or waveform of the synaptic currents. The series resistance can be monitored as a sharp onset capacitative component by applying a hyperpolarizing current pulse of 1–5 mV amplitude and 10–20 ms duration following the stimulation pulses.

7.2.4.6. Electrical Stimulation

To observe synaptic inputs in slices, metal or glass microelectrodes are commonly used to stimulate the neural elements. Monopolar or bipolar electrodes are used for metal electrodes. To make glass stimulating electrodes, glass micropipettes are pulled so they have a larger tip diameter than the ordinary recording patch pipettes; they are then filled with the extracellular solution. During recording, an appropriate position for inducing the aimed responses should be identified by changing the stimulation site (lightly touching the surface of the slice with the tip of the electrode) and the stimulation intensity. There are two opinions as to whether a constant current or constant voltage mode should be used for electrical stimulation. If metal electrodes are used and the stimulus intensity is weak, a constant voltage mode causes no problems. However, if glass electrodes are used, a constant current is preferable because the electrode impedance sometimes changes during the experiment owing to plugging of the electrode with tissue debris, which obstructs the current flow.

7.2.5. Staining the Cells

To visualize and observe the cells after whole-cell recordings, it is necessary to label them by adding a marker material to the intracellular pipette solution used for the recordings. Fluorescent markers such as Lucifer yellow and biocytin are commonly used. For a variety of reasons, biocytin is more often selected. Its advantages include the following: (1) biocytin is a small molecule (MW 372.5), and it can therefore diffuse over the cytosol quickly after injection; (2) with this method, the stained cell can be stored for a long time and it is possible to observe it under electron microscopy with particular procedures. Cell staining with biocytin can be done simply by adding biocytin (5 mg/ml) to the pipette solution in advance. It is important to detach the patch pipette gently from the cell after the whole-cell recording is terminated as the recorded cell can be broken or detached from the slice together with the recording pipette. If the electrode is detached successfully from the cell, the slice is picked up with a dropper pipette and moved to 0.1 M phosphate buffer solution containing 4% paraformaldehyde and fixed overnight or longer (4°C).

The avidin-biotin-horseradish peroxidase (HRP) complex (ABC) method, which utilizes the selective and strong chemical bonding of avidin and biotin, is commonly used to visualize cells labeled with biocytin. The stained cells are visualized by diaminobenzidine tetrahydrochloride (DAB), which is a substrate of HRP (Fig. 7.4). It is also possible to visualize biocytin with fluorescent markers. For this purpose, it is recommended to use fluorescence-labeled streptavidin. In addition, the expression of chemical markers in the recorded cells can be examined by combining biocytin staining with immunohistochemical (IHC) staining (double fluorescence-IHC staining). Moreover, by lowering the concentration of fluorescent streptavidin, it is also possible to stain the cells with

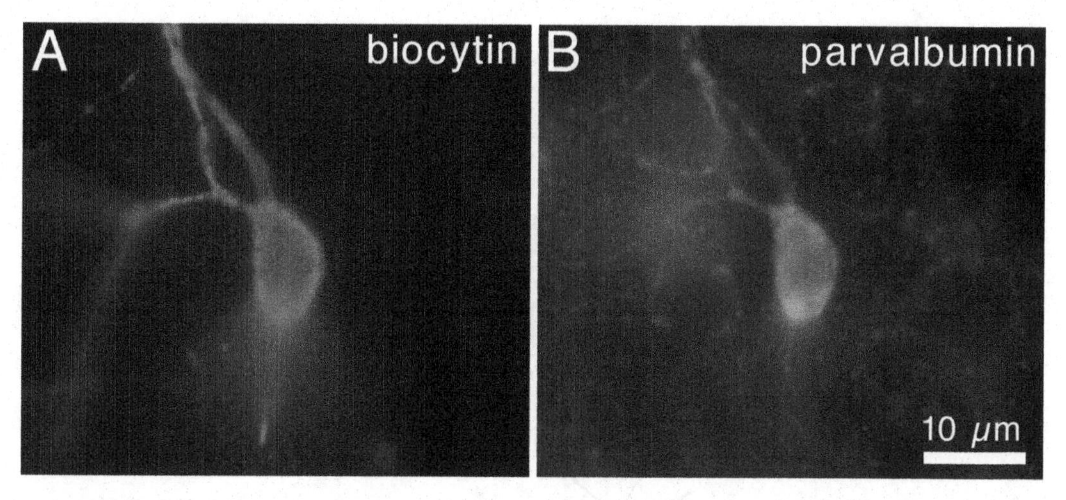

Fig. 7.5. Fluorescence double labeling method. (**a**) Cell stained with streptavidin. (**b**) Immunohistochemistry stain with the parvalbumin antibody.

the ordinary ABC method after the fluorescence observations and to maintain the biocytin staining permanently. In this case, it is supposed that some biocytin that was not bound to streptavidin remains in the cytosol and can be used for permanent visualization (Fig. 7.5). The standard procedure for biocytin staining is as follows:

1. Wash the slices with 0.05 M phosphate-buffered saline (PBS) solution (10 min ×2).
2. Soak in methanol (or PBS) containing 0.6–1.0% H_2O_2 (30 min).
3. Wash with PBS (10 min ×3).
4. Soak in PBS containing avidin-biotin-HRP complex (1%) and Triton X-100 (0.3–0.4%) (3 h).
5. Wash with PBS (10 min ×2).
6. Wash with 0.05 M Tris-buffered saline (TBS) (10 min ×2).
7. Soak in TBS containing 0.01% DAB and 0.3–1.0% nickel ammonium sulfate (20–30 min).
8. Add H_2O_2 (concentration 0.0003–0.01%) to the DAB–nickel ammonium sulfate solution and incubate slices (2–10 min).
9. Wash with TBS.

The entire reaction process can be performed at room temperature. Commonly the reaction procedures are conducted in culture plates.

It is sometimes necessary to re-cut the fixed slices for the detailed analysis of cell morphology (e.g., when the trajectories of dendrites and axons are complex or the focusing distance of the high-power objective lens is not sufficient). In such cases, slices

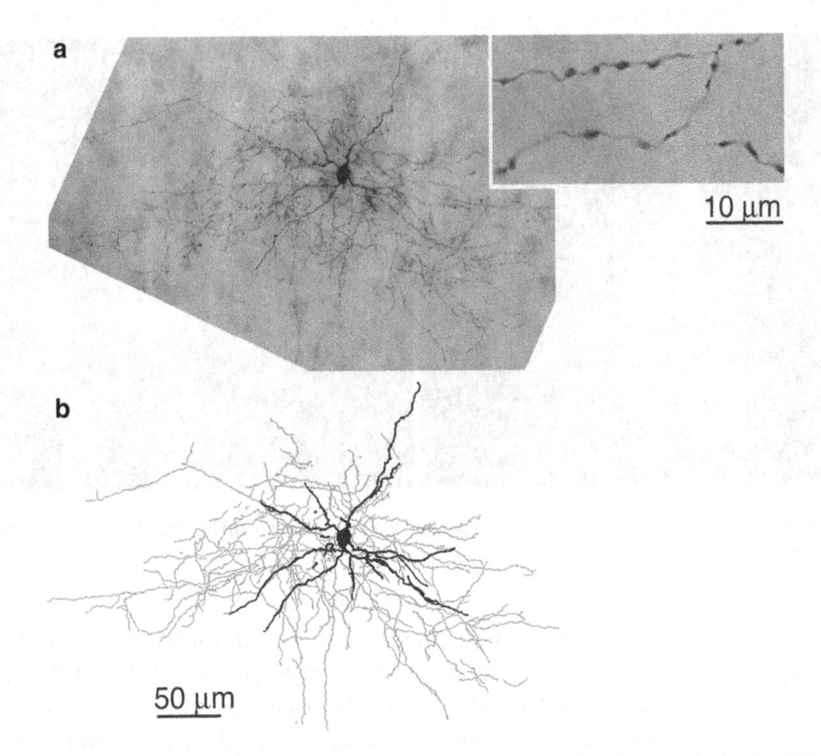

Fig. 7.6. Biocytin staining of a neocortical nonpyramidal cell (fast spiking cell). (**a**) Visualized with the ABC method after recutting to 50 μm thickness. *Inset.* Boutons on an axon are clearly observed with a 100× oil-immersion lens. (**b**) Reconstruction with Neurolucida software. Axons are *gray*; dendrites are *black*.

should be resectioned into 50-μm sections with the micro-slicer before performing the ABC reaction. It is recommended to re-section the slices embedded in either agar or gelatin and place them on a block of 4% agar. Figure 7.6 is a photomicrograph and reconstructed image (Neurolucida software; MBF Bioscience, Williston, VT, USA) of cortical nonpyramidal cells that were visualized by the ABC method after re-cutting. Re-cutting enables the clear-cut observation of axonal trajectories and terminal boutons on axons (Fig. 7.6a).

7.3. Supplement: Identification of the Axonal Projections of Recorded Cells by Retrograde Labeling

Intracellular staining with biocytin enables analysis of the axonal trajectories of recorded cells within a slice. It is difficult, however, to trace axonal projections to distant brain areas in ordinary slice preparations. To record from neurons whose target area is identified, a retrograde neural tracer is injected into the target region. For this purpose, tracers that can be observed with an epifluorescence microscope are suitable. Usually, the cholera toxin B subunit or fluorescent beads are used. For the injection of tracers, pressure injection is done with a pico-pump, or there is direct injection with a Hamilton microsyringe. The extent of the injection can be localized and the mechanical damage to the neural tissue at the injection site should be minimized if possible. This can be accomplished by using a glass micropipette pulled like a recording electrode and connected to a Hamilton microsyringe via a polyethylene tube.

As the time for the tracer to be transported to the targeted region depends on the distance between the injection site and the cell body, it is necessary to take steps to determine the appropriate survival time for individual experimental conditions with pilot experiments. If the retrograde labeling was performed appropriately, the experimenter should select those cells that are labeled with the fluorescent tracer, take photomicrographs, and make whole-cell recordings. If the recording pipettes are filled with biocytin, the identification and analysis of cell morphology is possible by visualizing biocytin after the experiments. In general, cells that are labeled intensely with the fluorescent tracer are not suitable for recording (they appear swollen when observed with a phase-contrast microscope). In addition, exposure to excitation light for a long time should be avoided because it not only causes bleaching of fluorescence but damages the cells. It is also important to select healthy cells quickly with clear fluorescent labeling.

References

1. Neher E, Sakmann B (1976) Single-channel currents recorded from membrane of denervated frog muscle fibres. Nature 260:799–802
2. Hamill OP, Marty A, Neher E, Sakmann B, Sigworth FJ (1981) Improved patch-clamp techniques for high-resolution current recording from cells and cell-free membrane patches. Pflugers Arch 391:85–100
3. Edwards FA, Konnerth A, Sakmann B, Takahashi T (1989) A thin slice preparation for patch clamp recordings from neurones of the mammalian central nervous system. Pflugers Arch 414:600–612

Chapter 8

Patch-Clamp Recording Method in Slices for Studying Presynaptic Mechanisms

Tomoyuki Takahashi, Tetsuya Hori, Yukihiro Nakamura, and Takayuki Yamashita

Abstract

The synapse plays pivotal roles in various neuronal functions, with its efficacy determining the opening and closing of neuronal circuits. Among the parameters that determine synaptic efficacy, presynaptic factors are of crucial importance. However, these factors are among those that are the least well studied. Patch-clamp recording from the calyx of Held presynaptic terminal, described in this chapter, has enabled us to address directly questions about the presynaptic regulatory mechanisms in synaptic transmission.

8.1. Introduction

The regulation of transmission efficacy at synapses plays a fundamental role in central neuronal function. Both presynaptic and postsynaptic mechanisms underlie the regulation of synaptic efficacy, but our knowledge concerning presynaptic mechanisms is relatively limited. This is because of the small structure of most presynaptic terminals, making it difficult to record electrophysiological signals directly, although presynaptic biophysical properties for transmitter release were thoroughly investigated at the squid giant synapse. Mammalian central nerve terminals, however, have a variety of regulatory mechanisms that can only be investigated by recording from mammalian presynaptic terminals.

In 1989, extending the microelectrode recording method from thin spinal cord slices (1) (Fig. 8.1A), the patch-clamp method from central nervous system (CNS) slices (Fig. 8.1B) was developed (2). For further extension, it was then hoped to establish a preparation in which direct recordings could be made from CNS presynaptic terminals. In 1994, Forsythe, at Leicester University

Yasunobu Okada (ed.), *Patch Clamp Techniques: From Beginning to Advanced Protocols*, Springer Protocols Handbooks, DOI 10.1007/978-4-431-53993-3_8, © Springer 2012

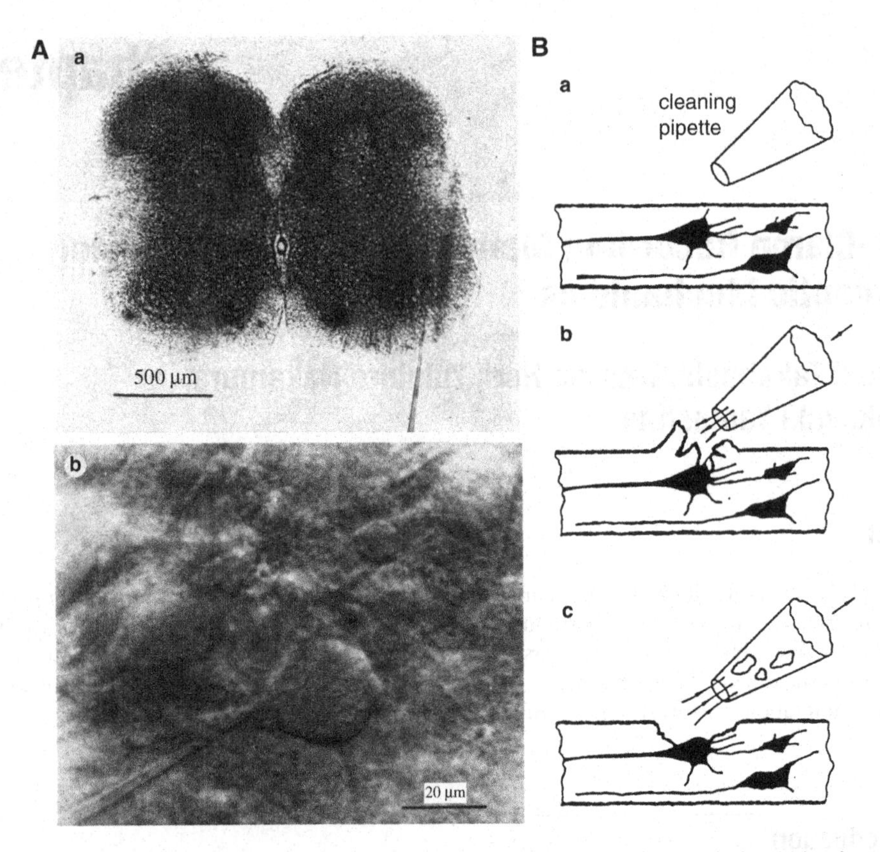

Fig. 8.1. Thin slice preparation and patch-clamp recording from neurons. (**A**) Thin spinal cord slice. (*a*) Lumbar spinal cord slice from P1 rat (130 μm in thickness). A microelectrode is seen on the right ventral horn. (*b*) A spinal motoneuron penetrated with a microelectrode (*from below left*) under upright Nomarski optics (40× water immersion lens). Adapted from Takahashi (1978) (1) by permission. (**B**) Tissue cleaning method for patch clamp recording from neurons in slices. (Adapted from Edwards et al. (1989) (2) by permission).

succeeded in whole-cell patch-clamp recording from a giant nerve terminal called the calyx of Held in the auditory brain stem slices of immature rats (3). During the next 2 years, Borst et al. (4) and Takahashi et al. (5) established methods for making simultaneous recordings from presynaptic terminal and postsynaptic neuron pairs. Hence, it became possible to address questions concerning presynaptic mechanisms in mammalian CNS preparations. In this chapter, we describe the practical details concerning this method.

8.2. Experimental Setup

8.2.1. Stage-Fixed Upright Microscope with Movable Objectives

After establishment of the slice patch-clamp method (2), the stage-fixed upright microscope became widely available from several manufacturers (Zeiss, Olympus, Nikon, Leica). We routinely use 40× to 63× water immersion objectives for presynaptic terminal identification. We find that the 60× objective from Olympus is handy for manipulation, with its long working distance and slender shape. Conveniently, it can also be used in Zeiss microscopes. The Nikon objective (60×) is bright with its high numerical aperture (NA, 1.0), but its "flat top" shape limits the angle of the electrodes to some extent. It should also be noted that soiled objective lenses often cause poor resolution. We typically use a 60× objective and a 1.6× magnifier to expand the picture by 1,000-fold.

As healthy presynaptic terminals are usually hidden behind postsynaptic cell bodies, it is necessary to change focus up and down while searching for the terminal. The microscope focus is ideally motorized, particularly for multielectrode manipulations, but for manual focusing we recommend extending the focus knob with an optional attachment. An infrared apparatus or Nomarski condenser is useful for enhancing contrast of synaptic structures, but a tilted condenser and placement of a narrow-band red filter in the light path provide equally good contrast to the picture (Fig. 8.2).

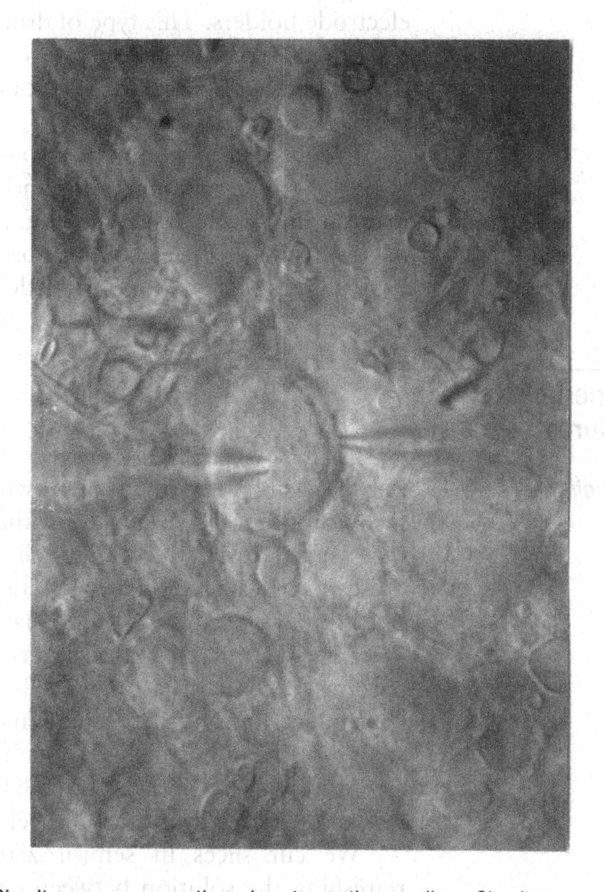

Fig. 8.2. Simultaneous presynaptic and postsynaptic recordings. Simultaneous recording from the calyx of Held nerve terminal (*right*) and postsynaptic medial nucleus of the trapezoid body (MNTB) neuron (*left*) under an upright microscope with tilted condenser illumination and a 60× water immersion lens.

Electrode manipulations can be made while viewing a video monitor attached to a CCD camera mounted at an eyepiece. A conventional CCD camera by itself is enough for many purposes, but a contrast enhancer (Argus C2741-62; Hamamatsu Photonics), if available, is a luxury that gives better pictures. We think that a black and white analog monitor gives the best resolution, but because it is rarely available nowadays a personal computer (PC) monitor can be used as a substitute. PC monitors also have additional advantages of less electrical noise and lower price.

8.2.2. Manipulators

In simultaneous presynaptic and postsynaptic recordings, it is critical to minimize the drift of manipulators because manual corrections of two electrode positions during experiments are troublesome. Any drift can increase whole-cell access resistance, thereby distorting stable recordings and voltage-clamp performance. We use motor-driven manipulators from Luigs & Neumann (Ratingen, Germany). Even when manipulators are stable, electrodes can drift owing to a change in room temperature and shrinking or swelling electrode holders. This type of drift can be reproduced by holding a warmed soldering iron near the holder. The magnitude of drift depends on the holder; we have found the holder from G23 instruments (UK) satisfactory.

Even after removing these sources of drift, a slight difference in osmolarity between external and pipette solutions can still cause drift during whole-cell recording. After membrane rupture, manual corrections of electrode positions are often necessary for 5–10 min as this type of drift settles.

8.3. Experimental Procedures

8.3.1. Slicing

For successful recording from presynaptic terminals, the most critical technical point is the quality of the slices. To obtain high-quality slices, it is necessary to keep the slicer in good condition, with little noise or vertical razor blade movements. We use two slicing devices, one from Dosaka EM (Kyoto, Japan) and the other from Leica (Wetzlar, Germany). For direct recording from presynaptic terminals, it is essential that many cells on the surface of the slices are kept intact after cutting. Concerning the cutting speed, for brain stem slices from postnatal day 14 (P14) rats, we find that the slowest speed gives the best results with the Dosaka slicer, and 0.4 mm/s seems optimal with the Leica slicer.

We cut slices in semi-frozen cutting solution, frequently refreshing the solution between cuttings to keep it cold and free from depolarizing molecules such as glutamate. Immediately after cutting, each slice is transferred to artificial cerebrospinal fluid kept at 35–37°C to let it recover quickly from cutting damage.

8.3.2. Cell Identification and Cleaning Tissues

Calyx of Held presynaptic terminals and postsynaptic principal cells of the medial nucleus of the trapezoid body (MNTB) are visualized in transverse brain stem slices (150–200 μm thickness) just outside of the pyramidal tract (6). The giant calyx terminal can be detected along the margin of the MNTB cell soma. Calyces of Held terminals are rarely found in slices with only a few MNTB cells on the surface. We usually find one or two patchable calyces on each side of a slice from P13–16 rats. Patchable calyces typically have 1.5–2.0 μm of exposed surface. Recording from smaller structures requires a finer-tip electrode of higher resistance, which limits voltage-clamp performance. After identifying a calyx visually, we free it from surrounding tissues by giving it gentle puffs of pipette solution from a cleaning pipette with a tip diameter of ~10 μm (Fig. 8.1B). This manipulation also informs us about the three-dimensional distribution of the calyx and often exposes a larger, hidden part of the calyx. In general, this cleaning procedure is useful only for calyces in rats or mice older than P13. In P7–10 rodents, before the onset of hearing, calyces are larger but relatively fragile. Thus, in the latter case, we advance a patch pipette with internal positive pressure and, after confirming a dimple formed on the calyx, a GΩ seal can easily be attained after releasing the pressure.

8.3.3. Simultaneous Presynaptic and Postsynaptic Recordings

After achieving a whole-cell recording configuration with a postsynaptic MNTB neuron, we advance a presynaptic patch pipette, which was previously positioned nearby, to the associated calyx and form GΩ seal, followed by whole-cell rupture. We sometimes fail to obtain a GΩ seal, presumably owing to a rough surface of a calyx or because of its unexpectedly smaller size. After establishing the whole-cell mode on a calyx terminal, a depolarizing current injection can generate an action potential in the calyx and an excitatory postsynaptic current (EPSC) in the postsynaptic neuron.

For stable EPSC recordings, it is essential to include glutamate at ~3 mM in the presynaptic pipette as its absence causes a time-dependent run-down of EPSCs (7). Conversely, loading the presynaptic pipette with 100 mM glutamate augments the EPSC amplitude with time. These phenomena arise from the fact that the presynaptic cytoplasmic glutamate concentration is in equilibrium with the vesicular glutamate content, and the endogenous glutamate concentration in the nerve terminal is estimated to be several millimoles (7).

In simultaneous presnaptic and postsynaptic recordings (Fig. 8.2), it is possible to evoke EPSCs by presynaptic Ca^{2+} currents elicited by depolarizing current pulses or action potential waveform command pulses. In these experiments, it is essential to keep the access resistance low and stable. For postsynaptic recording, we use patch pipettes with a resistance of 1–2 MΩ and keep the access resistance as low as 3–5 MΩ; for presynaptic recording, we use 4- to 6-MΩ electrodes with ~10 MΩ access resistance and compensate by up to 80% when necessary. Presynaptic pipettes with higher access resistances can be used if the presynaptic terminal is used only for action potential generation. It is often useful to keep the osmolarity of the presynaptic solution equal to (305–315 mOsm/kg) or slightly higher (by 5–10 mOsm/kg) than that of the external solution to keep the access resistance stable during whole-cell recording.

8.3.4. Presynaptic Terminal Perfusion

One of the features of whole-cell recording is that molecules in a patch pipette easily diffuse into the cell. This action can be demonstrated by loading the pipette with a small fluorescent molecule such as Lucifer Yellow, which quickly stains the entire presynaptic terminal. Taking advantage of this phenomenon, it is possible to introduce new molecules into a presynaptic terminal by patch pipette perfusion after obtaining control data (8). For infusing molecules into calyces, we started with previously reported pipette perfusion methods and simplified them.

We connect a fine glass tube (o.d. 300 μm, 10 cm long) (PT-030; Takao, Kyoto, Japan) to a fine plastic tube that has been heat-pulled from an Eppendorf "yellow tip." This plastic tube keeps its strength after pulling and can be cut to a tip diameter of 350–400 μm, allowing it to fit inside the glass tube (Fig. 8.3). We insert the plastic tube into the glass tube and seal the junction by gentle heating. We then mount the tube assembly onto a three-port (Y-shape) pipette holder (WPI) and heat-pull the glass pipette using a horizontal pipette puller (PN-30, Narishige, Japan), to achieve a tip diameter of 10–15 μm. The advantage of using the glass tube is that it can be inserted close to the tip of the patch

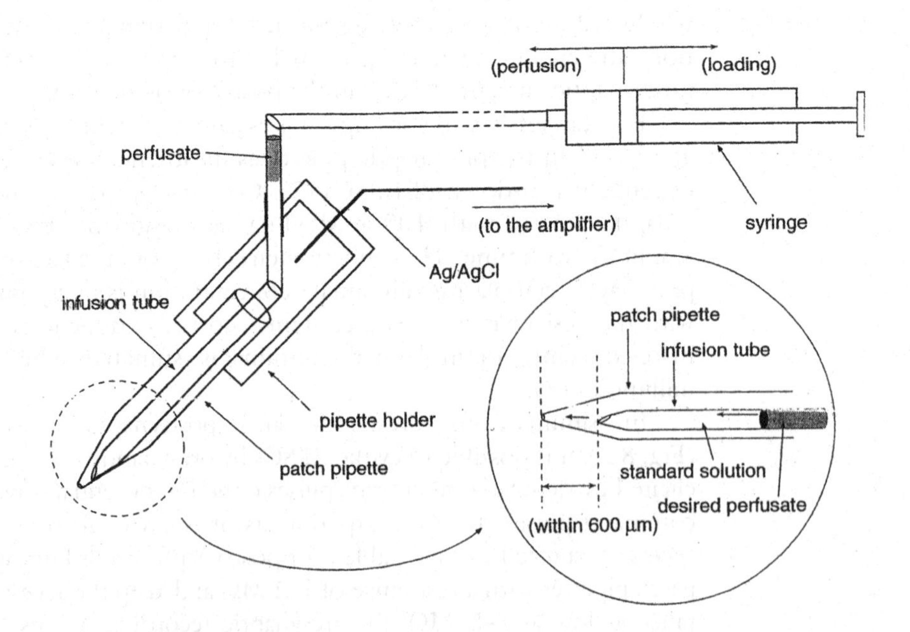

Fig. 8.3. Patch pipette perfusion system.

pipette (150–200 μm), thereby enabling quick infusion of molecules into a presynaptic terminal.

It takes about 2 min before a small molecule (molecular weight ~200) reaches its maximum concentration in a presynaptic terminal. This method allows one to compare control and test solutions at the same presynaptic terminal. After taking control data, we give positive gas pressure (8–10 psi) via a solenoid valve to the perfusion tube, thereby infusing a solution in a tube into a patch pipette. To minimize passive diffusion of the tube solution into the patch pipette solution, we fill the tip of the tube solution with patch pipette solution. Also, to minimize the effect of dilution by the control solution, we reduce the volume of the patch pipette solution to less than 3 times of that of the tube solution. This method is also useful for presynaptic infusion of large molecules, such as peptide fragments or antibodies, which often prevent formation of a GΩ seal.

8.3.5. Ca^{2+} Imaging at Presynaptic Terminals

Calcium imaging of presynaptic terminals can be made by whole-cell loading of Ca^{2+} indicator dyes into the terminal. Compared with imaging using membrane-permeable dyes, direct loading of membrane-impermeable dyes provides a better signal-to-noise (S/N) ratio. Among Ca^{2+} indicator dyes, those having high Ca^{2+} binding affinity (e.g., Fura-2) are useful for measuring Ca^{2+} at a low concentration range. However, this dye tends to saturate at high Ca^{2+} concentrations, and it cannot resolve fast Ca^{2+} signals. For monitoring Ca^{2+} transients in presynaptic terminals, we utilize the low-affinity dye Oregon Green BAPTA-5N (100 μM) and sample signals at high frequencies (>100 Hz). Presynaptic Ca^{2+} transients can be induced by a presynaptic action potential elicited by depolarizing current injection or by Ca^{2+} current pulses under voltage-clamp control.

8.3.6. Membrane Capacitance Measurements from Presynaptic Terminals

Cell membrane capacitance is proportional to the membrane area. Therefore, membrane capacitance of presynaptic terminals is increased by exocytosis of synaptic vesicles and decreased by endocytosis. Thus, measuring membrane capacitance is a powerful method for monitoring vesicle exo-endocytosis. To record membrane capacitance, a sine wave voltage command is applied to a presynaptic terminal in whole-cell voltage clamp mode, and the capacitance component is extracted from the current output (Lindau–Neher method (9)). We use the EPC9 (HEKA, Lambrecht/Pfalz, Germany) amplifier in combination with lock-in amplifier software (Patchmaster; HEKA). As an important advantage, capacitance measurements provide information as to whether a reagent has an effect on vesicle exocytosis, endocytosis, or both, as the latter is usually slower than the former (10).

However, application of capacitance measurements to presynaptic terminals has some limitations: (1) Capacitance change cannot be measured reliably for the initial 500 ms, when membrane conductance changes (e.g., by Ca^{2+} channel gating (11)). (2) Although in principle the method can resolve single-vesicle exocytosis, as exemplified in secretory cells, the S/N ratio in presynaptic terminal whole-cell recording is not high enough to resolve a small number of vesicle exo-endocytosis events. The S/N ratio can be improved in capacitance measurements using the cell-attached mode, however (11). At small nerve terminals in cultured neurons, where capacitance measurements are not feasible, styryl dyes or pH-sensitive green fluorescent protein (GFP)-tagged vesicle proteins have been widely used to monitor vesicle exo-endocytosis from a population of synaptic boutons.

8.4. Present Status of Recording from Other Presynaptic Terminals and Future Scope of the Calyx of Held Preparation

Direct recordings from large presynaptic terminals have been made at glutamatergic hippocampal mossy fiber terminals (12) as well as at γ-aminobutyric acid (GABA)ergic cerebellar basket cell terminals called "pinceau" (13). Whole-cell recording is also feasible at the end-bulb of Held primary sensory terminal-forming synapse on ventricular cochlear neurons (14). However, simultaneous presynaptic and postsynaptic recordings have been made successfully only at the calyx of Held. This synapse is a typical fast glutamatergic relay synapse specialized for the function of sound localization. Unlike hippocampal synapses, this synapse lacks long-term plasticity. Postsynaptic *N*-methyl-D-aspartic acid (NMDA) receptors, which play an induction role in long-term synaptic plasticity, decline in their expression as animals mature (15). Furthermore, the α-calmodulin kinase II, which is also thought to play an essential role in long-term plasticity, is scarce in the MNTB region, with its concentration being approximately 1% of that in the hippocampus (Naoto Saitoh, unpublished observation). Nevertheless, this synapse is similar to other synapses, having short-term plasticity, retrograde presynaptic regulatory mechanisms, and a variety of GTP-coupled presynaptic receptors.

An important aspect of the calyx of Held is its postnatal developmental changes. Rodents start to hear sound during the second postnatal week. During this period, dramatic changes are observed in morphology, molecular composition, and presynaptic functional properties (16–19). As this is also the period at which myelination starts, recording from the calyx becomes more difficult after the second postnatal week. Because of this technical reason, most studies at the calyx have been made during the prehearing period, typically at P8–10. Conclusions drawn from observations in prehearing calyces require reinvestigation at posthearing calyces as many of the conditions observed early do not persist after hearing is initiated (16–19). In this respect, the calyx of Held provides an interesting model for investigating the mechanism underlying synaptic maturation.

The slice patch-clamp method was developed as an extension of recording in visualized neurons in thin slices and has since reached a stage at which simultaneous presynaptic and postsynaptic recordings are possible. The next goal should perhaps be to visualize presynaptic molecules by real-time imaging to elucidate the presynaptic molecule–function relation.

Acknowledgments

We thank Ervin Johnson for helpful comments and English-language editing.

References

1. Takahashi T (1978) Intracellular recording from visually identified motoneurons in rat spinal cord slices. Proc Roy Soc Lond B 202:417–421
2. Edwards FA, Konnerth A, Sakmann B, Takahashi T (1989) A thin slice preparation for patch clamp recordings from neurones of the mammalian central nervous system. Pflugers Arch 414:600–612
3. Forsythe ID (1994) Direct patch recording from identified presynaptic terminals mediating glutamatergic EPSCs in the rat CNS, in vitro. J Physiol 479:381–387
4. Borst JGG, Helmchen F, Sakmann B (1995) Pre- and postsynaptic whole-cell recordings in the medial nucleus of the trapezoid body of the rat. J Physiol 489:825–840
5. Takahashi T, Forsythe I, Tsujimoto T, Barnes-Davies M, Onodera K (1996) Presynaptic calcium current modulation by a metabotropic glutamate receptor. Science 274:594–597
6. Forsythe ID, Barnes-Davies M (1993) The binaural auditory pathway: excitatory amino acid receptors mediate dual time-course excitatory postsynaptic currents in the rat medial nucleus of the trapezoid body. Proc R Soc Lond B Biol Sci 251:151–157
7. Ishikawa T, Sahara Y, Takahashi T (2002) A single packet of transmitter does not saturate postsynaptic glutamate receptors. Neuron 34:613–621
8. Hori T, Takai Y, Takahashi T (1999) Presynaptic mechanism for phorbol ester- induced synaptic potentiation. J Neurosci 19:262–7267
9. Lindau M, Neher E (1998) Patch-clamp techniques for time-resolved capacitance measurements in single cells. Pflugers Arch 411:37–146
10. Yamashita T, Hige T, Takahashi T (2005) Vesicle endocytosis requires dynamin-dependent GTP hydrolysis at a fast CNS synapse. Science 307:124–127
11. He L, Wu X-S, Mohan R, Wu L-G (2004) Two modes of fusion pore openings revealed by cell-attached recordings at a synapse. Nature 444:102–105
12. Geiger JRP, Jonas P (2000) Dynamic control of presynaptic Ca^{2+} inflow by fast-inactivating K^+ channels in hippocampal mossy fiber boutons. Neuron 28:927–939
13. Robertson B, Southan AP (1998) Patch-clamp recordings from cerebellar basket cell bodies and their presynaptic terminals reveal an asymmetric distribution of voltage-gated potassium channels. J Neurosci 18:948–955
14. Lin KH, Oleskevich S, Taschenberger H (2011) Presynaptic Ca2+ influx and vesicle exocytosis at the mouse endbulb of Held: a comparison of two auditory nerve terminals. J Physiol 589: 4301–4320
15. Futai K, Okada M, Matsuyama K, Takahashi T (2001) High-fidelity transmission acquired via a developmental decrease in NMDA receptor expression at an auditory synapse. J Neurosci 21:3342–3349
16. Iwasaki S, Momiyama A, Uchitel OD, Takahashi T (2000) Developmental changes in calcium channel types mediating central synaptic transmission. J Neurosci 20:59–65
17. Fedchyshyn MJ, Wang LY (2005) Developmental transformation of the release modality at the calyx of Held synapse. J Neurosci 25:4131–4140
18. Nakamura T, Yamashita T, Saitoh N, Takahashi T (2008) Developmental changes in calcium/calmodulin-dependent inactivation of calcium currents at the rat calyx of Held. J Physiol 586:2253–2261
19. Yamashita T, Eguchi K, Saitoh N, von Gersdorff H, Takahashi T (2010) Developmental shift to a mechanism of synaptic vesicle endocytosis requiring nanodomain Ca^{2+}. Nature Neurosci 13:838–844

Chapter 9

Analysis of Synaptic Plasticity with the Slice Patch-Clamp Recording Technique

Toshiya Manabe

Abstract

The practical details of the "blind" patch-clamp recording method and its usefulness and applications to the analysis of synaptic transmission and plasticity in brain slice preparations are described. The blind patch-clamp method displays its greatest power in the experiments in which long-term, stable recordings are required, such as those for the analysis of long-term potentiation (LTP) in hippocampal slices. Relatively thick slices, which have less damage during preparation, can be used with this method. Thus, the organization of neural tissues remains rather normal, which enables better, stable recordings. Recordings can be made from neurons located deeper in a slice, which are healthier than those near the surface. Furthermore, it is not necessary to clean and remove connective tissues around a recorded neuron, and the damage to a slice such as severance of the dendrite is minimal. These advantages make long-lasting, stable recordings possible during which the neuronal network is fairly well preserved. This method has been an indispensable tool in the analysis of synaptic plasticity.

Abbreviations

AMPA	α-Amino-3-hydroxy-5-methyl-4-isoxazolepropionate
CNS	Central nervous system
EPSC	Excitatory postsynaptic current
IPSC	Inhibitory postsynaptic current
LTD	Long-term depression
LTP	Long-term potentiation
mEPSC	Miniature EPSC
mIPSC	Miniature IPSC
NMDA	*N*-methyl-D-aspartate
sEPSC	Spontaneous EPSC
sIPSC	Spontaneous IPSC

Yasunobu Okada (ed.), *Patch Clamp Techniques: From Beginning to Advanced Protocols*, Springer Protocols Handbooks, DOI 10.1007/978-4-431-53993-3_9, © Springer 2012

9.1. Introduction

Patch-clamp recording techniques, including the blind patch-clamp recording, have made it possible to record miniature synaptic currents and single-channel currents; however, at the early stages, they were applicable only to quite simple preparations, such as the neuromuscular junction and cultured cells. The slice patch-clamp recording technique, developed later, enabled recording of whole-cell and single-channel currents from neurons in slice preparations in the central nervous system (CNS), and this technique has been an indispensable tool for the analyses of synaptic functions in the CNS.

There are mainly two methods for patch-clamp recording using the slice preparation. One is the "visualized" method, in which a recording patch pipette approaches neurons to be recorded by visualizing them with a high-performance microscope. The other is the "blind" method, in which a recording patch pipette blindly approaches neurons in a cell layer with a stereoscope and then identifies the neuron to be recorded with an electrophysiological method. The visualized method, which is applicable to relatively thin brain slice preparations, is described in detail elsewhere in this book. Here, the detailed method of the blind patch-clamp recording technique and its application to the analysis of synaptic plasticity are described. It is of note that the blind patch-clamp technique is applicable to relatively thick brain slice preparations (400–500 μm) that have been used conventionally for intracellular and extracellular field-potential recordings.

As indicated by its name, the blind method enables whole-cell patch-clamp and single-channel recordings without visualizing the neuron, which is the greatest merit of this method. The blind method was developed independently by two groups (1, 2) and has been used extensively for studying CNS functions in many laboratories (3–6). On the other hand, with the visualized method, fine manipulations are feasible because the recorded neuron can be seen under a microscope. However, the visualized method is somewhat unsuitable for analysis of synaptic plasticity because the recordings are usually made from neurons located near the surface of thin slices, and it is sometimes difficult to maintain a stable recording for a sufficiently long time.

9.2. Characteristics of the Blind Patch-Clamp Recording Method Using Slice Preparations

Although the advantages of patch-clamp recording in brain slices by the blind method are generally similar to those of the visualized method, there are some advantages peculiar to the blind method.

1. Because relatively thick slices can be used, they are less damaged during preparation, and the organization of neuronal tissues remains rather normal, enabling better, stable recordings.

2. Recordings can be made from neurons located relatively deep in a slice (e.g., 100–200 μm from the surface) in addition to neurons near the surface of the slice.
3. Because it is not necessary to clean and remove connective tissues around a recorded neuron, the damage to a slice (e.g., severance of a dendrite) can be avoided.
4. As it is not necessary to visualize a neuron under a microscope to form a giga-ohm seal, a high-performance microscope is unnecessary, and experiments can be performed at relatively low cost.
5. Experimental manipulations are easier because of a wider visual field and larger space.
6. In an experimental setup for the blind method, not only patch-clamp recordings but also extracellular field-potential and intracellular recordings can be made. It is also possible to perform whole-cell patch-clamp and extracellular field-potential recordings simultaneously.

One of the disadvantages of the blind method is a "washout" of intracellular components, which is commonly seen in whole-cell patch-clamp recordings. One of the disadvantages peculiar to the blind method is that it is difficult to record from a specified kind of cell in the tissue where different kinds of cells coexist because the cells are not identified under a microscope. Another disadvantage is technical difficulty in forming a giga-ohm seal that results from a characteristic of the blind method in which the attachment of a patch pipette to the cell surface is monitored only by a change in the resistance of the pipette. As for this point, this procedure is similar to that of intracellular recordings, and an experimenter who is familiar with the intracellular recording technique experiences little difficulty.

As a result of these properties, it is possible to make long-term, stable recordings with the blind method with the neuronal network remaining fairly well preserved. This fact makes the blind method a powerful tool for analyzing synaptic plasticity, which requires long-term, stable recording.

9.3. Practical Aspects of the Blind Patch-Clamp Recording

9.3.1. Tissues

For the blind patch-clamp recording method, tissue containing a single type of cell is suitable. The main tissues to which the blind method is applied in the CNS are the hippocampus, cerebral cortex, amygdala, and cerebellum. Slice preparations are most commonly used, but the blind method is sometimes applied also to in vivo preparations (7) as described in other chapters of this book.

9.3.2. Preparation of Brain Slices

The method for preparing brain slices for the blind patch-clamp recording, which are usually 400–500 μm thick, is basically the same as that for preparing thin slices (refer to other chapters in this book). Although the thicker slice can be cut with a slice chopper, there seems to be a higher success rate of proper giga-ohm seal formation in slices prepared by a vibrating slicer than in those prepared by a slice chopper.

9.3.3. Experimental Equipment

A pipette puller and electrophysiological equipment for recording a membrane current with the blind method are the same as those with the other types of patch-clamp recording. A high-performance microscope is unnecessary. A low-priced stereoscope is sufficient for the blind method as it can identify the cell layers or populations of the neurons in a slice.

9.3.4. Experimental Procedures

The procedure for patch-clamping with the blind method is similar to that for the other methods described in detail in other chapters. Usually, a recording pipette is not coated with Sylgard; however, if the Sylgard coating is necessary, it should be minimal because the pipette sometimes penetrates deep in a slice with the blind method, and the Sylgard around the pipette could press the tissue heavily, which may cause substantial damage.

Because the technique for searching a cell and making a giga-ohm seal with the blind method is different from that of the visualized method, it is described here in detail (Fig. 9.1).

1. Monitoring the resistance of a patch pipette with an oscilloscope allows determination of whether a pipette tip is attaching to the cell membrane. Put the pipette filled with internal solution into the bath of the recording chamber. Give square voltage waves (1 mV, 20–40 ms) to the pipette in the voltage-clamp mode through a patch-clamp amplifier to monitor the current in response to the waves with an oscilloscope. The resistance of the pipette can be calculated according to Ohm's law. In the case of standard internal solution, a pipette with the resistance of 4–8 MΩ is suitable for the blind method. Although this procedure can also be performed in the current-clamp mode, only the procedures in the voltage-clamp mode are described in the following material.
2. The pressure in the patch pipette is controlled by a ~10-ml syringe until the giga-ohm seal is accomplished. Fine control of the pressure can be achieved using the syringe. Generally, positive pressure applied to the pipette during the search for the cell with the visualized method is relatively weak; however, in the case of the blind method, considerably strong pressure is usually applied. In a standard electrophysiological setup, positive pressure is applied by pushing out about 1–2 ml of air from the

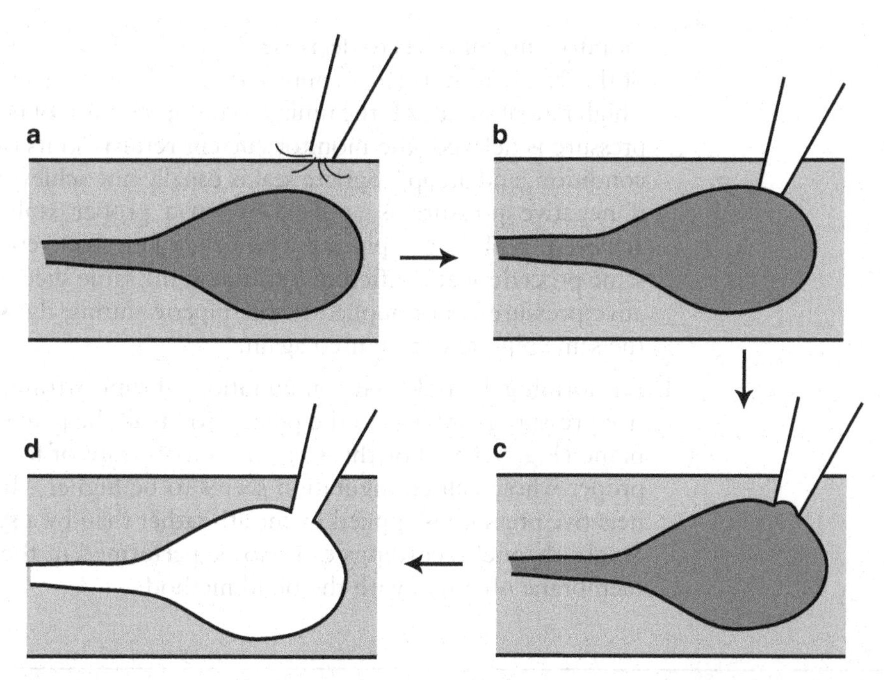

Fig. 9.1. Whole-cell patch-clamp recording with the blind method. (**a**) Patch pipette is advanced just above the surface of a slice. Because relatively strong positive pressure is applied to the pipette, intracellular solution is ejected from the tip of the pipette. (**b**) Pipette is further advanced in the slice above the cell layer and is about to attach to a cell body. (**c**) Just after detecting an increase in electrode resistance, fine negative pressure is applied to the pipette and a giga-ohm seal is achieved. (**d**) Whole-cell configuration is created by applying stronger negative pressure to break the membrane patch at the pipette tip.

syringe, but the volume of the air should be determined experientially. Then, move the pipette near the cell layer of the slice, and advance it step by step by a few micrometers using a micromanipulator while monitoring the pipette resistance with an oscilloscope (Fig. 9.1a). To ensure a higher success rate for establishing a giga-ohm seal, the pipette should be advanced parallel to the cell layer and/or in a direction such that the pipette encounters as many neurons as possible.

3. While advancing the pipette as described above, the recorded current exhibits a large, abrupt shift when the pipette attaches to a cell surface. Advance the pipette further while keeping the positive pressure in the pipette until the monitor current decreases (the pipette resistance increases when its tip attaches to the cell membrane) (Fig. 9.1b). In general, it is difficult to maintain stable recordings in the cell near the surface of slices; therefore, the cell encountered first during the search is sometimes broken by applying stronger positive pressure to the pipette. A giga-ohm seal between the tip of the pipette and the cell membrane is achieved within a few to a few tens of seconds by providing the pipette with fine negative pressure as soon as the

monitor current starts to decrease (Fig. 9.1c). If the conditions of the slices are normal, an appropriate seal can be achieved at a high rate of success. If the timing of the application of negative pressure is delayed, the monitor current returns to its original condition, and an appropriate seal is usually not achieved even if negative pressure is applied. When a proper seal is not achieved, replace the pipette with a new one and repeat the same procedure at a different location in the same slice. If negative pressure is not applied to the pipette during the search, the same pipette can be used again.

4. For forming a whole-cell configuration, abrupt, strong, negative pressure is applied to the pipette to break the patch membrane (Fig. 9.1d). For this step, the success rate of forming a proper whole-cell configuration seems to be higher when the negative pressure is applied by mouth rather than by a syringe. Single-channel recordings can also be performed in the patch membrane obtained with the blind method.

9.4. Equipment Required for the Blind Patch-Clamp Recording

- Faraday cage (shield cage)
- Vibration-control table: It is desirable to use a high-performance air-cushion table
- Stereoscope: A low-priced stereoscope, with which a cell layer is identified in the slice preparation, is sufficient for the blind method. When the cell layer is not clearly identified, the rate of success in forming a proper seal is quite low
- Lighting apparatus: It should be arranged so the tip of a recording pipette can be clearly seen
- Patch-clamp amplifier
- Electrical stimulator
- Isolator
- Oscilloscope
- Personal computer for acquiring and storing data
- Interface
- Manipulator: It is desirable to use a micromanipulator, with which a recording pipette is advanced step-by-step by a few micrometers. The micromanipulator is moved three-dimensionally by a few millimeters to a few centimeters with another manually operated manipulator. If the manipulators exhibit slippage during the recording, it is extremely difficult to perform long-term, stable recordings

9.5. Applications of Blind Patch-Clamp Recording to Analysis of Synaptic Plasticity

There are many advantages of applying the slice patch-clamp recording technique to the analysis of synaptic transmission. Representative advantages are noted here, and examples are described. Note that there are many other advantages that are not described in this section.

9.5.1. Lower Noise Level

The most remarkable advantage in the whole-cell patch-clamp recording is that the noise level is much lower than that in the conventional intracellular recording. It was almost impossible to record and analyze the spontaneous excitatory (sEPSC) or inhibitory (sIPSC) postsynaptic currents or the miniature EPSC or IPSC currents (mEPSC, mIPSC) with the conventional intracellular recording technique. However, using the whole-cell patch-clamp recording technique, these very small events can be recorded clearly, thereby enabling precise analyses of synaptic transmission in the CNS. In general, the sensitivity of the postsynaptic cell to neurotransmitters can be evaluated by analyzing the amplitude of mEPSCs or mIPSCs, and the release probability of neuro transmitters can be estimated by analyzing the frequencies of these events. However, it should be noted that this interpretation is a general principle, and there are exceptions.

An example of the analysis of synaptic plasticity using mEPSCs (sEPSCs) is as follows: There has been a long-lasting heated debate on whether long-term potentiation (LTP) in the CA1 region of the hippocampus is expressed by a change in the presynaptic terminal (an increase in the release probability) or in the postsynaptic cell (an increase in sensitivity to the neurotransmitter glutamate). As one of the approaches to solve this problem, we recorded mEPSCs before and after LTP induction and compared their amplitude and frequency (6). The size of mEPSCs (sEPSCs) is increased during LTP and short-term potentiation induced by *N*-methyl-D-aspartate (NMDA) application, which suggests that a postsynaptic change is involved in these types of synaptic potentiation (Fig. 9.2). This conclusion, derived from the electrophysiological examinations, is supported by recent studies using both molecular biological and morphological approaches. Although this was the first ever report of this electrophysiological method that was used for the analysis of synaptic plasticity, it has become one of the most frequently used standard methods for the analysis.

An example of the analysis of sEPSCs using Sr^{2+}, which originate in the peculiar synapses that are activated by electrical stimulation, is as follows: In the analysis of sEPSCs and mEPSCs described above, miniature events randomly appear from all the synapses on the recorded neuron if ingenuity (6) is not applied. However, when external Ca^{2+} is replaced with Sr^{2+}, miniature events that comprise EPSCs evoked by electrical stimulation appear asynchronously, and

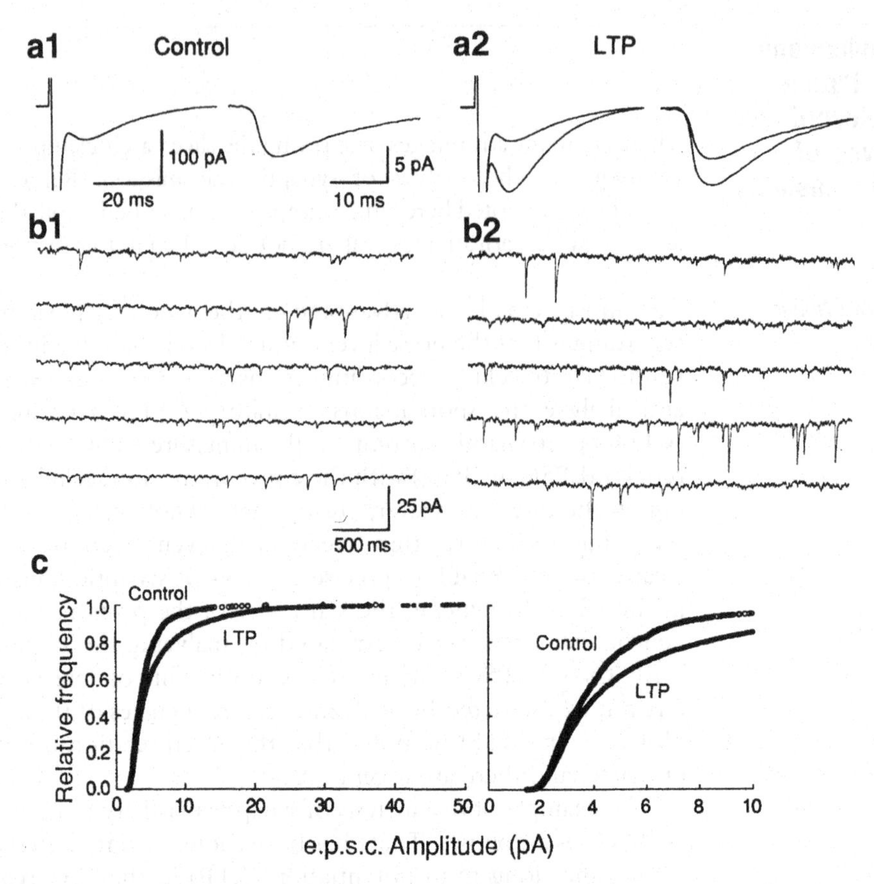

Fig. 9.2. Amplitude of spontaneous excitatory postsynaptic currents (sEPSCs) is increased during long-term potentiation (LTP) in the hippocampal CA1 region. (**a1**) Averaged traces of EPSCs evoked by electrical stimulation (*left*) and sEPSCs (*right*) before LTP induction. (**a2**) Averaged traces of EPSCs evoked by electrical stimulation (*left*) and sEPSCs (*right*) after LTP induction (larger traces). (**b1**) sEPSCs before LTP induction. (**b2**) sEPSCs after LTP induction. Larger sEPSCs appear that are rarely observed before LTP induction. (**c**) Cumulative histogram of the amplitudes of sEPSCs before and after LTP induction. Distribution is shifted to the right after LTP induction, indicating an increase in amplitude of the sEPSCs. The difference between the distributions is statistically significant by the Kolmogorov–Smirnov test (cited and modified from (6)).

sEPSCs originating in the activated synapses can be recorded and analyzed much more selectively (8). Using this method, it has been reported that LTP and long-term depression (LTD) in the hippocampal CA1 region are mediated by an increase and a decrease, respectively, in the sensitivity of the postsynaptic cell to the neurotransmitter (9).

9.5.2. Better Controlled Membrane Potential

Because of the extremely improved access to the cell with whole-cell patch-clamp recording, the membrane potential is much better controlled in the voltage-clamp mode.

The current–voltage (I–V) relation of synaptic responses can be evaluated more accurately. With the conventional intracellular recordings, it was difficult to estimate the reversal potential and examine the I–V relation of synaptic responses accurately because the cell was not efficiently clamped at positive membrane potentials. However, with whole-cell patch-clamp recordings, relatively fast responses such as EPSCs can be recorded much more accurately, and the reversal potential can be determined more precisely.

With the conventional intracellular recording method, it was difficult to clamp considerably slow responses of large amplitude (e.g., NMDA receptor-mediated synaptic responses), and the time course of the synaptic response was not accurately determined. These problems are overcome with whole-cell patch-clamp recording.

As an example of the application of this method to the analysis of synaptic plasticity (10, 11), α-amino-3-hydroxy-5-methyl-4-isoxazolepropionate (AMPA) receptor- and NMDA receptor-mediated synaptic responses can be recorded in the same cell. Also, we can determine which component of synaptic responses – AMPA or NMDA EPSCs – is modified during LTP or LTD. Again, although this electrophysiological method was first used for analyzing synaptic plasticity at the time of its publication, it has become one of the most frequently used standard methods for the analysis.

Another advantage of this method is that NMDA receptor-mediated synaptic responses can be recorded stably for a long time, making it possible to estimate the release probability of the neurotransmitter by using MK-801, an open-channel blocker of the NMDA receptor (5) (Fig. 9.3). This method was developed and applied independently by three groups (5, 12, 13), and it has been a standard method for this analysis.

9.5.3. Ability to Evaluate Changes in Activity of Single Channels

Another advantage in the whole-cell patch-clamp recording is that a change in the activity of single channels of the neurotransmitter receptor accompanying synaptic plasticity can be evaluated.

Single-channel properties can be examined directly in the slice patch-clamp recording. It has been proposed that a change in phosphorylation level of the AMPA and/or NMDA receptors is involved in the expression of synaptic plasticity, and changes in single-channel properties of the receptors accompanying synaptic plasticity can be examined with this method (14, 15).

With the nonstationary noise analysis of synaptic transmission, which is also applicable to the slice patch-clamp recording, the single-channel property of the receptor channels activated by released neurotransmitters at the synapse is examined. Because it is usually difficult to record directly single-channel currents originating in the receptor localized at the synapse in the CNS, the single-channel properties of the receptor are often estimated by noise analysis of synaptic currents in a nonstationary state. For more details of the noise analysis, refer to Chap. 6 of this book. It has been demonstrated with this method that an increase in the single-channel conductance of AMPA receptors is associated with LTP expression in the hippocampal CA1 region (16).

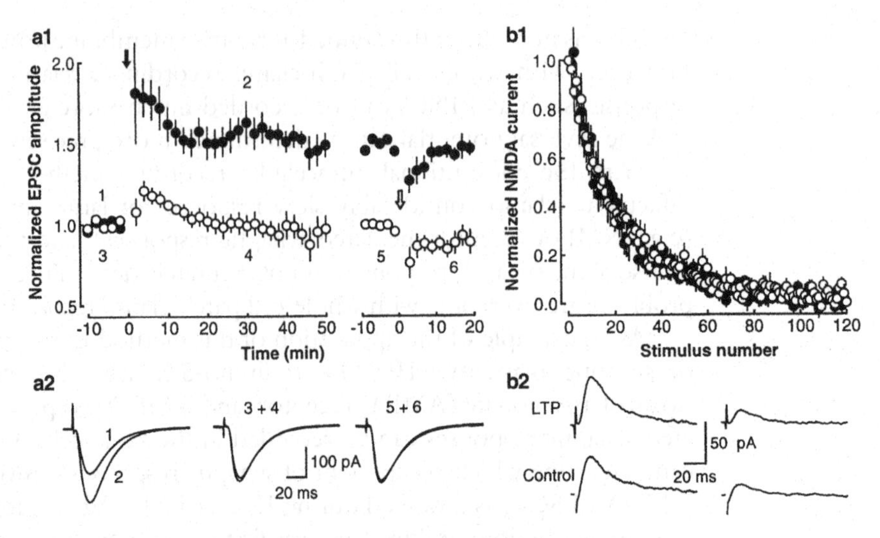

Fig. 9.3. Examination of a possible increase in probability of neurotransmitter release during LTP using MK-801. (**a1**) Two independent pathways are stimulated in the CA1 region of the hippocampal slices. LTP is induced in one pathway (*closed circles*) by pairing postsynaptic depolarization with electrical stimulation (*black arrow*), whereas LTP is not induced in the other pathway (*open circles*) by the second pairing (*open arrow*) due to "washout" of the LTP-induction mechanism caused by the prolonged (more than 1 h) whole-cell recording. Numbers 1–6 correspond to the traces in **a2**. (**a2**) Sample traces of evoked EPSCs. Numbers of the traces correspond to those shown in **a1**. (**b1**) *N*-Methyl-D-aspartate (NMDA) receptor-mediated synaptic responses are recorded in the presence of MK-801, an open-channel blocker of the NMDA receptor. The probability of neurotransmitter release is determined by measuring the rate of decay of NMDA EPSCs. The higher the release probability, the faster is the EPSC decay. The decay rate in the LTP pathway is the same as that in the control pathway, indicating that an increase in release probability is not associated with LTP. (**b2**) Sample traces of NMDA EPSCs. The left traces are averages of the first six NMDA EPSCs, and the right traces are averages of the six NMDA EPSCs around the 25th EPSC (cited and modified from (5)).

9.6. Conclusions

The practical details of the blind method, its usefulness, and its applications to the analysis of synaptic plasticity are described. The visualized slice patch-clamp method using thinner slice preparations displays its greatest power in the experiment in which the cell type must be identified under a microscope, whereas the blind patch-clamp method demonstrates its superb capability in the experiment in which the identification of cell types is not necessary and long-term, stable recordings are required. A method in which the advantages of the two methods are combined has been tried (17). Here, a cell is identified under a microscope and approached with the blind method, and a giga-ohm seal is achieved. This new method seems to have been successful in many cases. In experiments in which patch-clamp recordings are performed using slice preparations, the most important factor determining success or failure is the decision at the planning stage of the project as to which method will be used. The slice patch-clamp recording technique has already been an indispensable tool in the analysis of synaptic plasticity, and even in this post-genome era the technique continues to maintain its importance.

References

1. Blanton MG, Lo Turco JJ, Kriegstein AR (1989) Whole cell recording from neurons in slices of reptilian and mammalian cerebral cortex. J Neurosci Methods 30:203–210
2. Coleman PA, Miller RF (1989) Measurement of passive membrane parameters with whole-cell recording from neurons in intact amphibian retina. J Neurophysiol 61:218–230
3. Bekkers JM, Stevens CF (1990) Presynaptic mechanism for long-term potentiation in the hippocampus. Nature 346:724–729
4. Malinow R, Tsien RW (1990) Presynaptic enhancement shown by whole-cell recordings of long-term potentiation in hippocampal slices. Nature 346:177–180
5. Manabe T, Nicoll RA (1994) Long-term potentiation: evidence against an increase in transmitter release probability in the CA1 region of the hippocampus. Science 265:1888–1892
6. Manabe T, Renner P, Nicoll RA (1992) Postsynaptic contribution to long-term potentiation revealed by the analysis of miniature synaptic currents. Nature 355:50–55
7. Ferster D, Jagadeesh B (1992) EPSP-IPSP interactions in cat visual cortex studied with in vivo whole-cell patch recording. J Neurosci 12:1262–1274
8. Dodge FA, Miledi R, Rahamimoff R (1969) Sr^{2+} and quantal release of transmitter at the neuromuscular junction. J Physiol 200: 267–284
9. Oliet SHR, Malenka RC, Nicoll RA (1996) Bidirectional control of quantal size by synaptic activity in the hippocampus. Science 271:1294–1297
10. Sakimura K, Kutsuwada T, Ito I, Manabe T, Takayama C, Kushiya E, Yagi T, Aizawa S, Inoue Y, Sugiyama H, Mishina M (1995) Reduced hippocampal LTP and spatial learning in mice lacking NMDA receptor ε1 subunit. Nature 373:151–155
11. Manabe T, Aiba A, Yamada A, Ichise T, Sakagami H, Kondo H, Katsuki M (2000) Regulation of long-term potentiation by H-Ras through NMDA receptor phosphorylation. J Neurosci 20:2504–2511
12. Hessler NA, Shirke AM, Malinow R (1993) The probability of transmitter release at a mammalian central synapse. Nature 366:569–572
13. Rosenmund C, Clements JD, Westbrook GL (1993) Nonuniform probability of glutamate release at a hippocampal synapse. Science 262:754–757
14. Derkach V, Barria A, Soderling TR (1999) Ca^{2+}/calmodulin-kinase II enhances channel conductance of α-amino-3-hydroxy-5-methyl-4-isoxazolepropionate type glutamate receptors. Proc Natl Acad Sci USA 96:3269–3274
15. Yu X-M, Askalan R, Keil GJ II, Salter MW (1997) NMDA channel regulation by channel-associated protein tyrosine kinase Src. Science 275:674–678
16. Benke TA, Lüthi A, Isaac JTR, Collingridge GL (1998) Modulation of AMPA receptor unitary conductance by synaptic activity. Nature 393:793–797
17. Kato HK, Watabe AM, Manabe T (2009) Non-Hebbian synaptic plasticity induced by repetitive postsynaptic action potentials. J Neurosci 29:11153–11160

Chapter 10

Patch-Clamp Recordings from Neuronal Dendrites

Hiroshi Tsubokawa and Hiroto Takahashi

Abstract

Dendritic patch-clamp recordings have recently become popular as an electrophysiological technique to investigate the functional properties of ion channels in neuronal dendrites. During the past decade, experimentalists have pioneered recording from dendrites of neurons in brain slices under direct visual control (Davie et al., Nat Protoc 1(3):1235–1247, 2006). In this chapter, we describe advanced recording techniques including dendritic patching by the "blind approach."

10.1. Introduction

Neurons, the minimal units of the brain, have a variety of shapes. They are constructed of three components: cell body, axon, dendrites. Dendritic arbors extend their branches like a tree to maximize their ability to sample incoming inputs. Dendrites receive tens of thousands of synaptic inputs and are responsible for processing and integrating this information to determine spike output. Inputs can be processed and modified locally but can also be substantially altered during their propagation throughout the neuronal architecture. The hodgepodge of inputs descend toward the soma and axon hillock, where it is decided whether to transfer information to postsynaptic cells in the form of an all-or-none event called an action potential, or spike.

The distances between the distal ends of dendrites and the soma are substantial. The dendritic arbor of a layer V pyramidal neuron in the cortex ramifies and reaches more than 1 mm from its soma. Thus, some inputs must traverse great distances to affect spike output. Incoming inputs do not naturally maintain their integrity during this passage. During the propagation from dendrite to soma, the original information at the input sites is modified by

Yasunobu Okada (ed.), *Patch Clamp Techniques: From Beginning to Advanced Protocols*, Springer Protocols Handbooks, DOI 10.1007/978-4-431-53993-3_10, © Springer 2012

the dendrite's own active and passive properties. Classic theoretical work predicted that passive electrical parameters, including membrane resistance, membrane capacitance, and intracellular resistance, contribute to alter the electrical signal. Dendritic patch recording, in which a recording electrode is situated on a dendrite rather than at the more usual somatic location, revealed that channels and receptors have different expression patterns along the dendrites (1). The nonuniform distribution of conductances actively alters the electrical potential as it propagates.

Patch-clamp recording reports spatially restricted information and controls membrane potential in only a limited area. Using patch-clamp recordings from the soma, it is difficult to maintain tight electrotonic control over the elaborate and expansive dendritic arbor that is typical of cortical and hippocampal pyramidal neurons. This "space-clamp" issue has been confirmed experimentally, with the implication being that sites more and more remote from the recording site are under increasingly poor experimental control (2). As conventional somatic patch recordings cannot effectively clamp local dendrites, it is difficult to interpret synaptic activities. Indeed, although both local regenerative events and large synaptic events may be observed at the soma, many small synaptic events are likely to be filtered out completely during their passage from distal dendritic sites. Dendritic patch recordings decrease the distance from the recording site and the region of interest, allowing more precise control and more precise recording of synaptic conductances. The somatic patch technique allows vestiges of local events to be inferred but it cannot describe their original shape.

To explore synaptic events directly and faithfully, dendritic patch recordings are therefore imperative. During the past couple of decades, dendrite-exclusive events such as dendritic spikes and plateau potentials have been described. Additionally, application of a particular drug was shown to affect distal apical dendrites but not more proximal or basal dendrites (3–5). In particular, dendritic spikes and plateau potentials play an important role in intracellular signal transmission and are strongly associated with physiological phenomena such as synaptic plasticity. Those local dendritic activities could not have been detected with somatic patch recordings. Application of the dendritic patch technique provided the breakthrough to find these critical phenomena.

Here, we describe the techniques required for making whole-cell recordings from the dendrites of neurons. Several protocols have been published (6, 7), but we aim to describe the process in more detail and, in particular, focus on practical tips. We hope that it will be useful for researchers who are just starting or still struggling to obtain direct dendritic recordings.

10.2. Methods

10.2.1. Age of Animals for Dendritic Patch Recording

Although the age of the animal should be based on the particular research process, we mention two factors that particularly affect the ability to make dendritic patches. Brain tissue from juvenile mice (1–2 weeks postnatal) seems best able to tolerate oxygen deprivation and mechanical insults. Additionally, young animals have fewer tough fibers so there is less pulling of the tissue caused by blade vibration, making it easier to produce healthy brain slices. Unfortunately, the dendrites of juveniles are quite transparent and can be difficult to identify. In contrast, the dendritic arbor matures about 8 weeks after birth, increasing the contrast of the dendrites; thus, they are easier to locate in older animals though it is more difficult to make quality slices.

10.2.2. Anesthesia

Animals should be anesthetized according to the locally approved protocol. The inhalation of volatile anesthetics (e.g., with isoflurane) is a popular method for anesthetizing animals. Care should be used if ether is employed because it causes capillaries to dilate. After decapitation, and when the ether dissipates, the capillaries shrink and trap blood cells in vessels throughout the brain, leaving the brain colored pink by the remaining red blood cells. A ketamine/xylazine mixture is also used to maintain the condition of animals for long periods while performing cardiovascular perfusion. Cardiac perfusion increases the quality of slices from older animals in particular. Older animals have thick skulls, which prolong the time required to extract the brain. They also have stiff fibers that make it more difficult to cut. Cardiac perfusion ameliorates those difficulties. After perfusion with a cold cutting solution, the brain is chilled; thus, the additional time required to remove the skull does not have a negative impact. In addition to cooling the brain, the cutting solution rinses out fluids and debris from veins. Such clearance permits the blade to cut smoothly through the tissue.

10.2.3. Cutting Solution

Traumatic stress induces overactivation of neuronal metabolism, which promotes cell death. The cutting solution should therefore be chilled. Lowering the temperature of the cutting solution reduces the metabolic activity of neurons that occurs in response to trauma. If one uses a slushy solution, a potential osmolality mismatch between the cutting solution (~345) and artificial cerebrospinal fluid (ACSF) must be considered, which might affect the look and accessibility of the membrane. The iced cutting solution also has particles that can dent the edge of the blade, so care should be taken not to let ice particles bump a blade. Because the blade is vibrating with high speed, even small particles might abrade the edge of the blade. A blunt blade does not smoothly cut through slices, resulting in pulling of the tissue rather than slicing it. The pulling causes mechanical stress to neurons, yielding unhealthy slices.

The composition of the cutting solution should also be considered. Ringer's solution that substitutes Na^+ and Ca^{2+} ions with nonionic sucrose or cationic choline can be beneficial by helping to suppress neuronal overactivity. Antioxidants such as ascorbate, and pyruvate can also help to maintain cell viability.

The animal's genetic background should be considered, as some strains are sensitive to neuronal toxicity due to overactivation of *N*-methyl-D-aspartic acid receptors (NMDARs). If so, an NMDAR blocker can be added to the dissection solution during slicing.

10.2.4. Microtome

A quality Microtome (slicer) is a critical component for obtaining healthy slices. Several are available from different vendors. One feature of the slicer to consider is Z-compensation. Z-compensation mechanisms minimize the deviations of the blade in the *Z*-axis, which helps make high-quality slices more consistently. All good slicers permit control of the speed of the cut, the amplitude and frequency of the vibration, the angle of the blade, and the thickness of the slice. The best parameters are chosen on an empirical basis for specific preparations. Researchers are obliged to explore the best parameters based on their experience.

10.2.5. Material, Angle, and Direction of the Blades

Blades are made of a variety of materials, including sapphire, glass, ceramic, carbon, and stainless steel. Blades made of carbon and stainless steel are disposable and easy to handle. Although inexpensive and convenient, they must be handled with the same care as the expensive sapphire and ceramic blades to preserve the cutting side of the blade, which is extremely sensitive. Do not touch the edge of a blade when you handle it. When you take a blade from a container, hold the side of blade so you do not touch the edge; and do not remove the cover in a way that damages the cutting surface. As the diameter of a dendrite is quite small compared to the soma, you need to keep the cutting stress on the slice as low as possible. You can imagine the effect of a blunt or dented blade edge on slice quality. Blades are not necessarily clean when removed from the package. Industrial oil or other materials may coat the blade, which should be wiped clean with alcohol or acetone and rinsed with distilled water. Again, this cleaning process must be performed delicately. Sapphire blades provide the best performance but are the most expensive and require extra care.

The angle of the blade with respect to the tissue and cut direction is important for making high-quality slices (Fig. 10.1). An angle of 12°–20° is reasonable, although it depends on the type of blade. The exact shape of the edge of the blade depends on the type (e.g., taper, chisel, concave), and the blade should be installed so the taper of the blade has a slight clearance angle to the tissue, so it does not scrape along the slice during cutting.

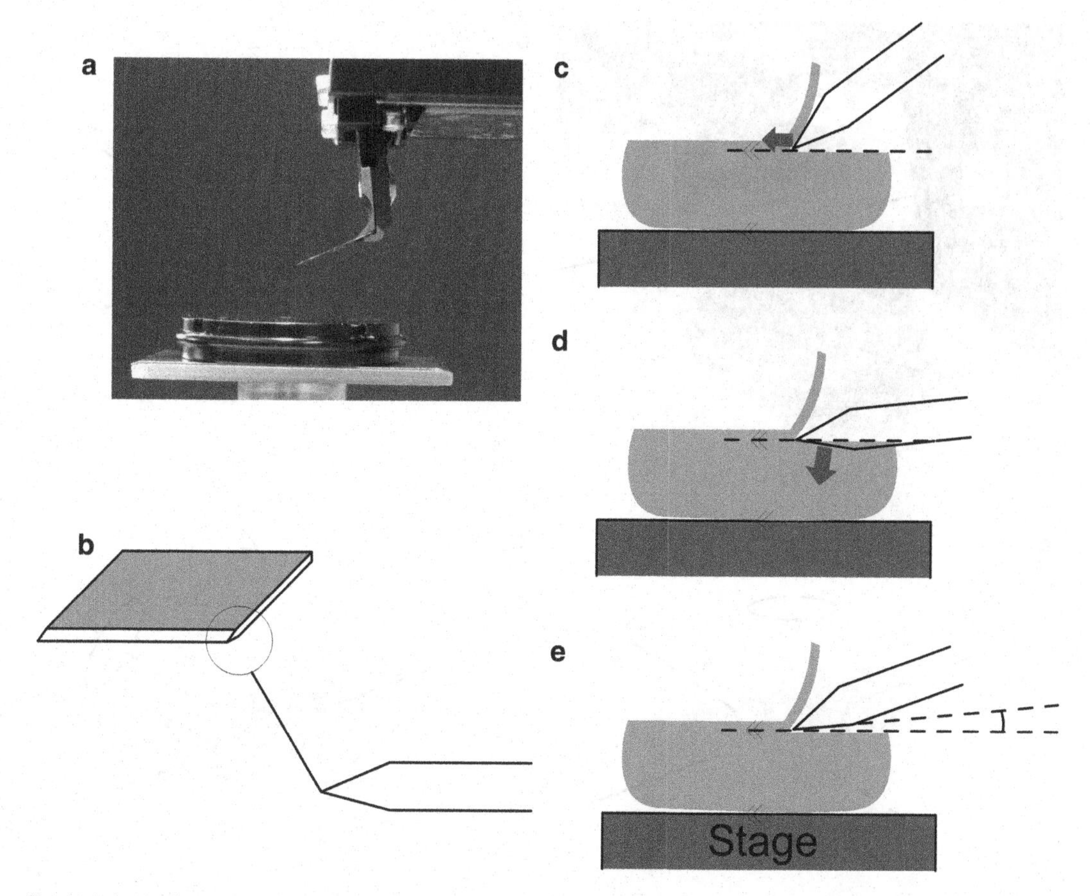

Fig. 10.1. Angle of the blade. The insertion angle of a blade should be optimized. (**a**) Blade attached to an attachment at an angle. (**b**) Blade has a taper that may considerably affect the slice condition. (**c**) Blade that has been set at too steep an angle pushes the tissue block, as shown by the arrow. (**d**) If the angle is too shallow, the blade scrapes the tissue and smashes the surface of the slice (*arrow*). (**e**) Proper angle of the blade provides some relief and minimizes stresses to the tissue.

The rotation of the blade axis also has to be considered when you install it in a holder (Fig. 10.2). Improper installation of a blade causes mechanical stress to the brain block, resulting in unhealthy slices.

10.2.6. Dissection

Dissection should be performed on ice to lower the rate of metabolic activity in the brain. The dissecting blade used to cut the brain should be inserted very gently. Mechanical twisting of the brain imposes stress on the neurons, damaging them. Try to minimize excessive movement of the brain. Spread a minimum amount of glue or tissue adhesive onto a prechilled stage. Excessive adhesive can soak into the tissue brain block, which prevents clean slicing. If this happens, wait for the glue to dry and remove it with forceps. Any extra glue that may be floating in the solution in the cutting chamber can be removed with a KimWipe.

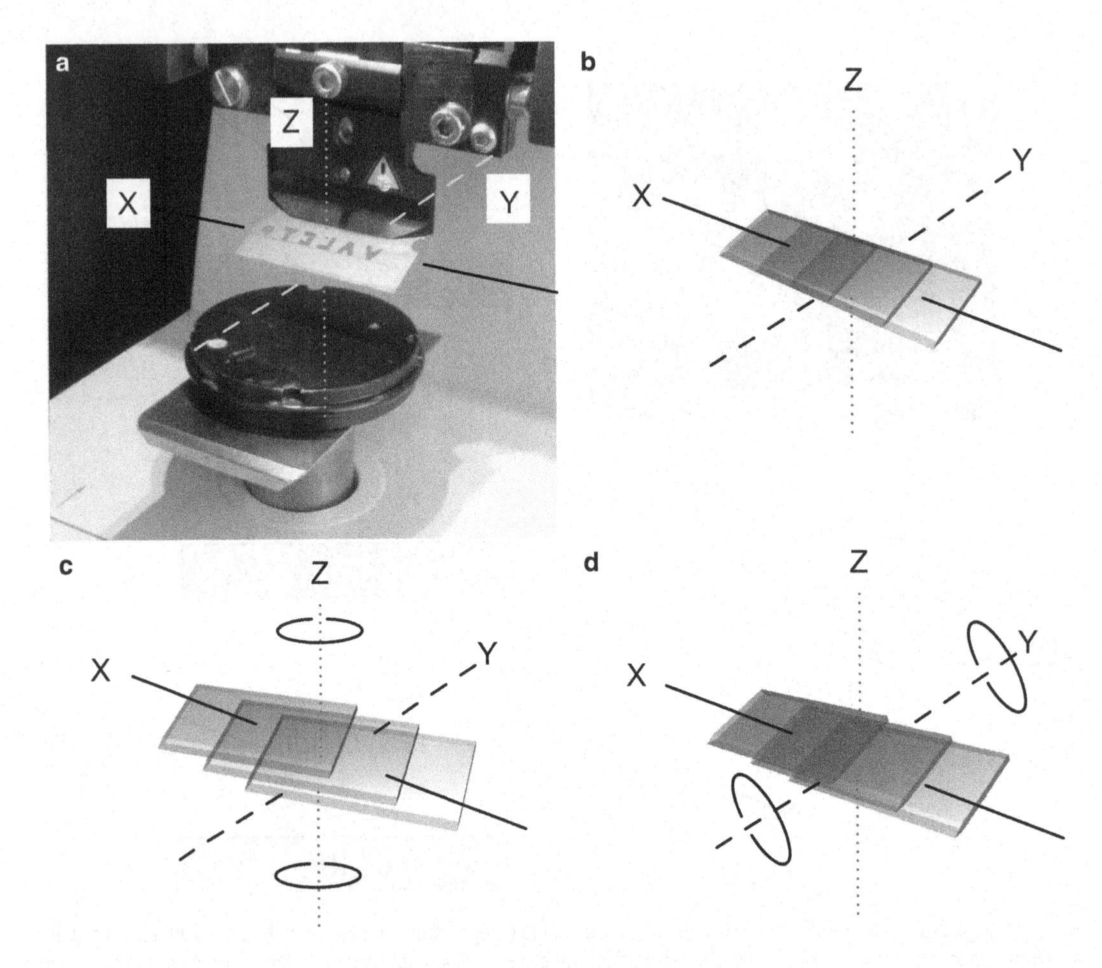

Fig. 10.2. (**a**) Blade installation and movement. Blade axes are illustrated in each panel. (**b**) Perfect alignment of the blade in every axis allows the blade to move only in the *X*-axis direction. (**c**) Poor installation in the *Z*-axis makes the blade insert into the brain block at an angle. (**d**) *Y*-axis skew leads to *Z*-distortion of the blade.

10.2.7. Slicing

The meninges should be removed from the surface of the brain before slicing. In particular, aged animals have tough fibers in/on the brain, which make slicing more challenging. This surface tissue can be removed with forceps, but strive to minimize contact with the brain. Pinch the fiber from the hemisphere to be discarded, and remove it cautiously one piece at a time. Magnifiers help during this procedure. In the hippocampus, there are stiff fibers running in the ventricle, between the cortex and hippocampus. Once the edge of the hippocampus is exposed, pause the slicer and remove the fibers before the blade is positioned at the region of interest. The fibers are difficult to cut, which results in the tissue dragging with the blade. Generally slices are made at ~400 μm thickness. It is said that thicker slices have less damage than thinner slices because thinner slices have less mechanical strength and tend to be distorted with blade vibration. In thicker slices, though, it is more difficult to see dendrites because of the lower light intensity and greater scattering.

10.2.8. Recovery

After the initial cutting, the slices are relocated to a recovery chamber. This solution contains warmed ACSF that is saturated with a carbon–oxygen mixture (95% O_2/5% CO_2). Right after slicing and because of damage from the blade, the neurons are particularly sensitive to environmental changes in temperature and osmolarity. Although 35–37°C is typical, in some cases lower temperatures (room temperature) may result in a better recovery (8). Direct transfer from a cutting solution to ACSF is possible, but a gradual exchange of the cutting solution with ACSF can also be used. Slices from the aged animal demand more oxygen, and in this case the use of a surface interface chamber can supply more oxygen to the tissue than a submersion chamber. After about an hour, the slices in the chamber are maintained at room temperature. It generally takes 1–2 h from the start of the procedure to be ready to start the experiments.

10.2.9. Looking at Dendrites

Quality optics are required to image thin dendrites. Differential interference contrast (DIC) coupled with infrared illumination (IR-DIC) generally works well and allows objects to be seen in high contrast. IR illumination is much better than visible light because it penetrates more deeply into the tissue with less scattering and lower absorption. IR filters are available from many manufacturers and can be installed between a condenser and the light source. An IR-specific camera may be needed to detect the IR image efficiently. An alternative is the Dodt system, although it may be more expensive than a regular IR-DIC system. The Dodt system does produce dendrite images with sharp contrast, which aids in successfully patching them. A much less expensive variant, sometimes used by the authors, is simply to use very strong oblique illumination. Here, the condenser polarizer of a typical DIC system is replaced with a crescent-shaped mask that blocks part of the illumination light.

To view thin dendrites, relatively high magnification is needed. We typically use a 40× objective with a 4× magnifier or a 60× objective with a 2× magnifier. High numerical aperture objective lenses (0.9–1.0) help collect as much light as possible and give high-resolution images. If the objective lens is equipped with a correction collar, it may be possible to sharpen the images with adjustment. Fluorescence images coupled with DIC have been used for dendritic patches of basal dendrites on pyramidal cells (9). Basal dendrites have a smaller diameter than apical dendrites, making them more difficult to observe. Even with quality optical systems it is difficult to obtain clear images. When the cells are filled with a fluorescent dye, it is much easier to trace the branches. Two-photon microscopes are usually necessary for this technique.

10.2.10. Patching Dendrites

10.2.10.1. Glass Pipette

Borosilicate glass capillaries are available from many manufacturers. Because the dendrite is much smaller than the soma, pipette drift can severely affect seal stability for dendrite patches. A slightly longer tapered pipette is used to avoid pushing. If you put large electrode into the slice, the tissue is initially pushed, but over time it creeps back toward the original position, which displaces the patched dendrite from the pipette. The pipette size for dendrite patching should be as small as 7–15 MΩ when measured with regular internal solution. The tip of the pipette is polished with a forge to make the seal smooth and help maintain the seal for a long period. The shape of the pipette also has an effect on the process of making the seal and breaking it. A good recording pipette is usually made by a 3- to 4-step pulling procedure. If the taper of the pipette is too curved, it takes time to achieve a seal, and it would be difficult to rupture the membrane. A straight taper seems to be better for making seals and breaking in. The pipettes used for making a patch on the basal dendrites are much smaller than one used for apical dendrites. After sealing and breaking into the cell, the access resistance is around 150 MΩ. Such high resistance distorts the electrical signal and in particular underestimates the membrane potential in the current-clamp mode (10). This fact must be considered when designing an experiment.

10.2.10.2. Manipulators

Sutter Instruments (Novato, CA, USA), Luigs & Neumann (Ratingen, Germany), and EXFO Burleigh (Quebec, Canada) all have manipulators with the degree of precision control necessary for patching dendrites. Piezo systems offer fine control, but the piezo movement is affected by temperature, and most have long travel, although some researchers make patches with these systems without a problem. In contrast, motorized systems have lower resolution but have long travel, can switch from large coarse motions to small fine motions with just one click, and have sufficient resolution for easy patching. Vibration has to be minimized for successful dendritic patching. A stiff post is needed to support the manipulators. Even with large, robust posts, the distance between the tip of the electrode and the axis of a post amplifies motion. Vibration and drift may come from cables, and they need to be properly restrained. Even small vibrations and drift are easily observable under the high magnifications used for dendritic patching.

10.2.10.3. Select a Healthy Dendrite

It is important to select a healthy dendrite to have success with dendrite patches. Transparent dendrites are usually a sign of good health and can be found inside the tissue, ~30 μm from the surface (Fig. 10.3). Dendrites that run on the surface of the slice are damaged by the cutting blade. Although these dendrites have high contrast and are easy to see, there is no guarantee that they are patchable. In our experience, the high contrast cells that appear clearly are almost always damaged. Once the pipette touches the

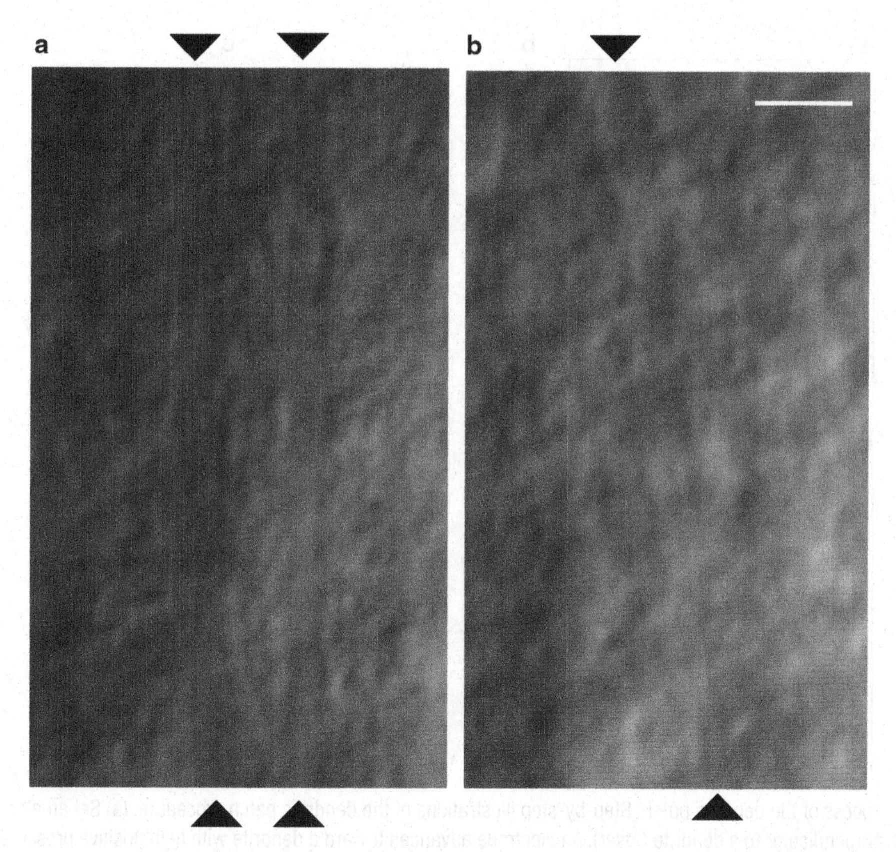

Fig. 10.3. Typical patchable dendrites. Images were captured using a 60× objective with a 2× magnifier. (**a**) Two dendrites of layer V pyramidal neurons in the mouse cortex, located 300 μm from the soma (*black arrowheads*). Despite having high contrast, they are patchable. More transparent dendrites should also be visible. (**b**) Distal dendrite at ~700 μm from the soma. *White bar* = 5 μm.

dendrite it can give a tactile impression that can be used to estimate dendritic health. Unhealthy dendrites have extra stiffness when a pipette touches the dendrites. If you "poke" a pipette into such a dendrite, a large portion of the dendrite moves with it. If this occurs, it is best to move to another dendrite. On the other hand, when a healthy dendrite is poked, it distorts only locally and seems to embrace the tip of the pipette. Such softness makes a tight seal, and the seal usually lasts because the dendrite sticks to and moves with the pipette, somewhat compensating for any displacement or drift.

10.2.10.4. Access to a Dendrite

The glass pipette is installed onto the head stage and moves down by action of the manipulator. The manipulator should be set at an angle such that the electrode can be directly inserted into the slice, straight toward the dendrite. During this process, positive pressure is applied to the glass pipette through a port on the pipette holder. A manometer or pressure gauge can be used to monitor the pressure. Too much pressure from the pipette can make patching impossible. Control the pressure carefully when the pipette is

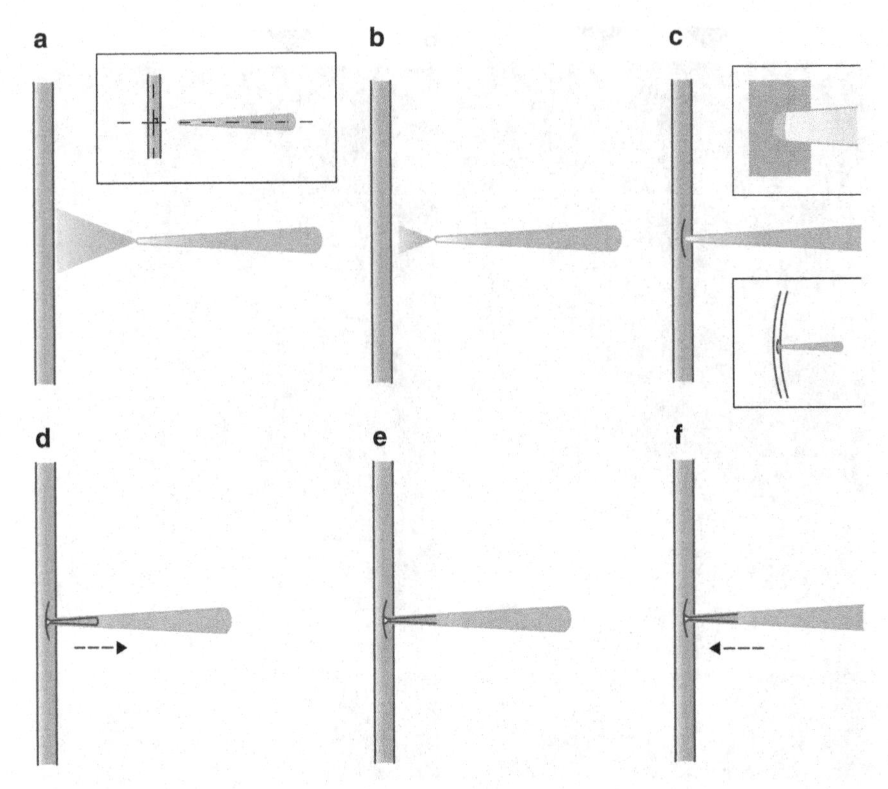

Fig. 10.4. Process of the dendrite patch. Step-by-step illustrations of the dendritic patch procedure. (**a**) Set an electrode so the axis is perpendicular to a dendrite (*inset*). An electrode advances toward a dendrite with high positive pressure, which cleans the neighborhood of dendrites. This process also makes a smooth path between the pipette and the dendrite. The disconnection of tissue can minimize drift caused by the pulling of adjacent tissues. (**b**) Lower the pressure when the pipette approaches the dendrite. High pressure makes the dendrite deflect, so the pipette cannot touch it. (**c**) When the electrode touches the dendrite, the dendritic membrane becomes distorted. If the membrane is healthy, it surrounds the tip of the pipette, and a dimple may be seen (*top inset*). An unhealthy dendrite is pushed by the pipette and does not form a dimple (*bottom inset*). (**d**) Gentle suction may be needed (*arrow*), and patience is needed until a giga-seal is formed. (**e**) Repetitive and gentle suction causes the membrane to break. (**f**) After initial entry, application of slight positive pressure (*arrow*) may be needed to clear membrane pieces that can slightly block the pipette, although it is not always necessary.

about to touch the dendrites, so it can clear tissue around the dendrite but not greatly disturb it (Fig. 10.4). Once the dendrite is touched by the pipette, there is a clear dimple/distortion made on the dendrite. Push the pipette a couple of microns deeper to make the contact tight and then release the positive pressure. If the cell is healthy enough, the giga-seal is made right after release of the pressure. Gentle sucking of the pipette can help make the seal. Sometimes it takes time to make the seal, so be patient. Compared to the soma, patching a dendrite needs more gentle suction to make a giga-seal. Strong suction can suck in the entire dendrite, including the opposite side of the membrane, breaking the seal. With good optics, the membrane that is drawn into the pipette can be visualized, and one can monitor the suction.

When the seal is greater than a giga-ohm, it is time to break in. Suck the pipette with short, gentle bursts and repeat it until the membrane breaks in.

10.2.10.5. Dual Patch to Soma and Dendrite

For a dual patch of the soma and dendrite, care should be taken during slicing to ensure that the soma and the dendrites are intact and preferably in the same place. Generally, at most one or two good slices are obtained from a hippocampus of one hemisphere. Manipulators should be positioned so both electrodes can have access to the soma and dendrite with opposite or perpendicular configurations. It is good to set both electrodes close to target sites, just above the slice. Do not set one by one because such a one-by-one procedure increases the risk of disturbing the first patch. Once set near the targets, choose either target for the first seal and break-in. Because they are usually far enough apart, making the seal and breaking in with the second pipette does not disturb the first one.

10.2.10.6. Blind Patch

Dendrites have many branches, and daughter dendrites are much smaller in diameter than the mother dendrites. It is difficult to see these thin dendrites even with IR-DIC optics. Under such circumstances, blind patching can be used to record from these small processes. By monitoring the resistance change reported from the test pulse while moving the pipette into the slice, it is possible to detect when the electrode encounters one of the daughter branches. If a fluorescent dye is added to the internal solution in the recording pipette, the patched neuron is visible after break-in. Blind patches can be made on basal and even distal apical dendrites with a high success rate in vivo. Researchers have even been able to make patches that hold in freely moving animals. Because the blind patch technique is applicable even to regions that are difficult to access with visual guidance, they are a strong tool for monitoring electrical signals from deep within the brain.

10.3. Advantages and Disadvantages of the Dendritic Patch

10.3.1. Local Events

Due to space-clamp issues, a patch on the soma is not able to clamp membrane potentials at distal dendrites accurately (2). The reverse is also true: The dendritic electrode is not able to clamp the soma, meaning that the electrodes accessing the soma and the dendrite have some degree of electrical isolation. Some researchers have taken advantage of this localization and succeeded in monitoring soma and dendrite events independently (11). A patch made on the dendrite clamps the voltage locally, and this restriction allows investigation of the local region of the dendrite without contaminating effects from the soma. Rather than clamping the entire cell, dendritic patches monitor and control electrical events in the local region.

10.3.2. Local Dialysis

Compared to the soma, the dendrite has a small volume. One concern that emerges is the strong dialysis that tends to occur between the pipette and the dendrite. The diffusion into the rest of the cell may take a long time because the thin dendrite offers resistance to quick dialysis. This can be used to the researcher's benefit – for example, by local, intracellular application of a drug. Because of the slow diffusion, one could look at the local effect of a drug before the drug reaches the entire cell.

10.3.3. Morphological Distortion

Once the patch is made on the dendrite, it is stable and even tolerates some drift. However, a stable patch or a stable electrical signal does not mean that morphological changes are not occurring. Thus, it is necessary to watch the morphology of the dendrite and any attached spines carefully.

References

1. Magee JC (2007) Voltage-gated ion channels in dendrites. In: Stuart G, Spruston N, Häusser M (eds) Dendrites. Oxford University Press, Oxford, pp 225–250
2. Williams SR, Mitchell SJ (2008) Direct measurement of somatic voltage clamp errors in central neurons. Nat Neurosci 11(7): 790–798
3. Golding NL, Spruston N (1998) Dendritic sodium spikes are variable triggers of axonal action potentials in hippocampal CA1 pyramidal neurons. Neuron 21(5):1189–1200
4. Takahashi H, Magee JC (2009) Pathway interactions and synaptic plasticity in the dendritic tuft regions of CA1 pyramidal neurons. Neuron 62(1):102–111
5. Tsubokawa H, Ross WN (1997) Muscarinic modulation of spike backpropagation in the apical dendrites of hippocampal CA1 pyramidal neurons. J Neurosci 17(15):5782–5791
6. Davie JT et al (2006) Dendritic patch-clamp recording. Nat Protoc 1(3):1235–1247
7. Poolos NP, Jones TD (2004) Patch-clamp recording from neuronal dendrites. Curr Protoc Neurosci, Chapter 6: Unit 6.19
8. Moyer JR Jr, Brown TH (1998) Methods for whole-cell recording from visually preselected neurons of perirhinal cortex in brain slices from young and aging rats. J Neurosci Methods 86(1):35–54
9. Larkum ME et al (2009) Synaptic integration in tuft dendrites of layer 5 pyramidal neurons: a new unifying principle. Science 325(5941):756–760
10. Zhou WL et al (2008) Dynamics of action potential backpropagation in basal dendrites of prefrontal cortical pyramidal neurons. Eur J Neurosci 27(4):923–936
11. Pouille F, Scanziani M (2004) Routing of spike series by dynamic circuits in the hippocampus. Nature 429(6993):717–723

Chapter 11

In Vivo Blind Patch-Clamp Recording Technique

Hidemasa Furue

Abstract

The methods for in vivo blind patch-clamp recording from the spinal cord, brain stem, and cortex neurons are described herein, including the (1) technique for in vivo preparations, (2) recordings of action potential and synaptic potentials under current-clamp conditions and excitatory and inhibitory synaptic currents evoked by natural physiological stimulation under voltage-clamp conditions, and (3) identification of recorded neurons and pharmacological characterization of synaptic currents. The signal-to-noise ratio and stability are good using this in vivo recording approach, and it is comparable to that found in vitro. The blind patch technique is a low cost method as it does not require expensive microscopes to visualize the recorded neurons. Furthermore, the relatively wide working space over the recording sites facilitates recordings from deep layers of the brain. Therefore, this blind in vivo technique is suitable for making neural recordings from any region in a wide range of animals from rodents to larger animals.

11.1. Introduction

The use of in vivo preparations is essential to permit recordings under physiological conditions, and the data obtained enable us to understand the functional significance of neuronal activity. In particular, the in vivo neuronal responses to natural physiological stimulation allow us to identify the key neuronal mechanisms that regulate a broad range of functions of the tissues and organs. Classically, action potentials elicited in neurons in vivo have been recorded predominantly by extracellular and intracellular recording techniques. In vivo intracellular recording techniques, with sharp electrodes, enable us to monitor both neuronal discharge and subthreshold synaptic membrane potentials. However, it is difficult to obtain stable recordings because cardiac and respiratory pulsations of the recording area in vivo can cause significant damage to the impaled neurons. Therefore, stable recordings of neuronal membrane potentials in vivo have tended to be limited to large neurons.

Yasunobu Okada (ed.), *Patch Clamp Techniques: From Beginning to Advanced Protocols*, Springer Protocols Handbooks, DOI 10.1007/978-4-431-53993-3_11, © Springer 2012

Furthermore, the high resistance of sharp electrodes makes it difficult to isolate synaptic ionic currents under voltage-clamp conditions (because of high series resistance and capacitative coupling), and so the ability to perform detailed quantitative analyses of ion currents and their kinetics (particularly those associated with synaptic transmission) is restricted.

In an attempt to solve these problems, the whole-cell patch-clamp recording technique has been adapted to allow in vivo recordings. Creutzfeldt's group reported the first recordings; they successfully recorded membrane potentials of cat visual cortex neurons using patch electrodes under whole-cell current-clamp conditions (1). Subsequently, in vivo whole-cell voltage recordings have been performed in various areas in mammals including the spinal cord, thalamus, cerebellum, olfactory bulb, visual cortex, barrel cortex, and motor cortex (see (2) for review). Furthermore, successful in vivo recordings of excitatory (EPSCs) and inhibitory (IPSCs) postsynaptic currents from the spinal cord (3, 4), cerebellum (5, 6), and auditory cortex (7) have been examined under voltage-clamp conditions. The in vivo whole-cell recording technique has thus become an extremely powerful tool for studying synaptic details of natural physiological responses.

The in vivo patch-clamp recording techniques described in this chapter relies on the use of the "blind" technique, whereby cells are sought and gigaseals obtained by monitoring the change in pipette resistance as the pipette is advanced through the neuropil. This technique allows us to obtain stable recordings of excitatory and inhibitory synaptic currents evoked in single neurons in vivo under voltage-clamp conditions and action potentials and synaptic potentials under current-clamp conditions. The signal-to-noise ratio is excellent and it is comparable to that recorded during in vitro slice patch experiments. The blind patch technique is a low cost method as there is no need for expensive microscopes (e.g., infrared-differential interference contrast (IR-DIC) or two-photon excitation microscopes) to visualize neurons in vivo. Because there is no need for a water immersion objective, there is a relatively wide working space above the recording sites available for in vivo blind recordings, which allows a greater degree of freedom. There is also no optical limitation placed on the depth of recordings, making it possible to obtain cells from deep layers of the brain or tissues. Therefore, the blind technique is suitable for recording from any deep neuronal nuclei not only in rodents but in larger animals.

This chapter addresses methods for in vivo whole-cell recording from spinal dorsal horn, brain stem, and cortex in the rodent. It shows synaptic responses evoked by natural physiological sensory stimulation and isolated synaptic currents at different holding potentials (Fig. 11.1) and illustrates how these responses can be blocked by drugs applied from the surface of the recording area. Thus, this method is also useful for elucidating changes in synaptic mechanisms that underlie the phenotypic changes seen in transgenic

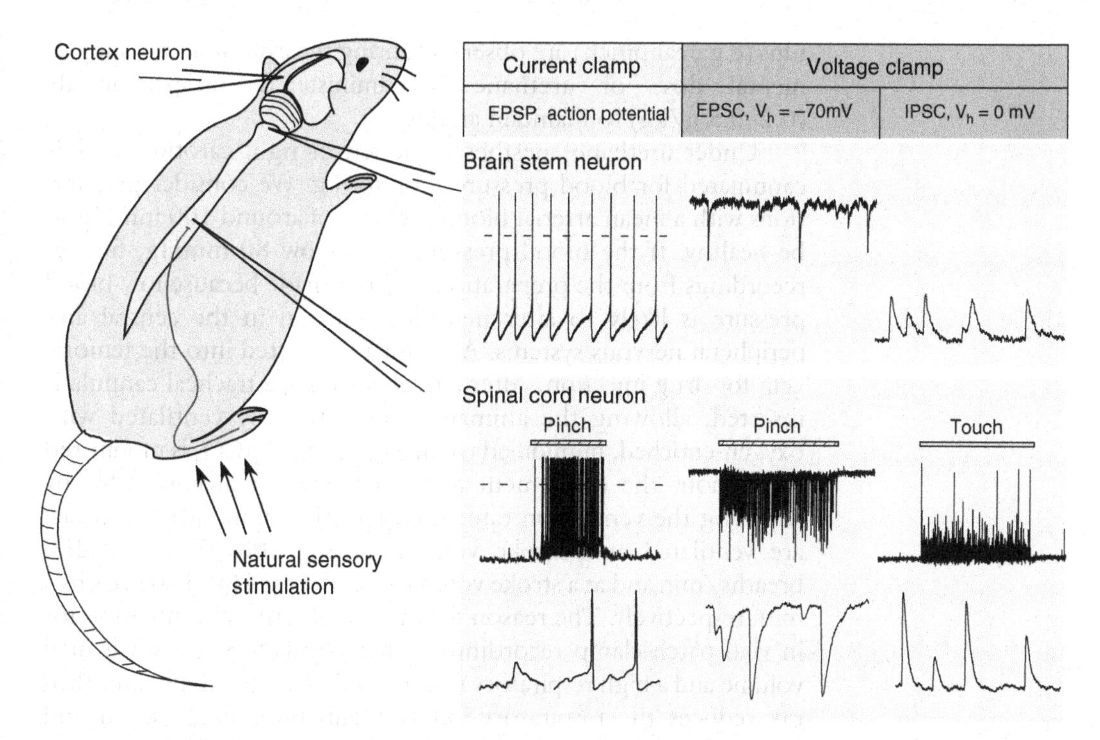

Fig. 11.1. In vivo blind patch-clamp recordings from the spinal cord, brain stem, and cortex. Under current-clamp conditions, spontaneous action potentials are elicited in brain stem (locus coeruleus) neurons, and excitatory postsynaptic potentials (*EPSP*) and action potentials are evoked in spinal dorsal horn (lamina II of the spinal cord, substantia gelatinosa) neurons by pinch stimulation applied to the skin of the hindlimb. Under voltage-clamp conditions, excitatory (*EPSC*) and inhibitory (*IPSC*) postsynaptic currents are recorded at a holding potential of −70 and 0 mV, respectively. Note the spontaneous EPSCs and IPSCs evoked in locus coeruleus neurons and EPSCs and IPSCs elicited in substantia gelatinosa neurons by cutaneous sensory stimulation applied to the hindlimb. Scale bars are not indicated in each trace. The in vivo blind technique enables quantitative analysis of synaptic responses, such as the amplitude and frequency of the evoked responses and their kinetics.

animals and pathological models. Detailed methods for these in vivo patch-clamp recording techniques, especially stimulation protocols for natural sensory stimulation and quantitative analysis of the evoked synaptic responses, have been described elsewhere (2).

11.2. In Vivo Preparations

11.2.1. Anesthesia and Artificial Ventilation

The methods for obtaining healthy in vivo preparations have been described elsewhere (2). Briefly, urethane is used as the general anesthetic for in vivo animal experiments because the anesthetic effect lasts longer than that of barbiturates and ketamine. Following a single intraperitoneal injection, urethane induces anesthesia within several minutes and typically lasts more than 8 h. In our experiments, rats (6–15 weeks old) and mice (6–12 weeks old) are deeply anesthetized with urethane injected intraperitoneally at 1.2–1.5 g/kg. If withdrawal reflexes in response to a noxious stimulus (e.g., ear pinch) are observed during the experiment, a supplemental dose of urethane is administered intraperitoneally (0.2–0.5 g/kg) to maintain areflexia.

Under urethane anesthesia, the left or right carotid artery is cannulated for blood pressure monitoring. We consider preparations with a mean arterial blood pressure of around 100 mmHg to be healthy. If the blood pressure falls below 80 mmHg, further recordings from the preparations are not made because low blood pressure is likely to alter neuronal function in the central and peripheral nervous systems. A cannula is inserted into the femoral vein for drug injection. After a tracheostomy, a tracheal cannula is inserted, allowing the animals to be artificially ventilated with oxygen-enriched, humidified room air. End-tidal pCO_2 is monitored throughout the experiment and maintained at around 3.8% by adjusting the ventilation rate and/or tidal volume. Mice and rats are ventilated at a stroke volume of 200–400 μl at 150–250 breaths/min and at a stroke volume of 2–3 ml at 100–150 breaths/min, respectively. The reason for the use of artificial ventilation for in vivo patch-clamp recordings is that ventilation at a small tidal volume and a high respiratory rate in combination with pneumothorax reduces the respiratory-induced pulsation of the brain and especially of the spinal cord and brain stem. Rectal temperature is monitored and maintained constant at 37–38°C by means of a circulating hot water blanket underneath the abdomen. All vital signs should be kept within the physiological range for optimal recording conditions.

11.2.2. Exposure of the Dorsal Surface of the Spinal Cord

The lumbar spinal cord is exposed by laminectomy because the primary afferent terminals from the neurons innervating the hindlimb make contact with spinal dorsal horn neurons in this region, allowing neuronal responses to be elicited by sensory stimulation applied to the ipsilateral hindlimb. After a posterior midline skin incision is performed over the thoracolumbar spine, local anesthetic (0.5% lidocaine with 1:100,000 epinephrine) is injected intramuscularly into both sides of the spine to decrease bleeding during surgery. A laminectomy (Th12-L2) is performed with fine bone scissors (no. 14077; Fine Science Tools, Heidelberg, Germany) to expose the dorsal surface of the lumbar enlargement of the spinal cord at the segmental level of L3-5. Micro-drills can also be used for this laminectomy. Following laminectomy, the dura matter is opened along the midline. The animal is then placed in a stereotaxic apparatus (model STS; Narishige, Tokyo, Japan), and the lumbar spine is fixed by clamp arms (Fig. 11.2a).

11.2.3. Exposure of the Dorsal Surface of the Brain Stem

Under anesthesia with urethane, a tracheal cannula is inserted to allow artificial ventilation. After bilateral thoracostomy, the animal is placed in a stereotaxic head frame (model SG or SR; Narishige) (Fig. 11.2b). Following exposure of the skull after the midline skin

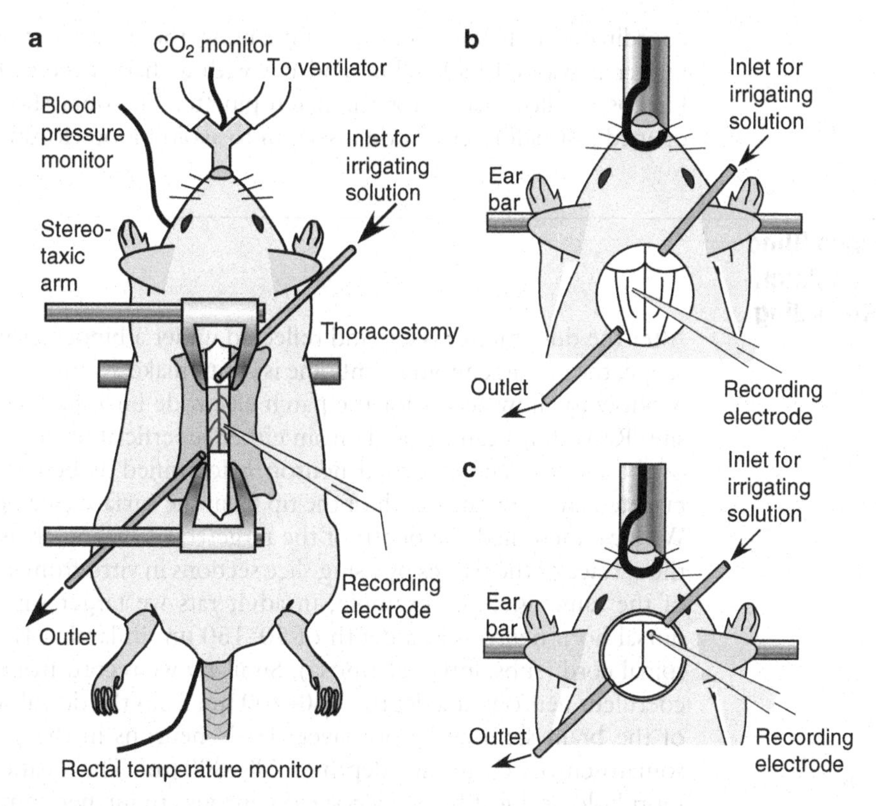

Fig. 11.2. In vivo preparations fixed in the stereotaxic apparatus. (**a**) Thoracolumbar vertebrae of the in vivo preparation are fixed by two clamps after laminectomy. The animal is intubated and artificially ventilated. End-tidal CO_2 partial pressure is maintained at around 3.8%. Bilateral pneumothorax is established to reduce spinal cord movement by respiration. Blood pressure is measured through a cannulated carotid artery. Rectal temperature is kept at 37–38°C by a circulating hot water blanket underneath the animal's abdomen. (**b**) After occipital craniotomy, the cerebellum is removed to expose the dorsal surface of the brain stem. (**c**) Craniotomy is performed to make a small hole above the somatosensory cortex based on the stereotaxic coordinates to allow insertion of a patch pipette into the cortex.

incision, we performed an occipital craniotomy area using a micro-drill. Under a binocular microscope, the mid-part of the cerebellum is removed by suction aspiration to expose the dorsal surface of the brain stem. Hemostasis from the cerebellum is achieved using small collagen sponges.

11.2.4. Exposure of Somatosensory Cortex

An anesthetized animal (rat) is fixed in a stereotaxic head frame as shown in Fig. 11.2c. Following a midline incision into the scalp, the sutures are identified, and the incisor bar is adjusted until the height of lambda and bregma are equal. After clearing the connective tissue on the skull, a plastic cylinder (1.2 cm diameter, 0.5 mm height) is mounted onto the center of the dorsal surface of the cranium and glued on using epoxy resin or dental cement to serve as a bath. Under visual guidance through a binocular microscope, a craniotomy is carefully performed using a dental drill. A hole is opened above the right or left somatosensory cortex based on the stereotaxic coordinates. Following removal of the bone, the surface of the dura matter is washed with saline and cut with a sharp needle or fine forceps to allow access for the patch pipette. The dura should be opened just before electrode insertion to avoid blood clotting.

11.3. In Vivo Blind Whole-Cell Patch-Clamp Recording

After the dura matter is cut and reflected under a binocular microscope, the pia–arachnoid membrane is cut to make a 1 mm diameter window to allow access for the patch electrode into the recording site. Recordings can be made from either superficial or deep layers of the tissues. The recorded neuron is identified as being in the targeted layers by the depth of the tip from the surface (see below). We have identified the depth of the targeted recording areas from the surface of the tissues by using slice sections in vitro from animals of the same age. For example, in adult rats we target superficial dorsal horn neurons at a depth of 50–150 μm in lamina II of the spinal cord (substantia gelatinosa). Similarly we record from locus coeruleus neurons at a depth of 50–200 μm from the dorsal surface of the brain stem and from layers IV–V neurons in the primary somatosensory cortex at a depth of 700–800 μm. The location and morphological features of recorded neurons are further confirmed by intrasomatic injection of biocytin or neurobiotin (Fig. 11.3). After finishing the experiments, the tissues are fixed with transcardial perfusion of formalin and then cut into slice sections. Morphological features of recorded single neurons, their axon terminations, and expressing neuronal markers can be identified with further immunohistochemical staining.

Electrodes are pulled from thin-walled borosilicate glass capillaries with resistances of about 10 MΩ. Potassium- and cesium-based patch-pipette solutions are used. The solutions have following compositions (mM): K gluconate 135, KCl 5, $CaCl_2$ 0.5, $MgCl_2$ 2, EGTA 5, ATP-Mg 5, Hepes-KOH 5 (pH 7.2); and Cs_2SO_4 110, TEA-Cl 5, $CaCl_2$ 0.5, $MgCl_2$ 2, EGTA 5, ATP-Mg 5, HEPES-CsOH 5 (pH 7.2). The potassium gluconate-based solution is used predominantly for current-clamp recordings. The cesium-based solution is used for voltage-clamp recordings, especially for recording IPSCs, because cesium and TEA stabilize the recordings of IPSCs at a holding potential of 0 mV due to inhibition of potassium channels. The surface of the brain–spinal cord is continuously irrigated with 95% O_2–5% CO_2-equilibrated Krebs solution with the following composition (in mM): NaCl 117, KCl 3.6, $CaCl_2$ 2.5, $MgCl_2$ 1.2, NaH_2PO_4 1.2, glucose 11, $NaHCO_3$ 25.

Using a fine, stable micromanipulator, the tip of a patch pipette is put into the irrigating Krebs solution. Electrode resistance is monitored continuously by applying a square voltage pulse (5 mV, 10 ms)

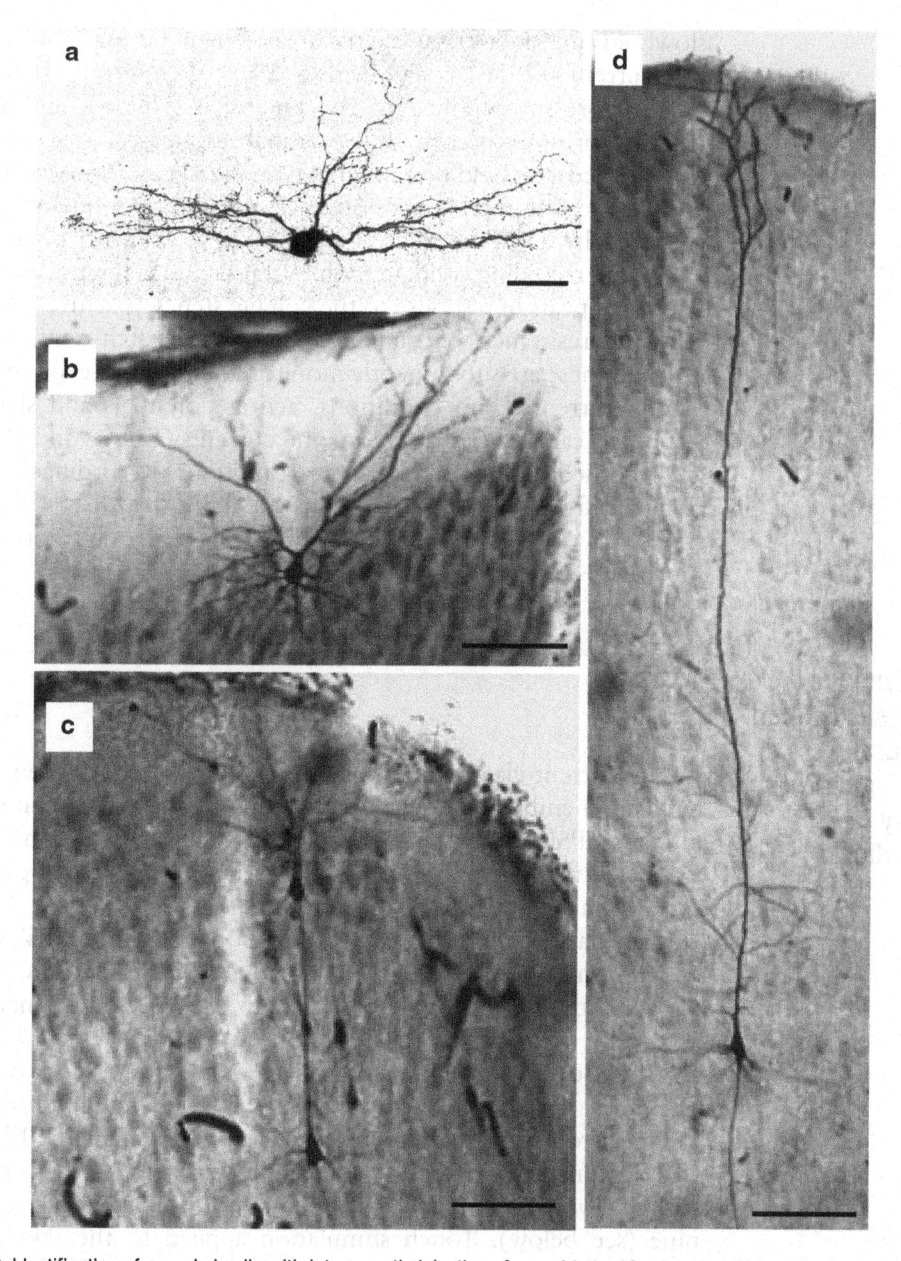

Fig. 11.3. Identification of recorded cells with intrasomatic injection of neurobiotin. After in vivo blind patch-clamp recordings, the preparations are fixed and the sections stained to reveal the neurobiotin. (**a**) Spinal dorsal horn neurons in laminae II–III. (**b–d**) Neurons recorded from the somatosensory cortex. Morphological features of recorded neurons can be analyzed after electrophysiological characterization in vivo. (Modified from Doi et al. (11), with permission)

to the electrode under voltage-clamp conditions at a holding potential of 0 mV. After applying a positive pressure to the patch pipette, it is advanced into the targeted layers or areas from the surface of the tissues. When the tip of the electrode contacts the tissue surface, a deflection is seen in the current baseline (useful for stereotaxic depth estimation). From this point, the electrode is slowly advanced into the targeted areas. When the amplitude of the step current response is slightly decreased (often by only 10%), the positive pressure applied to the pipette is released; and slight negative pressure is then added to make a gigaseal of 5–20 GΩ. Once sealed, the holding potential is changed to −70 mV, which is near the resting membrane potential; and the patch membrane is ruptured by applying additional negative pressure to obtain the whole-cell recording configuration. A transient membrane capacitive current appears in response to the voltage step. Capacitive currents and series resistance are compensated for with a patch-clamp amplifier; and these are subsequently monitored during the experiments. Under current-clamp conditions, action potentials and synaptic potentials can be recorded. Synaptic currents (EPSCs and IPSCs) are recorded under voltage-clamp conditions. Recordings can be obtained from in vivo preparations for more than 8 h. During a single day of recordings, 2–10 neurons may be thoroughly investigated.

11.4. Synaptic Responses Evoked by Natural Physiological Sensory Stimulation

Under current-clamp conditions, spinal dorsal horn neurons have a resting membrane potential of around−70 mV and exhibit spontaneous EPSPs. When pinch stimulation is applied to the skin of the ipsilateral hindlimb, a barrage of EPSPs are evoked, resulting in the generation of repetitive overshooting action potentials. Under voltage-clamp conditions at a holding potential of −70 mV, spinal dorsal horn neurons exhibit spontaneous EPSCs that are sensitive to the non-*N*-methyl-D-aspartic acid (NMDA) glutamatergic receptor antagonist 6-cyano-7-nitroquinoxaline-2,3-dione (CNQX, see below). Cutaneous pinch stimulation elicits a barrage of EPSCs. Spontaneous IPSCs are observed at a holding potential of 0 mV (to remove the driving force for cation-mediated EPSCs) that are sensitive to either a γ-aminobutyric acid-A ($GABA_A$) receptor antagonist bicuculline or a glycine receptor antagonist strychnine (see below). Touch stimulation applied to the skin elicits barrages of IPSCs. The frequency and amplitude of the spontaneous and evoked synaptic currents can be analyzed quantitatively (3, 4, 8, 9). Similarly, action potentials and EPSPs can be recorded from locus coeruleus (10) and somatosensory cortex (11, 12) neurons in vivo under current-clamp conditions, and EPSCs and IPSCs can be recorded under voltage-clamp conditions (see example traces in Fig. 11.1).

11.5. Drug Application In Vivo

Drugs can be administered by an intravenous, intraperitoneal, intramuscular, intradermal, or subcutaneous route (13). The in vivo blind patch-clamp technique also enables us to examine the actions of drugs applied systemically by intravenous injection on central synaptic currents (14). In addition, drugs can be locally applied to the recorded neurons from the surface of the tissues (2, 3, 8) in irrigating Krebs' solution in a way similar to that in the in vitro brain slices. When tetrodotoxin (TTX) is applied to the surface of the spinal cord, action potentials elicited by a current injection through the recording pipette are completely suppressed within tens of seconds, as shown in Fig. 11.4a. Similarly, CNQX applied to superficial dorsal horn neurons completely inhibits spontaneous EPSCs within 30 s, although the inhibitory action is washed out within several minutes (Fig. 11.4b). The CNQX action is dose-dependent. A half-maximum inhibitory concentration of CNQX applied to the surface of the spinal cord was 3 μM, which is consistent with that obtained in slice experiments (2). These data suggest that drug applied to the surface of the spinal cord rapidly diffuses into the neuropil to reach ion channels and receptors. These findings also indicate that the effect is due to a direct action in the spinal cord and is not achieved via systemic recirculation.

Pinch-evoked EPSCs are also reversibly inhibited by CNQX and TTX applied to the surface of the spinal cord (Fig. 11.4c). Superficial dorsal horn neurons receive monosynaptic inputs from primary afferent fibers (15). It is suggested that TTX acts on the dorsal root and axons of the afferent fibers that make synaptic contact with recorded neurons. Drug application from the surface of the tissues enables us to isolate synaptic responses (16). Application of bicuculline, strychnine, and CNQX completely inhibited spontaneous EPSCs at a holding potential of −70 mV and spontaneous IPSCs at a holding potential of 0 mV (Fig. 11.5a,b). At a holding potential of +40 mV in the presence of this antagonist cocktail, substantia gelatinosa (lamina II of the spinal cord) neurons exhibited spontaneous EPSCs with slower kinetics than those recorded at −70 mV in the absence of receptor antagonists. The supplemental addition of an NMDA receptor antagonist, D-APV, completely blocked these spontaneous slow EPSCs (Fig. 11.5c), suggesting that they are mediated through NMDA receptors.

Thus, the surface application of drugs can isolate individual synaptic currents in neurons of in vivo preparations. The kinetics of the onset and offset of these drug actions is dependent on the depth of the recorded neuron. With neurons that were located at

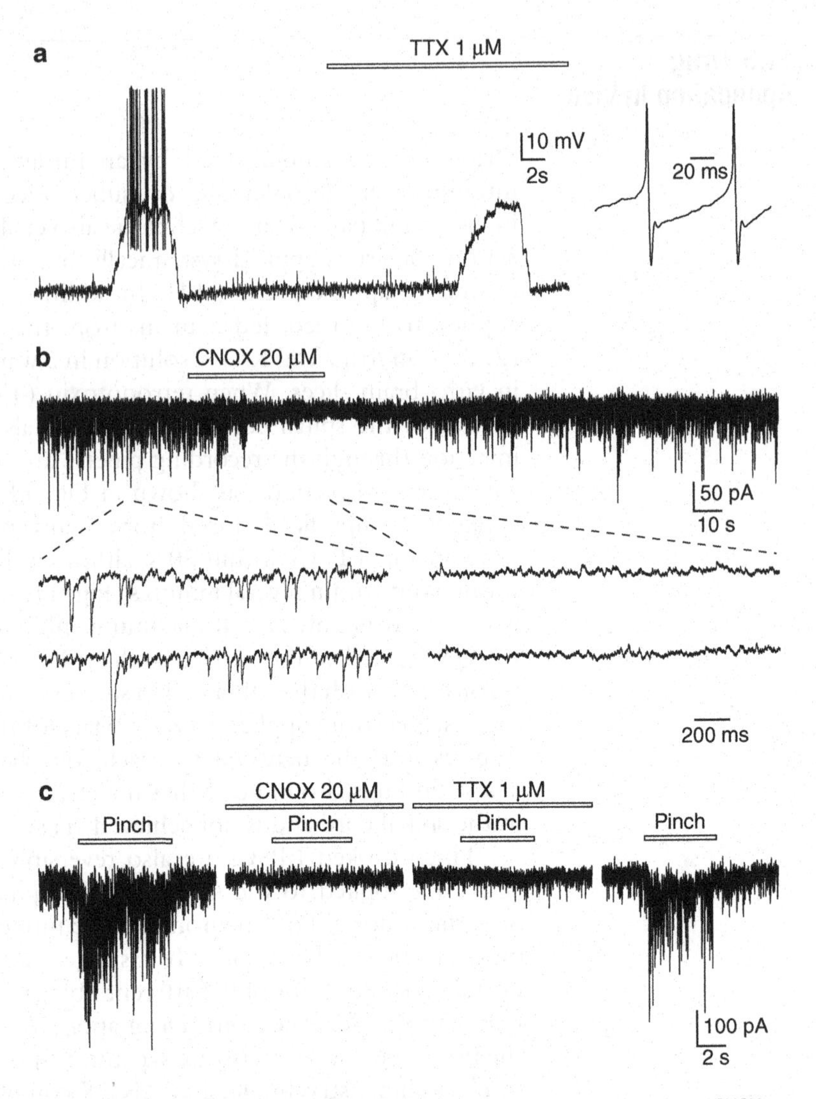

Fig. 11.4. Effects of tetrodotoxin (*TTX*) and 6-cyano-7-nitroquinoxaline-2,3-dione (*CNQX*) on action potentials and EPSCs evoked in spinal dorsal horn neurons. The drugs are applied via the surface of the spinal cord. (**a**) Action potentials elicited by current injection through the recording pipette and inhibition of the firing by TTX (*right trace*). *Left trace* shows action potentials shown in the *right trace* over an expanded time course. (**b**) Effect of CNQX on spontaneous EPSCs. *Lower* two traces show consecutive spontaneous EPSCs on an expanded time base. (**c**) CNQX and TTX blocked the EPSCs evoked by cutaneous pinch stimulation.

a depth of 250 μm from the surface, the CNQX (20 μM) took a longer time to inhibit the EPSCs (approximately twice as long as in neurons at a depth of 125 μm) (2). For neurons located at a depth of >500 μm, however, it is difficult to apply drugs topically because higher concentrations of drugs with a longer action onset time (>30 min) are needed before the drugs reach the neurons.

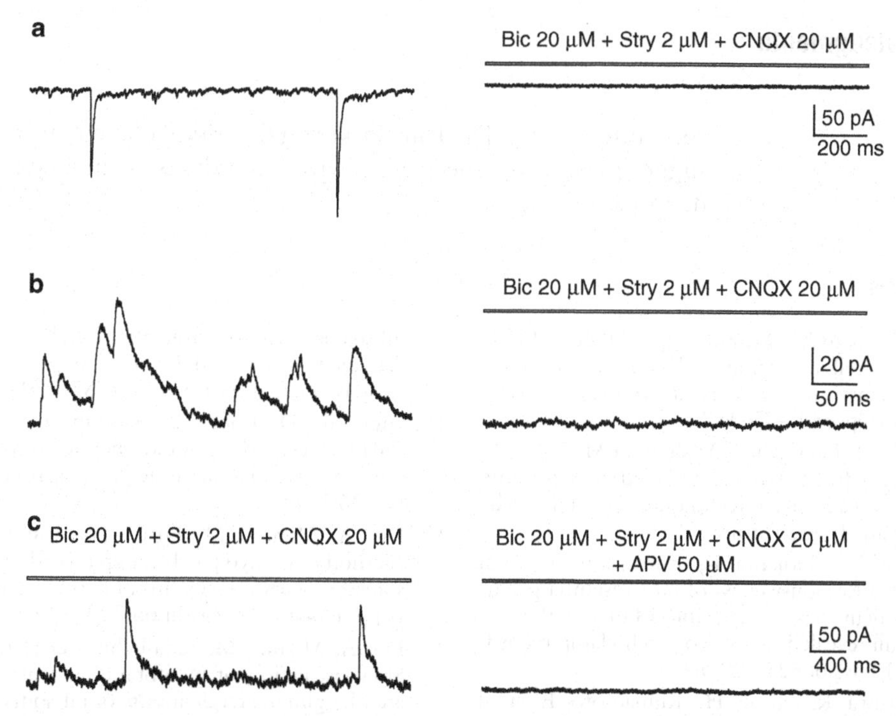

Fig. 11.5. Isolation of synaptic currents evoked in spinal dorsal horn neurons in vivo by receptor antagonists applied via the surface of the spinal cord. (**a**) Spontaneous EPSCs mediated through non-*N*-methyl-D-aspartic acid (NMDA) glutamate receptors at a holding potential of −70 mV (*left trace*). Application of bicuculline (*Bic*), strychnine (*Stry*), and CNQX completely inhibited spontaneous EPSCs (*right trace*). (**b**) Spontaneous γ-aminobutyric acid (GABA)ergic and glycinergic IPSCs recorded at a holding potential of 0 mV (*left trace*). Application of bicuculline, strychnine, and CNQX completely blocked the spontaneous IPSCs (*right trace*). (**c**) NMDA receptor-mediated EPSCs. In the presence of these three antagonists, substantia gelatinosa neurons exhibited spontaneous EPSCs with slow kinetics at a holding potential of +40 mV (*left trace*). Supplemental addition of an NMDA receptor antagonist, D-APV (*APV*), completely suppressed the EPSCs with slower kinetics. (Modified from Katano et al. (16), with permission).

11.6. Conclusion

The in vivo blind patch recording technique has become a powerful tool, not only for elucidating the functional significance of synaptic activities under natural physiological conditions but also for identifying abnormalities in synaptic transmission in pathological conditions or in genetically modified animals. This technique allows us to record from neurons located deep in the brain or tissues with a relatively wide working space over the recording sites, which allows the method to be applied to a wide range of animals, especially larger animals such as primates.

Acknowledgments

The author thanks Dr. Tony Pickering for critical comments, helpful suggestions, and especially in vivo recordings from brain stem neurons.

References

1. Pei X, Volgushev M, Vidyasagar TR et al (1991) Whole cell recording and conductance measurements in cat visual corztex in-vivo. Neuroreport 2:485–488
2. Furue H, Katafuchi T, Yoshimura M (2007) In vivo patch. In: Wolfgang W (ed) Patch-clamp analysis: advanced technique, 2nd edn. The Humana Press, Totowa
3. Furue H, Narikawa K, Kumamoto E et al (1999) Responsiveness of rat substantia gelatinosa neurones to mechanical but not thermal stimuli revealed by in vivo patch-clamp recording. J Physiol 521:529–535
4. Narikawa K, Furue H, Kumamoto E et al (2000) In vivo patch-clamp analysis of IPSCs evoked in rat substantia gelatinosa neurons by cutaneous mechanical stimulation. J Neurophysiol 84:2171–2174
5. Loewenstein Y, Mahon S, Chadderton P et al (2005) Bistability of cerebellar Purkinje cells modulated by sensory stimulation. Nat Neurosci 8:202–211
6. Chadderton P, Margrie TW, Hausser M (2004) Integration of quanta in cerebellar granule cells during sensory processing. Nature 428:856–860
7. Wehr M, Zador AM (2003) Balanced inhibition underlies tuning and sharpens spike timing in auditory cortex. Nature 426:442–446
8. Sonohata M, Furue H, Katafuchi T et al (2004) Actions of noradrenaline on substantia gelatinosa neurones in the rat spinal cord revealed by in vivo patch recording. J Physiol 555:515–526
9. Kato G, Yasaka T, Katafuchi T et al (2006) Direct GABAergic and glycinergic inhibition of the substantia gelatinosa from the rostral ventromedial medulla revealed by in vivo patch-clamp analysis in rats. J Neurosci 26:1787–1794
10. Sugiyama D, Imoto K, Kawamata M et al (2011) Descending noradrenergic controls of spinal nociceptive synaptic transmission. Pain Res 26:1–9
11. Doi A, Mizuno M, Furue H et al (2003) Method of in-vivo patch clamp recording from somatosensory cortex in rats (in Japanese). Nippon Seirigaku Zasshi 65:322–329
12. Doi A, Mizuno M, Katafuchi T et al (2007) Slow oscillation of membrane currents mediated by glutamatergic inputs of rat somatosensory cortical neurons: in vivo patch-clamp analysis. Eur J Neurosci 26:2565–2575
13. Waynforth HB, Flecknell PA (1992) Administration of substances. In: Experimental and surgical technique in the rat, 2nd edn. Academic, San Diego
14. Takazawa T, Furue H, Nishikawa K et al (2009) Actions of propofol on substantia gelatinosa neurones in rat spinal cord revealed by in vitro and in vivo patch-clamp recordings. Eur J Neurosci 29:518–528
15. Yoshimura M, Nishi S (1993) Blind patch-clamp recordings from substantia gelatinosa neurons in adult rat spinal cord slices: pharmacological properties of synaptic currents. Neuroscience 53:519–526
16. Katano T, Furue H, Okuda-Ashitaka E et al (2008) N-ethylmaleimide-sensitive fusion protein (NSF) is involved in central sensitization in the spinal cord through GluR2 subunit composition switch after inflammation. Eur J Neurosci 27:3161–3170

Chapter 12

Two-Photon Targeted Patch-Clamp Recordings In Vivo

Kazuo Kitamura

Abstract

The advent of two-photon microscopy has enabled us to visualize individual neurons in the intact brain. This technique, used in combination with whole-cell patch-clamp recordings, has facilitated targeted intracellular recording from particular neurons of interest. This chapter provides a practical guide for implementing in vivo two-photon targeted patch-clamp recording and describes potential outcomes using the technique.

12.1. Introduction

Understanding how neurons integrate the thousands of synaptic inputs into output action potentials and how the activities of single neurons contribute to the information processing in the intact brain is one of the most important questions in neuroscience. Addressing this issue requires high-resolution, reliable recordings of synaptic inputs and spike outputs in identified single neurons. Whole-cell patch-clamp recordings allow us to record synaptic and spike activity of neurons in vivo (1, 2) and thus have provided important insights into synaptic integration and interplay between a single neuron and network activity in various brain regions (3–5). As described in the previous chapter, the classic "blind" in vivo patch-clamp technique is simple and can be applied to deep brain tissues. In principle, however, it is impossible to target specific types of neuron and a particular site of cells with the blind method, and in many cases the success rate and quality of recordings are not as good as those of slice patch-clamp recordings.

To overcome these limitations, a targeted patch-clamp technique has been developed by combining in vivo two-photon microscopy with whole-cell patch-clamp recordings (two-photon targeted

Yasunobu Okada (ed.), *Patch Clamp Techniques: From Beginning to Advanced Protocols*, Springer Protocols Handbooks, DOI 10.1007/978-4-431-53993-3_12, © Springer 2012

patching, or TPTP) (6). TPTP enables targeted recordings from specific neurons by visualizing them using fluorescent proteins (FPs); it thus greatly enhances the success rate and quality of recordings. In addition, TPTP pioneered the way for recordings from neurons whose activity was manipulated by gene targeting (overexpression or knock-down of functional molecules) (7).

Another method, termed "shadow patching," has been developed for implementing targeted patch-clamp recordings in vivo (8). With shadow patching, neurons do not need to be fluorescently labeled by FPs or other cell-permeant fluorescent dyes. Instead, they can be visualized by negative contrast image created by loading the extracellular space with nonpermeant fluorescent dye. This approach facilitates making targeted patch-clamp recordings in intact and unmodified brains. Furthermore, this method can be used for targeted single-cell electroporation of dye or plasmid DNA (8, 9).

Although these methods can be applied only to brain regions that are available to two-photon imaging (e.g., neocortex, cerebellar cortex), they enable us to implement technically difficult experiments, including targeted recordings from minor types of neuron (10) and neuronal dendrites (8), simultaneous whole-cell recordings, and dendritic calcium imaging (11). Therefore, TPTP offers a new avenue for investigating functional properties of single neurons in vivo.

12.2. Practical Implementation of Two-Photon Targeted Patch-Clamp Recordings

To obtain significant results by performing experiments efficiently, the most important point is to maintain animals in good conditions by optimizing the surgical procedure and by monitoring the animal's vital signs (depth of anesthesia, breathing and heart beat rate, etc.).

12.2.1. Surgery

Surgical procedure for two-photon targeted patching is essentially the same as that for blind patching. To place an anesthetized animal (rat or mouse) under a two-photon microscope, one can use either a standard stereotaxic frame (ear bars and a nose clamp) or the metal head plate. However, use of the metal head plate is recommended because of the mechanical stability and the placement of the craniotomy relative to the focal plane of the two-photon microscope.

Place the anesthetized animal in a stereotaxic frame and perform craniotomy according to the standard procedure. A skin incision is made along the midline, and the skull is exposed. After removing all muscles and connective tissues, the skull is dehydrated and a custom-made small stainless steel head plate is glued onto the skull using superglue and dental cement. If dental cement is not firm enough to attach the head plate to the skull, the use of bone screws can be helpful. The head plate is then firmly screwed on a stainless rod, which is attached to the stage. In this way, the brain surface is placed in parallel to the imaging plane of the two-photon microscope. A round craniotomy is made using a dental drill with a small tip steel burr (0.5 mm in diameter). Drilling is performed very carefully and intermittently, and chilled HEPES-buffered saline is applied to avoid heating due to drilling. Any bleeding from the skull must be stopped immediately by using bone wax or Gelfoam. A circle (1–3 mm in diameter) is drilled until a thin piece of the skull that remains, and the piece of skull is then gently lifted out by a pair of fine forceps. Subsequently, the dura is cut by a fine syringe needle (30 gauge) and then removed by a pair of fine forceps (no. 5). Extra care must be taken not to touch the cortical surface during this procedure. Damage to the cortical surface makes it difficult to perform any imaging or recording experiments.

In contrast to the imaging/recording of preparations in vitro, movement artifacts due to pulsation and breathing provide serious problems for in vivo experiments. Application of agarose (1.5–3.0% in saline, ~36°C) onto the craniotomy is known to suppress this artifact. A small coverslip (10 × 5 mm, thickness #1) is placed over the agarose to cover approximately half of the cranial window. The coverslip is then held by wire springs or fixed by dental cement to further suppress movement artifacts. It is possible to suppress movement artifacts to 1–2 μm by this method.

12.2.2. Recording Electrode, Internal Solution, and Microscope

The patch electrode for two-photon targeted patching is basically the same as that for slice patching. We use a conventional thick-walled borosilicate glass capillary (o.d. 1.5 mm, i.d. 0.86 mm, with filament) and pull it with a simple two-step electrode puller. Although the patch electrode fabricated by this procedure normally has a shorter taper length than that of the classic sharp electrode for intracellular recordings, it does not cause problems during in vivo recordings because the depth of two-photon targeted patching is normally <500 μm from the brain surface. The resistance of the patch electrode is typically 4–7 MΩ for somatic recordings and 10–12 MΩ for dendritic recordings when filled with standard pipette internal solution (e.g., 133 mM K $MeSO_3$, 7.4 KCl, 10 HEPES, 3 MgATP, 0.3 Na_2GTP, 0.3 $MgCl_2$, pH 7.3). Although one can use the electrode with higher tip resistance, it is difficult to obtain low-access recordings with higher-resistance electrodes. On the other hand, the electrode with lower tip resistance (~2 MΩ) that is often used for slice patching is also difficult to use because of the difficulty of setting the appropriate pressure to obtain a sharp image of the tip of the electrode. Composition of the internal solution is the same as that for slice patching except that fluorescent dye (e.g., 50 μM Alexa594) must be added to visualize the electrode and the neurons using a two-photon microscope.

A conventional patch-clamp amplifier, interface board, and motorized manipulators can be used. A two-photon microscope for in vivo imaging is required. Commercial systems that are available from several companies can be used if the system provides the following: (1) image acquisition at 1–2 frames/s with 512 × 512 or 256 × 256 pixel resolution; (2) simultaneous multiple color imaging capability; (3) motorized specimen stage and focus. Most of the commercial instruments provide (1) and (2), but (3) is often optional. However, two-photon targeted patching requires manipulation of electrode and specimen position under the two-photon microscope, which should be covered by the blackout curtain or box. Therefore, it is highly recommended to have a motorized specimen stage and microscope focus for improving the success rate and efficiency of the experiment.

12.2.3. Two-Photon Targeted Patching

An advantage of TPTP is the ability to make targeted recordings from genetically identified cell types and from genetically modified neurons. Figure 12.1 shows an example of TPTP from a cerebellar Golgi cell. The cerebellar Golgi cell, one of the interneurons in the cerebellar cortex, is genetically labeled with green fluorescent protein (GFP) in this transgenic mouse (12). Thus, TPTP allows us to make reliable recordings from any minor neurons, which are normally difficult to record by blind patching. Although genetic modification to express FPs by making transgenic lines, viral infection, or electroporation of plasmid DNA is required to visualize

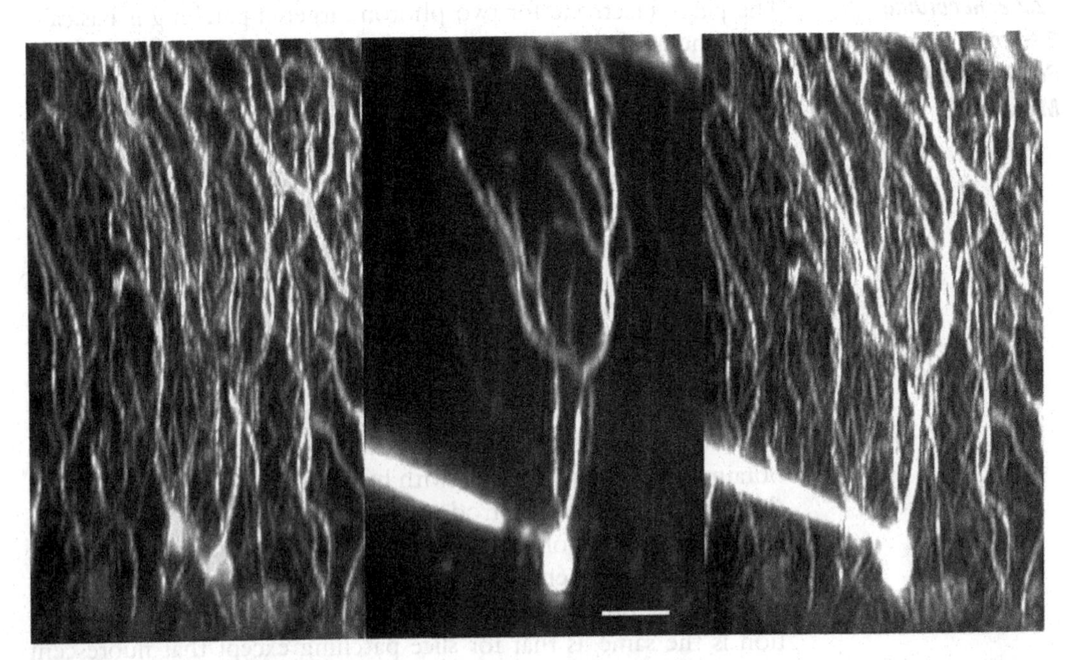

Fig. 12.1. Two-photon targeted patching from a cerebellar Golgi cell. *Left* GFP fluorescence in a Golgi cell. *Middle* Alexa fluorescence from the recorded cell. *Right* Overlay. *Bar* 20 μm.

the cells, recent progress in genetic engineering technology (e.g., Cre/loxP recombination system) makes it easier to obtain cell type-specific expression of FPs. In addition, many transgenic mice have been made available from comprehensive gene expression analysis by the GENSAT project (13).

The procedure for TPTP is as follows:

1. Prepare animals that express FPs in target neurons. Place the anesthetized animal in a stereotaxic frame and perform the craniotomy as described in the text. After craniotomy, place the animal under a two-photon microscope.
2. To visualize the tip of the electrode, nonpermeant fluorescent dye (e.g., 50 μM Alexa 594 hydrazide) is included in the pipette internal solution. Emission spectrum of the dye should be clearly distinguished from that of the FPs.
3. Exert relatively high positive pressure on the electrode (~200–300 mbar) with a 5-ml syringe, and place the electrode close to the brain surface over the region of interest, avoiding large blood vessels. Using a low-power objective lens (5×), visualize the tip of the electrode using either a video camera or observation through a binocular eyepiece. The electrode is then inserted in the brain and advanced to the appropriate depth with high pressure. The high pressure is required to avoid clogging of the electrode when it goes through the meninges. The pressure is then reduced to 15–25 mbar immediately. It is not essential to have a manometer, although it is useful to monitor the actual value, particularly for troubleshooting. Replace the low-power objective lens with a high-power lens (e.g., 40×), and turn to the two-photon imaging mode.
4. Adjust the pressure under two-photon imaging to obtain a sharp image of the tip of the electrode. Observe the tip of the electrode and the neuron of interest simultaneously in each color channel. The overlaid image of each channel is useful for positioning the tip of the electrode close to the neuron of interest. The wavelength of the excitation laser is ~850 nm for the GFP and Alexa594 combination.
5. Position the tip of the electrode on the neuron of interest. Care should be taken not to make too large lateral and vertical movement of the electrode within the brain to minimize damage to the brain tissue. When it is difficult to approach the target neuron without large electrode movement (~100 μm), retract the electrode to the outside of the brain and reposition the electrode. Precise positioning/repositioning is possible by trigonometric calculation based on the coordinate of the motorized microscope stage and manipulators (9).
6. Move the electrode toward the center of the soma of the target neuron. Contact of the tip of the electrode with the cell membrane

can be monitored by monitoring the tip resistance (2) or by observing the dimple in the cell membrane (8).

7. Remove the positive pressure to make a tight giga-ohm seal. Slight application of negative pressure is often helpful to achieve giga-ohm seal formation. After establishing the giga-seal, apply gentle suction ramp or suction pulses to rupture the cell membrane to obtain the whole-cell configuration. The initial access resistance should be ~20 MΩ for successful recording. Application of brief suction pulses just after establishing the whole-cell configuration is helpful for obtaining low access resistance.
8. It is advisable constantly to monitor and adjust the position of the tip of the electrode relative to the cell, particularly for long-duration recordings, because the brain tissue pushed by the electrode slowly drifts back to its original position.

12.2.4. Shadow-Patching

Shadow-patching is a variant of TPTP, but neurons do not need to be fluorescently labeled prior to the recordings. Neurons are visualized and identified in the intact brain by injecting nonpermeant dye into the extracellular space. When sufficient dye is present in the extracellular space, individual neurons become visible as dark "shadows" against a bright background because they do not take up the dye. We often use Alexa dye, because it is extremely bright and is not taken up by living neurons. Other dyes (e.g., calcium indicator Oregon Green BAPTA-1) can be used to perform shadow-patching. One can regulate perfusion by adjusting the pressure applied to the pipette to give a constant flow and relatively uniform distribution of dye in the surrounding neuropil. Under these conditions, the negatively stained "shadow" of individual neurons could be clearly visualized (Fig. 12.2).

Neuronal types can be identified by shadow-imaging. First, the depth of the pipette is adjusted to target neurons in a particular layer (e.g., 200–300 μm from the pia for layer 2/3 in the cerebral cortex, or 150–200 μm from the pia for the Purkinje cell layer). Second, preliminary identification of the neuronal type based on the size and shape of the cell body is carried out rapidly by scanning in a single plane of the optical section. This allows, for example, pyramidal cells to be tentatively discriminated from interneurons in the cortex and Purkinje cells to be readily identified by their size, circular somata, and dense packing in the Purkinje cell layer. This revealed features, such as the proximal dendrites, that further aid in the identification of cell type (Fig. 12.3). Thus, pyramidal cells can be readily identified by their prominent apical dendrite, and interneurons can be distinguished by their somatodendritic structure (round shape without obvious apical dendrites). The success rate for correct identification of Purkinje cells and molecular layer interneurons in the cerebellar cortex was 100%, and the rates for correct identification of interneurons and pyramidal cells in the

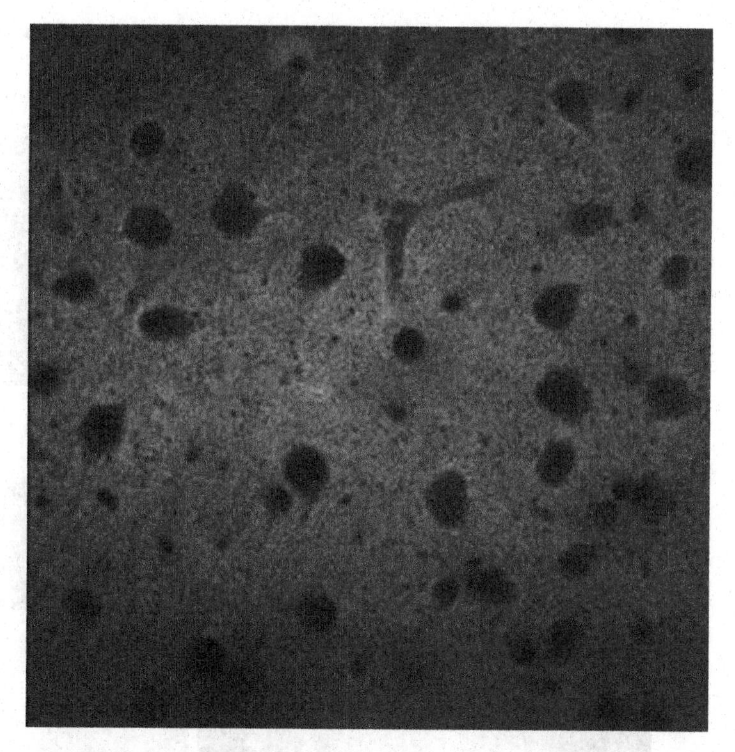

Fig. 12.2. Shadow-imaging of neurons in vivo. Neurons in layer 2/3 of the somatosensory cortex of the mouse. Extracellular space was filled with Alexa 594 to create negative shadows of the cells. Depth from the pial surface was ~200 μm. *Bar* 20 μm. (Modified from (8)).

cortical layer were 56% and 87%, respectively. These success rates are comparable to those achievable in brain slices using infrared differential interference contrast video-microscopy.

One concern is that the repeated scanning of the same area with relatively high laser power may lead to photodamage. The average power needed to visualize individual neurons with sufficient contrast to identify cell type was typically <30 mW at 200–300 μm from the brain surface, comparable to what is needed to visualize GFP-expressing neurons for TPTP. However, we never observed signs of photodamage (morphological changes or change in resting membrane potential) during our experiments, even when using much higher power than required for visualization (~50 mW). This may be related to the fact that with shadow-patching the dye is located extracellularly, and thus photodamage is likely to be less of concern.

Once a neuron has been identified, it can be targeted with the same pipette used for the dye perfusion to make a giga-seal patch-clamp recording under direct visual control with two-photon microscopy (Fig. 12.4). The details of the procedure for shadow-patching are the same as for TPTP (described above). Shadow-patching allows reliable whole-cell recordings to be made from the

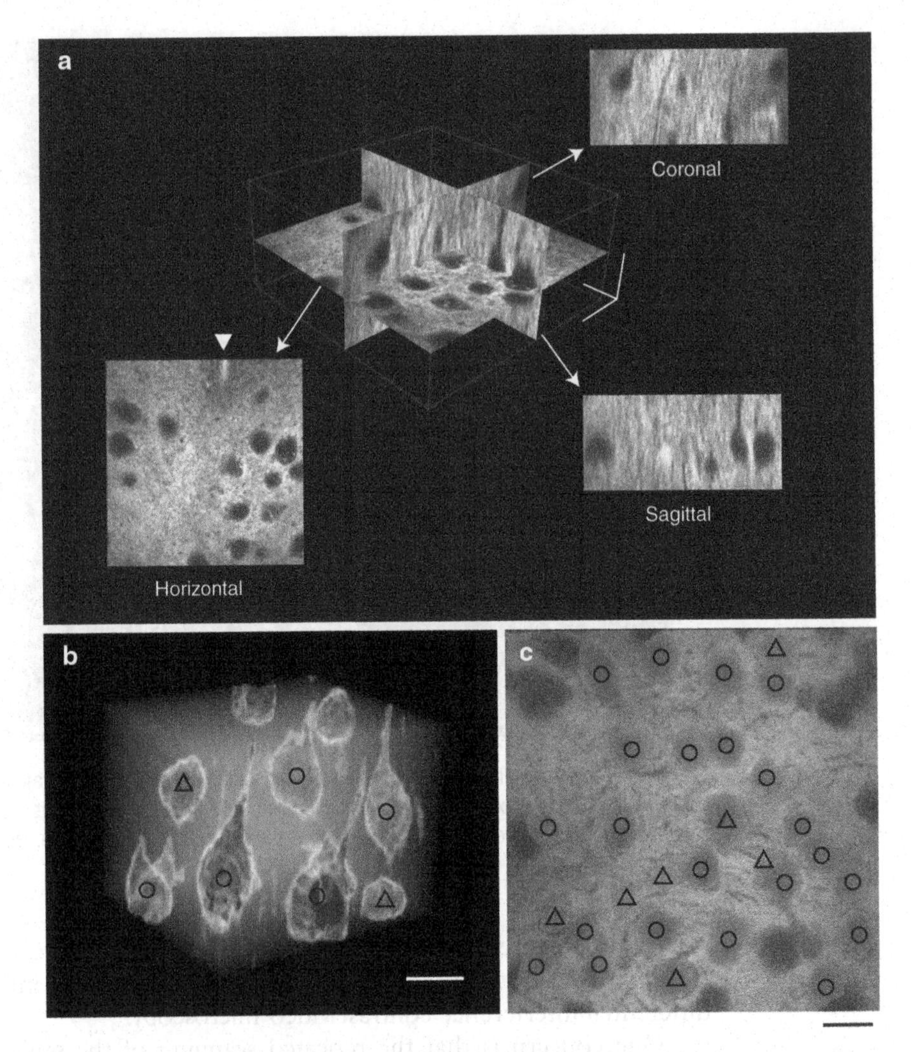

Fig. 12.3. Identification of neurons by shadow-imaging. (**a**) Reconstructed images of unlabeled neurons visualized by negative contrast in the mouse barrel cortex. Depth from the pial surface was ~200 μm. Fluorescent dye (50 μM Alexa594) was perfused via a patch pipette (*arrowhead*). (**b**, **c**) Identification of the cell type. Three-dimensional reconstructions of neurons (**b**) and minimum intensity projection of the stack (**c**) shown in (**a**). The cell type was identified based on the somatodendritic structure: *circles*, putative pyramidal cells (with apical dendrites); *triangles*, interneurons (without apical dendrites); and unidentified cells (no symbol). *Bars* 20 μm. (Modified from (8)).

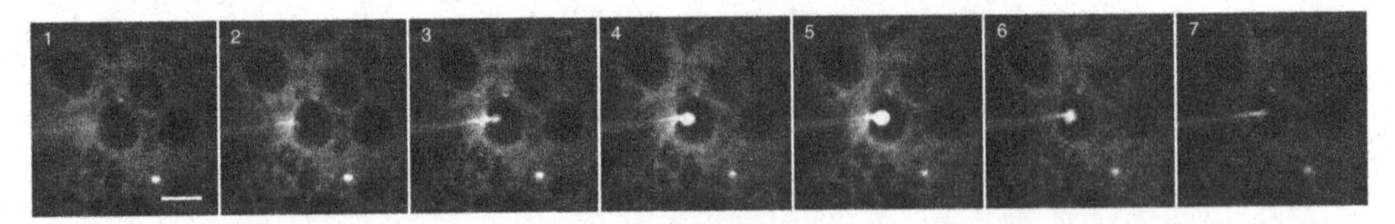

Fig. 12.4. Shadow-patching. Image sequence shows the patching process (2.5-s interval between images). Starting from the left, a pipette filled with Alexa 594 was used to identify the soma of a rat cerebellar Purkinje cell using the negative image. The pipette was then moved toward the cell until the dimpling of the somatic membrane caused by the flow of dye solution was clearly visible as an expanding bright "bubble" of dye (*4*, *5*). At this point (*6*), the pressure was released, and suction was applied to the pipette, which caused giga-ohm seal formation. At the same time, the dye concentration in the extracellular space began to fall rapidly as the extracellular perfusion (*7*) is terminated, and diffusion disperses the dye. *Bar* 20 μm. (Adapted from (8)).

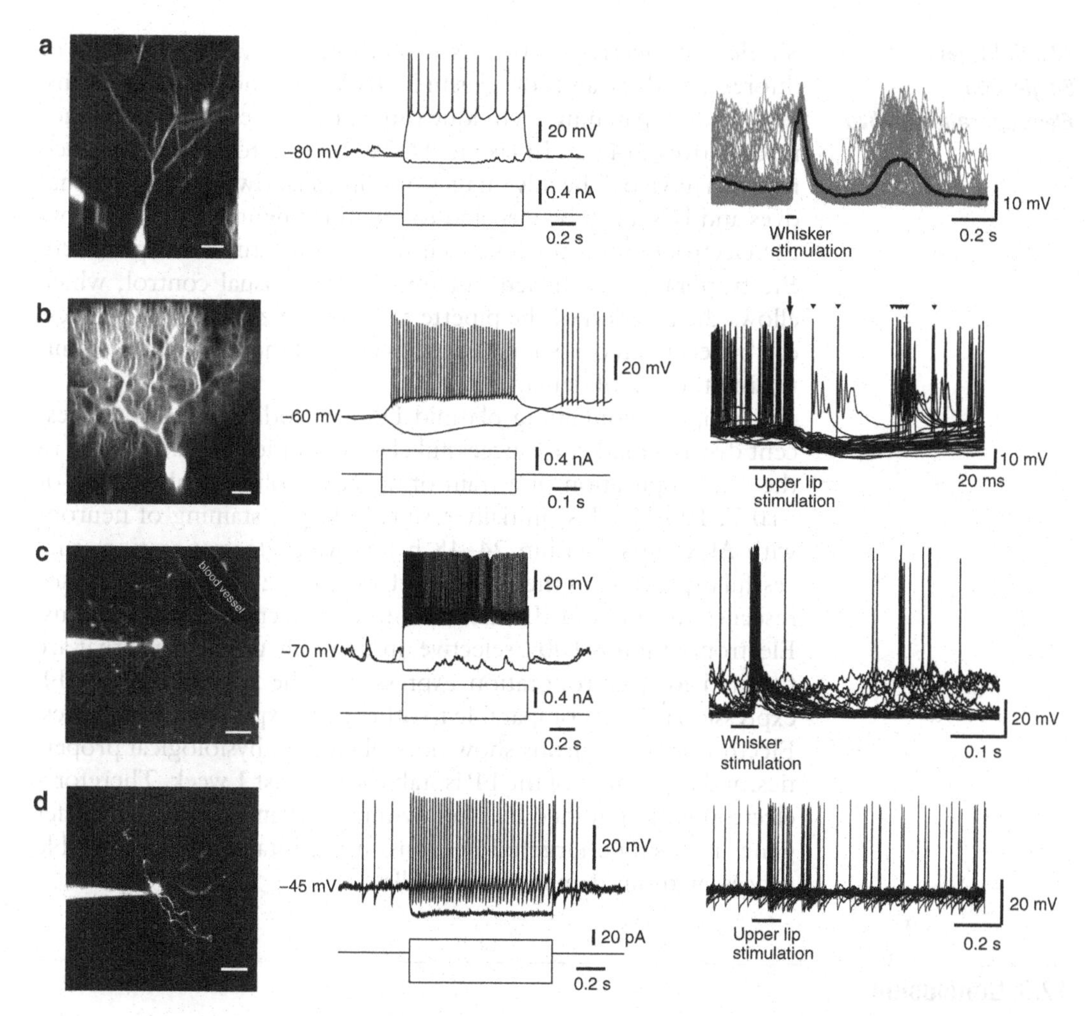

Fig. 12.5. Shadow-patching from neurons in the neocortex and cerebellar cortex in vivo. (**a**) Layer 2/3 pyramidal cell in the mouse barrel cortex. (**b**) Rat cerebellar Purkinje cell. (**c**) Fast spiking interneuron in the mouse barrel cortex. (**d**) Mouse cerebellar molecular layer interneuron (basket cell). *Bars* 20 μm. (Adapted from (8)).

somata of principal neurons (pyramidal neurons in layer 2/3 of the barrel cortex (Fig. 12.5a) and Purkinje neurons in the cerebellum (Fig. 12.5b), as well as the smaller interneurons in both regions (Fig. 12.5c,d). Because the Alexa dye diffuses rapidly from the extracellular space, patched neurons, once dialyzed with dye from the pipette, appear brightly fluorescent against the now dark neuropil. Under optimal conditions, the average success rate for the formation of giga-seals and subsequent whole-cell recording is ~70% for the trials with principal neurons. The access resistance is typically <20 MΩ, comparable to that obtained for whole-cell recordings in slices. The stability of shadow-patch recordings is also substantially improved; the recordings can be maintained for more than 30 min and sometimes up to 2 h.

12.2.5. Targeted Single-Cell Electroporation In Vivo

Single-cell electroporation is a method that allows loading of fluorescent dyes and/or plasmid DNA into individual neurons. The shadow-patching technique can be used to electroporate fluorescent dyes (14) and plasmid DNA (8, 9) into single identified neurons in vivo. The advantages of using shadow-imaging are that dyes and DNA can be targeted to identified neurons, and it allows the electroporation process itself to be monitored and optimized. Electroporation is carried out under direct visual control, which allows the position of the pipette to be optimized, and the process can be terminated as soon as dye has entered the cell, thereby minimizing the risk of damage.

A pipette containing plasmid DNA encoding FP and fluorescent dye as a marker of successful electroporation is placed close to the cell. Application of a train of negative voltage pulses (−15 or −10 V, 100 Hz, 1 s) initially results in bright staining of neurons with Alexa 594. Within 24–48 h following electroporation, successful expression of the FP plasmid is confirmed by bright FP fluorescence throughout the dendritic tree of electroporated neurons. Electroporation is 100% selective, in that only the neurons that are targeted for electroporation express FP. The success rate for FP expression is 75%, comparable to other gene expression techniques. Electroporated neurons show normal electrophysiological properties, and expression of the FP is stable for at least 1 week. Therefore, electroporation guided by shadow-imaging can be used to transfer genes into single targeted neurons in the intact brain in a stable process without damage to the cells.

12.3. Conclusion

The two-photon targeted patching technique has enabled us to obtain whole-cell recordings and single-cell electroporation in identified neurons in vivo by visualizing the target neurons under two-photon microscopy. The targeting technique allows us to make recordings from minor neurons, which have been difficult to obtain by blind patching. In addition, these techniques are particularly helpful for technically difficult experiments, such as combined whole-cell recording and calcium imaging experiments, where long-duration recordings with stable, low access resistances are important for efficient dye equilibration in the dendritic tree. Thus, two-photon targeted patching is expected to be a "standard" technique for in vivo recordings along with the increased use of the in vivo two-photon imaging technique.

References

1. Jagadeesh B, Gray CM, Ferster D (1992) Visually evoked oscillations of membrane potential in cells of cat visual cortex. Science 257(5069):552–554
2. Margrie TW, Brecht M, Sakmann B (2002) In vivo, low-resistance, whole-cell recordings from neurons in the anaesthetized and awake mammalian brain. Pflugers Arch 444(4):491–498
3. Brecht M, Schneider M, Sakmann B, Margrie TW (2004) Whisker movements evoked by stimulation of single pyramidal cells in rat motor cortex. Nature 427(6976):704–710
4. Chadderton P, Margrie TW, Häusser M (2004) Integration of quanta in cerebellar granule cells during sensory processing. Nature 428(6985):856–860
5. Crochet S, Petersen CC (2006) Correlating whisker behavior with membrane potential in barrel cortex of awake mice. Nat Neurosci 9(5):608–610
6. Margrie TW, Meyer AH, Caputi A, Monyer H, Hasan MT, Schaefer AT, Denk W, Brecht M (2003) Targeted whole-cell recordings in the mammalian brain in vivo. Neuron 39(6):911–918
7. Komai S, Licznerski P, Cetin A, Waters J, Denk W, Brecht M, Osten P (2006) Postsynaptic excitability is necessary for strengthening of cortical sensory responses during experience-dependent development. Nat Neurosci 9(9):1125–1133
8. Kitamura K, Judkewitz B, Kano M, Denk W, Hausser M (2008) Targeted patch-clamp recordings and single-cell electroporation of unlabeled neurons in vivo. Nat Methods 5(1):61–67. doi:nmeth1150 (pii) 10. 1038/nmeth1150
9. Judkewitz B, Rizzi M, Kitamura K, Hausser M (2009) Targeted single-cell electroporation of mammalian neurons in vivo. Nat Protoc 4(6):862–869. doi:nprot.2009.56 (pii) 10.1038/nprot.2009.56
10. Liu BH, Li P, Li YT, Sun YJ, Yanagawa Y, Obata K, Zhang LI, Tao HW (2009) Visual receptive field structure of cortical inhibitory neurons revealed by two-photon imaging guided recording. J Neurosci 29(34):10520–10532. doi:29/34/10520 (pii) 10.1523/J Neurosci 1915-09.2009
11. Jia H, Rochefort NL, Chen X, Konnerth A (2010) Dendritic organization of sensory input to cortical neurons in vivo. Nature 464(7293)):1307–1312. doi:nature08947 (pii) 10.1038/nature08947
12. Watanabe D, Nakanishi S (2003) mGluR2 postsynaptically senses granule cell inputs at golgi cell synapses. Neuron 39(5):821–829. doi:S0896627303005300 (pii)
13. Gong S, Zheng C, Doughty ML, Losos K, Didkovsky N, Schambra UB, Nowak NJ, Joyner A, Leblanc G, Hatten ME, Heintz N (2003) A gene expression atlas of the central nervous system based on bacterial artificial chromosomes. Nature 425(6961):917–925. doi:10.1038/nature02033 nature02033 (pii)
14. Nevian T, Helmchen F (2007) Calcium indicator loading of neurons using single-cell electroporation. Pflugers Arch 454(4):675–688

Chapter 13

Transporter Current Measurements

Mami Noda

Abstract

Na^+/K^+ pump, Na^+/Ca^{2+} exchanger, glutamate transporter and other transporters for dopamine, norepinephrine, GABA, and others, are all electrogenic. These transporters or exchanger transport Na^+ ions, with or without transmitters, in or out of the cells, while K^+ or Ca^{2+}, sometimes with anions, are transported to the opposite direction. Since more Na^+ is transported, it induces net positive charge movement, therefore producing membrane currents in the direction of Na^+ ion movement. Na^+/K^+ pump can work against ion gradient because it is an ATPase (Na^+/K^+-ATPase). In addition, under a certain condition, the Na^+/K^+ pump transforms into an ion channel so that it produces pump-channel currents. Other exchangers and transporters are generally dependent on electrochemical potential gradients and also voltage-dependent.

13.1. Introduction

The currents generated by electrogenic transporters can be recorded basically by the same method as membrane ionic channel currents. However, it is difficult to record single transporter currents, and the reversal potential of the transporter currents cannot be obtained. Another difference is that ions flow through ion channels only according to the elecrochemical gradient, whereas some transporters ions can be transported against the gradient. The measurements of typical transporter currents such as the Na^+/K^+ pump and Na^+/Ca^{2+} exchanger became possible through development of the patch-clamp technique using single cardiac myocytes (1, 2).

In principle, an ion channel needs no more than a single gate, but a pump requires at least two gates that open and close alternately to allow ion access from only one side of the membrane at a time (Fig. 13.1).

Yasunobu Okada (ed.), *Patch Clamp Techniques: From Beginning to Advanced Protocols*, Springer Protocols Handbooks, DOI 10.1007/978-4-431-53993-3_13, © Springer 2012

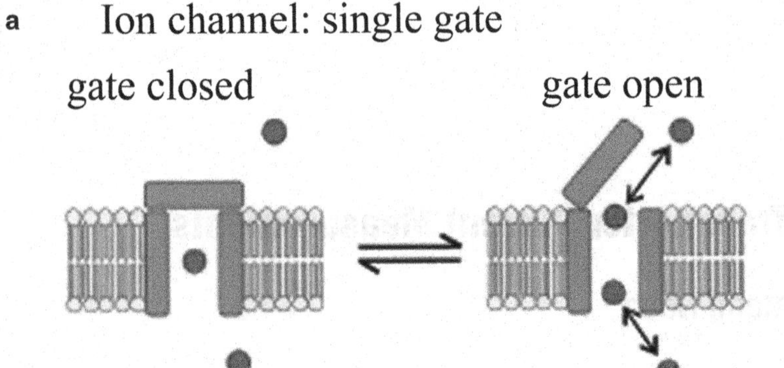

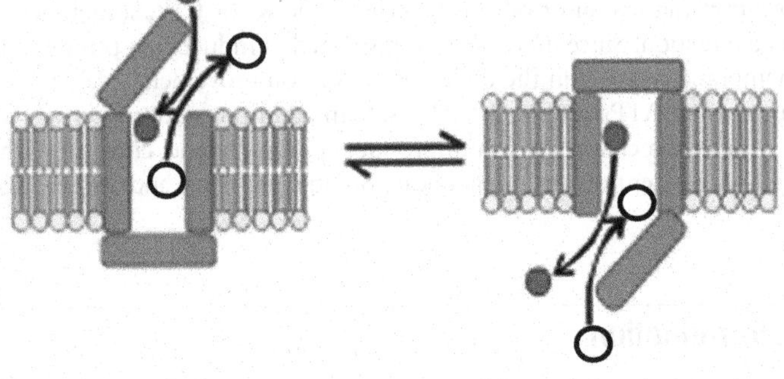

Fig. 13.1. Ion channels versus ion pumps. (**a**) Ion channels need only a single gate, which can be open or closed. When closed, ions (*filled circles*) cannot pass through the pore; but when open, ions can enter and flow through freely. (**b**) By contrast, ion-exchange pumps need two gates, which are never open at the same time. Each gate can open only if the other is closed, to prevent unrestricted ion flow. When one gate is open, one type of ion enters, and another leaves. (Modified from Gadsby et al. (3)).

13.2. Na^+/K^+ Pump Currents

13.2.1. Pump Currents

The Na^+/K^+ pump is a P-type ATPase crucial to the life of practically all animal cells. Its transport cycle is shown in Fig. 13.2 as a sequence of steps that give alternating access to ion-binding sites in an ion channel. The Na^+/K^+-ATPase pumps three sodium ions out of and two potassium ions into the cell for each ATP molecule. The $3Na^+/2K^+$-exchange transport cycle therefore produces current in the direction of Na^+ ion movement.

Although ion pumps and ion channels are both integral membrane proteins that mediate transport of ions across cell membranes, they traditionally have been viewed as very different, largely for kinetic reasons. Thus, up to 10^7–10^8 ions/s flow down their electrochemical gradient through an open channel, whereas a working pump moves only ~10^2 ions/s up their electrochemical gradient. Although

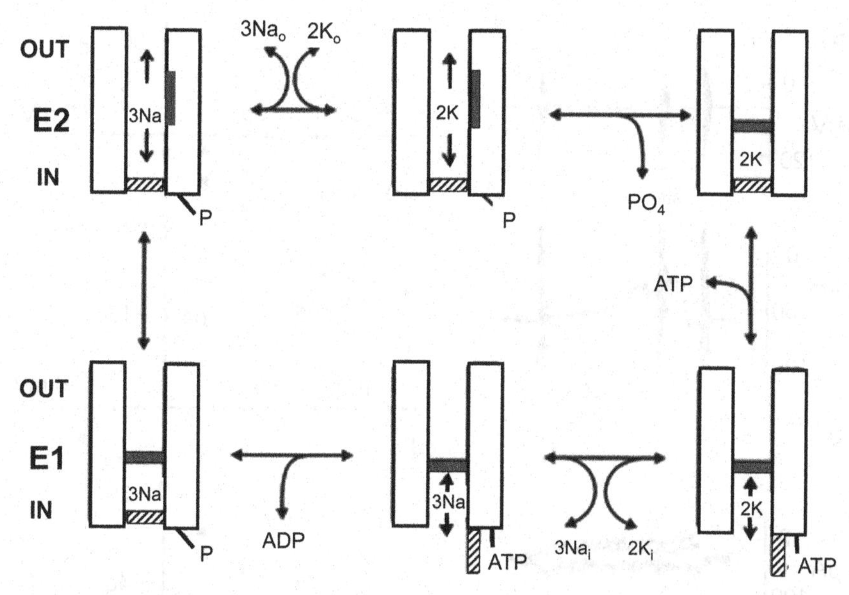

Fig. 13.2. Alternating-gate model of the Post-Albers transport cycle of the Na^+/K^+ pump represented in cartoon form as a channel with two gates never simultaneously open. Extracellular (OUT) and intracellular (IN) surfaces of the membrane are indicated, as are E2 states (*top*, external gate may open) and E1 states (*bottom*, internal gate may open). Occluded states with both gates shut (*top right* and *bottom left*) follow binding of two external K^+ and three internal Na^+ and subsequent phosphorylation, respectively. Adenosine triphosphate (ATP) acts with low affinity to speed opening of the internal gate and concomitant K^+ deocclusion and with high affinity to phosphorylate the pump. (From Artigas and Gadsby (7)).

the Na^+/K^+-ATPase pumps unequal numbers of Na^+ and K^+ ions in opposite directions across the cell membrane and so generates a net current, the current is too small (at most a few atto-amperes) to be measured for a single pump cycle. However, animal cells often contain many millions of Na^+/K^+-ATPase pumps; and although their transport cycles are not synchronized, their combined steady-state ion transport generates a readily measurable current (1, 4, 5). With saturating levels of intracellular ATP and Na^+, and of extracellular K^+, the component of cell membrane current generated by the pump remains outwards (net cation extrusion) over a broad (200 mV) range of membrane potentials, regardless of how high the extracellular $[Na^+]$ or $[K^+]$ is raised (6).

To drive the Na^+/K^+ pump reaction cycle backward, the transmembrane Na^+ and K^+ gradients are steepened by removing internal Na^+ and external K^+ and setting the $[Na^+]_o$ at 150 mM and internal K^+ concentration $[K^+]_i$ at 145 mM. Also, 5 mM ATP, 5 mM ADP, and 5 mM inorganic phosphate (Pi) are included in the pipette solution. Strophanthidin then causes an outward shift of holding current (Fig. 13.3a,b) reflecting abolition of the steady inward pump current. Steady-state, whole-cell current–voltage (I–V) relations determined before, during, and after the exposure to strophanthidin are shown in Fig. 13.3c. The resulting strophanthidin-sensitive I–V relation (Fig. 13.3d) reveals that inward current generated by the backward-running Na^+/K^+ pump also has a monotonic voltage

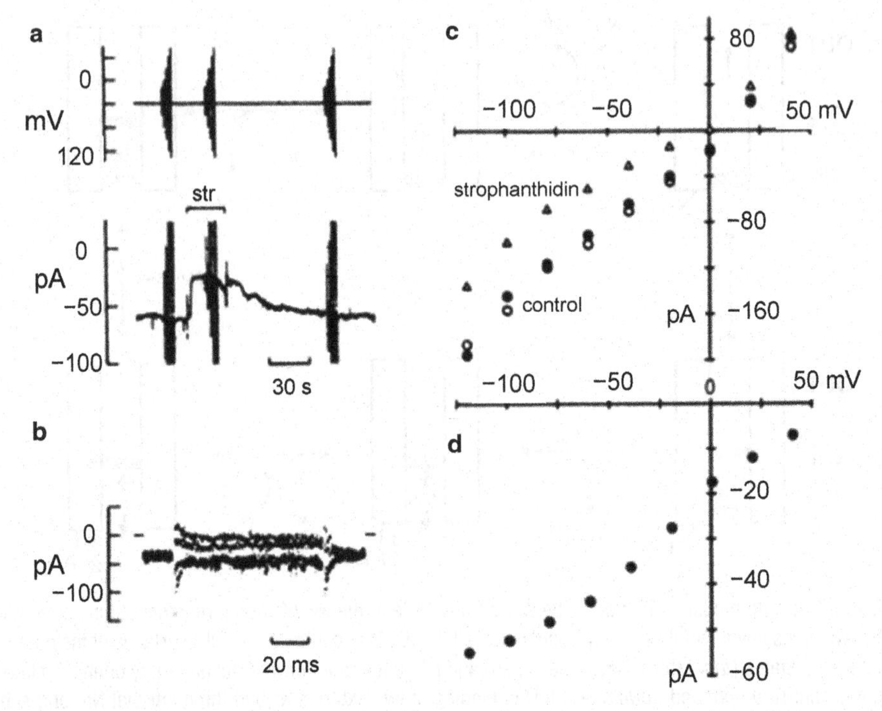

Fig. 13.3. Voltage dependence of inward current generated by the backward-running Na^+/K^+ pump. (**a**) Recordings of membrane potential (*top*) and membrane current (*bottom*); holding potential was −40 mV. *Horizontal bar* marks exposure to 0.5 mM strophanthidin (*str*). The K^+-free external solution contained 5 mM Ba^{2+}; the internal solution was free of Na^+, Cs^+, and creatine phosphate; it contained instead 145 mM K^+, 5 mM MgATP, 5 mM $Tris_2$ADP, and 5 mM phosphate. (**b**) Superimposed records of strophanthidin-sensitive currents for 80-ms pulses to +40, 0, −60, and −100 mV. They were obtained by subtracting each trace recorded in the presence of strophanthidin from the average of control traces recorded during pulses to the same potential just before and just after exposure to strophanthidin. (**c**) Whole-cell current–voltage relations from the experiment in (**a**), determined before (*open circles*), during (*triangles*), and after (*closed circles*) exposure to strophanthidin. Ordinate: steady current levels measured by averaging points over a fixed 8-ms period near the end of each pulse. Abscissa: membrane potential. (**d**) Current–voltage relation of the backward-running Na^+/K^+ pump. (From Bahinski et al. (8)).

dependence: It increases in amplitude as the membrane potential is made more negative, from an extremely small size at +40 mV toward an apparent plateau level near −100 mV.

13.2.2. Pump-Channel Currents

The two kinds of membrane protein that control movements of ions such as Na^+, K^+, and Cl^- into and out of cells have very different jobs. Ion channels let these charged atoms flow down chemical and electrical gradients; and pumps build up the gradients – hence the view that channels and pumps must be cut from different cloth. Nature, however, has produced a deadly tool that overcomes this stricture. It is called palytoxin, and is a complex 2.7-kDa molecule (with 63 stereogenic centers) originally extracted from *Palythoa* zoanthids collected from coral reefs. Palytoxin binds tightly to Na^+/K^+-ATPase pumps, transforming them into cation channels (Fig. 13.4).

Within a few seconds, a saturating concentration of palytoxin (100 nM) converts every one of the thousands of Na^+/K^+-ATPase

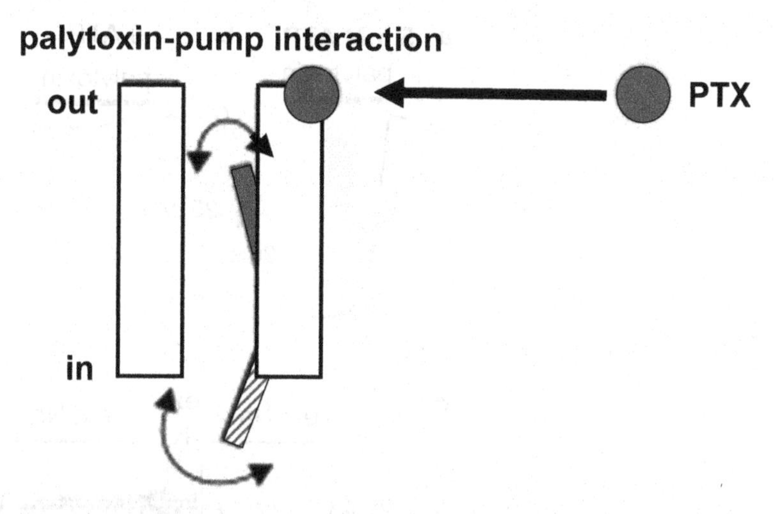

Fig. 13.4. Channel-like Na^+/K^+ pump with bound palytoxin (*PTX*), which allows the pump's two gates sometimes to be open at the same time. (Modified from Artigas and Gadsby (3)).

pumps in an excised outside-out patch into channels, as evident from the rapid development of inward (negative) current in the presence of an inward electrochemical potential gradient (Fig. 13.5a).

Another example is the cystic fibrosis transmembrane conductance regulator (CFTR) chloride channel, which is encoded by the gene mutated in patients with cystic fibrosis (CF). This channel belongs to the superfamily of ATP-binding cassette (ABC) transporter ATPases. ATP-driven conformational changes, which in other ABC proteins fuel uphill substrate transport across cellular membranes, in CFTR open and close a gate to allow transmembrane flow of anions down their electrochemical gradient. CF, an all-too-common deadly disease of young people, is caused by a defect in the expression or function of the CFTR protein. The gate that regulates chloride ion flow through the membrane-spanning pore in CFTR is thought to be opened when ATP binds to CFTR's two cytoplasmic nucleotide-binding domains (NBD1 and NBD2) and, acting as a molecular glue, holds them together. Hydrolysis of one of the ATPs disrupts the NBD1–NBD2 interaction and closes the channel gate, terminating anion flow. Kinase-mediated phosphorylation of a cytoplasmic regulatory domain in CFTR is needed for successful transmission of NBD events to the channel gate (11) (Fig. 13.6).

13.3. Na^+/Ca^{2+} Exchanger Currents

The Na^+/Ca^{2+} exchanger currents also became available for recording using the patch-clamp electrode (2, 13). The Na^+/Ca^{2+}exchanger is a sarcolemmal transporter that plays an important role in regulating

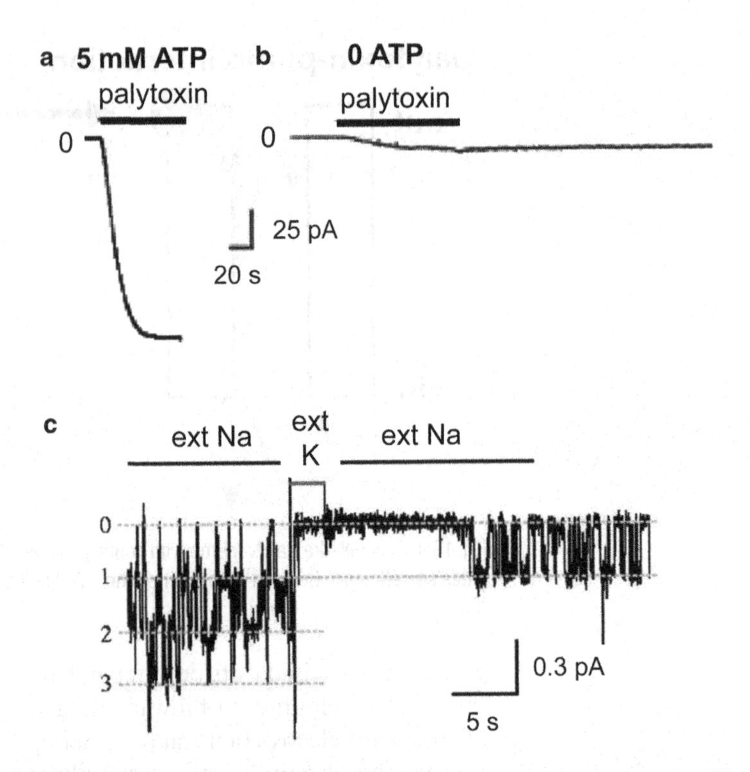

Fig. 13.5. Palytoxin (PTX) transforms Na^+/K^+-ATPase pumps into channels, one Na^+/K^+-ATPase pump at a time. (**a**, **b**) Macroscopic currents induced by the application of 100 nM PTX to outside-out patches excised from guinea pig ventricular myocytes, bathed in approximately 160 mM Na^+ solutions, and K^+ held at 40 mV, with 5 mM MgATP (**a**) or no ATP (**b**) present in the internal solution. Note the very different current scales in (**a**) versus (**c**). (**c**) Single palytoxin-bound pump-channel currents in an outside-out patch from a guinea pig ventricular myocyte, with a Na^+ internal solution without ATP. Palytoxin (100 pM) was applied for approximately 30 s at 1 min before the beginning of the trace. Holding potential was −50 mV. The 150 mM Na^+ in the external solution (*ext Na*) was replaced by 150 mM K^+ for the 2-s period indicated by the line. (Modified from Gadsby et al. (9) and Artigas and Gadsby (10)).

the intracellular Ca^{2+} concentration, especially in cardiac myocytes. The Na^+/Ca^{2+} exchange current is driven by the concentration gradients of both Na^+ and Ca^{2+} across the membrane. The exchange ratio is $3Na^+:1Ca^{2+}$ in cardiac myocytes and $4Na^+:(1Ca^{2+} + 1K^+)$ in retina and is therefore electrogenic. Under physiological conditions, the Na^+/Ca^{2+} exchange currents are inward (forward mode), pumping intracellular Ca^{2+} out and extracellular Na^+ in. However, the Na^+/Ca^{2+} exchange currents can be outward in depolarized conditions (reverse mode), causing extracellular Ca^{2+} to come in.

13.3.1. Inward Na^+/Ca^{2+} Exchange Currents

Without intracellular Ca^{2+}, brief exchange of the external solution from 140 mM Li^+ to 140 mM Na^+ did not induce any inward current (Fig. 13.7A,B). However, after loading internal Ca^{2+} with a 430 nM Ca^{2+} pipette solution (14–16), application of 140 mM Na^+ induced a large inward shift of the holding current, which decayed after a peak (Fig. 13.7A). Li^+ is used as a control because it is known

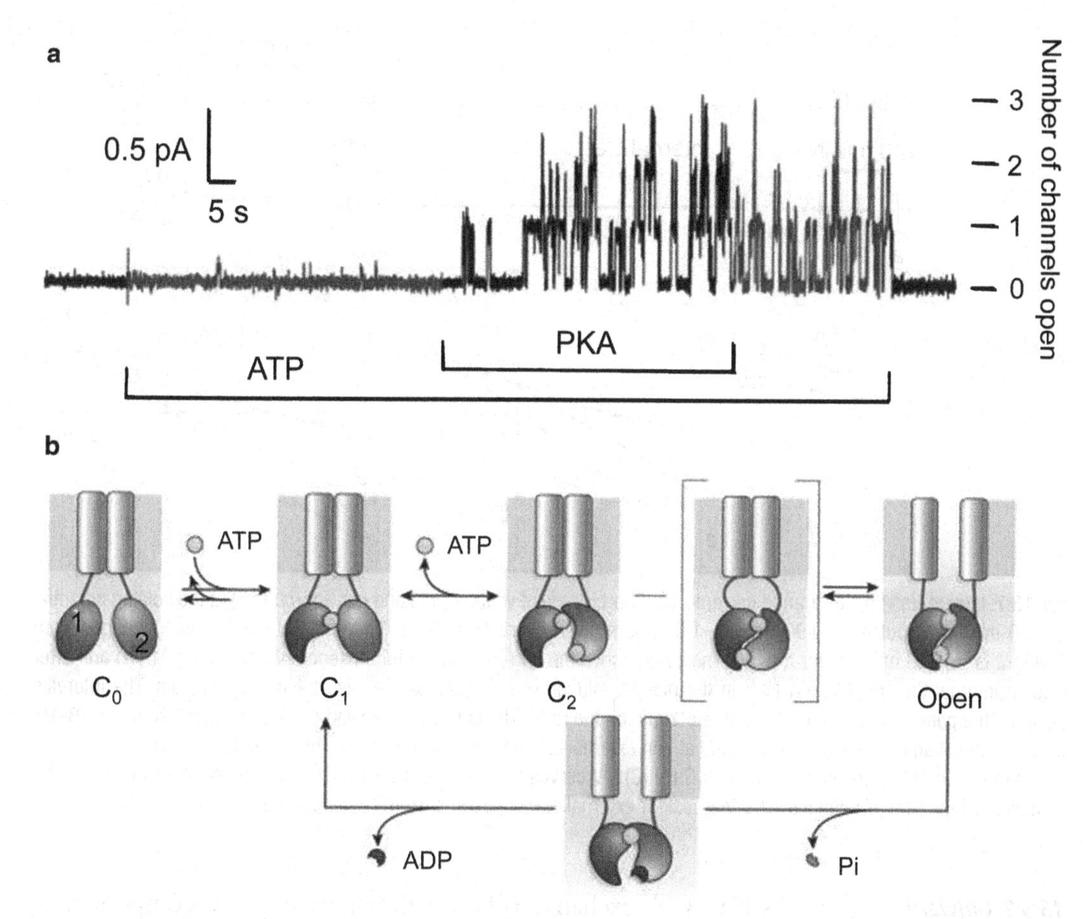

Fig. 13.6. Opening and closing of cystic fibrosis transmembrane conductance regulator (CFTR) channels. (**a**) CFTR's regulatory (R) domain must be phosphorylated by protein kinase A (PKA) before ATP is able to support channel opening. The recording shows chloride current flow through individual CFTR channels upon opening. Endogenous membrane-attached phosphatases partly dephosphorylate the R domain on PKA withdrawal, reducing the probability of finding an open channel. However, channels do not stop opening until ATP is removed; the number of simultaneously open channels is indicated at the right (for experimental condition; see (12)). (**b**) Present structural interpretation of ATP-dependent gating cycle of phosphorylated CFTR channels. The R domain is omitted. ATP (*circles*) remains tightly bound to nucleotide-binding domain 1 (NBD1) (*1*) Walker motifs for several minutes, during which time many closed–open–closed gating cycles occur. ATP binding to NBD2 (*2*) is followed by a slow channel opening step (C_2 to *Open*) that proceeds through a transition state (*square brackets*) in which the intramolecular NBD1–NBD2 tight heterodimer is formed but the transmembrane pore (*rectangles*) has not yet opened. The relatively stable open state becomes destabilized by hydrolysis of the ATP bound at the NBD2 composite catalytic site and loss of the hydrolysis product, inorganic phosphate (*Pi*). The ensuing disruption of the tight dimer interface leads to channel closure. (From Gadsby et al. (11)).

to substitute very weakly for Na^+ in the exchange mechanism. The internal solution contained no Na^+, and the external solution contained 1 mM external Ca^{2+}. The I–V relations obtained before and during the second external Na^+ perfusion in the presence of internal Ca^{2+} are shown in Fig. 13.7C,D (17). In addition, Na^+/Ca^{2+} exchange currents can be confirmed by using inhibitors: ex. mineral ions such as Ni, Mn and La; amiloride derivatives such as dichlorobenzamil; inhibitors such as KB-R7943, SN-6 and SEA0400; or inhibitory peptides such as XIP (intracellular loading).

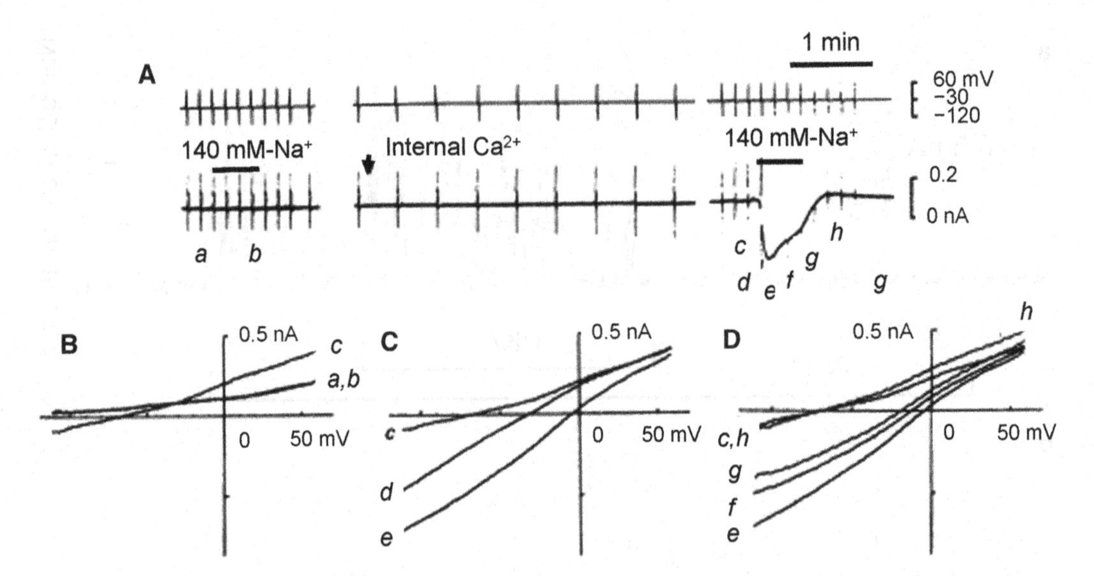

Fig. 13.7. Inward Na^+/Ca^{2+} exchange currents. (**A**) Chart record of voltage (*top*) and current (*bottom*) at the holding potential of −30 mV. Ramp pulses of +90 mV/0.5–1.0 s were given every 10 or 20 s. The external solution was changed from 140 mM Li^+ to 140 mM Na^+ for a period indicated above the current trace in the absence of internal Ca^{2+} (*left*) and after loading internal Ca^{2+} by 430 nM-Ca^{2+} in the pipette solution (*right*). External Ca^{2+} is at 1 mM throughout. The intervals between the panels are 3 min on the left and 1 min on the right. The current traces, indicated by *a–h*, are shown in (**B–D**). (**B**) Current–voltage (I–V) relations obtained at *a* in external Li^+ and *b* in external Na^+ in the absence of internal Ca^{2+} and at *c* in external Li^+ in the presence of internal Ca^{2+}. (**C**) I–V curves before (*c*) and during (*d* and *e*) the onset of external-Na^+-induced current. (**D**) I–V relations after the peak of external-Na^+-induced current (*e–g*). (From Kimura et al. (18)).

13.3.2. Outward Na^+/Ca^{2+} Exchange Currents

As Na^+/Ca^{2+} exchange is bidirectional, an outward component of the Na^+/Ca^{2+} exchange current should be induced by perfusing external Ca^{2+} in the presence of internal Na^+. For this purpose, the external solution consisted of 140 mM Li^+ without Na^+ and nominally free Ca^{2+} for the control and 1 mM Ca^{2+} for the test solution. Ba^{2+} (1 mM) was added to prevent the membrane from becoming leaky by eliminating all the external divalent cations from the control solution. The internal solution contained 30 mM Na^+ and 67 nM free Ca^{2+} (pCa 7.2) because it is known that Na^+/Ca^{2+} exchange does not operate in the absence of internal Ca^{2+}. The magnitude of the external Ca^{2+}-induced outward current became progressively larger when the concentration of external Ca^{2+} was increased from 0.1 to 20 mM (Fig. 13.8a). The control I–V curve obtained before the Ca^{2+} application (not shown) was subtracted from that at the peak response, and the difference I–V curves were superimposed in Fig. 13.8b. The dose–response relation to Ca^{2+} was obtained by plotting the current semilogarithmically against the external Ca^{2+} concentrations (Fig. 13.8c).

13.3.3. Bidirectional Na^+/Ca^{2+} Exchange

Since the Na^+/Ca^{2+} exchange is driven by the electrochemical potential gradients of Na^+ and Ca^{2+}, supposing a stoichiometry of $3Na^+$:$1Ca^{2+}$, the reversal potential is given by:

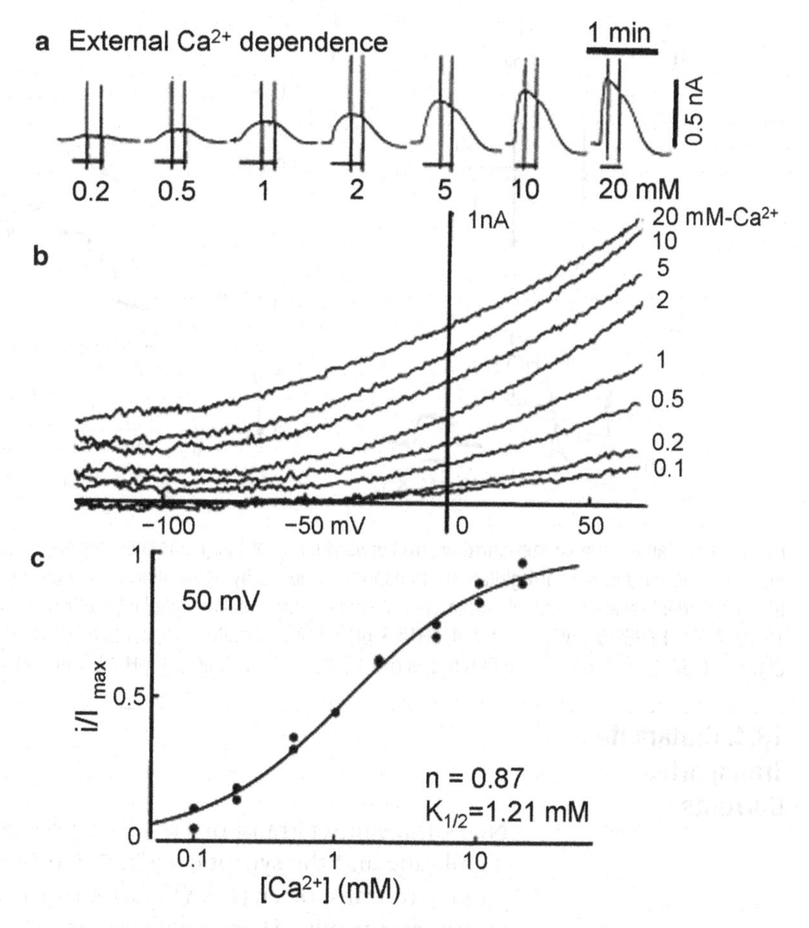

Fig. 13.8. (**a**) Current records on changing external Ca^{2+} from nominally free to various concentrations indicated below each trace between 0.2 and 20 mM. The holding potential is −30 mV. (**b**) I–V relations obtained by the difference between the peak of the activation (taken by the first ramp pulse on each trace in (**a**)) and control (not shown) for each trace. Ca^{2+} concentrations are indicated at the right of each trace. (**c**) Dose–response relation of the current at +50 mV. The current magnitude at each external Ca^{2+} concentration in (**b**) was normalized with reference to the current at 20 mM external Ca^{2+} (i/i_{max}) and plotted against external Ca^{2+} concentrations semilogarithmically. Note the Hill coefficient of 0.87 and the half-maximum ($K_{1/2}$) external Ca^{2+} concentration of 1.21 mM. (From Kimura et al. (18)).

$$E_{NaCa} = 3E_{Na} - 2E_{Ca}$$

where E_{NaCa} is the reversal potential of i_{NaCa}; and E_{Na} and E_{Ca} are the equilibrium potentials for Na^+ and Ca^{2+}, respectively. Under ionic conditions close to physiological ones – such as 140 mM external Na^+, 20 mM internal Na^+, 250 nM internal Ca^{2+}, and 1 mM external Ca^2 – the value of E_{NaCa} is −68 mV. When the holding potential were set at E_{NaCa}, the intracellular Ca^{2+} and Na^+ would be kept constant, and both inward and outward Na^+/Ca^{2+} exchange currents would be observed by applying hyperpolarizing or depolarizing pulses (19, 20).

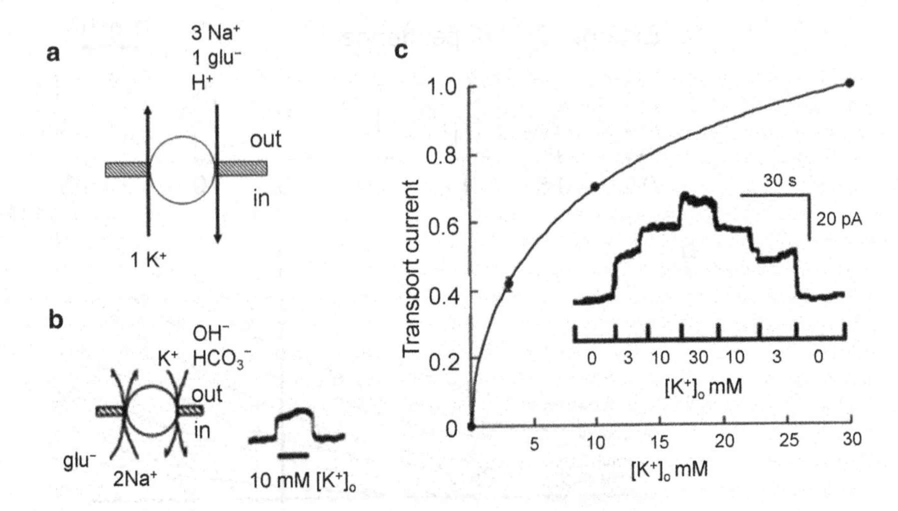

Fig. 13.9. Glutamate transporter currents. (**a**) Forward mode of the glutamate transporter. (**b**) Reverse mode of the glutamate transporter and outward glutamate transporter induced by application of extracellular K^+ in cultured microglia. (**c**) $[K^+]_o$-dependent reverse mode of glutamate transporter currents. Intracellular pipette solution (mM): NaCl 90, L-glutamate 10, Na_2ATP 3, HEPES 5, $CaCl_2$ 1, $MgCl_2$ 4, EGTA 5, pH 7.3 with *N*-methyl-D-glucamine (NMDG). Extracellular solution: choline Cl 80–110, $MgCl_2$ 0.5, $CaCl_2$ 3, HEPES 5, glucose 15, $BaCl_2$ 6, ouabain 0.1, pH 7.5 with NMDG. (From Noda et al. (22)).

13.4. Glutamate Transporter Currents

Neurotransmitter transporters are expressed on both the plasma membrane and the synaptic vesicle membrane. Plasma membrane glutamate transporter (EAAT) belongs to the Na^+/K^+-dependent transporter family. There are five known human EAAT subtypes; the glial carriers EAAT1 (GLT-1) and EAAT2 (GLAST) have the greatest impact on clearance of glutamate released during neurotransmission. Carriers expressed on neurons, Purkinje cells, and photoreceptor cells (EAAT3 (EAAC1), EAAT4, and EAAT5, respectively) suggest more subtle roles for these subtypes in regulating excitability and signaling.

Neuronal and glial glutamate transporters limit the action of excitatory amino acids after their release during synaptic transmission. To analyze their expression and function, there is a technique for measuring the glutamate transporter currents (21).

The current induced by an acidic amino acid such as glutamate is completely dependent on the presence of external Na^+. In fact, Li^+, Cs^+, choline$^+$, and TEA^+ were unable to substitute for Na^+. The relation between the external Na^+ concentration and the current amplitude can be explained if the binding of three Na^+ ions enabled transport. External D-aspartate, L-aspartate, and L-glutamate each induced a membrane current. This current seems to be to an electrogenic glutamate uptake carrier, which transports two or three Na^+ ions with every glutamate anion (and H^+ ions) carried into the cell (Fig. 13.9a).

Glutamate uptake is the forward mode, but the reverse mode (glutamate release) is also possible, especially when the extracellular

K^+ concentration becomes high, such as under ischemic conditions. In addition, it was also reported that in glial cells there is a carrier cycle in which two Na^+ ions accompanying each glutamate anion are transported, whereas one K^+ and one OH^- (or HCO_3^-) are transported at the same time (Fig. 13.9b). To measure the reverse mode of the glutamate transporter, extracellular K^+ is applied, which produces outward currents that are $[K^+]_o$-dependent (Fig. 13.9c) (22).

13.5. Other Transporters for Dopamine, Norepinephrine, GABA, and Others

Other amino acids are also transported together with Na^+. Na^+/Cl^--dependent transporters for dopamine, 5-hydroxytryptamine (serotonin), norepinephrine, γ-aminobutyric acid (GABA), and glycine have been reported. Among them, for example, uptake of dopamine by the cloned human dopamine transporter (DAT) has been shown to be electrogenic and voltage-dependent, with greater uptake observed at hyperpolarized potentials (23). Amphetamine elicits its behavioral effects by acting on the DAT to induce dopamine overflow into the synaptic cleft. Amphetamine binds to the DAT and is transported, thereby causing an inward current, making Na^+ more available intracellularly to the DAT. This enhances DAT-mediated reverse transport of dopamine, thereby producing an outward current (24).

Norepinephrine transporters have two functional modes of conduction: a classic transporter mode (T-mode) and a novel channel mode (C-mode). T-mode is putatively electrogenic because the transmitter and co-transported ions sum to give one net charge (25).

Another ion transporter, the H^+/Cl^- exchange transporter, was originally classified as a ClC Cl^- channel (CLC-ec1) (26). The ClC-0 channel is a "broken" Cl^-/H^+ antiporter in which one of the conformational states has become leaky for chloride ions. The electrical current through a single ClC-0 channel has been recorded (27).

See the references below for detailed methods for the individual transporter currents.

References

1. Gadsby DC, Kimura J, Noma A (1985) Voltage dependence of Na/K pump current in isolated heart cells. Nature 315:63–65
2. Kimura J, Noma A, Irisawa H (1986) Na–Ca exchange current in mammalian heart cells. Nature 319:596–597
3. Gadsby DC (2009) Ion channels versus ion pumps: the principal difference, in principle. Nat Rev Mol Cell Biol 10:344–352
4. Thomas RC (1972) Electrogenic sodium pump in nerve and muscle cells. Physiol Rev 52:563–594
5. Glitsch HG (2001) Electrophysiology of the sodium-potassium-ATPase in cardiac cells. Physiol Rev 81:1791–1826
6. Nakao M, Gadsby DC (1986) Voltage dependence of Na translocation by the Na/K pump. Nature 323:628–630

7. Artigas P, Gadsby DC (2003) Na^+/K^+ -pumpligands modulate gating of palytoxin-induced ion channels. Proc Natl Acad Sci USA 100:501–505
8. Bahinski A, Nakao M, Gadsby DC (1988) Potassium translocation by the Na^+/K^+ pump is voltage insensitive. Proc Natl Acad Sci USA 85:3412–3416
9. Gadsby DC, Takeuchi A, Artigas P, Reyes N (2009) Peering into an ATPase ion pump with single-channel recordings. Philos Trans R Soc Lond B Biol Sci 364:229–238
10. Artigas P, Gadsby DC (2002) Ion channel-like properties of the Na^+/K^+ pump. Ann NY Acad Sci 976:31–40
11. Gadsby DC, Vergani P, Csanády L (2006) The ABC protein turned chloride channel whose failure causes cystic fibrosis. Nature 440:477–483
12. Hwang TC, Nagel G, Nairn AC, Gadsby DC (1994) Regulation of the gating of cystic fibrosis transmembrane conductance regulator Cl channels by phosphorylation and ATP hydrolysis. Proc Natl Acad Sci USA 91:4698–4702
13. Mechmann S, Pott L (1986) Identification of Na–Ca exchange current in single cardiac myocytes. Nature 319:597–599
14. Fabiato A, Fabiato F (1979) Calculator programs for computing the composition of the solutions containing multiple metals and ligands used for experiments in skinned muscle cells. J Physiol Paris 75:463–505
15. Tsien RY, Rink TJ (1980) Neutral carrier ion-selective microelectrodes for measurement of intracellular free calcium. Biochim Biophys Acta 599:623–638
16. Tsien RY (1980) New calcium indicators and buffers with high selectivity against magnesium and protons: design, synthesis, and properties of prototype structures. Biochemistry 19:2396–2404
17. Watanabe Y, Koide Y, Kimura J (2006) Topics on the Na^+/Ca^{2+} exchanger: pharmacological characterization of Na^+/Ca^{2+} exchanger inhibitors. J Pharmacol Sci 102:7–16
18. Kimura J, Miyamae S, Noma A (1987) Identification of sodium-calcium exchange current in single ventricular cells of guinea-pig. J Physiol 384:199–222
19. Ehara T, Matsuoka S, Noma A (1989) Measurement of reversal potential of Na^+/Ca^{2+} exchange current in single guinea-pig ventricular cells. J Physiol 410:227–249
20. Yasui K, Kimura J (1990) Is potassium co-transported by the cardiac Na–Ca exchange? Pflugers Arch 415:513–515
21. Brew H, Attwell D. (1987) Electrogenic glutamate uptake is a major current carrier in the membrane of axolotl retinal glial cells. Nature 327:707–709. Erratum: Nature (1987) 328:742
22. Noda M, Nakanishi H, Akaike N (1999) Glutamate release from microglia via glutamate transporter is enhanced by amyloid-beta peptide. Neuroscience 92:1465–1474
23. Prasad BM, Amara SG (2001) The dopamine transporter in mesencephalic cultures is refractory to physiological changes in membrane voltage. J Neurosci 21:7561–7567
24. Khoshbouei H, Wang H, Lechleiter JD, Javitch JA, Galli A (2003) Amphetamine-induced dopamine efflux: a voltage-sensitive and intracellular Na^+-dependent mechanism. J Biol Chem 278:12070–12077
25. Galli A, Blakely RD, DeFelice LJ (1996) Norepinephrine transporters have channel modes of conduction. Proc Natl Acad Sci USA 93:8671–8676
26. Accardi A, Miller C (2004) Secondary active transport mediated by a prokaryotic homologue of ClC Cl^- channels. Nature 427:803–807
27. Lísal J, Maduke M (2008) The ClC-0 chloride channel is a 'broken' Cl^-/H^+ antiporter. Nat Struct Mol Biol 15:805–810

Chapter 14

Giant Patch and Macro Patch

Satoshi Matsuoka and Ayako Takeuchi

Abstract

The giant patch method was first developed in 1989. The major characteristic is the use of a pipette with a large tip diameter, which dramatically broadened the applications of the excised patch. The giant patch method enables (1) the current recording of transporters/channels with slow turnover rates and low expression levels, which had been impossible with the conventional method; (2) the rapid voltage-clamp; (3) rapid exchange of the bath and pipette solutions. We describe the devices and limitations for the application of the giant patch method to cardiomyocytes, *Xenopus* oocytes, and cultured cells. In addition, we describe another method, the macro patch, which we developed to overcome one of the limitations of the cardiac giant patch.

14.1. Introduction

The patch-clamp method, developed by Neher and Sakmann, has been continuously improved and widely applied to physiological research. Although this is a powerful method, the current recording of the excised patch is, in some instances, difficult because the area of excised patch membrane is so small. For example, the recording of electrical activity of transporters, such as Na^{+}-Ca^{2+} exchange and the Na^{+}-K^{+} pump, has been impossible because of their slow turnover rates. Low expression densities of ion channels, such as cardiac cystic fibrosis transmembrane conductance regulator (CFTR)-Cl^{-} channel, also makes the recording extremely difficult.

To overcome this problem, the giant patch method was developed by Hilgemann in 1989 (1). Characteristics of this method are as follows: (1) A glass pipette with a large tip is used. (2) The pipette tip is coated with a mixture of mineral oil and Parafilm, called "goop," to facilitate giga-ohm seal formation. (3) A bleb membrane of cardiomyocytes is used. The giant patch method is now applicable not only to cardiomyocytes but also to *Xenopus* oocytes and cultured cells (2–4).

Yasunobu Okada (ed.), *Patch Clamp Techniques: From Beginning to Advanced Protocols*, Springer Protocols Handbooks, DOI 10.1007/978-4-431-53993-3_14, © Springer 2012

The method has several advantages. First, it is applicable to recording the electrical activity of transporters with slow turnover rates and ion channels with low current densities because a larger excised patch membrane can be formed than when using the conventional method. Second, a faster voltage-clamp is expected because of lower electrical resistance of the patch pipette. Lastly, it is possible to exchange the pipette and bath solutions in a faster, easier way.

Originally, the giant patch membrane was excised from a bleb that protruded from a cardiomyocyte. It is uncertain, however, whether the property of a bleb membrane is the same as that of the intact sarcolemma of cardiomyocytes. Therefore, a method to form a larger excised patch from the intact cardiomyocytes has been long awaited. We have succeeded in creating a relatively large patch membrane from the intact cardiomyocyte and recording transporter currents; it is called a "cardiac macro patch" (5, 6).

In this chapter, we explain the original giant patch method for cardiac blebs and that of *Xenopus* oocytes and then describe the cardiac macro patch method. Several applications of these methods are also described.

14.2. Giant Patch Excised from a Cardiomyocyte Bleb

When ventricular myocytes are stored in a high K^+- and Na^+-free solution, one or several bulb-shaped bulges of the plasma membrane, or "blebs," appear. Hilgemann was the first to succeed in creating the excised membrane patch from blebs using glass pipettes with tip diameters of 10–22 μm and in recording Na^+ and Na^+-Ca^{2+} exchange currents (1, 7) (Fig. 14.1). Several researchers have modified and improved the original method. We describe our improved methods.

14.2.1. Preparation of Blebs

We have worked with bleb giant patches mostly from guinea pig ventricular myocytes. For giant patch experiments, a conventional collagenase method is enough to isolate the myocytes, with no specific methods required. After cell isolation, the ventricular myocytes and/or small tissue fragments are placed in a blebbing solution (150 mM KCl, 10 mM EGTA, 2 mM $MgCl_2$, 20 mM glucose, 15 mM HEPES, pH 7.2/KOH) and then stored overnight in the refrigerator. If a modified KB solution (25 mM KCl, 70 mM glutamate, 10 mM KH_2PO_4, 10 mM taurine, 0.5 mM EGTA, 11 mM glucose, 10 mM HEPES, pH 7.3/KOH) is used, larger blebs tend to grow.

Figure 14.1 demonstrates two blebs developed from a guinea pig ventricular myocyte. Blebs start to develop after 2–6 h and can be used for experiments about 12–48 h after formation. Early blebs

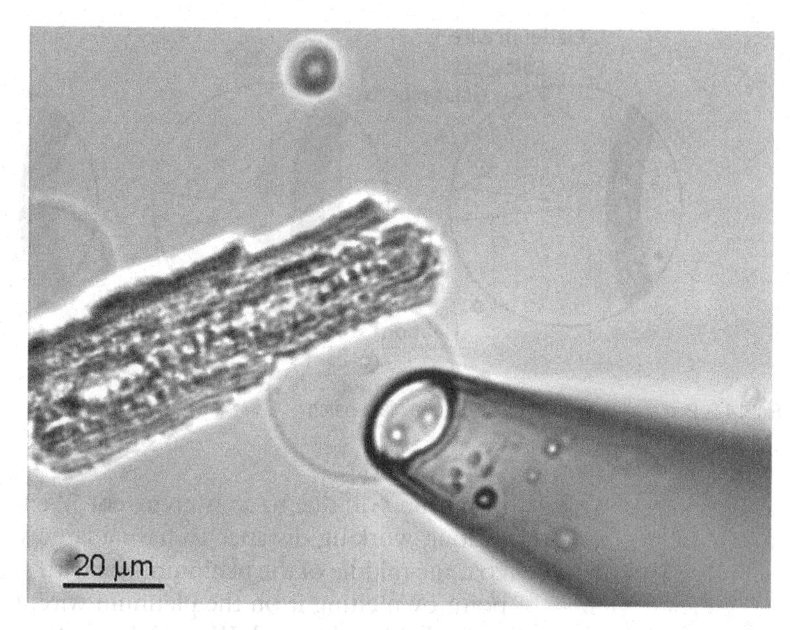

Fig. 14.1. Blebs developed in a guinea pig ventricular myocyte and a giant patch pipette (i.d. 18 μm). Giga-ohm seal is established on the bleb.

are small, and formation of the giga-ohm seal is difficult irrespective of the size. Usually, we carry out patch experiments after the overnight incubation. After bleb formation, some ventricular myocytes keep their original shape, but others do not. There seems no difference in electrophysiological properties of the blebs in cells with a rod-like shape or a round shape.

14.2.2. Preparation of Glass Pipette and Coating

For giant patch formation, a glass capillary with a relatively large outer diameter (o.d.) and a relatively thin wall is suitable. We use borosilicate glass capillaries provided by Hilgenberg GmbH, Malsfeld, Germany (o.d. 1.65 mm/i.d. 1.32 mm or o.d. 2.0 mm/i.d. 1.4 mm).

It is not straightforward to create a glass pipette with a clear, large tip. Figure 14.2 illustrates our method, which is essentially the same as that of Hilgemann (2, 3). First, we create a glass pipette used for a conventional whole-cell recording, whose tip diameter is about 2 μm. We use a pipette puller provided by Narishige, Tokyo, Japan (PB-7 or PC-10). A pipette with a relatively long tip facilitates the following manipulations.

To make the pipette tip wide, we use a microforge of our own making, which is mounted on the stage of an inverted microscope. The microforge consists of a relatively thick platinum wire (about 0.5 mm diameter) connected to a variable transformer, a foot switch to turn the current on and off, and two three-axis manipulators to control positions of the glass pipette and the platinum wire. The merits of using the thick platinum wire are that there is little heat expansion upon turning on the electricity and less possibility

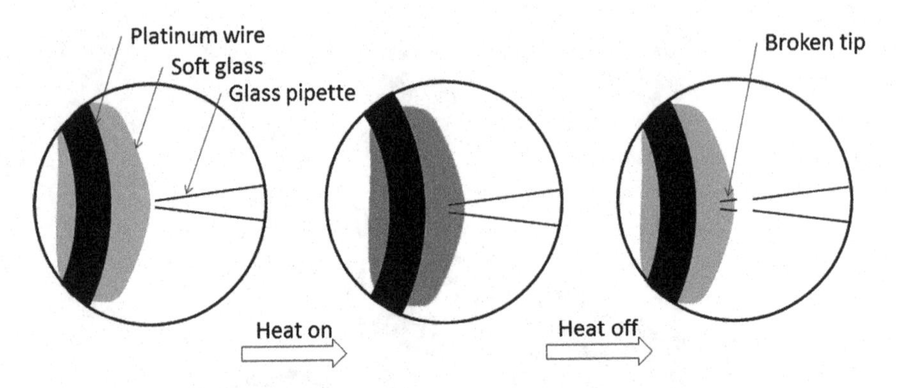

Fig. 14.2. Preparation of glass pipettes for a giant patch.

of burning off due to an overcurrent. We use a 100× objective lens of long working distance to have a large view of the glass tip. We coat the middle of the platinum wire with a soft glass of low melting point by melting it on the platinum wire. Glass containing lead or soda glass can be used. We mainly use hematocrit tubes (soda lime glass) produced by Chase Glass Co., which are provided by Iwaki Glass, Japan (currently Asahi Glass, Tokyo, Japan).

A glass pipette with a large tip diameter is made as follows. First, locate the glass pipette close to the coated platinum wire (Fig. 14.2, left). Second, heat the platinum wire by turning on the foot switch. The heating moves the platinum wire slightly toward the center. The power voltage may be the same as or slightly lower than that used for soft glass coating. Third, soon after starting the heat, move the platinum wire to the right using a manipulator so a small portion of the glass tip runs into the soft glass (Fig. 14.2, middle). Finally, turn off the switch. The platinum wire returns to the initial position, and the pipette tip breaks (Fig. 14.2, right). We repeat these procedures until the tip with a desired diameter is created. Lastly, apply heat to smooth the tip (heat polish).

We usually replace the soft glass on platinum wire every 2–3 days because it becomes difficult for the glass tip to run into the soft glass. Rather, the tip tends to melt down. This method needs practice, but making a perfectly round, smooth tip is not necessarily essential. It is possible to make a giant excised path with a glass pipette of irregular shape. It is the tip size that is important. Before adopting this method, we made pipettes by scratching the glass tip on the platinum wire (7, 8). Programmable and multistep micropipette pullers, such as products by Shutter Instrument, may be used to make a large pipette (9), although we do not have any experience with this method.

The next step is to coat the glass tip with a mixture of mineral oil and Parafilm. The mixture is made as follows. Cut Parafilm (Pechiney Plastic Packaging, Chicago, IL, USA) finely and then heat and mix well with a mineral oil of about the same weight as Parafilm, using a

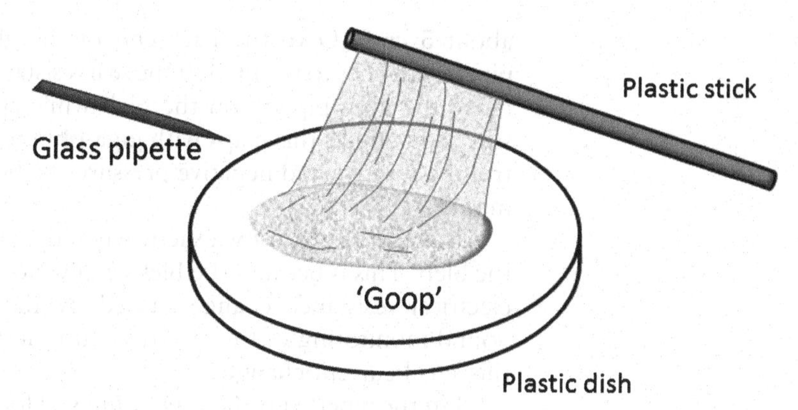

Fig. 14.3. Pipette coating with "goop".

glass stick or spoon until the lumps of Parafilm disappear. You may store the mixture in culture dishes at room temperature for a long time. Place a small amount of the mixture on the top of culture dish and then remix with a small amount of mineral oil using a plastic or glass stick until it becomes sticky and forms a thin film when the stick is elevated. Two types of mineral oil are available: light (light white oil M3516; St. Louis, MO, USA, Sigma) and heavy (heavy white oil 400–5; St. Louis, MO, USA, Sigma). The light mineral oil tends to facilitate giga-ohm seal formation and the heavy mineral oil to stabilize the patch membrane. Coat the glass tip by moving the glass pipette into a thin film of the mixture (Fig. 14.3). When using a liquid-type mixture made by adding more mineral oil, just dipping the glass pipette into the mixture is enough.

Fill a filtrated pipette solution into the glass pipette, and set it on a pipette folder. At this moment, the hole of the glass tip is sealed by the mixture. Blow out the sealed mixture by applying positive pressure with a syringe connected to the back-end of glass pipette. You will see a linear stream of pipette solution if the tip diameter is large enough.

You can use any composition of pipette and bath solutions for the giant patch experiments. Giga-ohm seal formation is easier when the pipette solution contains Ca^{2+}. In addition, at least 10 mM of Cl^- is necessary in both solutions to make the giga-ohm seal.

14.2.3. Formation of a Giga-Ohm Seal

Transfer the cells stored in the refrigerator to a culture dish or a recording chamber on the patch-clamp setup, and wait for awhile until the cells adhere to the bottom. Locate the glass pipette close to the bleb, applying weak positive pressure to the back-end. In most cases, blebs tend to move away from the pipette tip because of the solution flow from the pipette tip. Therefore, it is important to find cells that are stuck to the bottom. Gentle tapping of the microscope stage helps to find attached cells. You can usually form a giga-ohm seal within seconds by applying negative pressure of about 5 cmH_2O to the back-end of the glass pipette when the pipette attaches to the bleb. If the seal resistance is not high enough, move the glass pipette on the bleb while applying negative pressure. Sometimes this helps with giga-ohm seal formation. We control the positive and negative pressures by breathing in and out by mouth.

It is difficult to know exactly when the glass pipette attaches to the bleb. This is because the blebs easily move and also because the electrical resistance change caused by the attachment is small. Sound monitoring of the electrical current greatly helps detecting this small current change.

Lift the pipette up after giga-ohm seal formation, and an inside-out excised patch is easily formed. Sometimes a vesicle is formed at the pipette tip, although it is difficult to see. Quickly attaching the pipette tip to air bubbles in the bath chamber, if any, can effectively break the vesicle. If the desired membrane current is unable to be recorded, it is worth trying to break the vesicle.

If blebs are small and the giga-ohm seal formation is difficult, incubate the cells in the blebbing solution diluted to 70–80% for several minutes to make larger blebs. The addition of 5–8 mM Mg^{2+} to the bath solution sometimes facilitates the giga-ohm sealing. It is practically impossible to create an outside-out excised patch. The outside-out excised patch may be formed during the procedure of vesicle breaking, although it occurs on only rare occasions.

14.3. Giant Patch Excised from a *Xenopus* Oocyte

It is possible to create giant excised patches directly from *Xenopus* oocytes using glass pipettes with a tip diameter of about 20–35 μm (10, 11). Bleb formation is not required because of the large size of oocytes. This method is essentially applicable to any type of cloned channels/transporters if protein is expressed in the oocytes. Our method to form giant patches excised from oocytes is explained in the next sections.

14.3.1. Preparation of Xenopus Oocytes

For the giant patch experiments, you can prepare *Xenopus* oocytes with the conventional method and no special treatment. Place the cRNA-injected oocyte, whose follicles are removed with collagenase, in a high osmotic solution to shrink. Remove the clear vitelline membrane with fine tweezers under a stereo-microscope. Then, place the oocyte in a bath solution of physiological osmotic pressure. The composition of bath solution we use is the following: 100 mM KOH, 100 mM 2-(*N*-morpholino)ethanesulfonic acid, or MES (Sigma) or aspartate, 20 mM HEPES, 5 mM EGTA, and

5 mM $Mg(OH)_2$ (pH 7.0/MES). The high osmotic solution is prepared by adding 200 mM mannitol to the bath solution.

If you plan to change the cytoplasmic Ca^{2+} concentration, the endogenous Ca^{2+}-activated Cl^- current should be minimized. We add 100 μM niflumic acid or flufenamic acid to a pipette solution and replace Cl^- with MES or aspartate.

14.3.2. Preparation of Glass Pipette and Coating

The glass pipette is prepared as describe above. The tip diameter of our usual pipette is 20–30 μm. Giga-ohm seal formation is easier with a pipette tip that has an obtuse angle than one with an acute angle. Rounding the tip by heat polish aids giga-ohm seal formation. You can make a large pipette by cutting the glass in large diameter and then fire polishing strongly. Seal formation with such a large pipette, however, is quite difficult.

The pipette is coated in the same way as described above. Similar to the case with bleb giant patch, adding light mineral oil tends to facilitate giga-ohm seal formation, whereas heavy mineral oil helps stabilize the patch membrane. Addition of a small amount of decane to the mixture sometimes facilitates giga-ohm sealing of the oocyte patch.

14.3.3. Formation of Giga-Ohm Seal

The oocytes have two kinds of surface: animal (black) and vegetal (white) poles. The oocyte giant patches are usually excised from the animal pole. Figure 14.4 illustrates the process of patch formation.

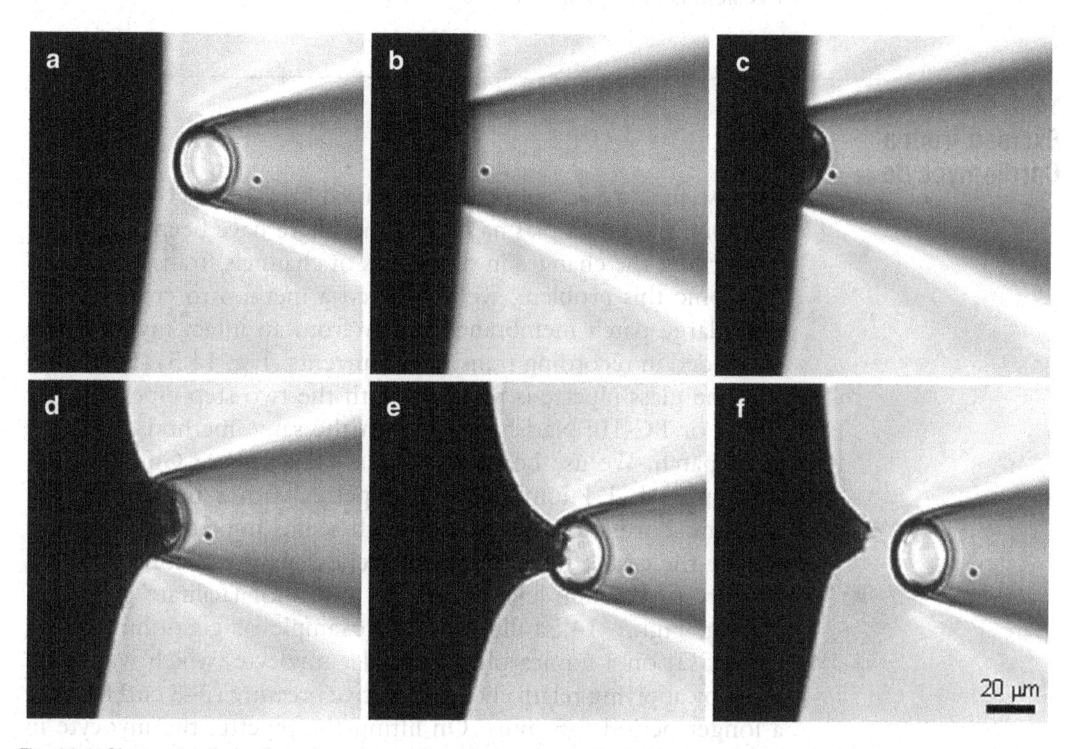

Fig. 14.4. Giant patch formation of an oocyte. The *black mass* on the *left* is part of the oocyte. The tip i.d. of the pipette (*right*) is 24 μm.

Place the glass pipette close to the oocyte (Fig. 14.4a) and then attach it to the oocyte surface by moving the pipette or the stage on which the oocyte is located (Fig. 14.4b). It is difficult to see the tip position at this point because we use an inverted microscope. You can detect the attachment by checking the increase in pipette resistance. The giga-ohm seal is usually formed within 1–5 min by applying negative pressure of ~10 cmH_2O (Fig. 14.4c), although sometimes the seal forms spontaneously without applying the continuous negative pressure.

Be careful when detaching the pipette. The conventional method of rapid pipette lifting usually fails. It is effective to move the pipette very slowly while watching the glass tip and checking the membrane current. The success rate of excision becomes higher when the pipette is pulled horizontally at an angle of 45°–60° than when it is lifted up. It could be possible to move the pipette horizontally at an angle of 90° (up or down direction in Fig. 14.4). As the pipette moves away, part of the oocyte membrane becomes stretched (Fig. 14.4d). Further pipette movement breaks the membrane within the glass tip, and the oocyte slowly backs away from the tip (Fig. 14.4e), resulting in formation of an excised patch (Fig. 14.4f).

In most cases, the excised patch is stable for 15–60 min. The possibility of vesicle formation is considerably lower than in the bleb giant patches. If a vesicle forms, the same method of breaking a vesicle is applicable.

14.4. Macro Patch Excised from a Cardiomyocyte

Because the cardiac giant patch is formed from the bleb membrane, not from the intact plasma membrane, there have been arguments concerning the changes in properties of channels/transporters. To overcome this problem, we developed a method to create a relatively large patch membrane directly from an intact myocyte and succeeded in recording transporter currents (Fig. 14.5) (5, 6).

The glass pipette is prepared with the two-step pipette puller (PB-7 or PC-10; Narishige) or with the same method as for the giant patch. We use borosilicate glass (Hilgenberg GmbH) (o.d. 2.0 mm, i.d. 1.4 mm) to make pipettes with a tip diameter of 3–8 μm. Formation of a larger patch seems impossible currently. Unlike the case of making giant patches, pipette coating with the mixture of Parafilm and mineral oil does not facilitate giga-ohm sealing. Figure 14.5a illustrates an example of giga-ohm sealing (2–4 GΩ) on a guinea pig ventricular myocyte, which was facilitated by applying relatively low negative pressure (3–8 cmH_2O) for a longer period (~5 min). On lifting the pipette, the myocyte in many cases lifts together with the pipette, failing the excised patch formation. Blowing the myocyte with a solution stream increases the success rate of the excised patch formation. To do that, we place the myocyte attached to the pipette close to the outlet of the solution from another glass capillary. The solution flow blows off the cell (Fig. 14.5b2). We use a double-lumen glass capillary (theta capillary) mounted on a piezo transducer. Applying an appropriate voltage makes the theta capillary move quickly to switch the solution rapidly (Fig. 14.5b3).

We have successfully recorded Na^{+}-Ca^{2+} exchange, Na^{+}-K^{+} pump, and CFTR-Cl^{-} currents with this method. Essentially the same current properties of the Na^{+}-Ca^{2+} exchange were obtained with bleb giant patches and macro patches, although the kinetics were different (6). Of the two methods – the bleb giant patch and the cardiac macro patch – the bleb giant patch is easier to form, and the larger excised patch membrane can be created once mastered. The success rate of current recording is lower with the cardiac macro patch.

14.5. Giant Patch Excised from a Cultured Cell

Excised patch analysis of channels/transporters expressed in cultured cells has been widely carried out. We found that inside-out patches were easily formed with conventional methods using glass pipettes of ~3 μm tip diameter. We could record Na^{+}-Ca^{2+} exchange

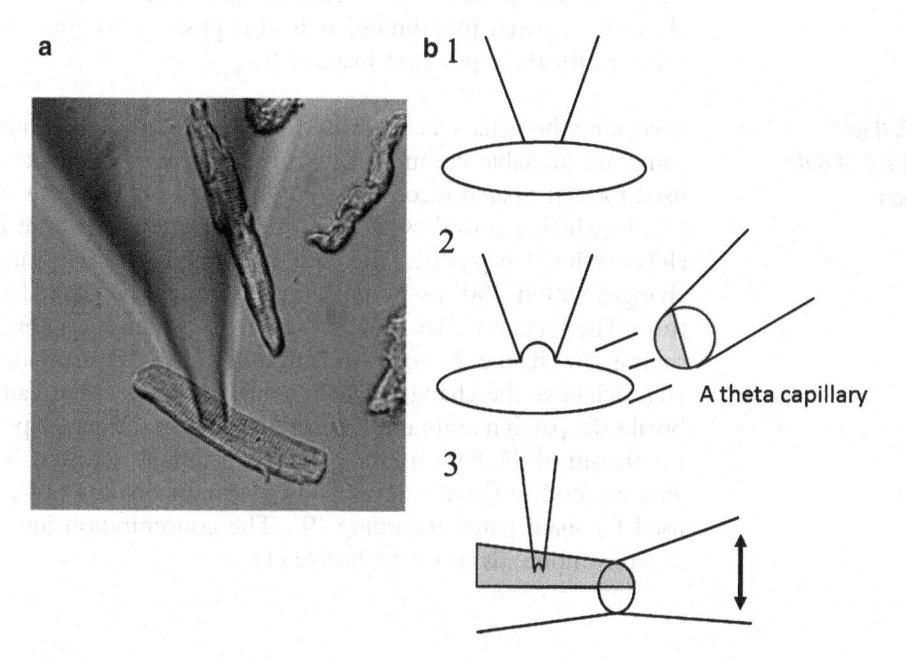

Fig. 14.5. Method to create a macro patch. (**a**) Guinea pig ventricular myocytes and a glass pipette for a macro patch (i.d. 7 μm). (**b**) Process for macro patch formation.

currents using the patches excised from the cultured cells (HEK 293 or COS cells) overexpressing Na^{+}-Ca^{2+} exchanger (NCX1). If the expression level of channels/transporters is high enough, the conventional method is sufficient. However, the giant patch method has substantial merit when a rapid solution change or a rapid voltage clamp is required (4). In those cases, a special device to excise the patch after giga-ohm sealing is necessary. The cell-blowing method described in Fig. 14.5 is one of the practical methods.

14.6. Applications

The giant patch method has developed more practical applications by combining it with other techniques. Some instances are introduced as follows. Details of the method are explained in the cited articles.

14.6.1. Perfusion of the Pipette Solution

Under the giant patch recordings, it is possible to perfuse the solution rapidly in the glass pipette. The pipette perfusion method is described in detail in Chap. 15. We applied the pipette perfusion method developed by Soejima and Noma (12) to the giant patch recording (13). The negative pressure for the perfusion has to be set lower (4–6 cmH_2O) than that used for whole-cell patch recordings because of the fragility of the giant patch membrane. However, the perfusion speed is faster because a larger inner tube can be set closer to a patch membrane. It is also possible to apply positive pressure for the pipette perfusion (2).

14.6.2. Rapid Exchange of Bath Solution

Several methods have been devised for changing bath solutions, and some are available commercially. Any of these instruments can be used for giant patch recordings. We modified Hilgemann's method (7), in which the outlets of three polyethylene tubes were located close to the glass pipette. The bath solution around the giant patch changes within 200 ms by pushing the syringe connected to each tube. The theta capillary mounted on a piezo transducer (Fig. 14.5) can rapidly change the solution within 30 ms (5). A larger tip diameter facilitates the change of bath solution, but a fast stream easily breaks the patch membrane. The angle between the glass pipette and the stream of the bath solution must be carefully adjusted. The oil-gate method, originally developed by Qin and Noma (14), can be used for giant patch recording (9). The concentration jump using caged compounds is also applicable (15).

14.6.3. Modification of Membrane Composition

The activity of channels/transporters is modified by the lipid composition of the membrane. Collins and Hilgemann (16) applied an oil droplet to the glass pipette of the giant patch and demonstrated that phosphatidylserine augments the Na^{+}-Ca^{2+} exchange current. Phosphatidylserine possibly diffuses in the mixture of Parafilm and mineral oil and thereby modifies the composition of the patch membrane. This method can be used for applying nonaqueous materials to patch membranes.

14.6.4. Single-Channel Recording

The giant patch method has an advantage in single-channel recording of channels whose expression density is low. Gadsby et al. (17) succeeded in making single-channel recordings of the CFTR-Cl^{-} channel in the giant patch excised from blebs of ventricular myocytes. This method is also applicable to single-channel recordings of cloned channels that express at low density in oocytes.

14.6.5. Fast Voltage-Clamp

The large tip diameter of the giant patch reduces the series resistance of the pipette. Therefore, a faster voltage-clamp is possible. Hilgemann (18) demonstrated the charge movement of the Na^{+}-K^{+} pump with a microsecond resolution. This method is applicable to other channels/transporters (4).

References

1. Hilgemann DW (1989) Giant excised cardiac sarcolemmal membrane patches: sodium and sodium-calcium exchange currents. Pflugers Arch 415:247–249
2. Hilgemann DW (1995) The giant membrane patch. In: Sakmann B, Neher E (eds) Single-channel recording, 2nd edn. Plenum, New York
3. Hilgemann DW, Lu CC (1998) Giant membrane patches: improvements and applications. Methods Enzymol 293:267–280
4. Couey JJ, Ryan DP, Glover JT, Dreixler JC, Young JB, Houamed KM (2002) Giant excised patch recordings of recombinant ion channel currents expressed in mammalian cells. Neurosci Lett 329:17–20
5. Fujioka Y, Komeda M, Matsuoka S (2000) Stoichiometry of Na^{+}-Ca^{2+} exchange in inside-out patches excised from guinea-pig ventricular myocytes. J Physiol 523:339–351
6. Fujioka Y, Hiroe K, Matsuoka S (2000) Regulation kinetics of Na^{+}-Ca^{2+} exchange current in guinea-pig ventricular myocytes. J Physiol 529:611–623
7. Collins A, Somlyo AV, Hilgemann DW (1992) The giant cardiac membrane patch method: stimulation of outward Na^{+}-Ca^{2+} exchange current by MgATP. J Physiol 454:27–57
8. Hilgemann DW, Collins A (1992) Mechanism of cardiac Na^{+}-Ca^{2+} exchange current stimulation by MgATP: possible involvement of aminophospholipid translocase. J Physiol 454:59–82
9. Doering AE, Lederer WJ (1994) The action of Na^{+} as a cofactor in the inhibition by cytoplasmic protons of the cardiac Na^{+}-Ca^{2+} exchanger in the guinea-pig. J Physiol 480:9–20
10. Matsuoka S, Nicoll DA, Reilly RF, Hilgemann DW, Philipson KD (1993) Initial localization of regulatory regions of the cardiac sarcolemmal Na^{+}-Ca^{2+} exchanger. Proc Natl Acad Sci USA 90:3870–3874
11. Matsuoka S, Nicoll DA, Hryshko LV, Levitsky DO, Weiss JN, Philipson KD (1995) Regulation of the cardiac Na^{+}-Ca^{2+} exchanger by Ca^{2+}: mutational analysis of the Ca^{2+}-binding domain. J Gen Physiol 105:403–420
12. Soejima M, Noma A (1984) Mode of regulation of the ACh-sensitive K-channel by the muscarinic receptor in rabbit atrial cells. Pflugers Arch 400:424–431
13. Hilgemann DW, Matsuoka S, Nagel GA, Collins A (1992) Steady-state and dynamic properties of cardiac sodium-calcium exchange: sodium-dependent inactivation. J Gen Physiol 100:905–932

14. Qin DY, Noma A (1988) A new oil-gate concentration jump technique applied to inside-out patch-clamp recording. Am J Physiol 255:H980–H984
15. Friedrich T, Bamberg E, Nagel G (1996) Na+, K+-ATPase pump currents in giant excised patches activated by an ATP concentration jump. Biophys J 71:2486–2500
16. Collins A, Hilgemann DW (1993) A novel method for direct application of phospholipids to giant excised membrane patches in the study of sodium-calcium exchange and sodium channel currents. Pflugers Arch 423:347–355
17. Nagel G, Hwang TC, Nastiuk KL, Nairn AC, Gadsby DC (1992) The protein kinase A-regulated cardiac Cl^- channel resembles the cystic fibrosis transmembrane conductance regulator. Nature 360:81–84
18. Hilgemann DW (1994) Channel-like function of the Na, K pump probed at microsecond resolution in giant membrane patches. Science 263:1429–1432

Chapter 15

Pipette Perfusion Technique

Minoru Horie

Abstract

The piette perfusion technique is a version of patch-clamp techniques and provides a greater intracellular access during electrophysiological recordings. This internal perfusion technique offers the intracellular change of not only ions but also substances with higher molecular weight such as enzymes, antibodies, and metabolites. As to the other versions of techniques, it requires dedicated time and personal training. This chapter describes the tips and tricks of the unique pipette perfusion technique.

15.1. Introduction and Historical Background

The introduction of patch-clamp techniques has revolutionized our understanding of ion channel physiology and diseases (1). Among variations of patch-clamp techniques, the whole-cell confi guration offers the standard recording mode. Because the mode clamps a single cell through the broken patch, it inevitably causes a loss of intracellular milieu, which then affects the physiological condition of the cell under voltageclamp (2). This weak point, however, gives a particular advantage to exchange the intracellular constituent across the cell membrane or to add drugs or metabolites, even a substance that is not physiological present in the cell, from the outside if one can change pipette solutions. A variety of approaches were then invented to accomplish a rapid exchange, and the pipette perfusion technique can be now applied to ion channel study (3–46). This chapter briefl y summarizes the historical background of the perfusion method and then introduces an up-to-date technique.

Yasunobu Okada (ed.), *Patch Clamp Techniques: From Beginning to Advanced Protocols*,
Springer Protocols Handbooks, DOI 10.1007/978-4-431-53993-3_15, © Springer 2012

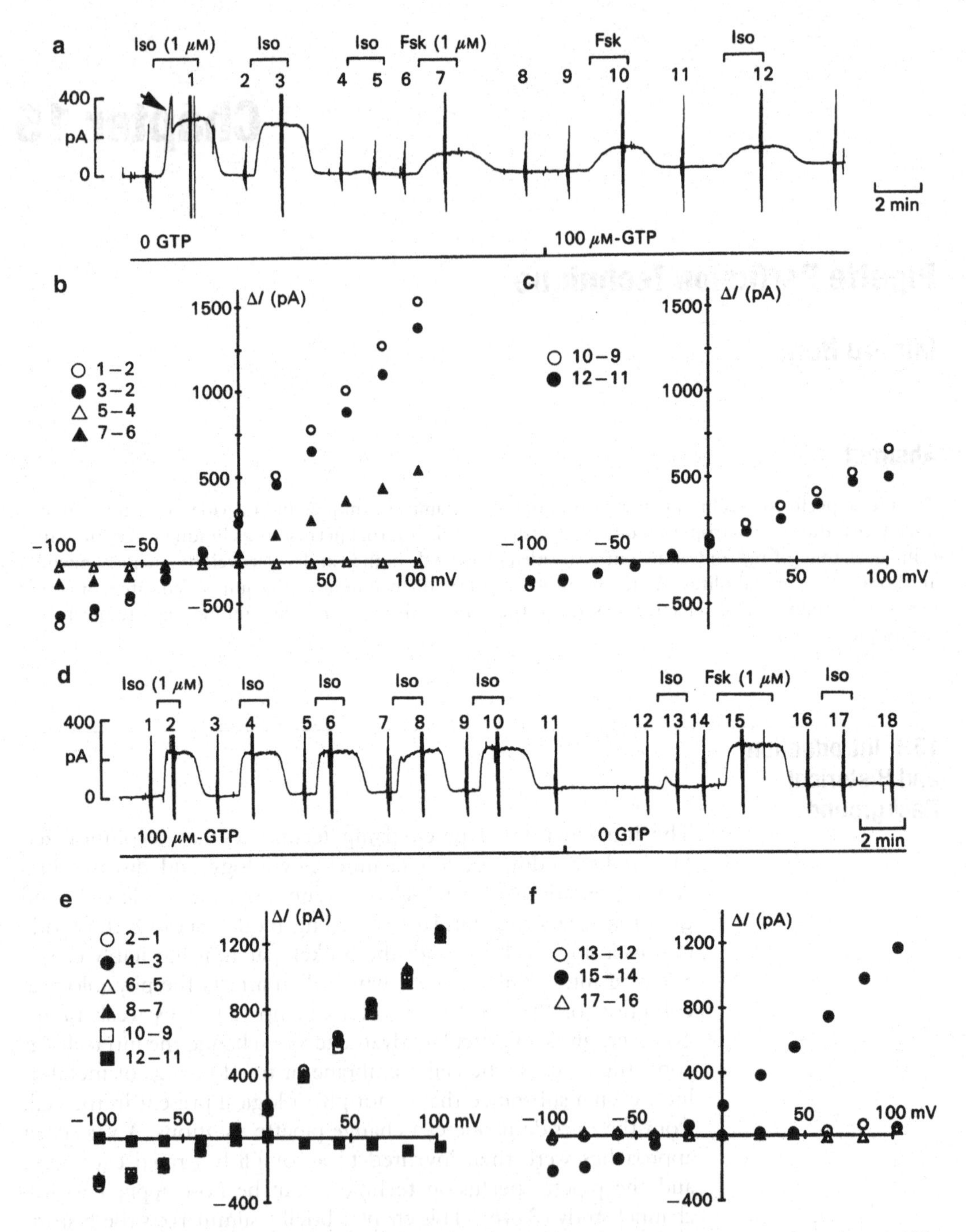

Fig. 15.1. Pipette guanosine-5′-triphosphate (GTP) 100 μM (*lower lines* in (**a**) and (**d**)) can prevent disappearance of the response to isoprenaline (**d**) and restore the response following its disappearance during dialysis with GTP-free pipette solution (**a**). (**a**) Chart record of whole-cell current. *Arrow* marks reduced amplifier sensitivity. *Thick vertical lines* (*1–12*) mark periods of application of voltage pulses to collect steady-state current–voltage (I–V) data. Holding potential was 0 mV; cell capacitance 143 pF; R_{pip} 0.6 MΩ; R_{acc} 1.3 MΩ. (**d**) Whole-cell current. Holding potential 0 mV; cell capacitance 210 pF; R_{pip} 0.7 MΩ; R_{acc} 2.1 MΩ. In contrast to the isoprenaline (*Iso*) responses, the Cl^- current responses to exposure to 1 μM forskolin (*Fsk*) were unaffected by addition (**a**) or withdrawal (**d**) of pipette GTP. (**b**), (**c**), (**e**), (**f**) Steady-state differences in the I–V relation. They were conductance-induced by agonists (isoprenaline or forskolin) and were obtained by subtracting control current levels from corresponding currents in agonists. Numbers correspond to the data collection episodes indicated in (**a**) and (**d**) (adapted from (2)).

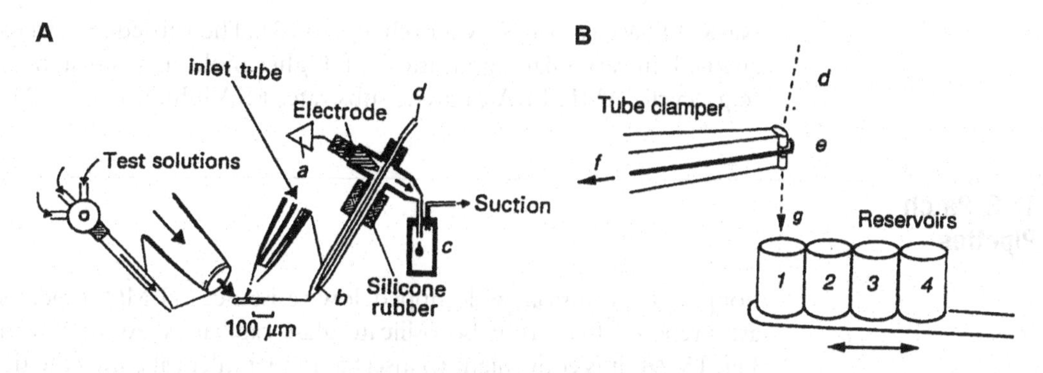

Fig. 15.2. Device for pipette perfusion. (**A**) Experimental setup. (**A**) From *left*: apparatus for bath solution exchange, magnified views of the apparatus, a single cardiac myocyte, a patch pipette with inlet tube, and a perfusion device with: *a*, connection to the patch-amplifier; *b*, inlet tube, pipette, and its holder; *c*, discharge tube of pipette solution and waste box; and *d*, a silicon tube connected to the supply tube. (**B**) From *left*, a tube-clamper mounted on a manipulator (not shown) and reservoir boxes (*1–4*) for new pipette solutions. The silicon tube (*d*) is connected to a respective reservoir box through a clamper hole (*e*), which is controlled by a thread wire (*f*). (Adapted from (3)).

Conventionally, in the whole-cell configuration, internal or pipette solutions are made to fill patch pipettes and usually contain ions and some other metabolites such as millimolar adenosine triphosphate (ATP), mimicking the intracellular milieu. To maintain a low Ca^{2+} concentration in the cytoplasm, Ca^{2+}-chelating agents are also employed in the pipette solution. These intracellular agents are especially needed during the receptor-mediated modulation of ion currents. During cellular signal transduction involving guanosine-5′-triphosphate (GTP)-binding proteins (G proteins), for example, intracellular GTP is definitely required for the maintenance of G proteins.

Figure 15.1 shows a chart of whole-cell chloride (Cl^-) current activated by cyclic adenosine monophosphate (cAMP)-dependent protein kinase A (PKA) that we recorded in a guinea pig ventricular myocyte (2). Using a pipette perfusion technique, increase in pipette GTP concentration (100 μM) rescued the response to isoprenaline (Fig. 15.1a). Conversely, pipette GTP (100 μM) was omitted, which completely abolished the current response to extracellular isoprenaline (1 μM) (Fig. 15.1c). In contrast, its response to forskolin, a direct PKA activator, remained intact, indicating that micromolar levels of GTP are essential for receptor-mediated regulation of this Cl^- conductance.

In cardiac electrophysiology, in which the patch-clamp method was applied to single acutely isolated myocytes during the early 1980s, the pipette perfusion method was first employed to measure single-channel activities of muscarinic acetylcholine-sensitive K^+ channels ($I_{K,ACh}$) in rabbit atrial myocytes (3). Figure 15.2 illustrates the perfusion device they employed. The device was then applied to the whole-cell current recording by their colleagues: isolation of L-type Ca current (4) conductance by the activation of

Na-K ATPase (5) or Na-Ca exchanger (13). The subsequent trials enabled intracellular application of higher-molecular substances (e.g., cyclic AMP, PKA catalytic substrate, PKA inhibitor.)(8–12).

15.2. Patch Pipettes

For pipette perfusion, wide-tipped, low-resistance (~1 MΩ) pipettes are prepared from thin borosilicate glass capillaries. As shown in Fig. 15.2A, it is convenient to insert a thinner inlet tube into the tip of the pipette as close as possible. As mentioned below, to inset the inlet tube as close as possible to the end of the pipette tip, the shape of the pipette should be acutely tapered to its wider tip (Fig. 15.2A). When used a Narishige double-step puller, for example, we employed a shorter distance for the first pulling. Pipettes with a lower resistance are preferable for efficient intracellular perfusion.

15.3. Device for Pipette Perfusion

Right panel of Fig. 15.2A shows a perfusion device made by plastic cross. The parts include a recording electrode, an inlet tube, a pipette and its holder, a discharge tube of pipette solution, and connection to a supply tube and reservoir boxes for new pipette solutions (see also Fig. 15.2B). These boxes are connected directly to the inlet tube via a thin silicon tube and a stainless thin pipe, which penetrates the perfusion device from top to bottom. The tip size of the inlet tube is ~50 μm (Fig. 15.2A). These tubes are made from a thin polyethylene tube with 2.5 mm o.d. Under a dissecting microscope, the tube was pulled on the platinum line heated electrically two times to obtain an acutely tapering tip of the inlet tube. As for the recording electrode, we used a small Ag/AgCl pellet connected by a short cable to a whole-cell patch-clamp amplifier (Fig. 15.2A, *a*). The end of the discharge silicon tube (Fig. 15.2A, *c*) is insulated by air in the waste box for old solutions, which largely reduces the electrical resistance of the perfusion device and the noise level during the recording.

Before each experiment, the inlet tube is connected to the perfusion device; and then the entire inside of the device is filled with normal Tyrode solution. The pipette is filled with the same solution and is connected to the pipette holder (Fig. 15.2A, *b*). Under a dissecting microscope, the tip of the inlet tube is placed as close as to the tip end of the pipette as possible (~200 μm) (Fig. 15.2A). This is a key point for obtaining efficient pipette perfusion.

15.4. Setting Up the Perfusion Device

The perfusion device is now mounted on the manipulator. To avoid loss of solution from the device, we clamp the discharge silicon tube (Fig. 15.2A, *c*) using a small clip normally used for vascular surgery. The tip end of the solution supply tube (also made of silicon) is directly dipped into one of the reservoir boxes and is filled with the desired pipette solution via a thin hole in the clamp (Fig. 15.2B, *e*). The solution supply tube can be temporarily stopped with the clamp device using a thin thread (Fig. 15.1B, *f*).

In the first solution reserve (Fig. 15.2B, *e*), we usually put normal Tyrode solution containing millimolar free Ca^{2+}, which facilitates the formation of a giga-seal against the myocyte. The solution supply tube is then changed to the second reservoir. At each point of solution exchange, the clamp must be used to tightly close the tube to avoid air leak. The reservoirs are placed on a short plastic bar that is connected to another manipulator, so we can move the reservoirs using the manipulator.

After all the connections are completed, we check for an absence of small air bubbles in the perfusion device because if the tip of the inlet tube traps air the device no longer works effectively. Under mild suction applied to the end of the waste reservoir (10–20 cmH_2O), pipette solution is drained from the solution supply reservoir toward the direction of the waste box through the perfusion device. Empirically, we know that the effective exchange of pipette solution is achieved when the waste solution rhythmically drips every 10 s. If the perfusion rate is faster than this speed, it is suspected that the tip of the inlet tube is far from the tip of the pipette or there is solution leak (usually at the connection site of the inlet tube and its holder).

15.5. The Experiment: Tips and Tricks

We start the experiment by obtaining a giga-seal on the cell membrane. The solution supply is stopped by clamping the silicon tube (Fig. 15.2A, *d*); and under the observation of myocytes with an inverted microscope (Nikon TMD), we usually choose small cells because they are more suitable for efficient exchange of intracellular milieu and employ relatively lower suction pressure (10–20 cmH_2O). After successful giga-seal formation, the suction is released once and the solution tube is moved to the next reservoir (Fig. 15.2B, *2*). This box is just for washing the end of the tube, which is needed because millimolar Ca^{2+} concentrations in the first box contaminate the intracellular solution, which usually contains only nanomolar levels of free Ca^{2+}. Therefore, the third box

(Fig. 15.2B, *3*) is a reservoir for the first desired pipette solution; the clamp for the solution tube is released, and suction is restarted. On continuing suction (10–20 cmH_2O), we count the number of wasted solution drops. Empirically, we know that four drops are enough for full exchange of the pipette solution, and we then clamp the supply tube again (although this number may vary depending on the device). At the completion of these processes, we can break the cell membrane by applying a higher suction pressure to the end of waste reservoir and form the whole-cell configuration.

There are several tips and tricks for successful experiments.

1. The most important player is the inlet tube. Once you get a good one, your experiments are 50% successful.
2. Training is required for quick, complete preparation of the perfusion device.
3. Checking for air bubbles is a key point to avoid loss of time.
4. Use a low suction pressure except for membrane breakup.
5. Obstruction of the inlet tube is always fatal. Clean it by using an ultrasonic cleaning machine immediately before each experiment.
6. After an experiment, wash the inlet tube and keep it in a clean box away from dust.

15.6. Typical Examples of Experiments with Pipette Perfusion

Figure 15.3a depicts a representative trace of whole-cell current recorded from an adult guinea pig ventricular myocyte. By changing the Cl^- concentrations in pipette solutions we attempted to confirm that the conductance activated by the PKA pathway is sensitive to Cl^- ions in a single myocyte. To induce the conductance in this specific experiment, we employed forskolin, which is known to activate directly the adenyl cyclase and thereby PKA. At the beginning, the pipette contained 24 mM Cl^-. The holding potential was 0 mV, and 100 nM forskolin applied to the bath solution increased the outward current (~180 pA) (Fig. 15.3a, *1*→*2*). In the continued presence of forskolin, the pipette Cl^- concentration was then changed to 109 mM, which reduced the outward current at 0 mV quickly (*2*→*3*). Subsequent washout of forskolin further decreased the current level (*3*→*4*).

Figure 15.3b shows the current–voltage relation for the forskolin-sensitive conductance obtained by a step-clamp method. The calculated reverse potentials were -40 mV and -2 mV, respectively. The values shown in Fig. 15.3b are very close to the Cl^- equilibrium potential calculated by the Nernst equation. Thus,

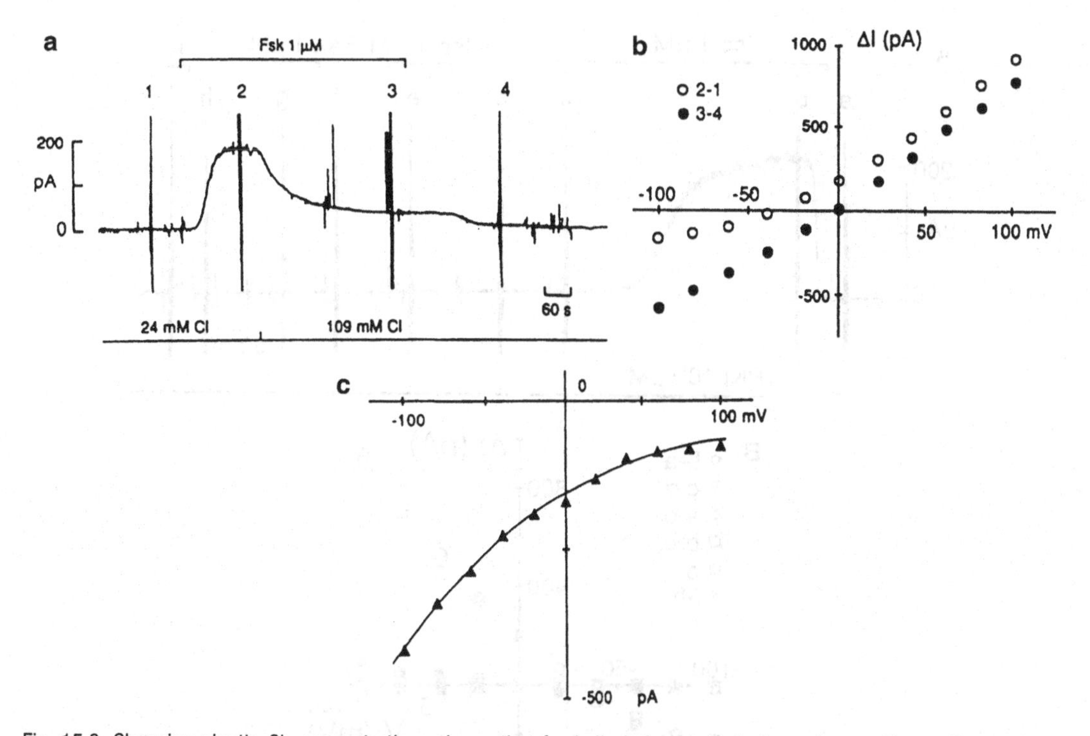

Fig. 15.3. Changing pipette Cl^- concentrations shows that forskolin-induced conductance is sensitive to Cl^-. (**a**) Chart record of the whole-cell current. Holding potential 0 mV, cell capacitance 106 pF. Pipette Cl^- concentrations were changed from 24 to 109 mM. (**b**) Steady-state difference *I–V* relations were obtained as in Fig. 15.1. *Open circles*, measured with 24 mM Cl^- in the pipette; *closed circles*, measured with 109 mM Cl^- in the pipette. (**c**) Difference currents induced by changing pipette Cl^- concentrations plotted against the membrane potentials. *Smooth line* was drawn based on the constant field equation of Goldman, Hodgkin, and Katz (for detail see text).

under the assumption that intracellular Cl^- concentrations correlate effectively with those in the pipette, we can estimate that the forskolin-induced conductance is selective to Cl^- or carried by Cl^--selective channels.

Depicted in Fig. 15.3c is the difference current–voltage relation (*3–2*), representing the conductance that was induced by the increase in pipette Cl^- concentrations. Supposed that there were no change in the current activation level by forskolin, the constant field equation by Goldmann, Hodgkin, and Katz (1) gives the relation between Cl^- permeability (P_{Cl}) and the difference current (ΔI) as:

$$\Delta I = P_{Cl}E_m VF^2\Delta[Cl^-]_i / RT\{1-\exp(E_m F/RT)\}$$

where $\Delta [Cl^-]_i$ is the difference in pipette (not intracellular) and is 85 mM (24–109 mM); and *F*, *R*, and *T* are the Faraday constant, gas constant, and absolute temperature, respectively. The cell membrane capacitance was measured as 185 pF in this specific experiment, and therefore the equation yielded the P_{Cl} value of 8×10^{-8} cm/s.

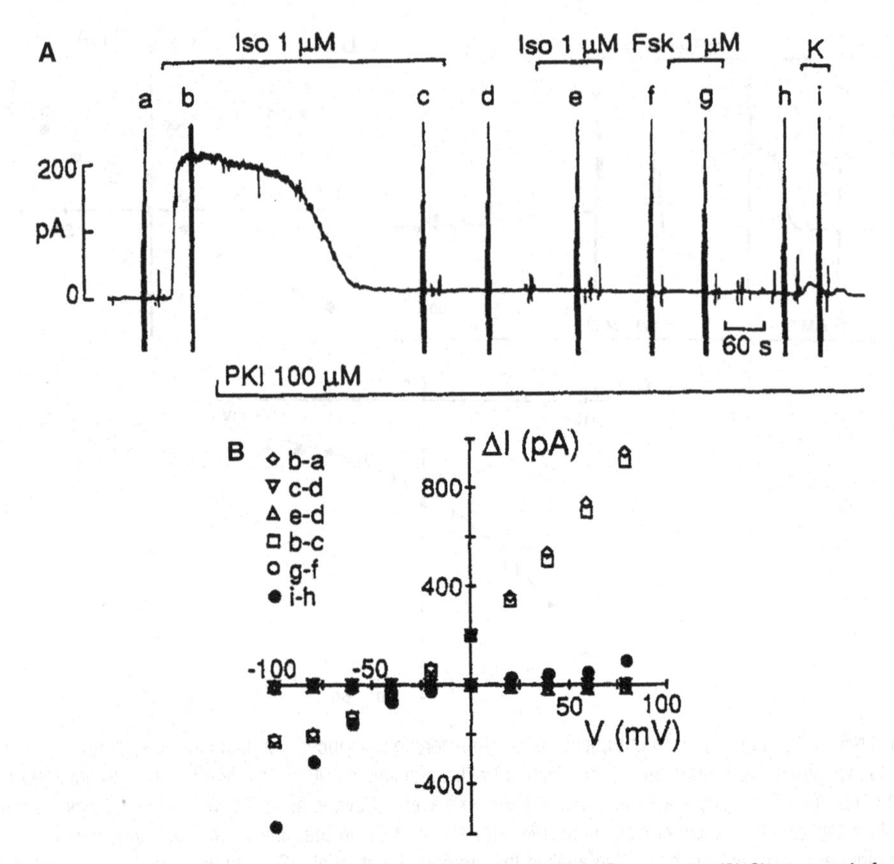

Fig. 15.4. Protein kinase A is exclusively involved during isoprenaline-induced Cl^- currents. (**A**) Chart record of whole-cell current at 0 mV shows complete abolition of Cl^- currents by intracellular application of the inhibitor peptide protein kinase inhibitor (*PKI*) 100 μM. Cell capacitance was 105 pF. (**B**) Difference I–V relations as indicated (numbers correspond to those in (**A**), showing that Iso-activated (*b–a*) and PKI-inhibited (in the continued presence of Iso, *b–c*) current–voltage relations are identical. PKI also prevented response to Iso-removal (*c–d*) or addition (*e–d*) and response (*g–f*) to froskolin. Activation of inward rectifier K^+ currents by exposure to 5 mM K^+ bath solution is shown by the relation *i–h*. (Adapted from (33)).

Figure 15.4 shows another modulation of the pipette perfusion technique, where we can introduce a substance that is not physiologically present in the cell. Again, a single guinea pig ventricular myocyte was voltage-clamped, and extracellular β-adrenergic stimulation by isoprenaline (1 μM) induced outward Cl^- currents at 0 mV holding potential. To determine if the signal transduction is mediated by PKA, in the continued presence of isoprenaline, we applied an inhibitor of PKA (Walsh inhibitor: molecular weight 2,221) (47) into the pipette solution using the technique. After loading the inhibitor, the membrane conductance was completely refractory to external isoprenaline or forskolin (1 μM), indicating that PKA is exclusively regulating this isoproterenol-dependent activation of Cl^- currents.

15.7. Conclusion and Future Aspects

Among a variety of configurations for the patch-clamp experiment, the pipette perfusion method requires fine training and mastering a delicate procedure, which may hamper its general usage. However, the technique offers a wide range of applications to research the intracellular regulating mechanisms in the physiological science.

References

1. Hille B (2001) Ion channels of excitable membranes, Third Edition. Sinaver Associates, 2001
2. Horie M, Hwang TC, Gadsby DC (1992) Pipette GTP is essential for receptor-mediated regulation of Cl^- current in dialysed myocytes from guinea-pig ventricle. J Physiol 455:235–246
3. Soejima M, Noma A (1984) Mode of regulation of the ACh-sensitive K-channel by the muscarinic receptor in rabbit atrial cells. Pflügers Arch 400(4):424–431
4. Matsuda H, Noma A (1984) Isolation of calcium current and its sensitivity to monovalent cations in dialysed ventricular cells of guinea-pig. J Physiol 357:553–573
5. Gadsby DC, Kimura J, Noma A (1985) Voltage dependence of Na/K pump current in isolated heart cells. Nature 315(6014):63–65
6. Fischmeister R, Shrier A (1989) Interactive effects of isoprenaline, forskolin and acetylcholine on Ca^{2+} current in frog ventricular myocytes. J Physiol 417:213–239
7. Sato R, Noma A, Kurachi Y, Irisawa H (1985) Effects of intracellular acidification on membrane currents in ventricular cells of the guinea pig. Circ Res 57(4):553–561
8. Kameyama M, Hofmann F, Trautwein W (1985) On the mechanism of beta-adrenergic regulation of the Ca channel in the guinea-pig heart. Pflügers Arch 405(3):285–293
9. Hartzell HC, Fischmeister R (1986) Opposite effects of cyclic GMP and cyclic AMP on Ca^{2+} current in single heart cells. Nature 323(6085):273–275
10. Hescheler J, Kameyama M, Trautwein W (1986) On the mechanism of muscarinic inhibition of the cardiac Ca current. Pflügers Arch 407(2):182–189
11. Kameyama M, Hescheler J, Hofmann F, Trautwein W (1986) Modulation of Ca current during the phosphorylation cycle in the guinea pig heart. Pflügers Arch 407(2):123–128
12. Kameyama M, Hescheler J, Mieskes G, Trautwein W (1986) The protein-specific phosphatase 1 antagonizes the beta-adrenergic increase of the cardiac Ca current. Pflügers Arch 407(4):461–463
13. Kimura J, Noma A, Irisawa H (1986) Na-Ca exchange current in mammalian heart cells. Nature 319(6054):596–597
14. Nakao M, Gadsby DC (1986) Voltage dependence of Na translocation by the Na/K pump. Nature 323(6089):628–630
15. Fischmeister R, Hartzell HC (1987) Cyclic guanosine 3′,5′-monophosphate regulates the calcium current in single cells from frog ventricle. J Physiol 387:453–472
16. Hartzell HC, Fischmeister R (1987) Effect of forskolin and acetylcholine on calcium current in single isolated cardiac myocytes. Mol Pharmacol 32(5):639–645
17. Hescheler J, Kameyama M, Trautwein W, Mieskes G, Soling HD (1987) Regulation of the cardiac calcium channel by protein phosphatases. Eur J Biochem 165(2):261–266
18. Gisbert MP, Fischmeister R (1988) Atrial natriuretic factor regulates the calcium current in frog isolated cardiac cells. Circ Res 62(4):660–667
19. Hescheler J, Trautwein W (1988) Modification of L-type calcium current by intracellularly applied trypsin in guinea-pig ventricular myocytes. J Physiol 404:259–274
20. Tseng GN (1988) Calcium current restitution in mammalian ventricular myocytes is modulated by intracellular calcium. Circ Res 63(2): 468–482
21. White RE, Hartzell HC (1988) Effects of intracellular free magnesium on calcium current in isolated cardiac myocytes. Science 239(4841 Pt 1):778–780
22. Bahinski A, Nairn AC, Greengard P, Gadsby DC (1989) Chloride conductance regulated by cyclic AMP-dependent protein kinase in cardiac myocytes. Nature 340(6236):718–721
23. Duchatelle-Gourdon I, Hartzell HC, Lagrutta AA (1989) Modulation of the delayed rectifier

potassium current in frog cardiomyocytes by beta-adrenergic agonists and magnesium. J Physiol 415:251–274

24. Gadsby DC, Nakao M (1989) Steady-state current-voltage relationship of the Na/K pump in guinea pig ventricular myocytes. J Gen Physiol 94(3):511–537
25. Hagiwara N, Irisawa H (1989) Modulation by intracellular Ca^{2+} of the hyperpolarization-activated inward current in rabbit single sino-atrial node cells. J Physiol 409:121–141
26. Horie M, Irisawa H (1989) Dual effects of intracellular magnesium on muscarinic potassium channel current in single guinea-pig atrial cells. J Physiol 408:313–332
27. Matsuoka S, Ehara T, Noma A (1990) Chloride-sensitive nature of the adrenaline-induced current in guinea-pig cardiac myocytes. J Physiol 425:579–598
28. Nakao M, Gadsby DC (1989) [Na] and [K] dependence of the Na/K pump current-voltage relationship in guinea pig ventricular myocytes. J Gen Physiol 94(3):539–565
29. Ono K, Trautwein W (1991) Potentiation by cyclic GMP of beta-adrenergic effect on Ca^{2+} current in guinea-pig ventricular cells. J Physiol 443:387–404
30. Duchatelle-Gourdon I, Lagrutta AA, Hartzell HC (1991) Effects of Mg^{2+} on basal and beta-adrenergic-stimulated delayed rectifier potassium current in frog atrial myocytes. J Physiol 435:333–347
31. Tareen FM, Ono K, Noma A, Ehara T (1991) Beta-adrenergic and muscarinic regulation of the chloride current in guinea-pig ventricular cells. J Physiol 440:225–241
32. Tseng GN, Boyden PA (1991) Different effects of intracellular Ca and protein kinase C on cardiac T and L Ca currents. Am J Physiol 261(2 Pt 2):H364–H379
33. Hwang TC, Horie M, Nairn AC, Gadsby DC (1992) Role of GTP-binding proteins in the regulation of mammalian cardiac chloride conductance. J Gen Physiol 99(4):465–489
34. Ono K, Tareen FM, Yoshida A, Noma A (1992) Synergistic action of cyclic GMP on catecholamine-induced chloride current in guinea-pig ventricular cells. J Physiol 453:647–661
35. Tseng GN (1992) Cell swelling increases membrane conductance of canine cardiac cells: evidence for a volume-sensitive Cl channel. Am J Physiol 262(4 Pt 1):C1056–C1068
36. Hanf R, Li Y, Szabo G, Fischmeister R (1993) Agonist-independent effects of muscarinic antagonists on Ca^{2+} and K^{+} currents in frog and rat cardiac cells. J Physiol 461:743–765
37. Hwang TC, Horie M, Gadsby DC (1993) Functionally distinct phospho-forms underlie incremental activation of protein kinase-regulated Cl^{-} conductance in mammalian heart. J Gen Physiol 101(5):629–650
38. Parsons TD, Hartzell HC (1993) Regulation of Ca^{2+} current in frog ventricular cardiomyocytes by guanosine 5′-triphosphate analogues and isoproterenol. J Gen Physiol 102(3):525–549
39. Oliva C, Cohen IS, Mathias RT (1988) Calculation of time constants for intracellular diffusion in whole cell patch clamp configuration. Biophys J 54(5):791–799
40. Tang JM, Wang J, Quandt FN, Eisenberg RS (1990) Perfusing pipettes. Pflugers Arch 416(3):347–350
41. Velumian AA, Zhang L, Carlen PL (1993) A simple method for internal perfusion of mammalian central nervous system neurones in brain slices with multiple solution changes. J Neurosci Methods 48(1–2):131–139
42. Lapointe JY, Szabo G (1987) A novel holder allowing internal perfusion of patch-clamp pipettes. Pflugers Arch 410(1–2):212–216
43. Byerly L, Yazejian B (1986) Intracellular factors for the maintenance of calcium currents in perfused neurones from the snail, *Lymnaea stagnalis*. J Physiol 370:631–650
44. Verrecchia F, Duthe F, Duval S, Duchatelle I, Sarrouilhe D, Herve JC (1999) ATP counteracts the rundown of gap junctional channels of rat ventricular myocytes by promoting protein phosphorylation. J Physiol 516(Pt 2): 447–459
45. Alpert LA, Fozzard HA, Hanck DA, Makielski JC (1989) Is there a second external lidocaine binding site on mammalian cardiac cells? Am J Physiol 257(1 Pt 2):H79–H84
46. Hattori K, Akaike N, Oomura Y, Kuraoka S (1984) Internal perfusion studies demonstrating GABA-induced chloride responses in frog primary afferent neurons. Am J Physiol 246(3 Pt 1):C259–C265
47. Cheng HC, Kemp BE, Pearson RB, Smith AJ, Misconi L, Van Patten SM et al (1986) A potent synthetic peptide inhibitor of the cAMP-dependent protein kinase. J Biol Chem 261(3):989–992

Chapter 16

Planar Lipid Bilayer Method for Studying Channel Molecules

Shigetoshi Oiki

Abstract

The planar lipid bilayer method is another way to examine channel molecules functionally at the single-molecule level. In contrast to patch-clamping, channel molecules are isolated from various biological resources and are reconstituted into an artificial membrane that has a defined lipid composition. Various techniques have been developed, from the conventional painting method to liposome-patch clamping. In this chapter, underlying principles and technical details for forming the planar lipid bilayer and methods for incorporating channel molecules into the bilayer are reviewed.

16.1. Introduction

The planar lipid bilayer (PLB) is a method of investigating the function of channel molecules electrophysiologically on an artificial membrane (1). The "artificial membrane" includes not only the PLB but also liposomes (or vesicles), which are comprised of specifically defined lipids. The PLB and liposomes differ in their geometrical configuration. Liposomes of different sizes, ranging from nanometers to micrometers, can be prepared. On the other hand, the PLB is "planar" in shape, and this planar geometry opens up the PLB method to manipulation by a variety of experimental procedures, such as applying a membrane potential across the membrane and free access to both sides of the membrane. The PLB method attained wide popularity in biological studies once a method for reconstituting membrane proteins had been established (2). Various types of channel proteins, including those of organelles, have been isolated and reconstituted into PLB (3–5). When reconstituted, some of the channels are functional, and others are not even though the structure of the PLB is similar to that of the biological membrane. The question thus arises: Why are the channel

Yasunobu Okada (ed.), *Patch Clamp Techniques: From Beginning to Advanced Protocols*, Springer Protocols Handbooks, DOI 10.1007/978-4-431-53993-3_16, © Springer 2012

molecules transferred in a potentially unfavorable environment? The answer to this question is addressed throughout in this chapter.

With patch-clamping of cell membranes, many kinds of molecules exist even on a small patch of membrane and may directly or indirectly affect the observed channels. In contrast, the PLB provides a simple, clean experimental platform. In a typical case, experiments can be performed with only minimum components (e.g., purified channel proteins, a membrane of defined lipid composition, and electrolyte solutions). The PLB method relies on the philosophy that the channel function can be reproduced from scratch through collecting channels of biological origin and membrane lipids. Using the PLB method, interaction between channels and membrane lipids can be studied in a simplified condition, and more complicated membrane function, such as the membrane fusion, can be reconstituted by collecting relevant components. The PLB serves as a sort of test tube for membrane proteins.

There are three practical reasons for using the PLB: first, to investigate those channels to which the patch-clamp method cannot be applied ("inaccessible" channels); second, to study structure–function relations of ion channels in simplified experimental conditions rather than by taking *in situ* measurements; third, to perform experiments for which the patch-clamp method is ineffective.

"Inaccessible" channels are those that reside in inaccessible anatomical locations and those on the membrane that are not amenable to a giga-ohm seal. Channels in inaccessible places include those in very small spaces, such as on microvilli and organelles (6). Also, channels expressed with high density on the membrane, such as the acetylcholine receptor channels on the electric organ of the torpedo fish, cannot be patch-clamped. Sometimes the anatomical difficulties can be circumvented, and patch-clamping is not absolutely impossible. For example, the failure of giga-seal patch-clamping of channels in the giant squid axon was remedied by approaching the patch-pipette from the inside of a cut-open axon (7).

Recently, the PLB method has acquired even greater popularity because the K^+ channel from *Streptomyces lividans* (KcsA), with its structural information (8), has become a flagship channel for studying structure–function relations (9). The KcsA channel, as a channel of bacterial origin, requires the PLB method for functional measurements. (Prokaryotic channels may not be expressed on eukaryotic cells without certain genetic modifications.) (10).

With increasingly widespread use of PLB, novel methods for PLB investigation have been developed (11). In this chapter, I present the basic methods for performing PLB experiments (12), including certain recently developed techniques. The chapter starts with historical episodes related to the physicochemical basis of PLB, after which methods for initiating PLB experiments are described. Among the PLB methods the chamber method is fundamental and a prerequisite for understanding all of the other PLB methods. In Sect. 16.4, the physicochemical background is described in some detail along with the methods for reconstituting channel proteins, which are described in greater detail in Sect. 16.5. The electronics employed in the PLB methods are outlined in Sect. 16.6, and examples of channel recordings are presented in Sect. 16.7. It is suggested that the readers start with the basic PLB experiments using the chamber method and familiarize themselves with how a stable and low-leakage membrane is formed.

16.2. What Is the Planar Lipid Bilayer?

The fluid mosaic nature of the biological membrane (13, 14) is made evident when a patch-pipette is pulled from a cell after gigaseal formation. The membrane threads elongate over a long distance. This experience gives an impression that the membrane is flexible and deformable. Most electrophysiologists, however, may not think that such membranes can be fabricated by simple manipulation. The following historical episodes are thus meant to provide a practical and realistic introduction to the concept and methods of obtaining a PLB.

In 1774, Benjamin Franklin formed a layer of molecular thickness by putting a small drop of olive oil on the surface of a pond. A thin layer spread across the large area of the surface (15). The thickness of the layer of molecular size could then be estimated.

In much more recent times, Langmuir (16) prepared a shallow trough and filled water, on which a small aliquot of amphiphilic molecules was spread. The amphiphilic molecules move freely on the surface, but as the available surface area was decreased the amphiphilic molecules gathered together and become oriented such that the hydrophilic moiety is anchored to the water surface; and the hydrophobic moiety faces the air, forming a monolayer (17). A glass plate is then slowly dipped into the water, and the monolayer is transferred onto the glass surface. Moving the glass plate up and down results in the formation of multiple layers (Langmuir–Blodgett film).

Mueller et al. put a small amount of membrane extract on a small hole, across which a membrane was formed (18). Microscopic observation revealed that the surface reflected an interference pattern. Subsequently, a dark area appeared and expanded on the surface. The relatively lower amount of reflected light indicated that this area was thinner than the rest of the area, with a thickness on the order of several tens of angstroms. This pioneering experiment was the origin of the PLB, which was known as the black membrane in earlier times. During this era of PLB investigation and development, the Japanese scientists Tetsuya Hanai (physical chemistry of PLB and characterization of the membrane doping substances) (19) and Masayuki Takagi (development of alternative folding methods) (20) made significant contributions.

In the short history of these developments, the implied story is that the researchers in this field attempted to understand the biological membrane by fashioning a membrane structure from scratch based on the knowledge of physical chemistry.

16.2.1. Planar Lipid Bilayer and a Soap Bubble

The PLB has a structure similar to that of the biological membrane and is constituted of the following layered structure (Fig. 16.1):

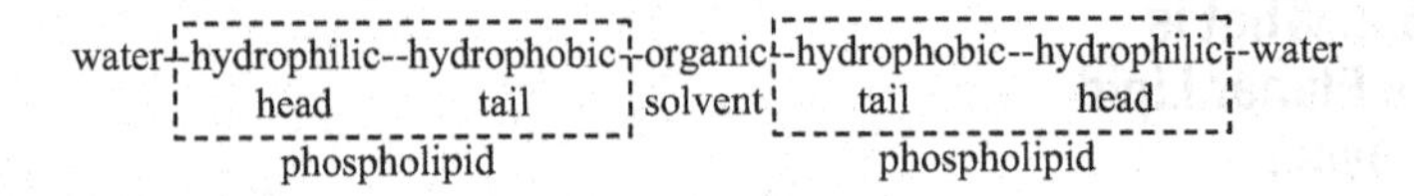

The amphipathic material encompasses bio membrane-derived and synthetic phospholipids. The PLB is different from the biological membrane by having a layer of organic solvent in the middle of the layered structure. This layered structure is reminiscent of a soap bubble, where detergent is used as an amphiphile. The layers of the soap bubble are as follows:

air – hydrophobic – hydrophilic – water – hydrophilic -- hydrophobic – air
tail head head tail
detergent detergent

In the soap bubble, there is a water layer in the middle. Since there is a hydrophobic air layer bounding both of the outer sides, this makes the amphiphilic monolayer orient in the opposite manner to that of the biomembrane and PLB.

The membrane of a soap bubble is rainbow-like in its appearance because the central water layer interferes with visible light, implying that the thickness of the layer is on the order of the wavelength of visible light. The color of the bubble surface changes gradually. Meanwhile, a spot without color appears at the top of the bubble, and spreads as a circular patch of increasing size down over the surface. This occurs by gravity-driven water movement down to the bottom, leading to a thinning of the water layer in the upper region. No visible light interferes at the top, and this area is blackened. The soap bubble is a metastable structure that eventually breaks. These features are shared with PLBs.

16.2.2. Planar Lipid Bilayer and Membrane Proteins

Once the PLB method was established, the PLB came into use as a reaction platform for membrane proteins. The ion transfer features of membrane-bound antibiotics were studied using the PLB. During the 1960s, single-channel current recordings were

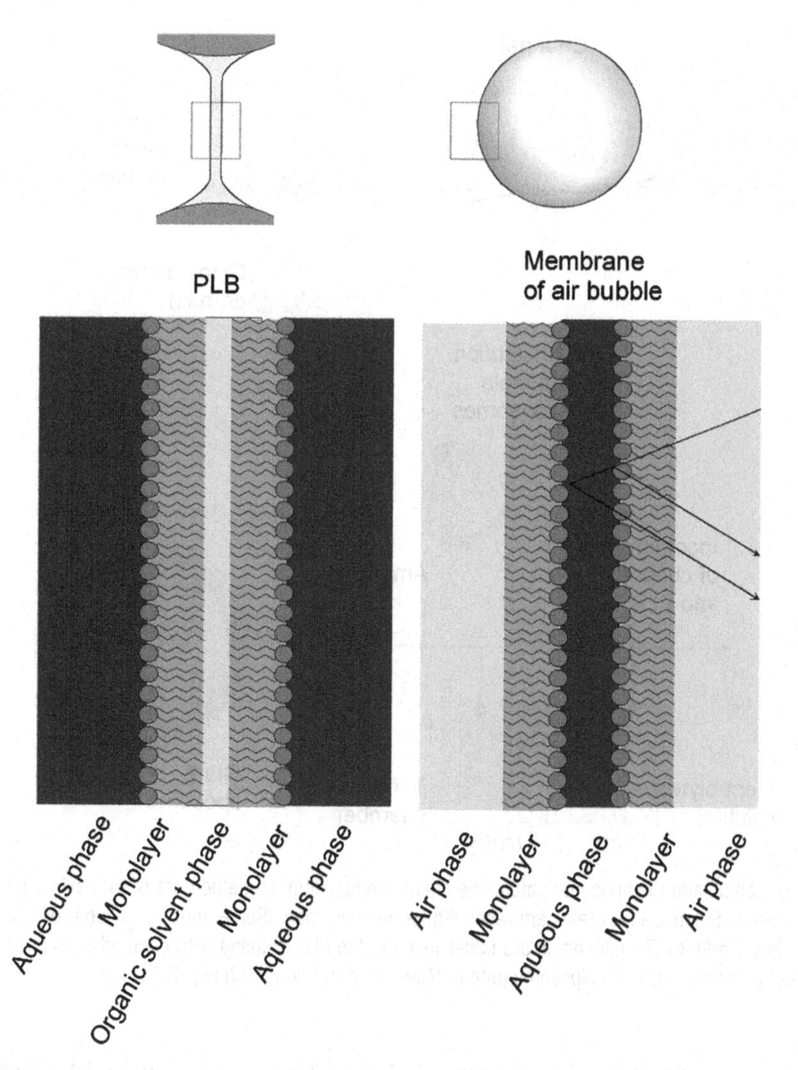

Fig. 16.1. Planar lipid bilayer (PLB) and the membrane of the air bubble. An enlarged view of the PLB formed on a small hole and the bubble membrane. The light shed on the membrane gives an interference pattern and is rainbow-colored.

successfully performed for excitability-inducing material (EIM) (21) and the antibiotic gramicidin (22). This was an epochal event, as they were the first recordings of single-molecule measurements. Neher conducted single-channel current measurements using PLB (23), and this experience led him to develop the patch-clamp method (24, 25).

During the late 1970s, a technique for reconstituting extracted channel proteins in the PLB was developed (2). A standard biochemical technique of protein reconstitution in liposomes was used to transfer membrane-embedded proteins into the PLB through membrane fusion processes (Fig. 16.2). Channel currents of organellar origin were readily measured. The membrane fusion

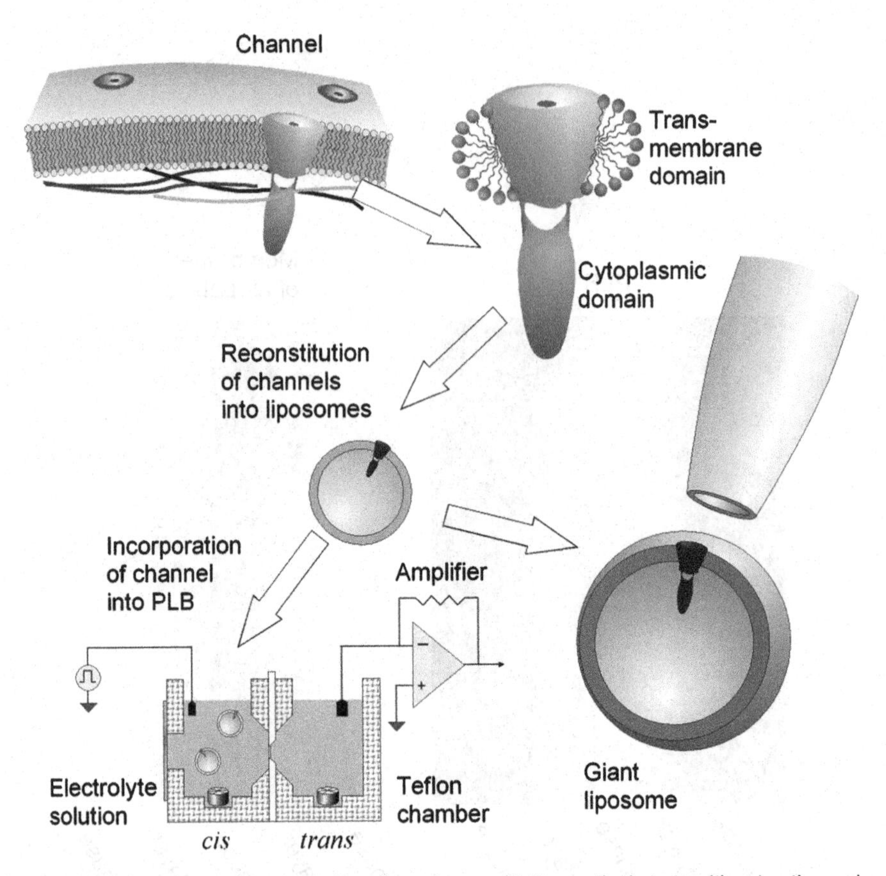

Fig. 16.2. PLB method and the liposome patch. The channel reconstitution method starts with extraction and purification of channel proteins. Solubilized proteins are reconstituted in liposomes. (Sometimes the solubilized proteins are directly incorporated into the PLB.) The liposomes are either fused to the PLB or fused with each other to form giant liposomes, which is useful for patch-clamping (liposome patch) (Modified from Oiki (2009) Fig. 1).

process itself is a subject of interest since the PLB provides a simple experimental environment. A target set of membrane proteins was reconstituted in a PLB to reconstruct a fission function of the membrane (26).

Crystallization of the prokaryotic potassium channel and subsequent elucidation of the three-dimensional structure have advanced the study of molecular mechanism of ion channels. As a counterpart of the structural study, the PLB method is recognized as the method of choice for functional measurements of prokaryotic channels.

Since the first single-channel current recordings during the 1960s, mechanistic understandings of channel function have advanced dramatically. During the long journey, the early pioneers struggled to understand the membrane by constructing and manipulating the structure. The passion and enthusiasm in those early days have carried over to the work being done in the channel field today.

16.3. Planar Lipid Bilayer Methods

The chamber method is the most fundamental of the PLB methods. In a chamber, two compartments separated by a septum having a small hole (e.g., 100 μm diameter) are filled with electrolyte solution. An aliquot of lipid-dispersed organic solvent is applied to the hole to form a lipid bilayer membrane (1). Channel molecules from a variety of resources were reconstituted in the PLB. The membrane is voltage-clamped, and currents through the channels are measured. In the PLB experiments formation of the membrane is the first and most important step (27).

16.3.1. Chamber Method

There are two methods of membrane formation: the painting method and the folding method. At the outset of studying PLB methods, one should first become acquainted with the painting method. Two compartments of a chamber are filled with an electrolyte solution of several milliliters or less. A septum separating the two compartments has a small hole that is several tens of micrometers in diameter. Before discussing formation of the membrane, we start with construction of the chamber.

16.3.1.1. Chamber Construction

The chamber is constructed of Teflon. In the method presented here, the Teflon plate is used for the septum, and on it a small hole is made each time the experiment is performed. The fresh hole has a sharp edge, rendering the ease of membrane formation and the stability of the formed membrane (28). Teflon is highly hydrophobic and is thus optimal supporting material for the membrane; on the other hand, it is mechanically fragile, and fresh renewal of the septum and fabrication of the hole on the septum at the time of each experiment makes the system clean, facilitating membrane formation. The following shaving method (29) is a simple way to make a small hole on the Teflon plate (Fig. 16.3a).

1. A square-shape Teflon plate (0.5–1.0 mm in thickness; 1 × 1 cm) is cut from a Teflon sheet or plate.
2. One end of a stainless-steel rod (3 mm diameter) milled to a conical shape with a smooth surface is heated with a gas burner.
3. The tip of the rod is pushed against the Teflon plate until a slight bulge appears on the other side of the plate.
4. The rod is removed from the Teflon plate, leaving a cone-shaped pit.
5. The slight bulge that appears on the other side of the plate is shaved with a razor, resulting in a hole with a sharp edge. The hole should have a precise round shape. The diameter is measured under microscopy.
6. The Teflon plate is washed with chloroform/methanol.

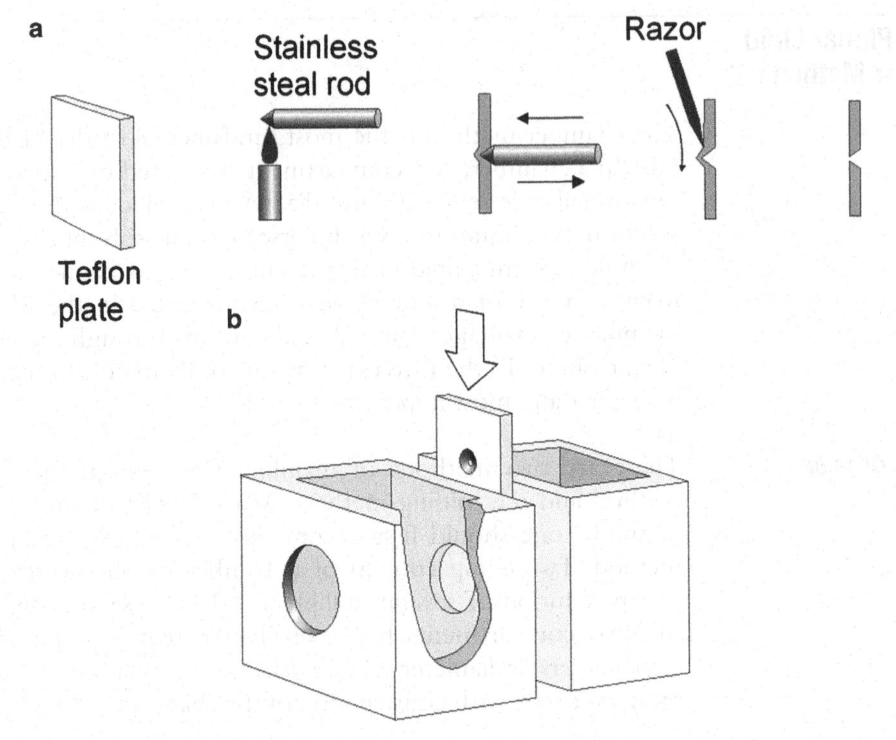

Fig. 16.3. Chamber method. (**a**) Method for making a hole in a Teflon plate. One end of a stainless-steel rod is heated and pushed into a Teflon plate until a bulge appears. The bulge is shaved once with a razor, and a small hole of conical profile is made. (**b**) Assembling the chamber. A Teflon plate is sandwiched by two Teflon chambers. The septum is sealed to prevent leakage of current between the two compartments.

During this process, the size of the hole is adjusted by controlling the degree of bulge and the depth of scraping by the razor. Holes >100 μm are easily made, and holes of 30 μm diameter can be made successfully.

This shaving method is applicable to the painting, folding, and punch-out methods. The hole made by this method fulfills the following requirements for stable and low-noise recordings (see Sect. 16.6.2) (29): (1) low stray capacitance; (2) low access resistance; (3) high membrane stability; and (4) low dielectric loss. For example, a thin Teflon sheet with a thickness of 25 μm can be used as the septum on which a hole is made, but such a thin septum significantly contributes to the stray capacitance. If the septum is thick, this thickness reflects the length of the hole and an increase in access resistance. The hole made by the shaved method is mechanically tough and has an open, conical shape, thus minimizing the access resistance. Also, the edge of the hole is very sharp, which facilitates both the membrane thinning process and stability.

Assembly of the chamber proceeds as follows:

1. The chamber is comprised of three parts: two compartments and a septum with a hole (Fig. 16.3b).
2. Except for the central part of the septum where the hole is located, vacuum grease (silicon grease; Dow Corning, Corning, NY, USA) is lightly applied to prevent an electrical current leak.
3. The septum is sandwiched by two blocks of the chamber and fixed.
4. The assembled chamber is set in a Faraday box.

16.3.1.2. Planar Lipid Bilayer Formation

Phospholipids

Phospholipids are dispersed in organic solvents (alkane, such as decane, hexadecane, and hexane) in a concentration range of 4–40 mg/ml (hereafter called "lipid solution"). In PLB experiments, pure or mixed lipids can be used. For example, a synthetic phosphatidylcholine diphytanoylphosphatidylcholine (DPhPC, Mr. 846.26; Avanti Plar Lipids, Alabaster, AL, USA) (Fig. 16.4) has branched and saturated acyl groups and has come into widespread use because of its high purity and chemical stability (30). On the other hand, the bacterial membrane contains phosphatidylethanolamine (PE), and these components of the native membranes are used as the lipid composition of PLB. When

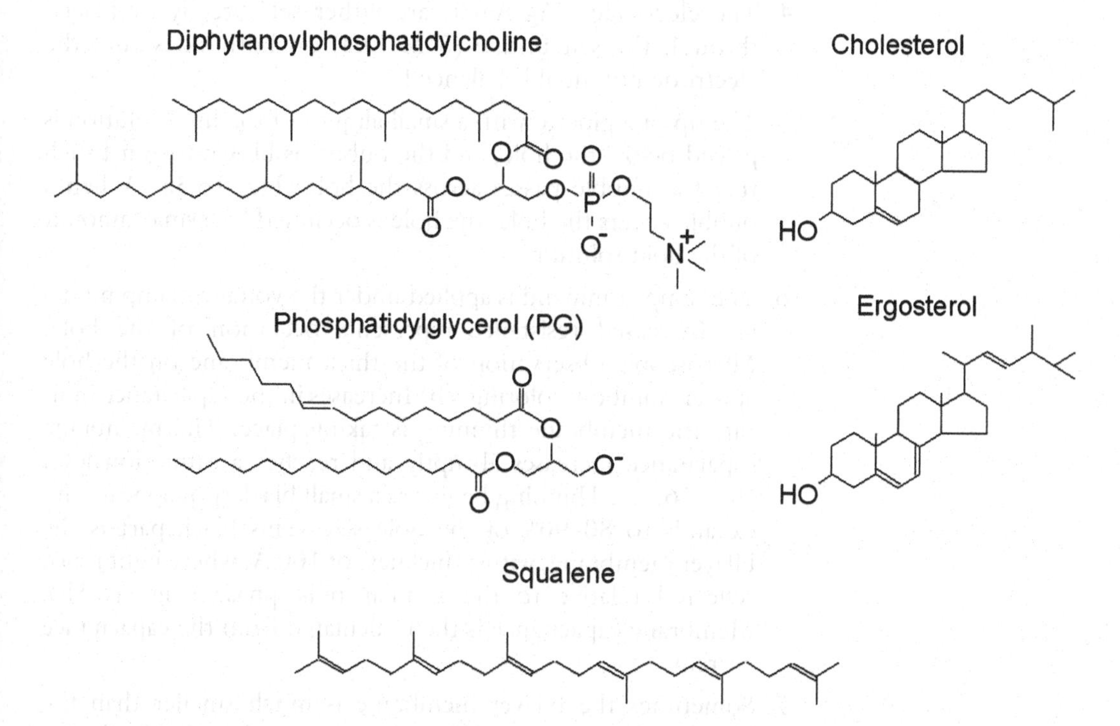

Fig. 16.4. Chemical formula of molecules in the PLB.

charged lipids are required, phosphatidylserine (PS) is used for channels of eukaryotic origin (e.g., PC/PS = 1:1; PE/PS = 1:1) and phosphatidylglycerol (PG) for those of prokaryotic origin (PE/PG = 3:1). For the KcsA potassium channel (see Sect. 16.7.2), PG is necessary for channel activity (31). On the other hand, the nicotinic acetylcholine receptor channel requires cholesterol for channel activity in addition to the phospholipids (32). In case that lipid information is not available, experiments should be started with mixed lipids extracted from soybeans (L-α-phosphatidylcholine type IV-s) that include various types of phospholipid.

Painting Method

Painting a small aliquot of phospholipid solution on a small hole leads to formation of the PLB (the physical processes underlying the formation are shown in Sect. 16.4.1). With this method, decane is used as an organic solvent. The name "painting" method has a historical derivation, in which a paint brush was used to apply lipid solution to the hole. We use a bubbling method.

1. Phospholipids are maintained in chloroform solution in a deep freezer. Chloroform is completely evaporated under the flow of nitrogen gas. By adding decane, a lipid solution of 20 mg/ml is prepared.
2. Precoating: A small aliquot of the lipid solution (2 μl) is added around the hole, and the solvent is dried under nitrogen gas.
3. Electrolyte solution fills both compartments.
4. The electrodes (Ag-AgCl) are either set directly or placed through the salt bridge (3 M KCl) on both sides, and the electrode potential is balanced.
5. The tip of a pipette with a small aliquot of the lipid solution is placed below the hole, and the bubble is blown forward such that the bubble passes across the hole (Fig. 16.5a). When a bubble covers the hole, the hole is occluded by a small amount of the lipid solution.
6. The ramp command is applied under the voltage-clamp mode. (a) Increased resistance represents occlusion of the hole. Microscopic observation of the thick membrane on the hole appears rainbow coloring. (b) Increases in the capacitance indicate the membrane thinning is taking place. The membrane capacitance increases abruptly and reaches a saturation level (Fig. 16.5c). Thinning begins as a small black spot appears and expands to 80–90% of the hole size. This black part is the bilayer membrane, with a thickness of 100 Å, where light is not reflected relative to the annular bulk phase (Fig. 16.5b). Membrane capacitance is then calculated from the capacitance current.
7. Sometimes the bilayer membrane is much smaller than the hole size because a considerable amount of the lipid solution

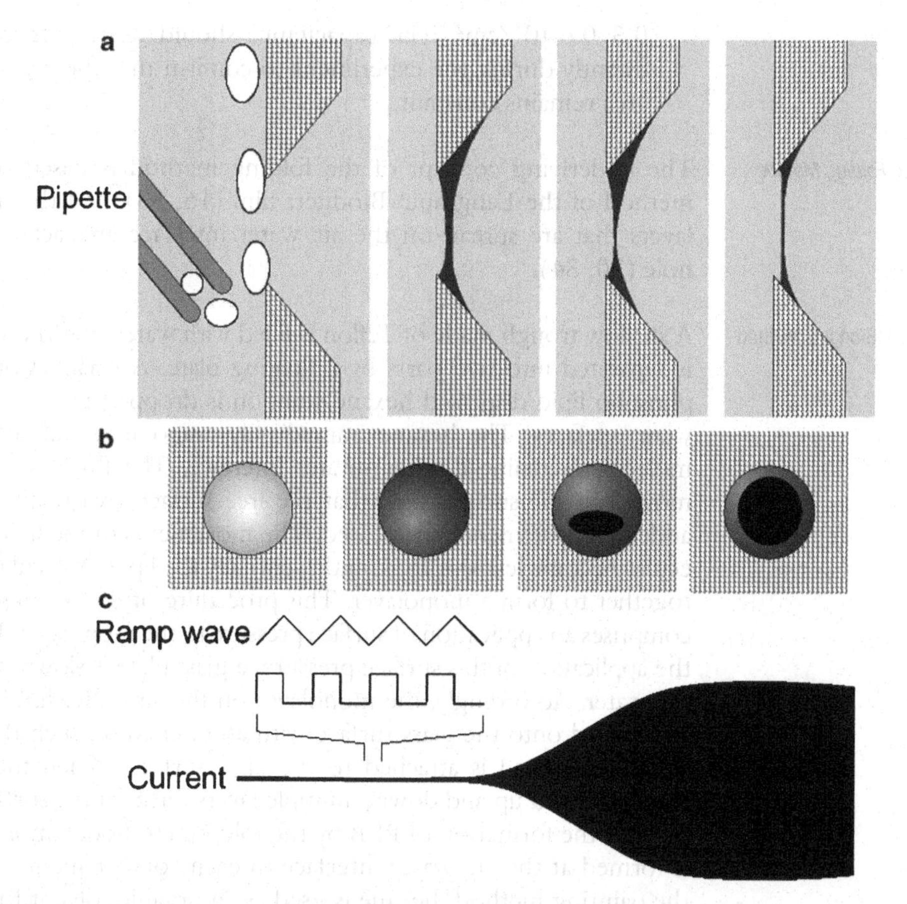

Fig. 16.5. Membrane-forming processes of the painting (bubble-blowing) method. (**a**) Cross section of a Teflon plate at the hole. The hole is occluded by an aliquot of lipid solution delivered by bubbling. Organic solvent escapes along the surface of the Teflon septum, and membrane thinning takes place. The bilayer is formed at the narrowest portion of the hole. (**b**) Front view of the hole. Thick membrane interferes with the light and is thus rainbow-like in coloring. A small spot appears and expands toward the edge of the hole. At equilibrium, the size of the black membrane is 90% that of the hole. (**c**) Current response to the ramp potential. As thinning proceeds, the capacitance increases and reaches a maximum. The current response to the ramp command is shown on an expanded scale, from which the membrane capacitance and resistance can be evaluated.

accumulates at the hole. In such cases, the membrane must be broken and re-formed several times until the capacitance increases to an optimal value.

8. A true membrane breaks when a very large membrane potential (e.g., 400 mV) is applied. If it does not, the hole is filled with the thick membrane and the capacitance is low. In this case, the "membrane" is broken by strong bubbling.
9. The membrane resistance and capacitance are then evaluated. The resistance should be >100 GΩ. The specific membrane capacitance is calculated from the measured capacitance and the diameter of the bilayer. When decane is used for the organic solvent, the specific capacitance of the membrane should be

0.3–0.6 μF/cm². The capacitance should be monitored frequently during the experiment to confirm that the membrane area remains constant.

16.3.1.3. Folding Method

The underlying concept of the folding method is based on the method of the Langmuir–Blodgett film (16, 33), and the monolayers that are spread on the air–water interface are faced at the hole (20, 34).

Langmuir–Blodgett Method

A shallow trough made of Teflon is filled with water, and the surface is separated into two parts by a floating plate. A small amount of phospho lipid-dispersed hexane solution is dropped on one of the water surfaces. The hexane immediately evaporates, and the lipid molecules remain at the air–water interface. The floating plate is moved on the surface such that the free surface area of the lipid-added compartment is decreased. The movements of the lipid molecules become less vigorous, and eventually the lipid molecules pack together to form a monolayer. This procedure of the floating plate comprises an application of surface pressure to the monolayer. Under the application of this surface pressure, a glass plate is slowly soaked the water. Accordingly, the monolayer on the air–water interface is transferred onto the glass surface with an orientation such that the hydrophilic head is attached to the glass surface. When the glass plate is moved up and down, multiple layers form on the surface.

For the formation of PLB by the folding method, a monolayer is formed at the air–water interface in each compartment. Unlike the painting method, hexane is used as an organic solvent for lipid dispersion, and a drop of the hexane solution is added to the water surface. Hexane evaporates rapidly and leaves a phospholipid monolayer at the air–water interface. Upon raising the water level by adding electrolyte solutions, the monolayer runs up on the Teflon surface and passes through the hole, where the two monolayers from both sides encounter and form a bilayer.

The membrane formed by this method has been referred to as a "solvent-free" membrane. The great advantage of this method over the painting method is the capacity to form an asymmetrical membrane in which the lipid composition of either leaflet is arbitrarily defined (35). Once formed, the membrane retains this asymmetry for a long time because the flip-flop or exchange of the lipid molecules between both leaflets of the membrane is very slow (36).

1. For precoating, highly hydrophobic squalene (Fig. 16.6a) is dispersed in hexane and added around the hole from both sides. It is then dried under the flow of nitrogen gas.
2. The electrolyte solution is filled in both compartments to the level below the hole. Ag-AgCl electrodes are set.
3. Phospholipids are dispersed in hexane (4 mg/ml), and a 50 μl aliquot is added on the electrolyte solution of both sides.

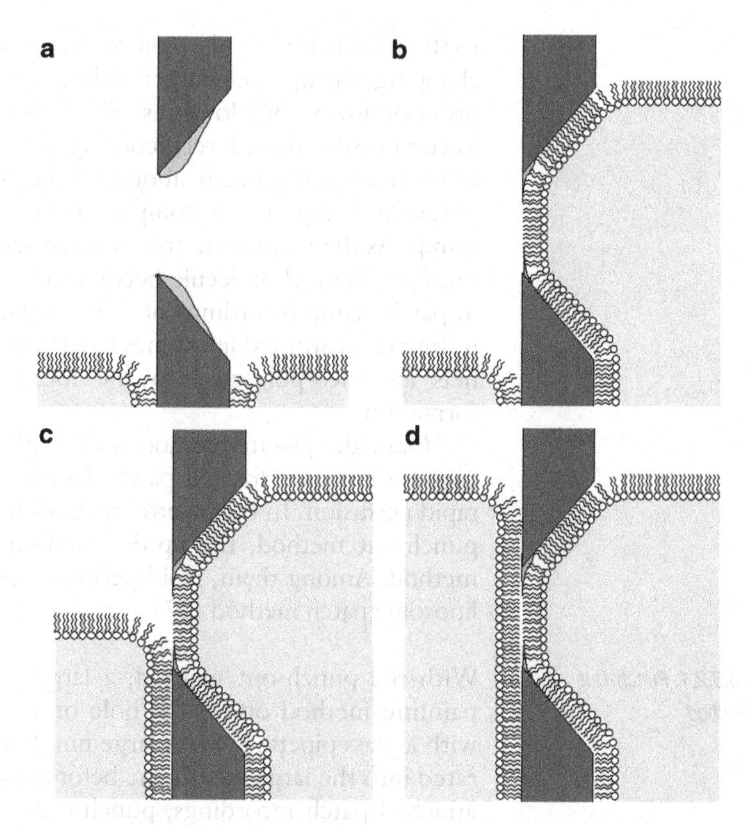

Fig. 16.6. Folding method. (**a**) Hole is precoated with squalene (*gray*). (**b**) By adding electrolyte solution on one side, the monolayer creeps up the Teflon surface. (**c**) Water level is raised on the other side. (**d**) Two monolayers are apposed at the hole. Squalene at the edge of the hole supports the bilayer. With this method, different kinds of lipids can be spread as a monolayer in each compartment, and an asymmetrical bilayer can be formed.

4. Leave the solutions for several minutes to allow complete evaporation of the hexane and formation of a monolayer (Fig. 16.6a).
5. The ramp potential is applied under the voltage-clamp mode.
6. The water levels of both sides are slowly raised by adding the electrolyte solution. The capacitance immediately increases when the water level passes across the hole and forms the bilayer.

The specific membrane capacitance (see Sect. 16.4.2) of the membrane formed by the folding method is 0.6–0.8 μF/cm². This value is close to the native biological membrane, suggesting that practically no solvent layer exists between the two monolayers.

16.3.2. Pipette Methods

Forming a membrane with a diameter of <30 μm is not easy to achieve with the chamber method because thinning of the membrane does not occur. On the other hand, a small membrane can be readily formed at the tip of the glass pipette, the size of which is generally <30 μm. The noise level is significantly reduced relative to the chamber methods. Compared to the patch pipette for patch-clamping, the tip size is larger and the electrode resistance is on the order of dozens of kilo-ohms. This low series resistance allows the lowest possible noise level recordings (see Sect. 16.6.2). Also, there is no need for concern about voltage errors of the membrane potential or capacitance compensation for attaining a fast voltage-clamp. As the membrane area is decreased, the probability of picking up a channel molecule becomes lower, similar to what occurs in patch-clamp recordings of cells. In particular, it is not practical to fuse reconstituted liposomes to a small PLB. Alternatively, channels are incorporated into the monolayer before the bilayer formation.

Once the pipette method is established for PLB, various techniques developed for cell patch-clamping can be applied, such as rapid perfusion. In the pipette method, there are three options: the punch-out method, the tip-dip method and the liposome patch method. Among them, rapid perfusion can be performed with the liposome patch method (37).

16.3.2.1. Punch-Out Method

With the punch-out method, a large membrane formed by the painting method on a large hole of a chamber is patch-clamped with a glass pipette (38). A large number of channels are incorporated into the large membrane beforehand; and similar to the cell-attached patch recordings, punch-out of the PLB with a pipette allows single-channel current measurements. In the punch-out method, the glass pipette differs from the cell patch clamp in two respects. The tip size is approximately 30 μm, and the glass surface is rendered hydrophobic by silanization.

The glass pipette (Fig. 16.7) can be fabricated as follows:

1. A borosilicate glass capillary is washed with a strong acid, followed by chloroform and methanol.
2. The glass pipette is pulled to make a pipette with a fine tip. The platinum heater for the puller may deposit platinum on the pipette surface and hinder the stable sealing between the membrane and the glass. In contrast, pulling the glass pipette under the heating with a gas burner or a laser puller circumvents this problem (Fig. 16.7b).
3. For fabrication of the glass pipette, a microforge is used. The platinum wire of the microforge is thickly coated with borosilicate glass to prevent platinum evaporation. This coating is replenished on every experimental day (Fig. 16.7a).
4. The platinum wire is heated below the temperature at which it turns red. The pipette tip is touched to the glass-coated platinum wire at the diameter of approximately 30 μm (Fig. 16.7c).
5. Turning off the heater switch shrinks the heater wire slightly and pulls the glass pipette at the attached site (Fig. 16.7d). The

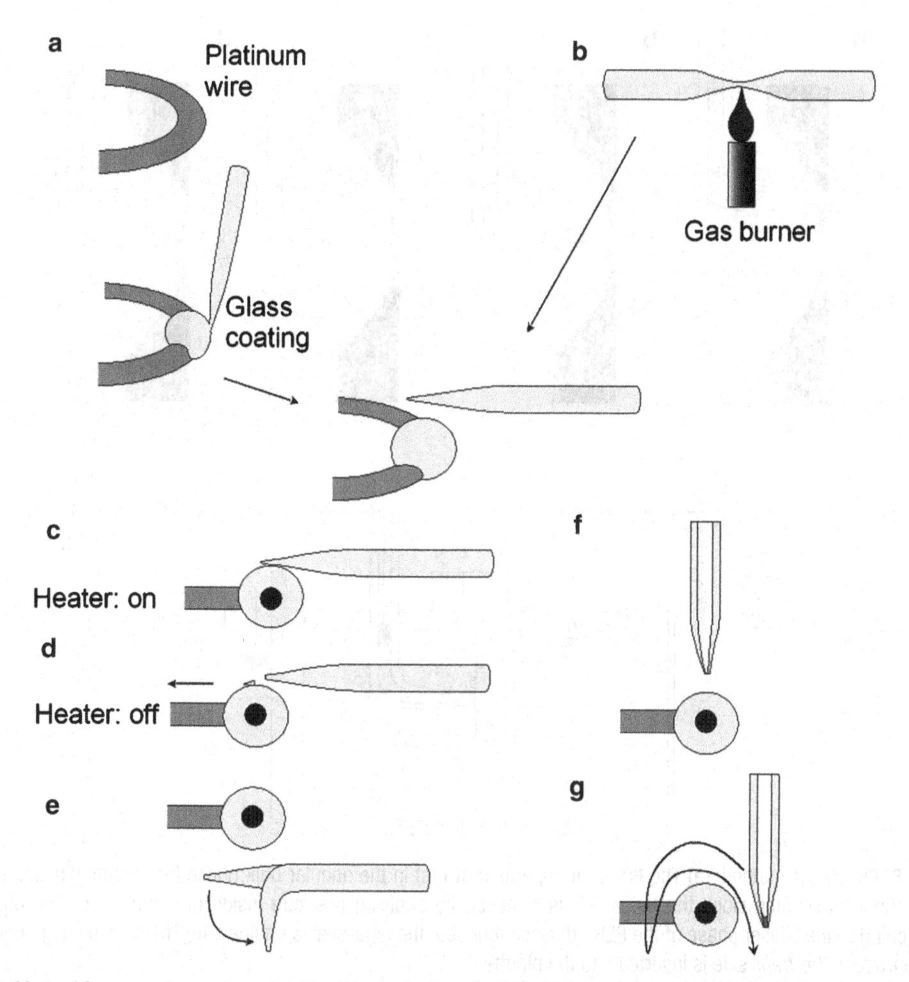

Fig. 16.7. Method for preparing a glass pipette with silanization. (**a**) The platinum is heavily coated with the glass. (**b**) A glass pipette is pulled using a conventional puller or a gas burner. (**c**) Cross section of the heater with the coated glass. The heater is on, and the tip of a pipette is touched to the glass surface lightly. (**d**) When the heater is turned off, it retracts slightly toward the left and pulls the pipette, breaking the tip rectangularly. The tip of the pipette should be 30 μm in diameter and have a proper circular shape without any defects. (**e**) For the punch-out method, the pipette is bent by the heater. (**f**, **g**) After filling the silane solution at the tip, the surface of the pipette is heated for silanization.

small tip of the pipette is pared from the rest of the pipette, leaving a right circular tip approximately 30 μm in diameter.

6. The tip of the pipette is slightly heat polished (Fig. 16.7f).
7. Trioctylsilane is diluted by benzene as a 10% solution.
8. The tip of the pipette is soaked in the silicone solution, and the tip is approached toward the heater at low temperature. Scanning the tip around the heater yields surface siliconization (Fig. 16.7g).
9. The pipette is bent at approximately 1 cm from the tip (Fig. 16.7e).
10. The pipette is washed with chloroform and methanol.

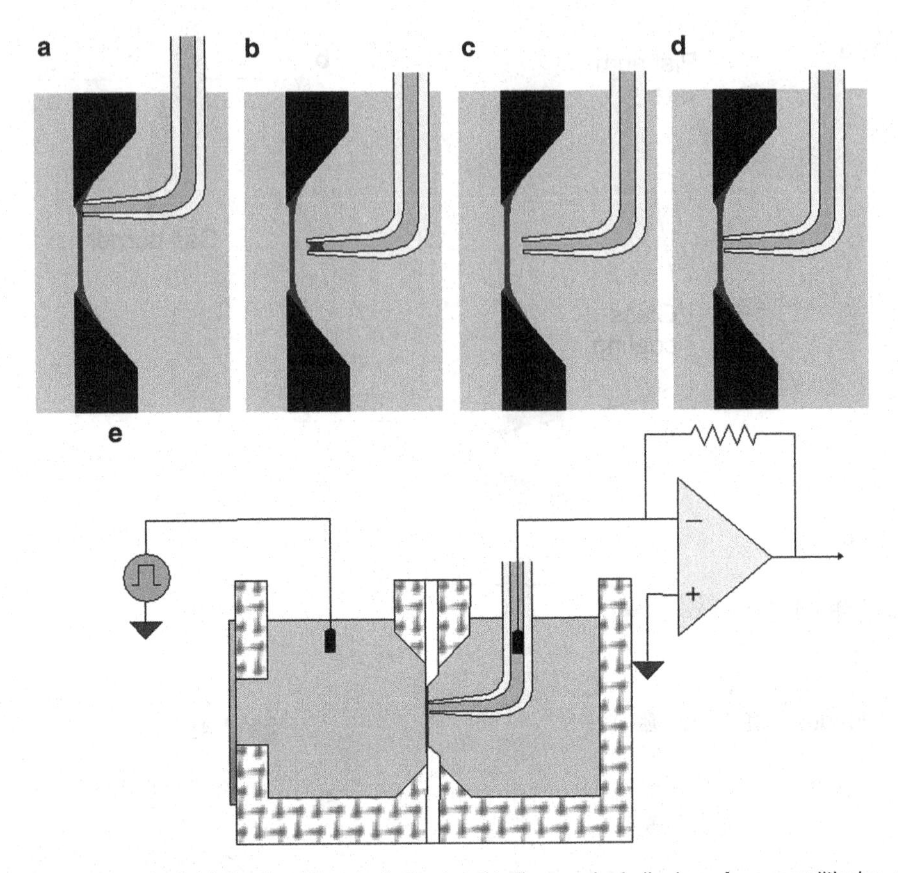

Fig. 16.8. Punch-out method. (**a**) The tip of the pipette is at rest in the annular bulk phase for preconditioning. (**b**) When pulled, the lipid solution clogs the tip, which is removed by applying pressure inside the pipette. (**c**) The open tip is approaching to the bilayer phase of the PLB. (**d**) Once touched, the giga-seal is attained. (**e**) The *cis* side is grounded, and the electrode in the *trans* side is inserted into the pipette.

The setting for the punch-out method (Fig. 16.8) is as follows:

1. A glass pipette is set in the electrode holder with aside outlet for applying positive or negative pressure. The pipette is set on the *trans* side of the chamber with the tip facing the hole.
2. The tip of the pipette is viewed through the hole with a microscope. The tip is approached right behind the hole.
3. A membrane is formed on the hole by the painting method (a big hole is preferable to observe the tip of the pipette in the *trans* side).
4. The tip of the pipette is moved to the annular phase on the formed membrane and left there for a few minutes. This results in precoating the pipette (Fig. 16.8a).
5. The pipette is withdrawn from the annular phase, and the membrane on the pipette tip is broken by applying pressure to the pipette (Fig. 16.8b,c).

6. The bilayer is patch-clamped (punched out). Once the tip is attached to the bilayer, a patch membrane is formed (Fig. 16.8d), which is indicated by abrupt decreases in the membrane noise. If the "precoating" is insufficient, the membrane on the hole is broken.

In the punch-out method, the electrode balance can be adjusted even during the course of an experiment without breaking the large membrane. The procedure is as follows. The tip of the glass pipette is passed through the large PLB without breaking it. A patch membrane is broken by applying positive pressure. The tip of the patch electrode is now in the *cis* side and electrically connected to the *cis* side electrode. When the pipette without the membrane is withdrawn toward the *trans* side, a new membrane is formed at the tip, a process similar to forming the outside-out patch membrane.

16.3.2.2. Tip-Dip Method

The PLB is formed on the tip of a glass pipette with a similar way of the folding method. The pipette tip is passed through a monolayer formed on an air–water interface. In earlier studies, the size of the tip was similar to that of a typical patch pipette; the formed membrane was not stable, and the seal resistance was not high (39–41). Here a modified tip-dip method is shown (42, 43).

Lessons learned from PLB formation and the punch-out method indicate that the formation of a stable seal takes place when the supporting material is hydrophobic, rather than hydrophilic. A patch pipette ~10 μm in size was fabricated by the previously mentioned method with the glass surface rendered hydrophobic. The thickness of the glass at the tip is much thicker than that for the pipette used for conventional patch clamping, and the contribution of the pipette to the total capacitance is small. With the low series resistance and the small membrane capacitance, this configuration may provide a system with the lowest noise level.

1. A total of 5 μl phospholipid solution (20–40 mg/ml in hexadecane) is added to the electrolyte solution (5 ml) in a glass beaker. The monolayer expands at the air–water interface, and the less volatile hexadecane stays on the monolayer in a lens shape.
2. A glass pipette having a tip size of 10 μm, made by the above-described method (i.e., not in an L shape) is set into the pipette holder with aside outlet for the application of pressure.
3. The tip of the pipette is left in the hexadecane phase (Fig. 16.9a). This is a process of precoating the pipette.
4. The clogged hexadecane is flushed out, and the tip is passed across the monolayer. At the beginning, the amount of hexadecane at the tip may be too much to form the bilayer. In this case, hexadecane in the tip is flushed out, and the tip is again passed through the interface.
5. Once the bilayer is formed, it can be re-formed repeatedly.

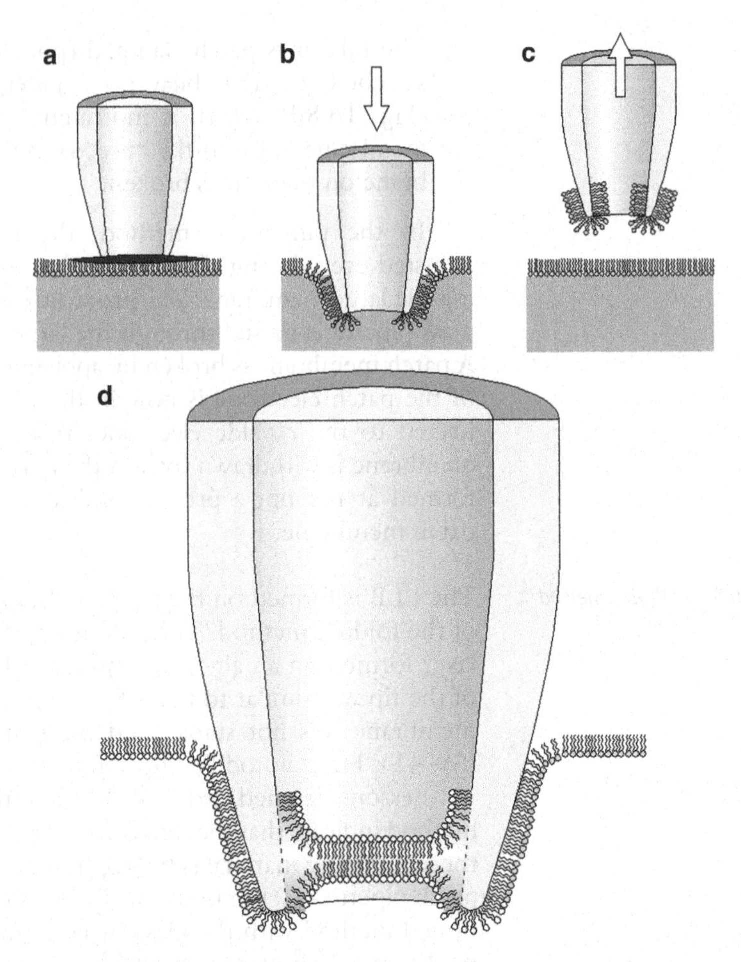

Fig. 16.9. Modified tip-dip method. (**a**) On the monolayer at the air–water interface, the organic solvent shapes a lens, toward which the tip of the pipette is approached (pre-coating). (**b**) The tip is immersed in the water phase. (**c**) Lipids are attached to the hydrophobic surface of the silanized glass pipette. (**d**) Two monolayers are apposed upon dipping the pipette into the water phase. Organic solvent on the monolayer forms the annular phase at the tip.

In this method, a thin layer of the organic solvent is laid over the lipid monolayer at the air–water interface. This organic layer prevents the channel protein from being exposed to the air. The PLB is formed at the tip of the pipette by a mechanism similar to the PLB in the chamber. In fact, if the tip size is larger, the thinning process of the PLB is observed as an incremental change of the membrane capacitance. The annular bulk phase interacting with the hydrophobic surface of the glass stably supports the bilayer (Fig. 16.9d), and the seal resistance is very high, on the order of 1 TΩ (1,000 GΩ), and a membrane potential of up to 400 mV can be applied. With this method, another solvent, such as hexadecane, can be used, and the organic solvent phase in the membrane becomes negligible.

16.3.2.3. Liposome Patch Method

The liposome patch method is another way of forming lipid bilayer at the tip of the pipette. Similar to the punch-out method, preformed lipid bilayer is patch-clamped. There are however some technical tricks, and stable patch membranes with the giga-seal are attained recently. Once established, the liposome patch allows the application of some of the various experimental techniques that the conventional patch clamp methods have developed, such as rapid solution exchange. Unlike other PLB methods, the liposome-patch membrane is completely solvent-free, since the liposome membrane does not contain organic solvents (37, 44, 45).

Giant unilamellar vesicles (GUVs) is prepared by a procedure shown in Fig. 16.10 (46). For the liposome patch, L-α-phosphatidylcholine type IV-s (Sigma-Aldrich) is used exclusively as the phospholipid composition. This lipid, despite its label, is a mixed lipid containing only 30% PC. GUV is formed in an electrolyte solution of low ionic strength, and liposomes with a diameter of up to 10 μm can be formed.

GUV containing solution is added to a chamber on the microscope stage, and giga-seal formation is performed using a procedure similar to that used for typical patch-clamping (Fig. 16.11). The tip of the patch pipette is treated with silane to render the glass surface hydrophobic. Slight negative pressure facilitates a giga-ohm

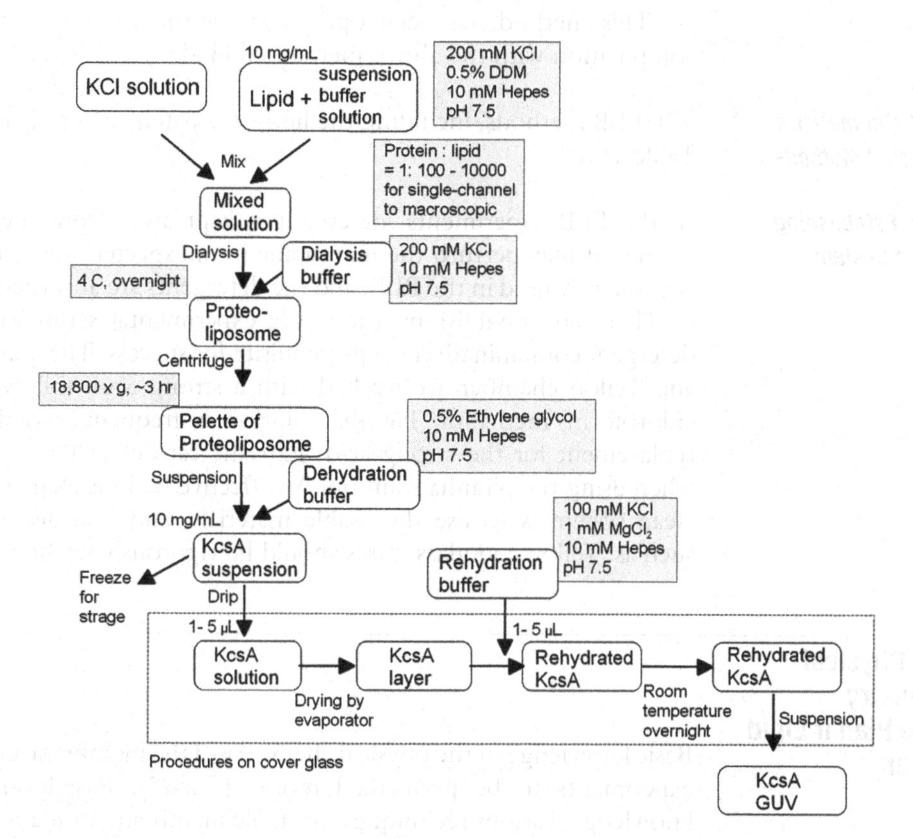

Fig. 16.10. Preparation of giant unilamellar vesicles (*GUV*). *DDM*, n-dodecyl β-D-maltoside; *KcsA*, K^+ channel from *Streptomyces lividans*.

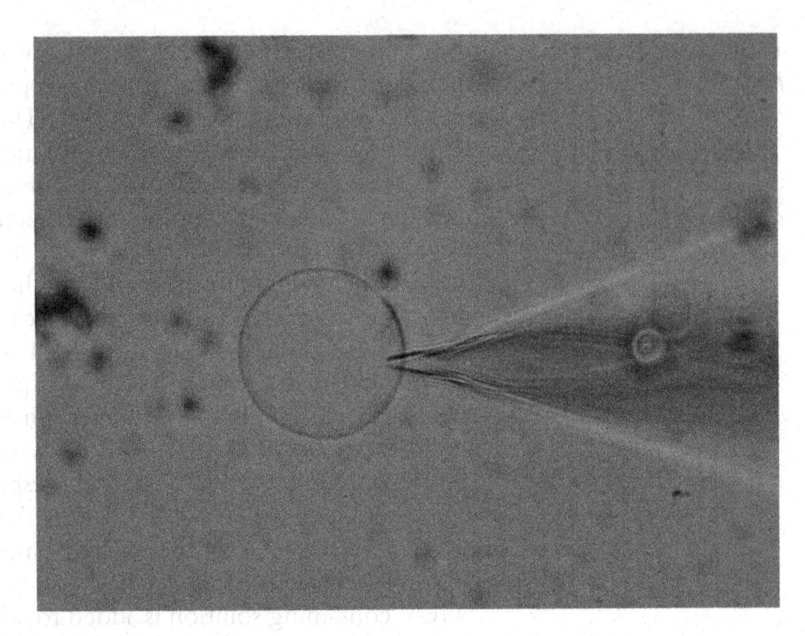

Fig. 16.11. Liposome patch.

seal. Once the giga-ohm seal is attained, the patch pipette is withdrawn to obtain an excised patch.

This method has been optimized for the KcsA channel, in combination with the above-mentioned lipid.

16.3.3. Comparison of the PLB Methods

The PLB methods, including the liposome patch, are compared in Table 16.1.

16.3.4. Establishing a Clean System

In the PLB experiments, detergent is kept away from the PLB because it may perturb the membrane in unexpected ways. Teflon is frequently used in the PLB, and the detergents are absorbed onto it. Therefore, establishing the entire experimental setup without detergent contamination is a prerequisite for success. The glassware and Teflon chamber are washed with a strong acid, followed by chloroform/methanol. Piranha solution is frequently used as a replacement for the strong acid. Extreme care must be exercised when using the piranha solution. An effective basic strategy for the clean system is to use disposable materials; and non-disposables such as, Teflon and glass wares should be vigorously washed.

16.4. Physical Chemistry of the Planar Lipid Bilayer

Basic knowledge of the physical chemistry of the membrane enables experiments to be performed more efficiently. Based on such knowledge, various techniques for stable membrane formation and membrane fusion have been developed.

Table 16.1
Comparison of planar lipid bilayer methods

		Procedure	Noise level	Membrane stability	Organic solvent	Perfusion	Sample preparation
Chamber	Painting	Easy	High	Stable	Present	Slow	Easy
	Folding	Not difficult	Medium	Stable	Trace amount	Slow	Easy
	Punch-out	Not difficult	Medium	Stable	Present	Slow	Easy
Pipette	Modified tip-dip	Not easy	Low	Stable	Present	Difficult	Easy
	Liposome patch	Not easy	Low	Less stable	Absent	Easy and rapid	Not easy

16.4.1. Formation and Structure of the Planar Lipid Bilayer

The formation of the PLB by the painting method proceeds with the following processes. A small aliquot of lipid solution is deposited on the hole. In the lipid solution occluding the hole, the phospholipids distribute at the water–organic solvent interface and form monolayers. The bulk organic solvent sandwiched by the two monolayers escapes toward the more hydrophobic Teflon surface, an event facilitated by buoyancy of the organic solvent. The driving force for the thinning is the Plateau–Gibbs boundary attraction force at the earlier time point and then van der Waals force for the middle and late periods (28). The layer of the organic solvent becomes progressively thinner until an equilibrium is reached. A thin layer of the organic solvent remains behind between the two monolayers (solvent-containing membrane).

Similar thinning processes take place in the formation of the "solvent-free" membrane by the folding method (47). For this method, the monolayer formed at the air–water interface does not contain organic solvent because hexane evaporates rapidly. However, the hole is precoated with squalene, which is sandwiched between the two apposing monolayers. As squalene has a long molecular length, it is excluded from the inter-monolayer space, and the membrane is thinner than that formed with decane using the painting method. Even in a "solvent-free" membrane, the annular phase is present at the edge of the hole.

The PLB is comprised of a lipid bilayer phase and an annular bulk solvent phase (or torus) (Fig. 16.12). The annular phase, which consists of phospholipids and a huge amount of organic solvents, surrounds and supports the bilayer on the one hand and mediates fixation to the supporting materials on the other.

Microscopic observation of the membrane reveals a distinct boundary line between the lipid bilayer phase and the annular phase. The flat shape in the bilayer phase becomes concave across the border, where a contact angle can be definable.

In the PLB, the lipid bilayer phase consists of the two monolayers and a sandwiched organic solvent phase. The thickness of the lipid bilayer phase is determined by (1) the length of the phospholipid molecules, (2) the presence or absence of cholesterol (the membrane thickens in the presence of cholesterol) (48), and (3) the thickness of the organic solvent phase (the longer the molecular length of the organic solvent, the thinner the membrane thickness).

The annular and PLB phases are chemically in equilibrium, although the volume of the annular phase is much larger. Thus, the annular phase predominantly governs the equilibrium status. Phospholipids are amphipathic molecules; and above a certain concentration known as the critical micelle concentration (CMC), phospholipids form micelles. A bilayer is formed only above the CMC. In the annular phase, which has a large pool of organic solvents, reverse micelles of phospholipids are found.

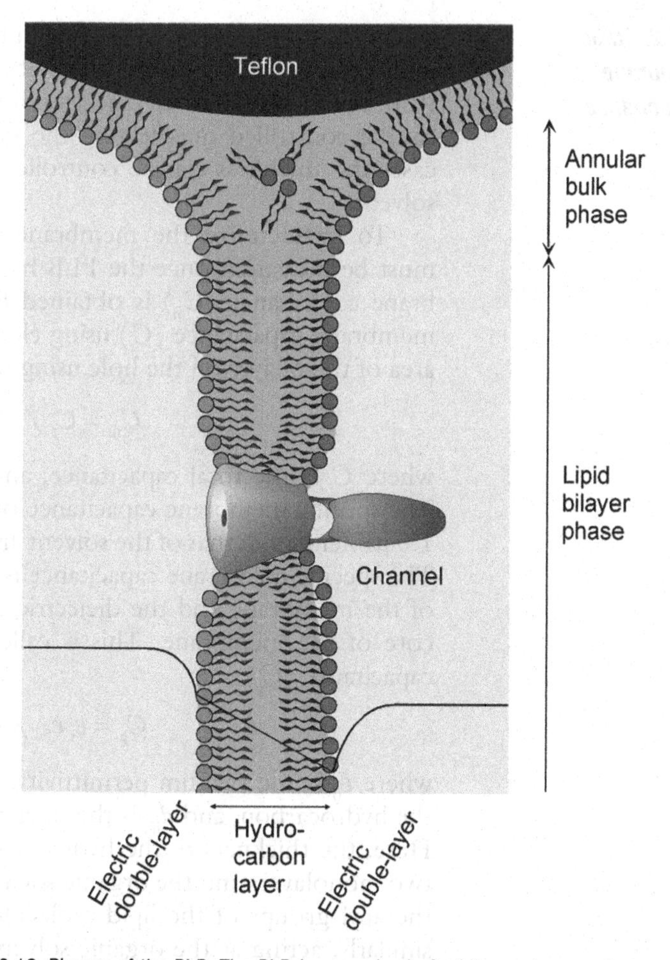

Fig. 16.12. Phases of the PLB. The PLB is comprised of a bilayer phase and an annular bulk phase near the supporting material. The bilayer has three layers: two monolayers of phospholipids and a layer of organic solvent. In the annular bulk phase, a huge amount of the organic solvent is present in which reversed micelles of the phospholipid are found. The distinct contact angle is defined at the border between the bilayer phase and the annular bulk phase.

When we measure electrical currents from either the PLB or a patch membrane, we expect that measured currents originate from the channels on the membrane rather than from leakage between the two compartments. The leakage frequently occurs at the interface between the membrane and the supporting material. Insulation of the two compartments is referred to as the "seal" (see Chap. 1). In the PLB, the annular phase keeps the tight connection between the bilayer membrane phase and the supporting Teflon material of the septum. The seal resistance can be as high as several hundreds of giga-ohms in PLB.

16.4.2. Static Membrane Capacitance

There are fundamental differences in the membrane electric capacitance of the PLB compared to that of the cell membrane. In the PLB, the membrane area (the area of the bilayer membrane phase) can be controlled by selecting the size of the hole; and in some cases the thickness can be controlled by selection of the organic solvent.

To characterize the membrane, the membrane capacitance must be measured once the PLB has formed. The specific membrane capacitance (C_m) is obtained from a determination of the membrane capacitance (C_t) using electrical measurements and the area of the bilayer on the hole using optical measurements

$$C_m = C_t / A_m \tag{16.1}$$

where C_t is the total capacitance, and A_m is the membrane area. The specific membrane capacitance of the biological membrane is 1.0 μF/cm², and that of the solvent-free membrane is 0.8 μF/cm². The specific membrane capacitance is determined by the thickness of the membrane and the dielectric constant of the hydrophobic core of the membrane. This is called the geometric membrane capacitance (C_g).

$$C_g = \varepsilon_o \varepsilon_{hc} / d_{hc} \tag{16.2}$$

where ε_0 is the vacuum permittivity, ε_{hc} the dielectric constant of the hydrocarbon, and d_{hc} is the thickness of the hydrocarbon core. Here, the thickness of the hydrocarbon core includes that of the two monolayers and the organic solvent layer (Fig. 16.12). In fact, the acyl groups of the lipid molecules and hydrocarbons behave similarly, acting as the organic solvent and are thus regarded as a homogeneous phase, with a dielectric constant of 2.

There are also electrical double layers at the membrane surface (C_{dl}: the Gouy–Chapman–Stern layer) that contribute to the specific membrane capacitance. The double layers in both sides of the membrane are electrically connected in series; thus,

$$\frac{1}{C_m} = \frac{1}{C_g} + \frac{2}{C_{dl}} \tag{16.3}$$

where C_{dl} is expressed by an equation similar to 16.2. In this case, E_{hc} represents the dielectric constant of the double layer, and d_{dl} represents the Debye length, which changes as a function of the surface charge density and ionic strength of the electrolyte solution.

In the case of the decane membrane, the measured membrane capacitance varies significantly by changing the membrane area rather than the membrane thickness. When a large amount of lipid solution is painted on the hole, the annular bulk phase occupies a large area of the hole, and the capacitance is relatively small.

The PLB is stable when the bilayer area occupies ~90% of the area of the hole. Sometimes the area of the bilayer estimated from the capacitance measurements is larger than that of the hole. In such a case, the membrane bulges to one side because of the hydrostatic unbalance, and the membrane is unstable.

The membrane capacitance increases when the absolute value of the membrane potential is increased.

$$\Delta F_V = -\frac{\varepsilon_0 \varepsilon}{2d} V^2 \tag{16.4}$$

where ΔF_v represents the free energy at a given voltage. This is the electrostriction, changing the equilibrium between the bilayer and annular phases through altering the contact angle; it leads to an increase in the bilayer area.

16.4.3. Solvent-Containing and Solvent-Free Membranes

The PLB membrane contains more or less organic solvent except for the liposome patch membrane. It has been reported however that the channel activities on both membranes are qualitatively similar, and are not distinguished from that on the biological membrane. How are the channel proteins accommodated to the PLB? The membrane protein has a hydrophobic region that surrounds the transmembrane domain and interfaces with the phospholipids. To be integrated into the membrane, the hydrophobic region must have a width of approximately 30 Å to match the thickness of the hydrophobic core of the membrane. When a membrane protein is incorporated into the PLB, the PLB must be deformed around the protein such that the two monolayers come closer together until they match the width of the hydrophobic region of the transmembrane domain, and the organic solvent is eliminated from between the two monolayers (Fig. 16.12). From the viewpoint of membrane proteins, they are surrounded by bilayers and might not have contact with the organic solvent.

16.4.4. Lipid Composition

Phospholipids are characterized by a hydrophilic head group and hydrophobic tails. Each chemical feature contributes to physicochemical characteristics of the bilayer membrane. The head groups confer charges onto the surface of the membrane, whereas the tail group defines the thickness of the membrane by the tail length and the fluidity of the membrane by the presence or absence of double bonds and branched structures. When charged lipids (e.g., PS, PI, PG) are used, the surface charge density causes the membrane capacitance to increase (17, 49, 50). In addition, the surface charge accumulates ions of the opposite sign near the membrane, and this alters the single-channel conductance (49, 51). When an asymmetrical membrane is formed, the surface potentials on the two sides differ significantly, and the voltage-dependent gating property is shifted because the intramembrane electrical field is affected by the asymmetrical surface potential.

In addition to the chemical contributions of the head and tail groups, the shape of phospholipid molecules is important for the physical features of the membrane (Fig. 16.13) (52, 53). Phospholipids form a curved or flat sheet when assembled. For example, PC has a cylindrical shape and fill a flat surface when assembled (bilayer phase, L_α). In contrast, PE has a small head and a reversed xz conical shape, assembling into a concave sheet (hexagonal phase H_{II}). Polymorphisms of the monolayer structure contribute significantly to the processes undergoing changes in the shape and curvature of the membrane sheet, such as in the membrane fusion (Fig. 16.12).

Lipid compositions alter the phase transition temperature (52). The membrane fluidity differs below and above this temperature, and thus temperature changes may alter the channel activity (54).

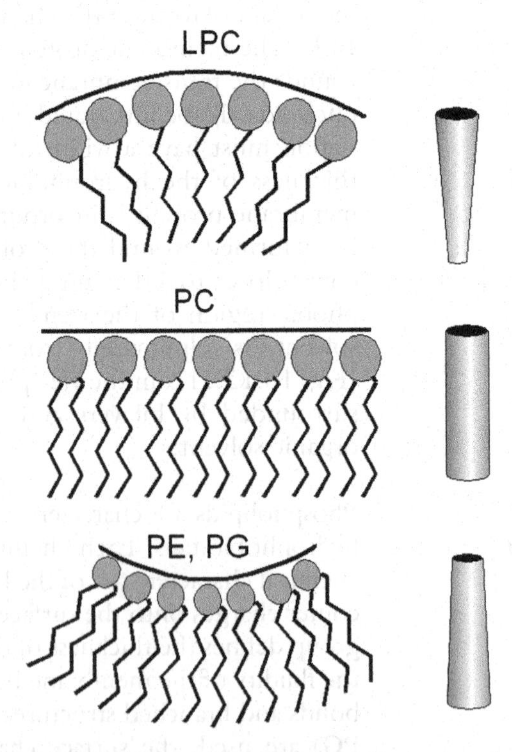

Fig. 16.13. Molecular configuration of phospholipids and the shapes and curvature of the self-assembled structure. Lysophosphatidylcholine (*LPC*) has one tail and shows a cone shape. In phosphatidylcholine (*PC*) the head and tail are of similar size and display a columnar shape. In phosphatidylethanolamine (*PE*) and phosphatidylglycerol (*PG*), the head group is small and assumes a reversed cone shape.

16.5. Reconstitution of Membrane Proteins

16.5.1. Purification of Membrane Proteins and Reconstitution in Liposomes

Various methods for purifying various types of membrane proteins have been reported. Here, a procedure for purifying and reconstituting the KcsA potassium channel in liposomes is presented (Fig. 16.14) (55). The histidine-tagged KcsA channel (at the N-terminal end) was used for this procedure. The result of sodium dodecyl sulfate polyacrylamide gel electrophoresis (SDS-PAGE) for the purified KcsA demonstrates that the channel is in the tetrameric state even in the presence of a detergent.

After purifying KcsA with an affinity column and gel filtration column, KcsA is reconstituted in liposomes. Initially, solubilized KcsA is covered with detergent on the hydrophobic portion of the transmembrane domain. As the detergent is removed, phospholipids replace this portion and form liposomes. The orientation of the reconstituted membrane protein is determined during this process.

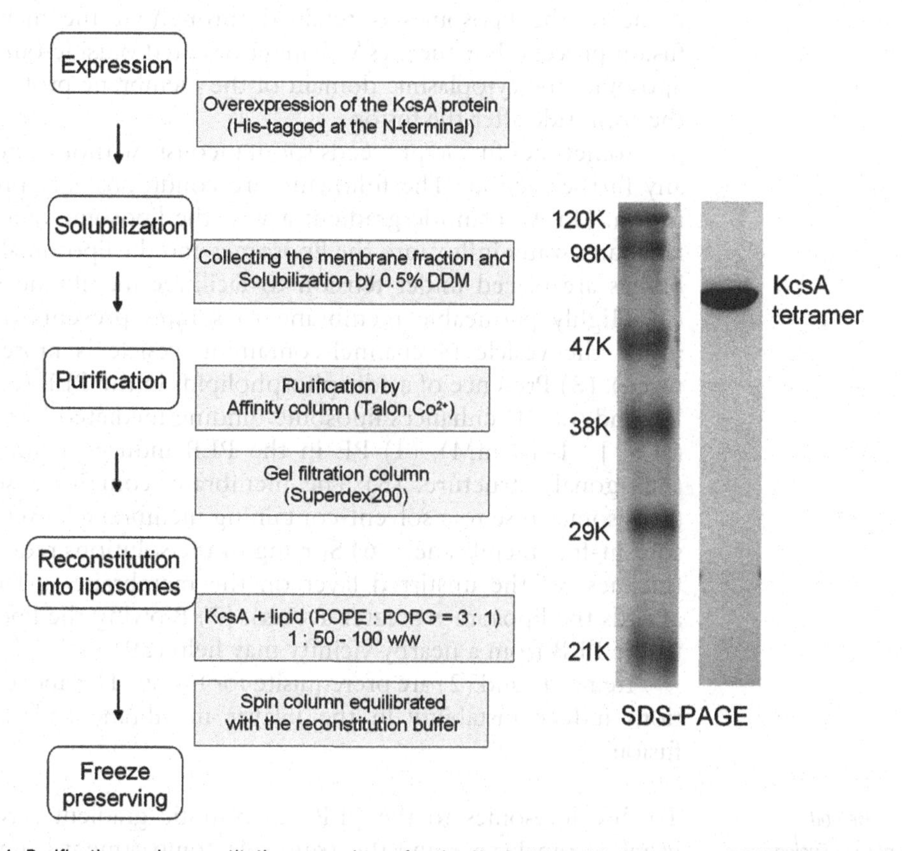

Fig. 16.14. Purification and reconstitution procedures for the KcsA channel in liposomes. *POPE:POPG*, 1-palmitoyl-2-oleoylphosphatidylethanolamine: 1-palmitoyl-2-oleoylphosphatidylglycerol; *SDS-PAGE*, sodium dodecyl sulfate polyacrylamide gel electrophoresis.

Various factors affect the orientation upon reconstitution. Among them, the shape of the transmembrane domain (i.e., cylindrical, conical, inversely conical) contributes to the orientation substantially. For KcsA, having the inversely conical transmembrane domain, the orientation is always in the outside-out with the cytoplasmic domain retained on the inside of the liposome (46, 56).

The reconstituted channels on the liposome are transferred to the PLB after membrane fusion. An intraliposomal hyperosmotic condition facilitates liposome fusion to the PLB (57, 58). Thus, nonelectrolyte solute is added to the reconstitution buffer solution to load the nonelectrolyte inside of the liposome upon reconstitution. The number of channels incorporated into the PLB can be controlled by changing the protein/lipid ratio upon reconstitution (37).

16.5.2. Reconstitution in the Planar Lipid Bilayer Via Membrane Fusion

In most of the PLB experiments, membrane proteins are added to the front compartment of the chamber, which is referred to as the *cis* side (with the other being the *trans* side). The orientation of the membrane proteins on the membrane fraction of the native membrane or the liposomes is retained throughout the membrane fusion process. For the KcsA channel oriented outside-out in the liposome, the cytoplasmic domain of the membrane protein faces the *trans* side after the fusion.

Sometimes fusion proceeds spontaneously without a need for any further action. The following are conditions that promote fusion. (1) An osmotic gradient across the liposome membrane facilitates water influx into the liposomes and the liposomal membranes are placed under tension to facilitate membrane fusion. (2) Highly permeable membrane to solutes prevents dilution inside the vesicle (a channel-containing vesicle is more easily fused). (3) Presence of acidic phospholipids in the PLB (e.g., PS, PG and/or PI) enhances liposome binding mediated by Ca ions ($[Ca^{2+}]_{cis}$ 1–10 mM). (4) PE in the PLB induces a nonbilayer (hexagonal) structure. (5) The membrane contains a solvent. (Liposomes fuse to a solvent-containing membrane better than a solvent-free membrane.) (6) Stirring of the solutions reduces the thickness of the unstirred layer on the membrane surface and affords the liposomes greater access). (7) Blowing the liposomes to the PLB from a nearby vicinity may help (29).

Items (1) and (2) are prerequisite for fusion. The above conditions induce instability in the bilayer membrane and facilitate fusion.

16.5.2.1. Method for Membrane Fusion

To fuse liposomes to the PLB, an osmotic gradient across the intraliposomal space and the *trans* side compartment is imposed. The followings are ways to facilitate the fusion.

1. The PLB is formed in symmetrical electrolyte solution, and nonelectrolyte is added to the *cis* side to impose an osmotic gradient across the PLB. An osmotic gradient such as 3:1 is a good starting point. The osmolarity of the *trans* side should be lower than that inside the vesicle.
2. Membrane vesicles are added to the *cis* side at the vicinity of the hole while stirring the solution. Higher intravesicular osmolarity than that on the *cis* side induces vesicle swelling and facilitates membrane fusion.
3. Stirring the solution reduces the thickness of the unstirred layer near the membrane and facilitates vesicle access to the PLB.

An abrupt manifestation of current followed by the usual single-channel activity indicates that fusion has just occurred. Further fusion events are observed as spontaneous increases in the number of active channels or current amplitudes. With this functional criterion for membrane fusion, the fusion of liposomes lacking channels is ignored. Thus, if no current is recorded from a sample for an extended period of time, it is not clear whether fusion has not taken place or the channels incorporated into the PLB are inactive. To circumvent this issue, the nystatin method was developed (see Sect. 16.5.3).

Among the experimental steps in the course of PLB study, efficient control of liposome incorporation into the PLB is the most difficult. Even using the same sample of liposomes, sometimes fusion does not occur. To simply reduce possible risks that the liposome may not approach to the PLB, the pressure-pulse method has been applied: Liposome containing solution is loaded at the tip of a pipette similar to that used for the punch-out method, and a brief pulse of pressure is applied to flush out the solution. A brief pulse does not break the PLB, but the liposomes gain closer access to the PLB, which leads to fusion.

16.5.2.2. Termination of Membrane Fusion

Stopping the stirring may terminate fusion. When Ca^{2+} is used for the fusion process, ethyleneglycol-bis-(2-aminoethylether)-*N*,*N*,*N′*,*N′*-tetraacetic acid (EGTA) is added to reduce the free Ca^{2+} concentration. Adding osmolyte to the *trans* side, equilibrating the osmotic gradient, is also effective.

16.5.2.3. Mechanism of Membrane Fusion

The lipid bilayers by themselves have the ability to produce membrane fusion without assistance from other molecules (59). Membrane fusion occurs when the hydrophobic moiety of phospholipid molecules is exposed to the surface. The hydrophobic exposure in two apposed membranes generates attractive force and two membranes approach closely until they are interdigitated (Fig. 16.15) (60). This hydrophobic exposure happens when the hydrophilic head groups of the membrane phospholipids are separated from

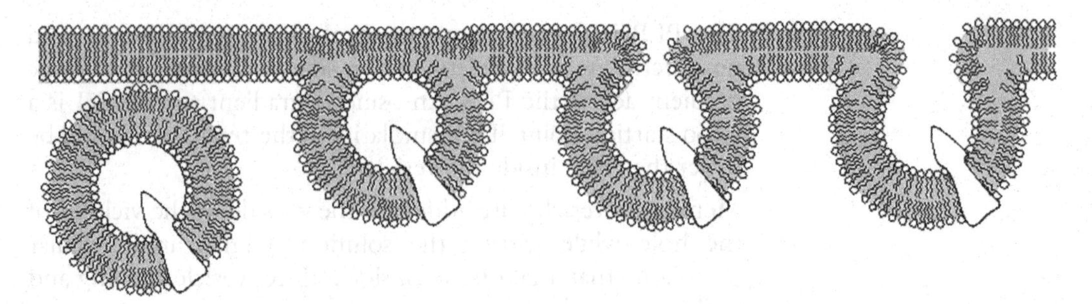

Fig. 16.15. Membrane fusion of liposomes to the PLB. The membrane fusion proceeds through a process of hemi-fusion, where the upper leaflet of the PLB and the inner leaflet of the vesicle form the bilayer. The fusion pore appears on the bilayer, and the membrane fusion proceeds.

each other by imposing high membrane tension or rendering the membrane having a high degree of curvature. The underlying mechanism of membrane fusion is lipid–lipid interaction, and fusion-inducing proteins simply deform the membrane to facilitate hydrophobic interactions (60, 61).

During membrane fusion, a hemi-fusion state is commonly observed. As the two bilayer membranes approach, the two touching monolayers from the two bilayers deform, and the exposed hydrophobic parts from the two monolayers become fused. Thus, the outer leaflet of the liposomes and the nearby (or *cis*) leaflet of PLB attain continuity. Only the two monolayers comprised of the inner leaflet of the liposome and the *trans* leaflet of PLB remain at the contact site, and they form a bilayer. This process is referred to as "hemi-fusion" (62). In the hemi-fused membrane, further deformation generates perturbation of the bilayer structure, and the lipids in the outer and inner leaflets become exchangeable. A small pore that is created there is known as the "fusion pore." Recently, these fusion processes have been generated using computer simulations (63, 64).

Osmotic differences are used to apply tension to the vesicle membrane. When the vesicles bind to the PLB, the osmotic gradient between the intravesicular solution and the *trans* solution drives water flux into the vesicle, and the intravesicular pressure increases (Fig. 16.16). Intravesicular volume expansion extends the membrane, and membrane fusion occurs. If there are no channels on the vesicle membrane, the intravesicular water flows out into the *cis* solution as the intravesicular pressure increases. On the other hand, if there are channels on the vesicle membrane, ions flow into the vesicle as the intravesicular concentration is diluted. This influx may maintain the volume expansion of the vesicle.

In the PLB, membrane fusion occurs without any contribution of "fusion" proteins. The PLB serves as a simple system to help us better understand the mechanism of membrane fusion.

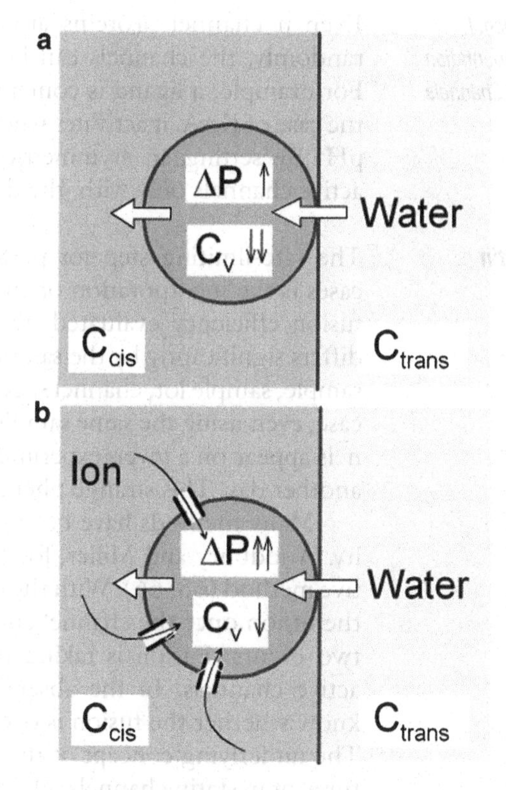

Fig. 16.16. Membrane fusion is facilitated by an osmotic gradient and the presence of channels on the vesicle membrane.

16.5.2.4. Orientation of the Channel Proteins Incorporated in the Planar Lipid Bilayer

The orientation of the membrane proteins on the PLB after fusion is predetermined by the orientation of the proteins in the vesicles to be fused. For example, a microsomal fraction is collected from cell fractionation, and the microsome is fused to the PLB to record the channel of the endoplasmic reticulum (ER) membrane. In this case the channels are oriented such that the cytoplasmic side protrudes outside of the microsome. After fusion of the microsomes to the PLB, the cytoplasmic side of the channel faces the *cis* side.

When purified channels are reconstituted in liposomes, the orientation of the membrane proteins is generally random. (In the case of KcsA, the orientation is right side-out.) For the channels oriented in the same direction on PLB, measurement of the macroscopic current becomes feasible (37).

The relation between the orientation of the membrane proteins in the vesicles and PLB is maintained so long as the PLB is not broken after the addition of channel vesicles. Once broken, the channel vesicles are distributed around the hole; and upon the reforming of the PLB, the channel vesicles are located on both sides of the membrane. Fusion of vesicles from the two sides results in a random orientation of the reconstituted channels.

16.5.2.5. Functional Control of the Orientation of Incorporated Channels

Even if channel proteins are incorporated into the membrane randomly, the channels can be functionally oriented in one way. For example, a ligand is contained on one side of the chamber. In the case of KcsA, it activates when the cytoplasm side becomes acidic pH. By setting an asymmetrical pH for the two compartments, active channels only with the desired orientation are recorded.

16.5.3. Nystatin Method

The rate-limiting step for performing PLB experiments in most cases is the incorporation of channel molecules into the PLB. The fusion efficiency evaluated by the appearance of active channels differs significantly by the sample preparation, heterogeneity of the sample, sample lot, channel species, and other factors. In an extreme case, even using the same sample, one may observe that many channels appear on a given experimental day but no channels appear on another day. This strange phenomenon still remains elusive.

Many methods have been proposed to overcome this variability. Woodbury and Miller, for example, developed a logical, effective method (65, 66). With the conventional method, we recognize the fusion once the channel current appears. This criterion implies two events: Fusion is taking place, and these liposomes contain active channels. In the absence of channel currents, we do not know whether the fusion is occurring or if the channels are active. The underlying concept of the nystatin method exploits three features of nystatin channels. (1) Nystatin is a hydrophobic substance and thus is readily transferred to the membrane. (2) Nystatin channel permeates both cations and anions (slightly more permeable to anions) and thus allows massive solute influx, leading to vesicle expansion. (3) Nystatin loses channel activity in the absence of a sterol (67). With the nystatin method, sterol is exclusively included in the vesicle, while no sterol is added to the PLB (Fig. 16.17).

Sterols are a subgroup of the steroids and include cholesterol and ergosterol (Fig. 16.4). Ergosterol maintains the channel activity of nystatin more effectively than cholesterol. Nystatin forms oligomers (decamers), and it is proposed that ergosterol holds them together.

When a vesicle fuses to the PLB, the nystatin channels on the vesicle are transferred to the PLB, and ionic currents through the nystatin channels are rendered observable. This appears as a jump of current at the moment of the vesicle fusion. This current originates from all the nystatin channels on the vesicle (Fig. 16.17c,f). After fusion, ergosterol molecules, which previously had been present exclusively on the vesicle membrane, diffuse toward the ergosterol-free region of the PLB. The ergosterol around the nystatin channels decreases, and the channels fall into inactivity. The conductance of the nystatin channel is too small to be resolved as stepwise conductance changes, so a decrease in the number of active channels appears as a decay in current amplitude (Fig. 16.17e,f). This type of current change, with its sudden appearance followed

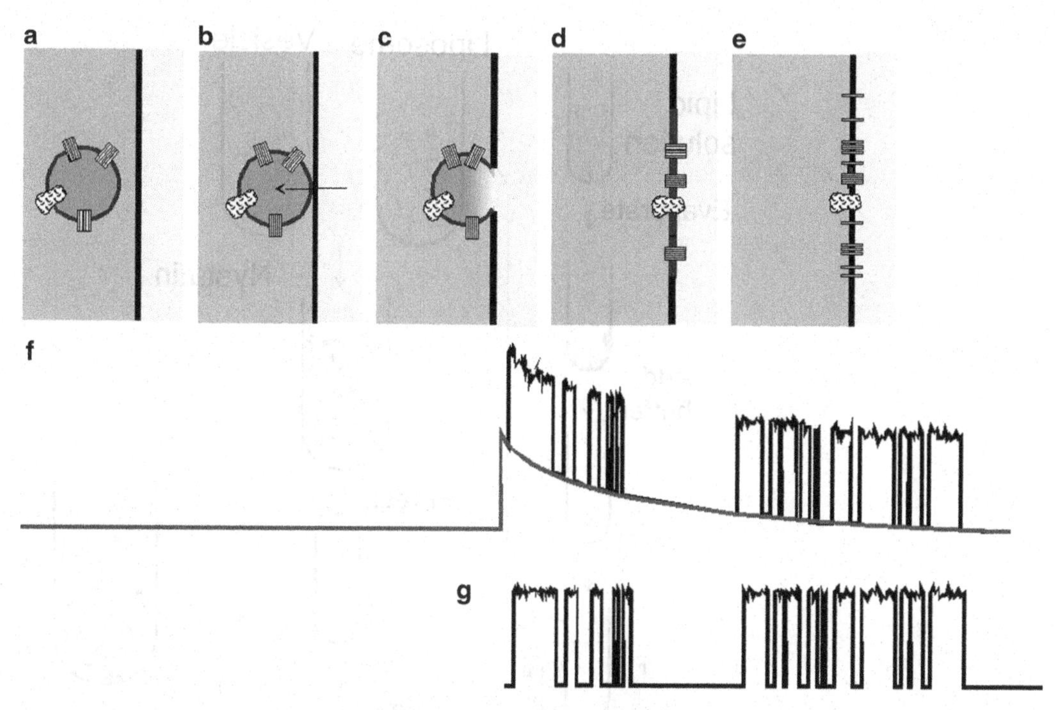

Fig. 16.17. Nystatin method: membrane fusion process. (**a–c**) Nystatin channels are active on the vesicle membrane because ergosterol is plentiful. (**c**) Once fused, a current jump is elicited by the action of many nystatin channels. When a channel having a measurable single-channel current exists on a vesicle, the channel activity appears upon membrane fusion. (**d, e**) Gradually, ergosterol escapes from the region of the previous vesicle membrane. As the concentration of ergosterol decreases, the nystatin channels disassemble. (**f**) A representative time course of the current upon a vesicle fusion containing nystatin and protein channels. The slowly decaying current represents the macroscopic currents through the nystatin channels. (**g**) Single-channel current of the reconstituted channel protein. From a current trace of (**f**), the nystatin current was subtracted. After disassembly of the nystatin channel, the single-channel activity of the incorporated channel remains.

by decay, represents vesicle fusion to the PLB, and each fusion event is visualized in the current recordings.

When a channel protein is included on a fused vesicle, a single-channel current of the channel is overridden on the nystatin decaying current, and the single-channel current continues irrespective of the disappearance of the nystatin current. Accordingly, all of the fusion events, regardless of whether the vesicle contains a channel, are visualized. Hence, this method allows separate evaluations of fusion events and channel activity.

Preparation of nystatin containing liposomes (Fig. 16.18) requires the following steps:

1. The lipid composition (mole percent) is as follows: PE 50%, PC 10%, PS 20%, ergosterol 20%.
2. The mixed lipids are dissolved in chloroform (1 mg/100 μl), and 2 μl of nystatin (stock solution 10 μg) is added.
3. After evaporation of chloroform under nitrogen gas, 0.2 ml of NaCl solution is added.

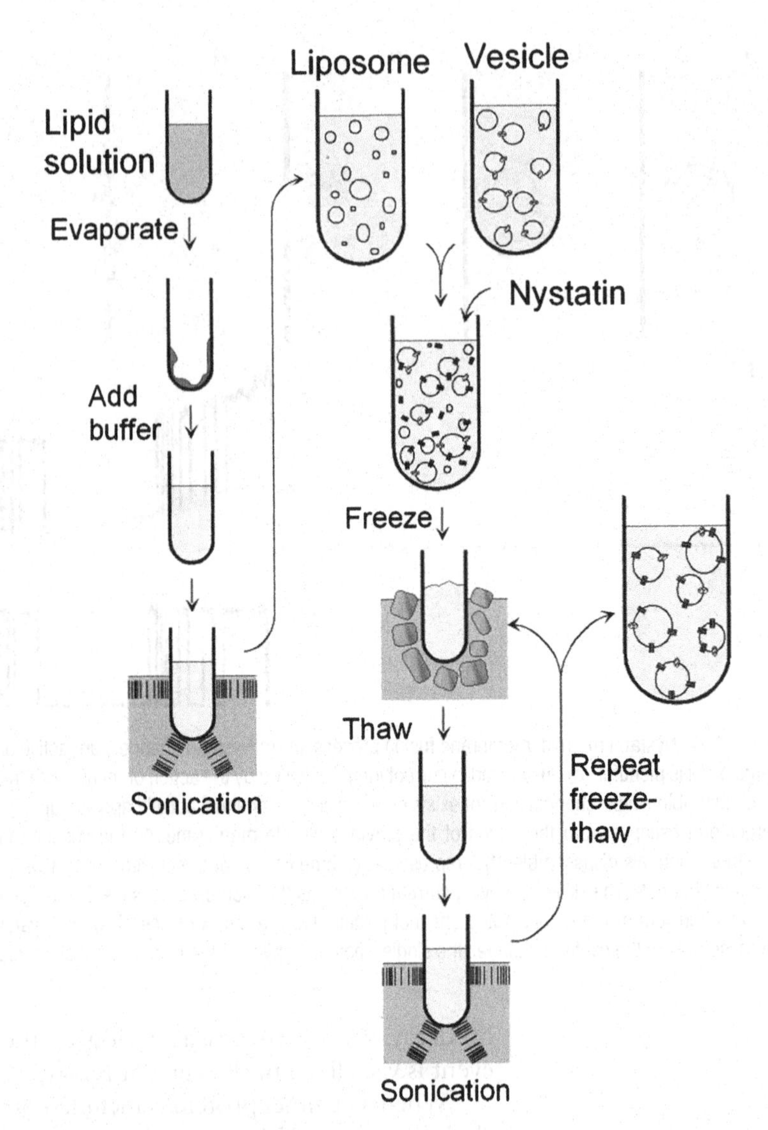

Fig. 16.18. Preparation of nystatin-incorporated liposomes.

4. Sonicate for 10–100 s. The test tube is dipped in the sonicator bath such that the water level of the sample and that of the sonicator bath are on the same level.
5. These liposomes are frozen by dry ice/ethanol.
6. Thaw the sample immediately before the experiment, and sonicate it for 5–15 s to yield giant unilamellar liposomes.

The final concentration of nystatin added to the liposome solution is 50 μg/ml. The nystatin channel activity is highly concentration-dependent, with a Hill coefficient of 10. This sharp concentration dependence makes the optimal concentration range narrow (50–60 μg/ml). Below this range the channel activity is too low; and above it nystatin forms a cation-selective channel (i.e., with double-sided action), rather than a nonselective channel. Moreover, the activity of the channel becomes independent on ergosterol (68).

The nystatin method is highly useful. Assuming a homogeneous distribution of nystatin on liposomes, the liposomal size and its distribution may be estimated from the amplitudes of the current jumps. Generally, the size of the liposome is inversely related to the time of the last sonication.

There is a case in which the nystatin method is not feasible. The acetylcholine receptor channel requires cholesterol to maintain its channel activity. The nystatin method cannot be used in such a case because incorporating cholesterol into the PLB prevents disassembly of the nystatin channels after vesicle fusion.

16.5.4. Direct Insertion Method

Various channel-forming substances have been reported. They involve small molecules with a variety of chemical characteristics and from various biological resources ranging from bacteria to human cells, and peptides or proteins of various sizes. The function of those channel-forming substances has been studied almost exclusively using the PLB. Most of these substances exhibit toxicity toward specific target cells, and called as channel-forming antibiotics.

The channel-forming substances are released from a cell into aqueous solution and are then inserted into a target cell membrane. Thus, these toxins can be readily incorporated into the PLB once they are added to one side of the chamber. In most cases, these substances are inserted into the membrane with the same orientation. Therefore, channel current measurements are easy in the PLB. In fact, one can readily start the single-channel current measurements in the PLB using the antibiotic gramicidin channel (69, 70).

In parallel with these channel-forming substances, membrane proteins have been inserted into the PLB directly from the aqueous phase without passing through the fusion process. When solubilized proteins are added to the aqueous solution, they are spontaneously transferred to the PLB and exhibit channel activity. The mechanism underlying the process is elusive. However, prokaryotic membrane proteins are incorporated into the cell membrane after translation of the protein (post-translational translocation) (71) in contrast to eukaryotic membrane proteins, in which they are incorporated into the membrane in step with translation (co-translational translocation).

16.5.5. Monolayer Incorporation

The channel molecules are incorporated into the monolayers formed at the air–water interface. The channel-incorporated monolayers are appositioned to form a bilayer with the folding method or the tip-dip method.

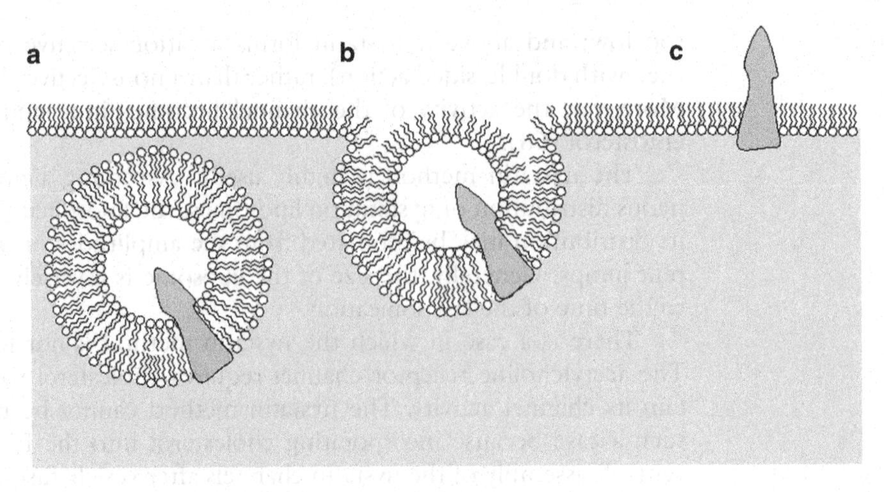

Fig. 16.19. Incorporation of a membrane protein into a monolayer.

Membrane protein is transferred by the following steps:

1. The phospholipid solution used for the folding method (4 mg/ml hexane) is added onto the water surface, where it forms a monolayer at the air–water interface.
2. Vesicle containing solution is added slowly to the surface.
3. The vesicle is "fused" to the monolayer, and there is transfer of the membrane protein to the monolayer (Fig. 16.19).

The vesicles in the aqueous solution move to the air–water interface, where the vesicle membrane is extended. Accordingly, channel proteins in the vesicle are transferred to the surface monolayer. With this process, the orientation of the membrane proteins is random. When incorporated into the monolayer, the membrane proteins are at least partly exposed to the air.

This method is an alternative method in the event that the fusion method is not successful. For example, nicotinic acetylcholine receptor (nAChR) channel has been elaborated in an effort to reconstitute it in the PLB through the vesicle fusion method, but the effort mostly failed. In contrast, the reconstituted nAChR channel was extended as a monolayer, and the tip-dip method was performed successfully.

16.6. Electronic Aspects of the Planar Lipid Bilayer System

The electrical setup for the PLB experiment is simple, partly because of the less complicated configuration. Ag-AgCl electrodes are sometimes dipped into the electrolyte solutions directly, and the series resistance is as low as 10 kΩ or less. Thus, the potentially serious issue of high series resistance (>1 MΩ) encountered in the conventional patch-clamping technique can be ignored in the PLB system. The issues that arise exclusively in PLB experiments are described here (29, 72, 73).

16.6.1. Definition of Membrane Polarity

The connection of the ground electrode to either the *cis* or *trans* compartment and the polarity of the membrane potential can be set arbitrarily in PLB experiments. In most cases, channel-reconstituted liposomes are added to the *cis* side, and these manipulations on the *cis* side is permitted with the *cis* side grounded. The membrane potential of the biological membrane is defined relative to the extracellular space. Similarly, the membrane potential for organelles is defined relative to the intraorganelle space (74).

There is an additional issue related to the orientation of channel proteins on the membrane. Depending on the source and methods of extraction used for the membrane fraction, the vesicles exhibit either a outside-out or inside-out configuration. Furthermore, the channels on the reconstituted liposome are either randomly or orderly oriented. Thus, specifying the orientation of the incorporated channel molecules in the PLB, and accordingly, defining the polarity of the membrane potential, are additional concerns when performing PLB experiments. For example, when the vesicle of a microsomal fraction is fused to the PLB from the *cis* side, the orientation of the membrane protein in the ER membrane is transferred such that the cytoplasmic side faces to the *cis* side. In this case, the membrane potential should be defined relative to the *trans* side.

16.6.2. Minimizing Background Noise

A patch-clamp amplifier is a highly sensitive current–voltage converter. For single-channel current measurements, the noise originating from either the input of the operational amplifier or the field effect transistor (FET) must be taken into consideration.

With conventional patch-clamping, the thermal noise originating from the feedback resistor (50 GΩ) of the head stage and the seal resistance are predominant in the overall noise. On the other hand, in the PLB, the noise that originates from clamping the voltage across the large capacitance membrane is the most significant background noise. One readily grasps this relation by reminding oneself of the following equation:

$$I = C\frac{\mathrm{d}v}{\mathrm{d}t} \tag{16.5}$$

At high-frequency ranges of voltage fluctuation, the current noise level is augmented. The spectrum density of the noise linearly increases as the bandwidth and the capacitance increase. Thus, the recording bandwidth is limited for large-capacitance PLB. The source of the voltage noise is (1) the FET input of the patch-clamp amplifier, (2) the voltage command circuit, and (3) the access resistance.

The resistance components in the recording chamber are arranged in a series from the electrode to the channel: they include (i) the interface between the electrode and the electrolyte solution, (ii) the salt bridge, (iii) the electrolyte solution and (iv) the geometrical configuration around the membrane (75). Among these components, the resistance around the membrane is the only factor that can be manipulated. In contrast to the conventional patch-clamping technique, which can have a pipette resistance as high as several mega-ohms, the PLB method using a chamber has a resistance on the order of several dozen kilo-ohms around the hole.

16.6.2.1. Input Voltage Noise

The input voltage noise of the head stage generates current noise through the total capacitance, C_t. The spectrum density, $S^2_{i(f)}$ follows Ohm's law.

$$S^2_{i(f)} = \frac{e_n^2}{X_c^2} \tag{16.6}$$

where e_n is the voltage rms (root mean square) noise of the head stage input, and X_c is the capacitance reactance of the PLB. The capacitance reactance is

$$1/\left(2f\pi C_t\right) \tag{16.7}$$

where f is the frequency, and

$$C_t = C_b + C_s + C_{FET} \approx C_b \tag{16.8}$$

where C_b is the capacitance of the PLB, C_s is the stray capacitance, and C_{FET} is the input capacitance. Thus,

$$S^2_{i(f)} = e_n^2\left(2\pi f C_b\right)^2 \tag{16.9}$$

The internal voltage noise can be evaluated by connecting the capacitance of the PLB to the amplifier. When an external voltage command is used, especially using a digital to analog (D/A) converter, the noise level must be taken into consideration.

16.6.2.2. Access Resistance Thermal Noise

The spectrum density for the current noise generated under voltage-clamping of the access resistance is as follows.

$$S^2_{i(f)} = 4kT\mathrm{Re}\{Y(f)\} \tag{16.10}$$

Where k is the Boltzmann constant, T is the absolute temperature, and Re[Y(f)] is the real part (conductance) of the admittance for the series R_a and C_b.

$$\mathrm{Re}\{Y(f)\} = \frac{(2\pi f C_b)^2 R_a}{1+(2\pi f C_b R_a)^2} \tag{16.11}$$

This becomes

$$\mathrm{Re}\{Y(f)\} = R_a(2\pi f C_b)^2 \tag{16.12}$$

at the lower extreme of R_a, similar to equation 16.7. Thus, three factors – internal noise of the head stage, voltage noise generated in the membrane capacitance, and access resistance in series – are expressed in a similar manner. At the upper extreme of R_a,

$$\mathrm{Re}\{Y(f)\} = 1 / R_a \qquad (16.13)$$

The noise is maximum when the dependence of R_a on the frequency changes from linear to reciprocal dependence.

16.6.2.3. Evaluation of Noise

Integrating $S_{i(f)}$ subscript over the bandwidth gives the variance of current. Figure 16.20 shows the current noise as a function of the access resistance and the membrane capacitance. The current level is shown as the peak-to-peak value from direct current (DC) to 1 kHz. The noise is represented as the sum of the input voltage noise of the amplifier (16.9, where $e_n = \sqrt{6\ \mathrm{nV/Hz}}$) and the voltage noise of the access resistance. When the access resistance is greater than several kilo-ohms, the input voltage noise of the amplifier can be ignored, and the voltage noise of the access resistance

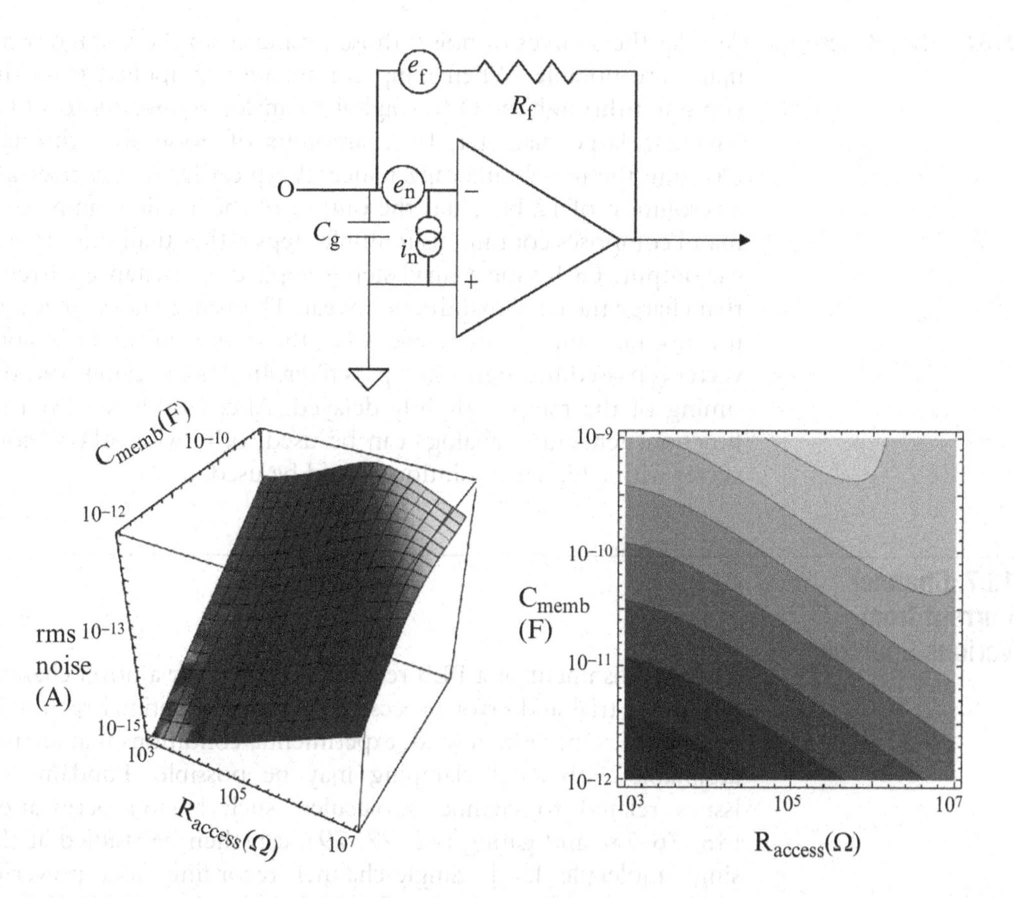

Fig. 16.20. Noise from the operational amplifier and the noise level as a function of the series resistance and membrane capacitance. The noise sources are located on the circuit of the operational amplifier in two positions.

becomes the predominant source of the noise. Reducing either the access resistance or the membrane capacitance is effective for reducing the noise. However, one may use a large membrane to facilitate the access to the fusing vesicles. In this case, reducing the access resistance is a practical alternative to reducing the noise. When the membrane capacitance is >100 pF, the noise exhibits the maximum value even if the access resistance is <1 MΩ. The effect of access resistance and membrane capacitance on the noise level is readily observed on the contour plot (Fig. 16.20b). Note that the noise level observed in a standard conditions of patch-clamping (e.g., several mega-ohms and several picofarads) can be achieved in the PLB experiment under an access resistance of several kilo-ohms and a membrane capacitance of several tens of picofarads.

Summarizing the above considerations, a useful strategy for recording with low levels of noise is the following: (1) small membrane capacitance, (2) use of a low-noise amplifier, (3) use of a low noise voltage command, (4) small access resistance, (5) choice of materials used for the chamber.

16.6.2.4. Ramp Voltage

Among the sources of noise, those arising from the voltage command are notable. When ramp commands are applied from the computer through the D/A (digital-to-analog) converter to a PLB having a large diameter, large amounts of noise arise through charging the membrane capacitance. A typical D/A converter has a resolution of 12 bits, and the output of the applied ramp command comprises continuous multiple steps rather than smooth linear output. Each time a small step is applied, capacitance currents that charge the large membrane appear. This is not "noise" because it is not random. To attenuate noise, the output of the D/A converter is passed through a low-pass filter. In this case, however, the timing of the ramp is slightly delayed. Alternatively, an external function generator (analog) can be used; otherwise a D/A converter with a higher resolution should be used.

16.7. Channel Current from Various Sources

The establishment of a PLB recording system for a novel channel requires a trial and error process to determine optimal recording conditions. Once established, experimental conditions that are not attainable with patch-clamping may be possible. Fundamental issues related to channel molecules, such as ion permeation (45, 76–78) and gating (42, 77, 79), can then be studied at the single-molecule level. Single-channel recording is a powerful method of examining issues of overwhelming importance that go beyond the scope of the usual physiological studies to reach chemical physics (80). Two examples follow.

16.7.1. Channel-Forming Toxin

Incorporating channel molecules into a target cell membrane is a common strategy for attaining cytotoxicity; it is referred to as a channel-forming toxin (81, 82). Various molecules, encompassing small antibiotic substances, polyenes, peptides, and proteins (even of large size), exhibit cytotoxic activity toward a broad spectrum of target cells. The channel-forming toxin is released into the extracellular space and is incorporated into a target membrane, where it exhibits channel activity. In addition to having membrane-targeting features, the molecules are designed to retain a stable structure in both aqueous solution and the membrane environment. This amphipathic nature of the substances is fundamental to their toxic effects. Some toxins form cation-selective channels and result in deterioration of the resting membrane potential. Here we present a recent example of the channel-forming toxin polytheonamide B (pTB), which originates from a marine sponge (83) (Fig. 16.21).

The pTB is a 48-mer peptide comprised of unusual amino acids. Both D- and L-chiral forms are alternatively aligned linearly, and the side chains are modified, such as methyl asparagine and norleucine (84). This strange structure must be synthesized through nonribosomal processes. The alternative D- and L-chiral forms suggest the formation of a β-helix structure like that of the gramicidin A channel (69). The β-helix is a hollow pore with an inner diameter of 4 Å. Similar to the gramicidin A channel (70), it is expected that pTB also exhibits a monovalent cation selectivity.

Adding pTB in a concentration as low as picomolars to the *cis* compartment of the chamber immediately results in channel activity that is cation selective. In contrast to the gramicidin A channel, pTB forms a membrane-spanning channel with a single molecule, which is evident from the Hill constant obtained from the concentration dependence of the current amplitudes pTB exhibits asymmetrical current-voltage (I–V) curves with slight rectification. When pTB is added from one side of the chamber, the channel is inserted into the membrane in the same direction. Using macroscopic measurements, voltage-dependent gating is elucidated.

16.7.2. KcsA Potassium Channel

The KcsA channel is a flagship channel for studying the structure–function relations of ion channels (85–91). The KcsA channel originated from the bacterium *Streptomyces lividans*, and the number of the amino acid residues is 160. Channel activity is attained with the formation of a homotetramer. Cytoplasmic acidic pH is sensed by a pH sensor in the cytoplasmic domain, and the channel becomes active below pH 5. Based on its high-resolution structure, various aspects of the functional features have been elucidated. The KcsA channel does not express as an active channel in eukaryotic cells without genetic modification (92), and the PLB or liposome patch methods have been used to characterize the channel function. For example, Chakrapani et al. (37) reconstituted the KcsA channel in giant liposomes and examined the kinetics of the pH dependence using a pH jump method.

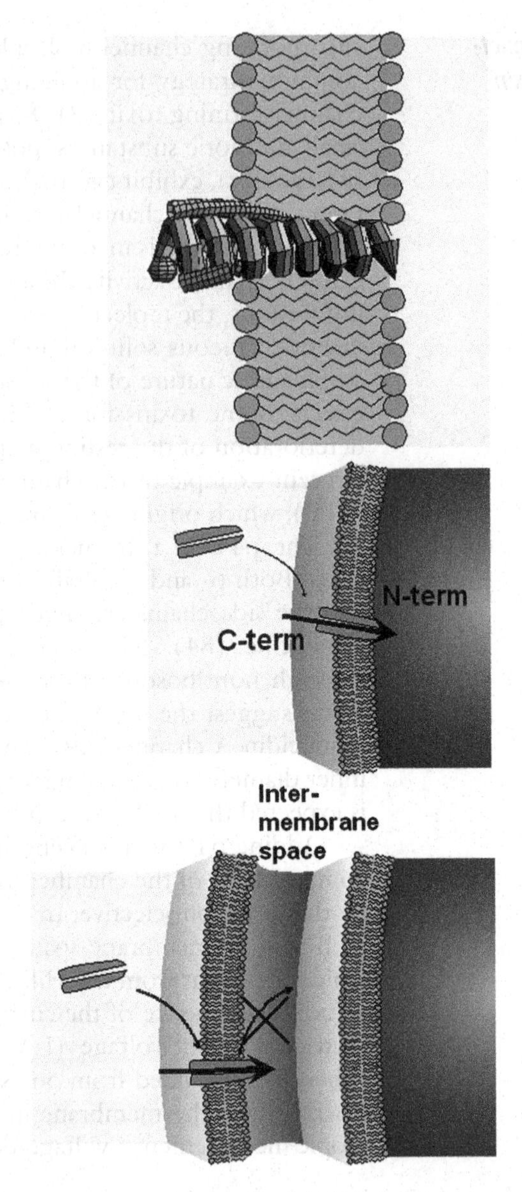

Fig. 16.21. Cytotoxic peptide polytheonamide B (pTB) is inserted into the membrane in the same orientation and never exits from the membrane. Thus, once encountered, pTB stays at the membrane and does not attack the next membrane. (Modified from Iwamoto et al. (83) Fig. 3).

The channel activities of KcsA were examined under various compositions of bilayer membranes, and it was revealed that channel activity is high in the presence of acidic phospholipids, such as PG (31). Like other potassium channels, KcsA exhibits inactivation. By introducing a mutation near the selectivity filter, the channel becomes noninactivating (93). On the other hand, mutations of the pH sensor render the channel pH-insensitive (94, 95). Finding hot spots on the sequence relevant to the specific function of the channel allows channels to be designed with specifically desired features (96).

16.8. Conclusion and Future Perspectives

The PLB method has undergone significant advances during the past decade. Previously, the PLB method had been almost exclusively used to study single-channel current measurements. The achievement of an oriented reconstitution enabled meaningful macroscopic current measurements, such as measurements of the current responses on the voltage steps in voltage-gated channels (97). The ability to examine both macroscopic and single-channel behavior by simply changing the number of incorporated channels in the PLB parallels the patch-clamp counterparts, including whole-cell and single-channel recordings. The oriented reconstitution cannot be achieved with any types of membrane proteins, and to date we have only a few successful examples (37, 97, 98). However, more effective technologies continue to be developed. For the membrane proteins such as transporters and pumps, functional characterization can be performed more intensively using the PLB once the massive oriented reconstitution becomes available.

The PLB is a sort of test tube for the study of membrane proteins. This method serves as the gold standard for studying channel molecules, including various channel-forming substances and proteins. We have employed the term "reconstitution" for the step by which extracted membrane proteins are incorporated into a clean artificial membrane. The channel current recorded in the simplified experimental conditions serves for analytical investigations. On the other hand, the term "reconstitution" can be expanded to mean that a specific membrane function is regenerated on the PLB by organizing several types of membrane proteins. The PLB can be used as a platform to integrate membrane proteins exhibiting certain functional aspects of the native membrane. Several types of functional protein are incorporated to reproduce membrane processes "from scratch." For example, physiological processes of vesicle formation were reconstituted (26).

The PLB method provides a unique experimental platform for exploring the frontiers of molecular investigation in membrane proteins including ion channels.

References

1. Miller C (1986) Ion channel reconstitution. Plenum Press, New York
2. Miller C, Racker E (1976) Ca^{++}-induced fusion of fragmented sarcoplasmic reticulum with artificial planar bilayers. J Membr Biol 30(3): 283–300
3. Montal M (1987) Reconstitution of channel proteins from excitable cells in planar lipid bilayer membranes. J Membr Biol 98(2):101–115
4. Favre I, Sun YM, Moczydlowski E (1999) Reconstitution of native and cloned channels into planar bilayers. Methods Enzymol 294:287–304

5. Morera FJ, Vargas G, Gonzalez C, Rosenmann E, Latorre R (2007) Ion-channel reconstitution. Methods Mol Biol 400:571–585
6. Sattsangi S, Wonderlin WF (1999) Isolation of transport vesicles that deliver ion channels to the cell surface. Methods Enzymol 294:339–350
7. Bezanilla F (1987) Single sodium channels from the squid giant axon. Biophys J 52(6): 1087–1090
8. Doyle DA, Morais Cabral J, Pfuetzner RA, Kuo A, Gulbis JM, Cohen SL, Chait BT, MacKinnon R (1998) The structure of the potassium channel: molecular basis of K^+ conduction and selectivity. Science 280(5360):69–77
9. MacKinnon R (2003) Potassium channels. FEBS Lett 555(1):62–65
10. LeMasurier M, Heginbotham L, Miller C (2001) KcsA: it's a potassium channel. J Gen Physiol 118(3):303–314
11. Demarche S, Sugihara K, Zambelli T, Tiefenauer T, Voros J (2011) Techniques for recording reconstituted ion channels. Analyst 136:1077–1089
12. Hanke W, Schlue W-R (1993) Planar lipid bilayers. In: Methods and applications. Academic Press, London
13. Singer SJ, Nicolson GL (1972) The fluid mosaic model of the structure of cell membranes. Science 175(23):720–731
14. von Heijne G, Rees D (2008) Membranes: reading between the lines. Curr Opin Struct Biol 18(4):403–405
15. Tanford C (2004) Ben Franklin stilled the waves: an informal history of pouring oil on water with reflections on the ups and downs of scientific life in genera. Oxford University Press, New York
16. Langmuir I (1917) The Shapes of Group Molecules Forming the Surfaces of Liquids. Proc Natl Acad Sci USA 3(4):251–257
17. Israelachvili JN (2011) Intermolecular and surface forces, 3rd edn. Academic, Amsterdam
18. Mueller P, Rudin DO, Tien HT, Wescott WC (1962) Reconstitution of cell membrane structure in vitro and its transformation into an excitable system. Nature 194:979–980
19. Hanai T, Haydon DA, Taylor J (1965) Polar group orientation and the electrical properties of lecithin bimolecular leaflets. J Theor Biol 9(2):278–296
20. Takagi M, Azuma K, Kishimoto U (1965) A new method for the formation of bilayer membranes in aqueous solution. Annu Rep Biol Works Fac Sci Osaka Univ 13
21. Bean RC, Shepherd WC, Chan H, Eichner J (1969) Discrete conductance fluctuations in lipid bilayer protein membranes. J Gen Physiol 53(6):741–757
22. Hladky SB, Haydon DA (1970) Discreteness of conductance change in bimolecular lipid membranes in the presence of certain antibiotics. Nature 225(5231):451–453
23. Neher E, Stevens CF (1977) Conductance fluctuations and ionic pores in membranes. Annu Rev Biophys Bioeng 6:345–381
24. Neher E, Sakmann B (1976) Single-channel currents recorded from membrane of denervated frog muscle fibres. Nature 260(5554):799–802
25. Sakmann B, Neher E (1995) Single-channel recording, 2nd edn. Plenum, New York
26. Tabata KV, Sato K, Ide T, Nishizaka T, Nakano A, Noji H (2009) Visualization of cargo concentration by COPII minimal machinery in a planar lipid membrane. EMBO J 28(21):3279–3289
27. Kapoor R, Kim JH, Ingolfson H, Andersen OS (2008) Preparation of artificial bilayers for electrophysiology experiments. J Vis Exp (20)
28. White SH (1986) The physical nature of planar bilayer membranes. In: Ion channel reconstitution. Plenum, New York
29. Wonderlin WF, Finkel A, French RJ (1990) Optimizing planar lipid bilayer single-channel recordings for high resolution with rapid voltage steps. Biophys J 58(2):289–297
30. Redwood WR, Pfeiffer FR, Weisbach JA, Thompson TE (1971) Physical properties of bilayer membranes formed from a synthetic saturated phospholipid in n-decane. Biochim Biophys Acta 233(1):1–6
31. Heginbotham L, Kolmakova-Partensky L, Miller C (1998) Functional reconstitution of a prokaryotic K^+ channel. J Gen Physiol 111(6):741–749
32. Addona GH, Sandermann H Jr, Kloczewiak MA, Miller KW (2003) Low chemical specificity of the nicotinic acetylcholine receptor sterol activation site. Biochim Biophys Acta 1609(2):177–182
33. Zasadzinski JA, Viswanathan R, Madsen L, Garnaes J, Schwartz DK (1994) Langmuir-Blodgett films. Science 263(5154):1726–1733
34. Montal M, Mueller P (1972) Formation of bimolecular membranes from lipid monolayers and a study of their electrical properties. Proc Natl Acad Sci USA 69(12):3561–3566
35. Montal M (1973) Asymmetric lipid bilayers: response to multivalent ions. Biochim Biophys Acta 298(3):750–754
36. Sherwood D, Montal M (1975) Transmembrane lipid migration in planar asymmetric bilayer membranes. Biophys J 15(5):417–434

37. Chakrapani S, Cordero-Morales JF, Perozo E (2007) A quantitative description of KcsA gating I. macroscopic currents. J Gen Physiol 130(5):465–478
38. Andersen OS (1983) Ion movement through gramicidin A channels: single-channel measurements at very high potentials. Biophys J 41(2):119–133
39. Coronado R, Latorre R (1983) Phospholipid bilayers made from monolayers on patch-clamp pipettes. Biophys J 43(2):231–236
40. Ehrlich BE (1992) Planar lipid bilayers on patch pipettes: bilayer formation and ion channel incorporation. Methods Enzymol 207: 463–470
41. Suarez-Isla BA, Wan K, Lindstrom J, Montal M (1983) Single-channel recordings from purified acetylcholine receptors reconstituted in bilayers formed at the tip of patch pipetts. Biochemistry 22(10):2319–2323
42. Oiki S, Koeppe RE 2nd, Andersen OS (1995) Voltage-dependent gating of an asymmetric gramicidin channel. Proc Natl Acad Sci USA 92(6):2121–2125
43. Oiki S, Koeppe RE II, Andersen OS (1997) Voltage-dependent gramicidin channels. In: Towards molecular biophysics of ion channels. Elsevier, Amsterdam
44. Delcour AH, Martinac B, Adler J, Kung C (1989) Modified reconstitution method used in patch-clamp studies of Escherichia coli ion channels. Biophys J 56(3):631–636
45. Iwamoto M, Oiki S (2011) Counting ion and water molecules in a streaming file through the open-filter structure of a K channel. J Neurosci 31:12180–12188
46. Cortes DM, Cuello LG, Perozo E (2001) Molecular architecture of full-length KcsA: role of cytoplasmic domains in ion permeation and activation gating. J Gen Physiol 117(2): 165–180
47. Niles WD, Levis RA, Cohen FS (1988) Planar bilayer membranes made from phospholipid monolayers form by a thinning process. Biophys J 53(3):327–335
48. Bretscher MS, Munro S (1993) Cholesterol and the Golgi apparatus. Science 261(5126): 1280–1281
49. Green WN, Andersen OS (1991) Surface charges and ion channel function. Annu Rev Physiol 53:341–359
50. Lakshminarayanaiah N (1984) Equations of membrane biophysics. Academic, Orlando
51. Latorre R, Labarca P, Naranjo D (1992) Surface charge effects on ion conduction in ion channels. Methods Enzymol 207: 471–501
52. Small DM (1986) The physical chemistry of lipids. From Alkanes to Phospholipids, Plenum, New York
53. Gruner SM, Cullis PR, Hope MJ, Tilcock CP (1985) Lipid polymorphism: the molecular basis of nonbilayer phases. Annu Rev Biophys Biophys Chem 14:211–238
54. Boheim G, Hanke W, Eibl H (1980) Lipid phase transition in planar bilayer membrane and its effect on carrier- and pore-mediated ion transport. Proc Natl Acad Sci USA 77(6): 3403–3407
55. Iwamoto M, Shimizu H, Inoue F, Konno T, Sasaki YC, Oiki S (2006) Surface structure and its dynamic rearrangements of the KcsA potassium channel upon gating and tetrabutylammonium blocking. J Biol Chem 281(38) :28379–28386
56. Williamson IM, Alvis SJ, East JM, Lee AG (2003) The potassium channel KcsA and its interaction with the lipid bilayer. Cell Mol Life Sci 60(8):1581–1590
57. Labarca P, Latorre R (1992) Insertion of ion channels into planar lipid bilayers by vesicle fusion. Methods Enzymol 207:447–463
58. Cohen FS, Niles WD (1993) Reconstituting channels into planar membranes: a conceptual framework and methods for fusing vesicles to planar bilayer phospholipid membranes. Methods Enzymol 220:50–68
59. Chernomordik LV, Kozlov MM (2008) Mechanics of membrane fusion. Nat Struct Mol Biol 15(7):675–683
60. Helm CA, Israelachvili JN (1993) Forces between phospholipid bilayers and relationship to membrane fusion. Methods Enzymol 220: 130–143
61. Efremov RG, Nolde DE, Konshina AG, Syrtcev NP, Arseniev AS (2004) Peptides and proteins in membranes: what can we learn via computer simulations? Curr Med Chem 11(18):2421–2442
62. Chernomordik LV, Kozlov MM (2005) Membrane hemifusion: crossing a chasm in two leaps. Cell 123(3):375–382
63. Marrink SJ, Mark AE (2003) The mechanism of vesicle fusion as revealed by molecular dynamics simulations. J Am Chem Soc 125(37):11144–11145
64. Tieleman DP, Leontiadou H, Mark AE, Marrink SJ (2003) Simulation of pore formation in lipid bilayers by mechanical stress and electric fields. J Am Chem Soc 125(21):6382–6383
65. Woodbury DJ, Miller C (1990) Nystatin-induced liposome fusion: a versatile approach to ion channel reconstitution into planar bilayers. Biophys J 58(4):833–839

66. Woodbury DJ (1999) Nystatin/ergosterol method for reconstituting ion channels into planar lipid bilayers. Methods Enzymol 294:319–339
67. Singer MA (1975) Interaction of amphotericin B and nystatin with phospholipid bilayer membranes: effect of cholesterol. Can J Physiol Pharmacol 53(6):1072–1079
68. Finkelstein A (1987) Water Movement Through Lipid Bilayers, Pores, and Plasma membranes. In: Theory and reality. Wiley-Interscience, New Jersey
69. Andersen OS (1984) Gramicidin channels. Annu Rev Physiol 46:531–548
70. Andersen OS, Koeppe RE 2nd, Roux B (2005) Gramicidin channels. IEEE Trans Nanobiosci 4(1):10–20
71. Johnson AE, van Waes MA (1999) The translocon: a dynamic gateway at the ER membrane. Annu Rev Cell Dev Biol 15:799–842
72. Sherman-Gold R (1993) The axon guide for electrophysiology and biophysics laboratory techniques. Axon, Foster City
73. Sigworth FJ (1995) Electronic design of patch clamp. In: Sakmann B, Neher E (eds) Single-channel recording. Plenum, New York
74. Bertl A, Blumwald E, Coronado R, Eisenberg R, Findlay G, Gradmann D, Hille B, Kohler K, Kolb HA, MacRobbie E et al (1992) Electrical measurements on endomembranes. Science 258(5084):873–874
75. Armstrong CM, Gilly WF (1992) Access resistance and space clamp problems associated with whole-cell patch clamping. Methods Enzymol 207:100–122
76. Morais-Cabral JH, Zhou Y, MacKinnon R (2001) Energetic optimization of ion conduction rate by the K^+ selectivity filter. Nature 414(6859):37–42
77. Sigworth FJ (1985) Open channel noise. I. Noise in acetylcholine receptor currents suggests conformational fluctuations. Biophys J 47(5):709–720
78. Sigworth FJ, Urry DW, Prasad KU (1987) Open channel noise III. High-resolution recordings show rapid current fluctuations in gramicidin A and four chemical analogues. Biophys J 52(6):1055–1064
79. Oiki S (2010) Single-channel structure-function dynamics: the gating of potassium channels. In: Cell signaling reactions: single-molecular kinetic analysis. Springer, New York
80. Hille B (2001) Ion channels of excitable membranes, 3rd edn. Sinauer Associated, Inc, MA
81. Shin YK, Levinthal C, Levinthal F, Hubbell WL (1993) Colicin E1 binding to membranes: time-resolved studies of spin-labeled mutants. Science 259(5097):960–963
82. Menestrina G, Serra MD, Prevost G (2001) Mode of action of beta-barrel pore-forming toxins of the staphylococcal alpha-hemolysin family. Toxicon 39(11):1661–1672
83. Iwamoto M, Shimizu H, Muramatsu I, Oiki S (2010) A cytotoxic peptide from a marine sponge exhibits ion channel activity through vectorial-insertion into the membrane. FEBS Lett 584:3995–3999
84. Hamada T, Matsunaga S, Yano G, Fusetani N (2005) Polytheonamides A and B, highly cytotoxic, linear polypeptides with unprecedented structural features, from the marine sponge, Theonella swinhoei. J Am Chem Soc 127(1):110–118
85. Shimizu H, Iwamoto M, Konno T, Nihei A, Sasaki YC, Oiki S (2008) Global twisting motion of single molecular KcsA potassium channel upon gating. Cell 132(1):67–78
86. Perozo E, Cortes DM, Cuello LG (1998) Three-dimensional architecture and gating mechanism of a K^+ channel studied by EPR spectroscopy. Nat Struct Biol 5(6):459–469
87. Takeuchi K, Takahashi H, Kawano S, Shimada I (2007) Identification and characterization of the slowly exchanging pH-dependent conformational rearrangement in KcsA. J Biol Chem 282(20):15179–15186
88. Uysal S, Vasquez V, Tereshko V, Esaki K, Fellouse FA, Sidhu SS, Koide S, Perozo E, Kossiakoff A (2009) Crystal structure of full-length KcsA in its closed conformation. Proc Natl Acad Sci USA 106(16):6644–6649
89. Schrempf H, Schmidt O, Kummerlen R, Hinnah S, Muller D, Betzler M, Steinkamp T, Wagner R (1995) A prokaryotic potassium ion channel with two predicted transmembrane segments from Streptomyces lividans. EMBO J 14(21):5170–5178
90. Chill JH, Louis JM, Miller C, Bax A (2006) NMR study of the tetrameric KcsA potassium channel in detergent micelles. Protein Sci 15(4):684–698
91. Kelly BL, Gross A (2003) Potassium channel gating observed with site-directed mass tagging. Nat Struct Biol 10(4):280–284
92. Gao L, Mi X, Paajanen V, Wang K, Fan Z (2005) Activation-coupled inactivation in the bacterial potassium channel KcsA. Proc Natl Acad Sci USA 102(49):17630–17635
93. Cordero-Morales JF, Cuello LG, Zhao Y, Jogini V, Cortes DM, Roux B, Perozo E (2006) Molecular determinants of gating at the potassium-channel selectivity filter. Nat Struct Mol Biol 13(4):311–318

94. Thompson AN, Posson DJ, Parsa PV, Nimigean CM (2008) Molecular mechanism of pH sensing in KcsA potassium channels. Proc Natl Acad Sci USA 105(19):6900–6905
95. Cuello LG, Cortes DM, Jogini V, Sompornpisut A, Perozo E (2010) A molecular mechanism for proton-dependent gating in KcsA. FEBS Lett 584(6):1126–1132
96. Cuello LG, Jogini V, Cortes DM, Pan AC, Gagnon DG, Dalmas O, Cordero-Morales JF, Chakrapani S, Roux B, Perozo E (2010) Structural basis for the coupling between activation and inactivation gates in K^+ channels. Nature 466(7303):272–275
97. Schmidt D, Cross SR, MacKinnon R (2009) A gating model for the archeal voltage-dependent K^+ channel KvAP in DPhPC and POPE:POPG decane lipid bilayers. J Mol Biol 390(5):902–912
98. Yanagisawa M, Iwamoto M, Kato A, Yoshikawa K, Oiki S (2011) Oriented reconstitution of a membrane protein in a giant unilamellar vesicle: Experimental verification with the potassium channel KcsA. J Am Chem Soc 133:11774–11779

Chapter 17

Patch-Clamp Capacitance Measurements

Takeshi Sakaba, Akaihiro Hazama, and Yoshio Maruyama

Abstract

Not only electrical conductance but also electrical capacitance of the cell membrane can be measured by patch-clamp techniques. Exocytotic events can be detected by recording changes in membrane capacitance. The membrane capacitance, which reflects the surface area of the plasma membrane, increases during an exocytotic process by fusion of secretory granules to the plasma membrane. In this chapter, we describe the patch-clamp method for measuring capacitance.

17.1. Capacitance Changes Associated with Exocytosis and Endocytosis

The capacitance of the cell membrane is proportional to the surface area, with a constant of 1 μF/cm^2. Therefore, changes in the membrane surface area can be monitored by measuring the cell capacitance (Fig. 17.1). Exocytosis of secretory granules results in an increase of the cell surface area that is compensated by membrane retrieval processes (e.g., endocytosis) that consequently reduce the cell surface area. Time-resolved capacitance measurements were developed by Neher and Marty (1) in 1982 to resolve the kinetics of exocytosis and endocytosis of secretory granules in a single cell. The conventional patch-clamp technique is used, in whole-cell or cell-attached configuration. Because of the low-noise and high temporal resolution, this method enables fast quantitative measurements of changes in membrane capacitance (2, 3). In addition, the intracellular solution can be potentially controlled through a measuring patch pipette.

In the whole-cell voltage-clamp mode (4), capacitance can be measured in two ways, each of which relies on the fact that the capacitive transient is proportional to the differentiation (temporal changes) of the membrane potential: (1) In response to a step voltage pulse, capacitive currents can be observed at the onset and offset of

Yasunobu Okada (ed.), *Patch Clamp Techniques: From Beginning to Advanced Protocols*, Springer Protocols Handbooks, DOI 10.1007/978-4-431-53993-3_17, © Springer 2012

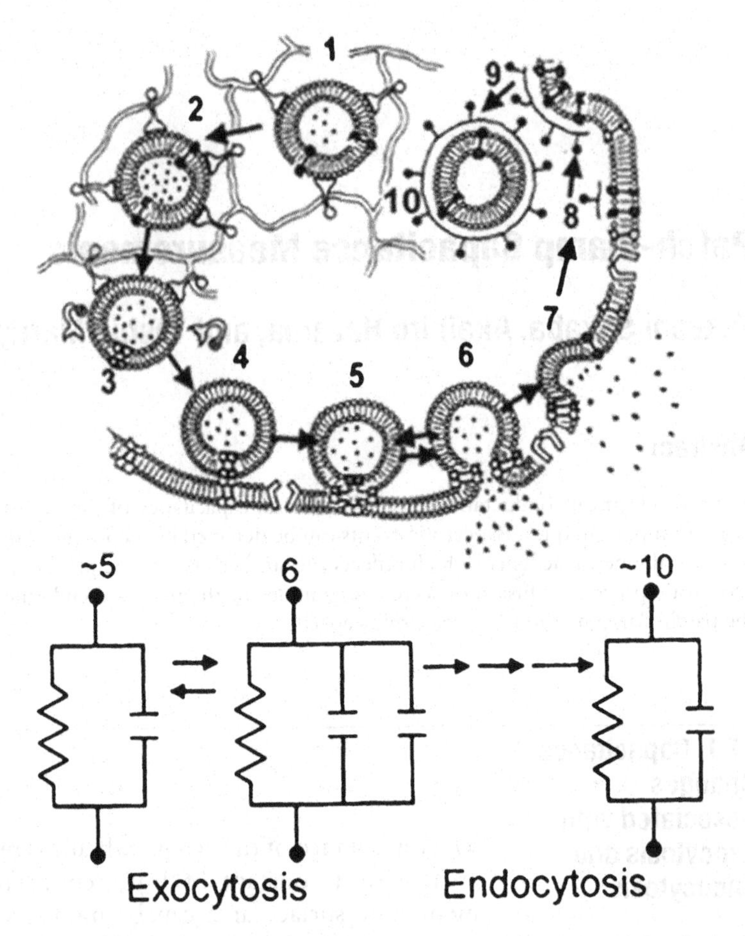

Fig. 17.1. Process for fusion of secretory granules to plasma membrane (*top*) and associated changes in capacitance (*bottom*). Secretory granules are transported to the plasma membrane (1–4), primed (5), fused to the plasma membrane (6) and endocytosed (8–10).

the pulse. (2) When the membrane potential is changed in a sinusoidal manner, the capacitive current component exhibits a phase delay of 90°. One can extract the cell capacitance in both cases.

Capacitance of the patch pipettes can be several picofarads when the pipette is immersed in the extracellular bath solution. Changes in pipette capacitance, mainly due to a change in the fluid level, may cause artificial capacitance changes. Usually patch pipettes are coated with Sylgard or dental wax to reduce any stray capacitance. Also, care must be taken not to change the fluid level when exchanging the extracellular solution.

17.2. Patch-Clamp Capacitance Measurements

17.2.1. Using a Step Voltage Pulse

The step voltage pulse method is intuitively easy to understand. Figure 17.2a shows an electrical circuit of an electrically single compartment cell under whole-cell voltage clamp. When a square

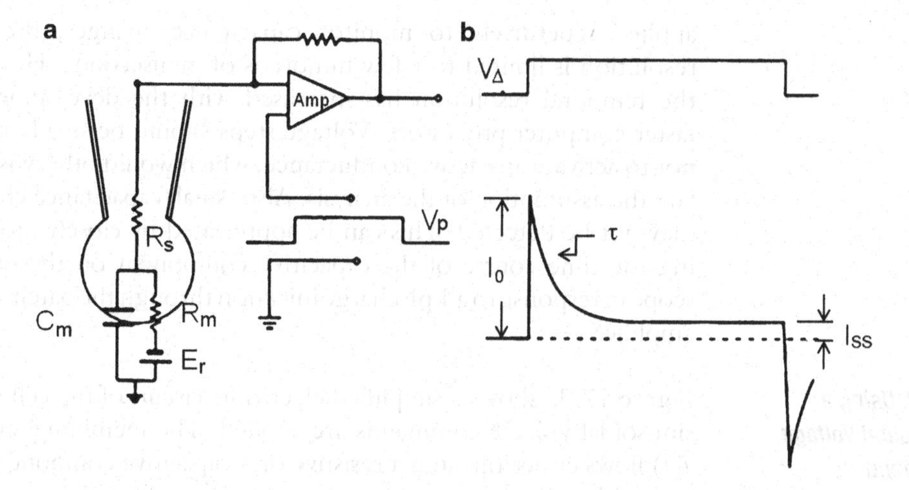

Fig. 17.2. (**a**) Electrical circuit of the cell under whole-cell voltage-clamp. (**b**) Capacitance currents are elicited in response to a step voltage pulse. R_m membrane resistance, C_m membrane capacitance, R_s series resistance, E_r reversal potential, V_Δ a step depolarizing pulse, I_0 instantaneous current, I_{ss} steady-state current, τ time constant of the current decay.

voltage pulse is applied to the cell membrane, capacitive currents can be seen (Fig. 17.2b). Membrane current $I(t)$ following the onset of the pulse can be described as:

$$I(t) = (I_0 - I_{ss})\exp(-t / \tau) + I_{ss}$$

where I_0 is the initial current amplitude, I_{ss} is the steady-state current amplitude following the capacitive transient, and τ is the time constant of capacitive decay.

Immediately following the pulse, the impedance of the capacitive component C_m is almost 0, and therefore described as

$$I_0 = \Delta V / R_s$$

where ΔV is the amplitude of the voltage step.

The steady-state current (I_{ss}) following the capacitive transient is described as:

$$I_{ss} = \Delta V / (R_s + R_m)$$

The time constant of the capacitive current τ is described as:

$$\tau = R_p C_m$$

where $R_p = R_s R_m / (R_s + R_m)$.

Then,

$$C_m = \tau / R_p$$

This method of capacitance measurement is implemented in the patch-clamp software provided by Axon (Sunnyvale, USA) and HEKA (Lambrecht, Germany). Because a small voltage step is applied repetitively to monitor capacitance changes, the time resolution is limited to a few hundreds of milliseconds. However, the temporal resolution has increased with the development of faster computer processors. Voltage steps should be small enough not to activate any active conductance, which would otherwise violate the assumption of the analysis. Also, small capacitance changes may not be detected. This can be appreciated by closely monitoring the time course of the capacitive component on the oscilloscope in response to a 1 pF charge injection through the patch-clamp amplifier.

17.2.2. Using a Sinusoidal Voltage Command

Figure 17.3a shows a simplified electronic circuit of the cell where sinusoidal voltage commands are applied. The membrane current $I(t)$ flows either through a resistive or a capacitive component and can be described as:

$$I(t) = G_m V + C_m (\mathrm{d}V/\mathrm{d}t)$$

where V is the command voltage, G_m is the membrane conductance, and the C_m is the cell capacitance. When a sinusoidal voltage

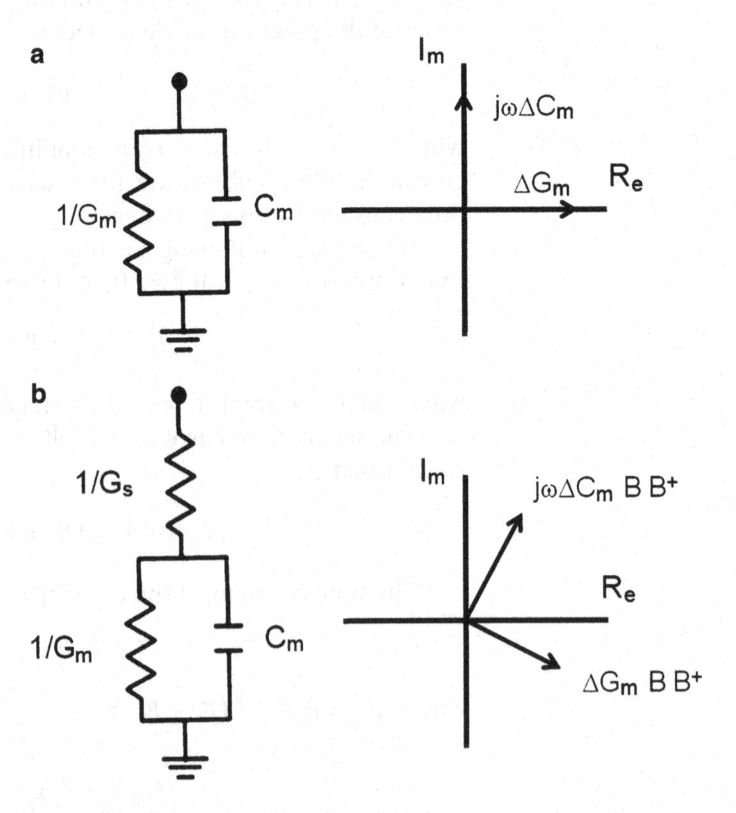

Fig. 17.3. Equivalent circuit of the cell under whole-cell voltage-clamp. (**a**) Ideal circuit without series resistance (R_s). I_m (membrane currents) and $1/G_m = R_m$ (membrane resistance) in the phase plane correspond to the real and imaginary part of the lock-in amplifier, respectively. (**b**) In reality, series resistance is added in addition to that in (**a**). Because of series resistance, conductance and capacitance changes are phase shifted, while they are orthogonal to each other.

command, $V = V_0 \sin \omega t$ (where ω is the frequency), is applied to the cell,

$$I(t) = G_m V_0 \sin \omega t + C_m V_0 \cos \omega t$$

This can be also written as:

$$Y(\omega) = G_m + j\omega C_m$$

where $Y(\omega)$ is the admittance and j is the imaginary unit.

There is a phase shift of 90° between the resistive and the capacitive current components. In the phase plane, G_m and C_m are orthogonal to each other. Determination of these two current components is necessary to measure C_m and G_m. Usually lock-in amplifiers are used for this purpose, but they are also implemented in the software of the patch-clamp amplifiers (e.g., HEKA Pulse and Patchmaster).

Two outputs of the lock-in amplifiers (usually 0° and 90° of the input phase) are in principle proportional to the changes in G_m and C_m. However, there is a series resistance in the real circuit that causes a further phase shift. Because of the series resistance, changes in G_m and C_m can no longer be easily detected. In several cases, experimental conditions need to be optimized to nullify the series resistance, or else the series resistance is taken into account, as described below.

In principle, three parameters (G_s, G_m, C_m) have to be determined experimentally (see also ref. (5)). In the Lindau–Neher method (2), three independent equations – one formulated from the direct current (DC) component of the membrane currents and the two from application of the sinusoidal voltage command – are numerically solved to estimate the three parameters. In the Neher–Marty method (1, 3), either G_s or G_m is assumed to be constant during the recording period. It is then possible to detect changes in the two other parameters, C_m and G_m or G_s; but the absolute value of C_m cannot be determined.

17.2.2.1. Lindau–Neher Technique

Figure 17.4a shows the system. For calculating the three parameters (G_s, G_m, C_m), two independent outputs of the lock-in amplifier and the DC component of the current value are required. Here, the DC component is driven by the voltage command at $V_{DC.}$

From two outputs of the lock-in amplifiers (A and B) and the DC current component, the following three equations are obtained.

Admittance Y can be described as:

$$Y = A + Bj$$

$$A = \left(1 + \omega^2 R_m R_p C_m^{\ 2}\right) / \left\{R_t \left(1 + \omega^2 R_p^{\ 2} C_m^{\ 2}\right)\right\}$$

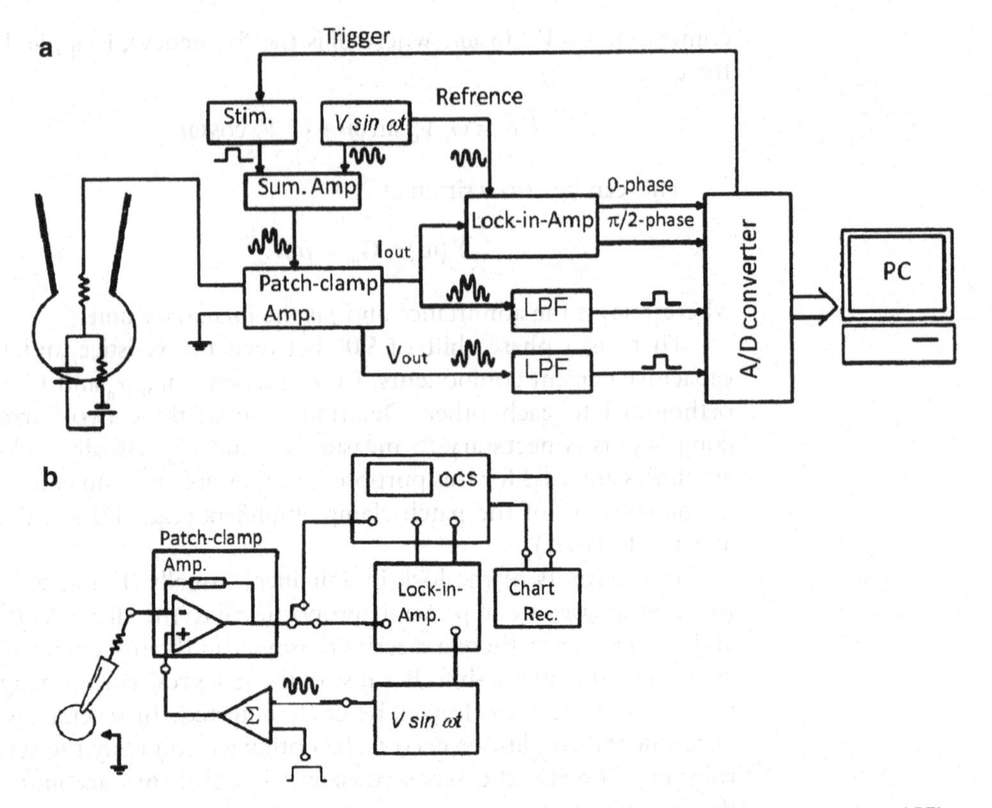

Fig. 4. Schemes for the Lindau–Neher (**a**) and Neher–Marty (**b**) methods. *A/D* analog to digital, I_{out} current out, *LPF* low-pass filter, *OCS* oscilloscope, *PC* personal computer, *Stim* stimulator, V_{out} voltage out.

$$B = \omega R_m{}^2 C_m / \left\{ R_t{}^2 \left(1 + \omega^2 R_p{}^2 C_m{}^2 \right) \right\}$$

where

$R_t = R_s + R_m; R_p = R_s R_m / (R_s + R_m);$

$R_s = 1/G_s$; and $R_m = 1/G_m$.

From the DC component, we can derive the DC current I_{dc} as:

$$I_{dc} = (V_{dc} - E_r) / R_t$$

E_r is the reversal potential of the component. As a variant of this method, a small step voltage is applied after each sampling period of the sinusoidal wave. Unknown parameters are C_m, R_m, and R_s; the other parameters are known. By solving the equations, we obtain

$$R_s = (A - G_t) / (A^2 + B^2 - AG_t)$$

$$R_m = \{(A - G_t)^2 + B^2\} / (G_t A^2 + B^2 - AG_t)$$

$$C_m = (A^2 + B^2 - AG_t)^2 / \omega B\{(A - G_t)^2 + B^2\}$$

where $G_t = 1/R_t$.

Because computer processing gets faster with time, this method, although somewhat computationally exhaustive, allows fast measurements of capacitance. Temporal and signal resolution rather depends on the used frequencies and amplitudes of the sine wave. Because the method is integrated into commercial software, it does not require additional computation and so is more widely used today.

17.2.2.2. Neher–Marty Method

Fusion of secretory granules with a diameter of 1 μm and surface area of 0.8 μm² would produce a capacitance increase of 80 fF. However, secretory granules can be even smaller; in the extreme case, fusion of synaptic vesicles with a diameter of 30–50 nm would produce a capacitance increase of only tens of abfarads. To capture such small capacitance changes, lock-in amplifiers are used in combination with the whole-cell patch-clamp or cell-attached mode, the latter of which increases the signal-to-noise ratio significantly. Here, we deal only with capacitance measurements in the whole-cell configuration. A sinusoidal voltage command is applied to the cell, and the output signals in-phase and out-of-phase (with a 90° delay) with the inputs isolated using lock-in amplifiers (phase-sensitive detection). With this method, the measured values are not C_m itself but, rather, the change in C_m would be amplified in proportion to the input frequency ω. This method requires a lock-in amplifier.

With the Neher–Marty method, the whole-cell capacitance is first canceled out by feedback compensation circuitry of the amplifier. Then, ΔC_m is measured by the lock-in amplifier. The compensation circuitry of the patch-clamp amplifier is used for calibrating the capacitance traces. Figure 17.3b shows two orthogonal outputs of the lock-in amplifier, one mainly reflecting ΔC_m, and the other mainly reflecting ΔG_m. ΔC_m and ΔG_m project on these two outputs. As described above, ΔG_m and ΔC_m are not exactly at 0° and 90° in phase with the input sine wave but are shifted owing to the series resistance. The error factors associated with ΔG_s compromise the accuracy of the measurements. By setting a phase angle of the lock-in amplifier, error factors can be minimized. To minimize the errors, four factors should be considered: (1) Cell capacitance should be compensated by the feedback circuitry of the amplifier. (2) R_s itself should be low. (3) The lock-in amplifier should compensate the phase shift. (4) If possible, appropriate intracellular and extracellular solutions should be selected.

In the original method of Neher and Marty (1), C_m and G_m can vary whereas G_s remains constant during the measurement. In the circuit of Fig. 17.3, the total admittance Y is a function of C_m and G_m and is described as:

$$Y(C_m, G_m) = (G_m + j\omega C_m)B$$

where B is a function of C_m, G_m, ω, and

$$B(C_m, G_m, \omega) = 1/\{1 + G_m/G_s + j\omega C_m/G_s\}$$

When C_m and G_m are changed, B becomes a function of $C_m + \Delta C_m$ and $G_m + \Delta G_m$.

$$B^+ = B(C_m + \Delta C_m, G_m + \Delta G_m, \omega)$$
$$= 1/\{1 + (G_m + \Delta G_m)/G_s + j\omega(C_m + \Delta C_m)/G_s\}$$

The change in the admittance ΔY is then

$$\Delta Y = Y(C_m + \Delta C_m, G_m + \Delta G_m) - Y(C_m, G_m)$$
$$= \{(G_m + \Delta G_m) + j\omega(C_m + \Delta C_m)\}B^+ - (G_m + j\omega C_m)B$$
$$= (\Delta G_m + j\omega\Delta C_m)BB^+$$

In the experiments, C_m and G_s should be canceled out by using the feedback circuitry of the amplifier. If $G_s \gg (G_m + \Delta G_m)$, B and B^+ approach 1. In reality, this is not perfectly achievable, and as a result the two outputs of the lock-in amplifier do not coincide with the exact phase of ΔG_m and ΔC_m. Therefore, the outputs are influenced by changes in both ΔG_m and ΔC_m. To track the changes of these two parameters accurately, the phase of the lock-in amplifier should be set properly. One way is to move the capacitance toggle of the patch-clamp amplifier back and forth, so ΔC_m is simulated artificially. Then, the phase of the lock-in amplifier is adjusted such that the change at 0° of the lock-in output becomes minimal whereas the change at 90° of the lock-in output becomes maximum. The phase of the lock-in amplifier is then set such that interference due to the incorrect setting of the phase can be minimized. Error factors associated with this method are mentioned in other articles in the literature.

Figure 17.5 shows exemplar traces of the capacitance transients at a single acinar cell (6–8). These traces were obtained from whole-cell patch-clamping, and $G_s\Delta(G_m + \Delta G_m)$ as well as cancelation of G_s and C_m by the amplifier was correctly achieved. Because there was no baseline drift, G_s should not have changed significantly during the measurement period. Acetylcholine 50 nM was applied using a local puff pipette, and care was taken not to change the fluid level. Patch pipettes were coated with Sylgard to reduce stray capacitance. At the beginning of the trace, a 200-fF change of the capacitance was introduced through the feedback circuit of the patch-clamp amplifier, which calibrates the capacitance. If the phase is set correctly, this capacitance change does not cause any changes in the conductance trace (G_m).

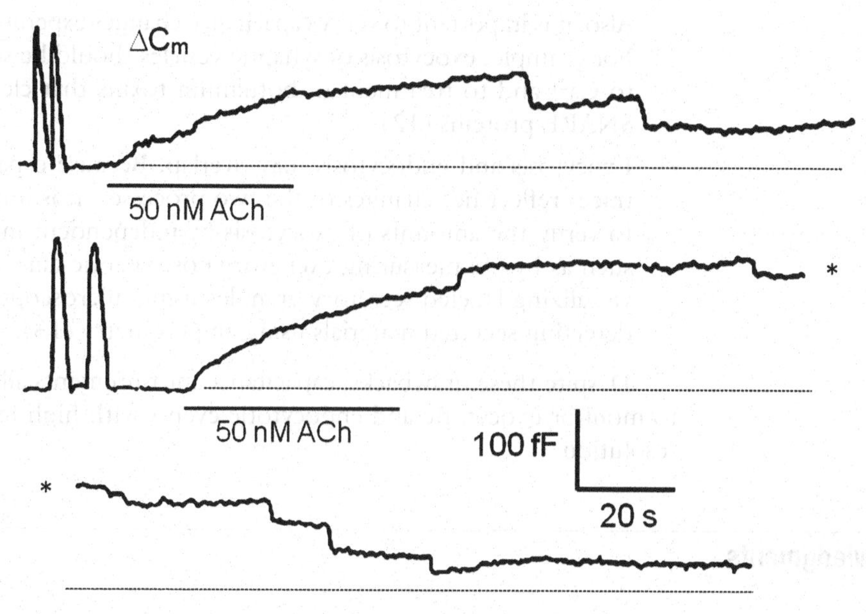

Fig. 17.5. Example traces of capacitance measurement using the Neher–Marty method. Capacitance changes are in response to application of acetylcholine (*ACh*) at a single acinar cell.

17.3. Other Considerations

There are other variants of capacitance measurements, which are described elsewhere (5, 9, 10). Capacitance measurements are usually implemented with patch-clamp amplifier software (e.g., Patchmaster; HEKA), and the manual describes the detailed procedure. It is convenient to use such a software-based semiautomatic approach, although it makes it difficult for users to realize the drawbacks of the method, of which there are three major ones.

- Because the method models a cell as a single electrical compartment, it may not be appropriate to apply it to cell types with complex morphology. For example, the method has been used at mossy fiber boutons in hippocampus (11), which has a small terminal with a very long axon. In this case, stimulation frequency must be optimized such that capacitance changes due to exocytosis from a terminal should be detected without any contamination from the axonal currents.
- The Neher–Marty method assumes that the capacitance, series resistance, and conductance changes do not change considerably. Nevertheless, it is important to monitor not only the capacitance but also the conductance and the series resistance simultaneously. If two or three traces show associated changes, it may be an artifact due to cross-talk among the parameters.

Also, it is important to verify capacitance changes experimentally. For example, exocytosis of synaptic vesicles should be sensitive to Ca^{2+} and to treatment of botulinum toxins that cleave the SNARE proteins (12).

- Exocytosis and endocytosis may overlap. Because capacitance traces reflect net changes of the two processes, it is important to verify the amounts of exocytosis by independent methods, such as by (1) measuring excitatory postsynaptic currents, (2) visualizing labeled secretory granules using microscopy, or (3) detecting secreted materials using amperometry (13).

Despite these drawbacks, capacitance measurements allow one to monitor exocytotic and endocytotic events with high temporal resolution.

Acknowledgments

We thank Andreas Neef and Raunak Sinha for their comments.

References

1. Neher E, Marty A (1982) Discrete changes of cell membrane capacitance observed under conditions of enhanced secretion in bovine adrenal chromaffin cells. Proc Natl Acad Sci USA 79:6712–6716
2. Lindau M, Neher E (1988) Patch-clamp technique for time-resolved capacitance measurements in single cells. Pflugers Arch 411:137–146
3. Gillis KD (1995) Techniques for membrane capacitance measurements. In: Sakmann B, Neher E (eds) Single channel recording, 2nd edn. Plenum Press, New York, pp 155–198
4. Sakmann B, Neher E (1995) Single-channel recording, 2nd edn. Plenum Press, New York
5. Okada Y, Hazama A, Hashimoto A, Maruyama Y, Kubo M (1992) Exocytosis upon osmotic swelling in human epithelial cells. Biochim Biophys Acta 1107:201–205
6. Maruyama Y (1988) Agonist-induced changes in cell membrane capacitance and conductance in dialysed pancreatic acinar cells of rats. J Physiol 406:299–313
7. Maruyama Y, Petersen OHP (1994) Delay in granular fusion evoked by repetitive cytosolic Ca^{2+} spikes in mouse pancreatic acinar cells. Cell Calcium 16:419–430
8. Maruyama Y (1996) Selective activation of exocytosis by low concentration of ACh in rat pancreatic acinar cells. J Physiol 492: 807–814
9. Fidler N, Fernandez JM (1989) Phase tracking: an improved phase detection technique for cell membrane capacitance measurements. Biophys J 56:1153–1162
10. Rohlicek V, Schmid A (1994) Dual frequency method for synchronous measurement of cell capacitance, membrane conductance and access resistance on single cells. Pflugers Arch 428:30–38
11. Hallermann S, Pawlu C, Jonas P, Heckmann M (2003) A large pool of releasable vesicles in a cortical glutamatergic synapse. Proc Natl Acad Sci USA 100:8975–8980
12. Yamashita T, Hige T, Takahashi T (2005) Vesicle endocytosis requires dynamin-dependent GTP hydrosis at a fast CNS synapse. Science 307:124–127
13. Haller M, Heinemann C, Chow RH, Heidelberger R, Neher E (1998) Comparison of secretory responses as measured by membrane capacitance and by amperometry. Biophys J 74:2100–2113

Chapter 18

Brief Guide to Patch-Clamp Current Measurements in Organelle Membranes

Yoshio Maruyama and Akihiro Hazama

Abstract

The standard giga-seal patch-clamp techniques opened the way to monitor ion-channel activity in the native organelle membrane in situ. One has to have an organelle preparation, covered with the intact membrane carrying rightly oriented channel proteins and suitable for proper patch-clamping. Here we deal with the intact "native" organelle, the nucleus with an inner and outer nucleic membrane (nuclear envelope), and mitochondria without the outer membrane but an intact inner membrane (mitoplast). We collect them by rupturing the cell using a variety of procedures including cell homogenization, cell swelling with a hypotonic solution, cell shaking in a Na citrate solution, and their appropriate combination, usually modified by individual laboratories.

18.1. Patch-Clamping the Inner Mitochondrial Membrane

18.1.1. Mitoplast Preparation

After homogenization of cells or a piece of tissue, we can collect mitochondria (1) from any type of cell with the standard sucrose density-gradient centrifugation procedure. It is a routine procedure and has no immediate problems. Rupturing the outer mitochondrial membrane by repeating hypotonic and hypertonic solution exposures, we have mitoplasts suitable for patch-clamping.

18.1.1.1. Solutions

- A solution (in mM): 250 sucrose, 1 EGTA, 5 HEPES/KOH (pH 7.2)
- B solution (in mM): 340 sucrose, 1 EGTA, 5 HEPES/KOH (pH 7.2)
- C solution (in mM): 150 sucrose, 1 EGTA, 5 HEPES/KOH (pH 7.2)
- Hypotonic solution (in mM): 1 EGTA, 5 HEPES/KOH (pH 7.2)

Yasunobu Okada (ed.), *Patch Clamp Techniques: From Beginning to Advanced Protocols*, Springer Protocols Handbooks, DOI 10.1007/978-4-431-53993-3_18, © Springer 2012

- Hypertonic KCl solution (in mM): 750 KCl, 1 EGTA, 5 HEPES/KOH (pH 7.2)

18.1.1.2. Procedure 1: Collection of Mitochondria

1. Transfer isolated tissues of 1–2 g to a 15-ml Dounce homogenizer on ice and add ice-cold A solution to give a 10% (tissue wet wt/vol) mixture.
2. Homogenize the tissue pieces on ice using seven or eight strokes with the pestle.
3. Place the homogenate (0.75 ml) on an equal amount ice-cold B solution in a 1.5-ml microfuge tube.
4. Centrifuge at 600*g* at 4°C for 10 min.
5. Transfer the supernatant, containing the mitochondria fraction, to a microtube, and centrifuge at 9,000*g* at 4°C for 10 min.
6. Aspirate and discard the supernatant. Subsequently, wash the pellet surface with ice-cold A solution once or twice.
7. Add ice-cold A solution and centrifuge again at 9,000*g* at 4°C for 10 min.
8. Aspirate and discard the supernatant. Suspend the pellet with 50–100 μl ice-cold C solution. Keep it at 4°C for later experiments.

18.1.1.3. Procedure 2: Making Mitoplasts (Rupturing the Outer Membrane by Osmotic Shock)

1. Add a few drops of the mitochondria preparation with ice-cold hypertonic KCl solution, and leave it for 10–60 min (hypertonic shock).
2. Dilute the above solution with the appropriate amount of ice-cold hypotonic solution to settle the tonicity at about 50 mosm/kg H_2O.
3. Recover the solution tonicity to 300 mosm/kg H_2O by adding hypertonic KCl solution.
4. Resuspend a drop of the above sample in the experimental solution, and subject it to the electrical experiments.

18.1.2. Giga-Seal Formation in the Mitoplast

Several procedures are reported to establish the giga-seal formation in mitoplasts. Operating a manipulator under the inverted microscope, some pursue a floating mitoplast and catch it at the tip of the electrode with brief suction (2). For mitoplasts tightly attached to the bottom, others first carry out an extensive wash of their surface with an experimental solution before accessing the patch pipette (3). To obtain giga-seal formation, we recommend use of a high-resistance electrode (10–20 MΩ with KCl solution), made from thick-walled glass pipettes. Figure 18.1 shows 108 pS Cl^- channel activity in the mitoplast from mouse pancreatic acinar cells using this procedure. In previous reports, mitoplasts manifested all of the major ion pathways across the inner membrane

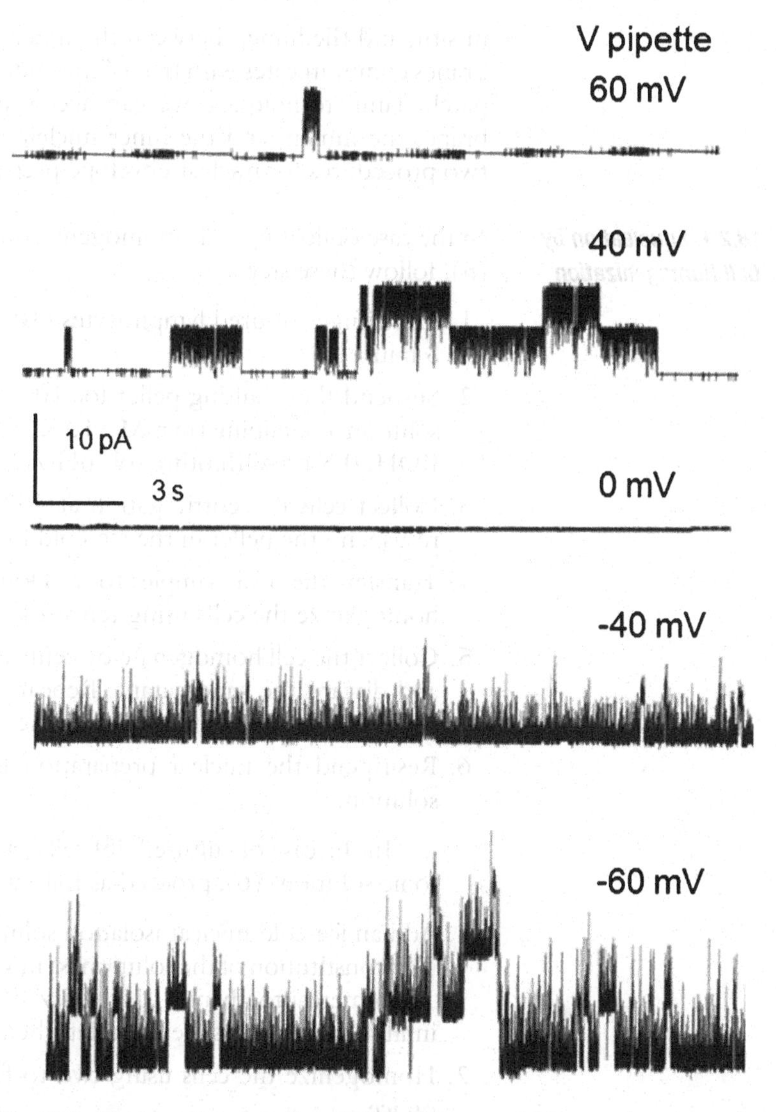

Fig. 18.1. Cl^- channel currents in mitoplasts from exocrine pancreatic acinar cells. Voltages in the patch pipette (*V pipette*) are indicated

– K^+, Cl^-, and Ca^{2+} channels (4) – which may relate to energy metabolism, Ca^{2+} signaling, apoptosis, and oxygen-sensing mechanisms.

18.2. Patch-Clamping the Nuclear Membrane or Nuclear Envelope

The nuclear envelope consists of two membranes, the inner and outer nucleic membranes, covering the eukaryotic cell nucleus from either the nucleoplasm or cytoplasm. The outer nucleic membrane continues to the perinuclear endoplasmic reticulum membrane in situ, and the lumen between the inner and outer nucleic membranes communicates with that of the endoplasmic reticulum. With patch-clamp techniques, we can access the outer nucleic membrane, the lumen, and the inner nucleic membrane. We describe two procedures for nuclear envelope preparation.

18.2.1. Enucleation by Cell Homogenization

In the case of floating cells (homogenization in hypotonic solution) (5), follow these steps.

1. Centrifuge cultured lymphocytes (10^6/ml) at 400*g* at 4°C for 3 min.
2. Suspend the resulting pellet for 10 min in ice-cold hypotonic solution, containing (in mM): 10 KCl, 1.5 $MgCl_2$, 10 HEPES/KOH, 0.5-D,L-dithiothreitol (pH 7.2).
3. Collect cells by centrifugation at 400*g* at 4°C for 3 min and resuspend the pellet in the ice-cold hypotonic solution.
4. Transfer the cell sample to a Dounce homogenizer, and homogenize the cells using ten strokes with the pestle on ice.
5. Collect the cell homogenate by centrifugation at 4°C for 3 min and discard the supernatant. Repeat the procedure twice and collect the pellet, which contains the nuclear envelopes.
6. Resuspend the nuclear preparation in isotonic experimental solution.

In the case of cultured Sf9 cells (homogenization in hypertonic solution) (6), proceed as follows:

1. Add an ice-cold nuclear isolation solution to the cell container. The constitution of the solution is (mM): 140 KCl, 250 sucrose, 1.5-β-mercaptoethanol, 10 Tris/HCl, and some protease inhibitors. The cells are then detached by gentle scraping.
2. Homogenize the cells using two to four strokes of the pestle on ice.
3. Add a drop (20–30 μl) of the homogenate to the isotonic KCl experimental solution (1 ml) in the experimental chamber on the stage of the inverted microscope.

18.2.2. Enucleation by Hypotonic Swelling and Cell Surface Operation

The approach based on enucleation by hypotonic swelling and cell surface operation is useful for studying the nuclear envelope from freshly isolated cells, particularly when the cell density is low, or when one has to identify the cell type in a heterogeneous cell population (7). We depict the approach in pancreatic acinar cells in some detail. The pancreatic tissue contains acini, ducts, and blood vessels, from which we select the acinar cells. First, we digest the tissue into single cells by serial treatment with collagenase and trypsin. We transfer the dispersed cells to a chamber and allow them to attach to its bottom. After the cells settle, we select the single acinar cells and carry out enucleation. We conduct all of the procedure in a small chamber on the stage of the inverted microscope under magnification of 600× to 800×. The setup would be familiar to every patch-clamper.

18.2.2.1. Procedure

1. In the vision field of the microscope, set up a glass pipette, its tip diameter the same size as that of a single cell (about 10–20 μm), and a fine glass needle to cut the cell surface. The pipette contains a hypotonic solution (20–50 mosm/kg H_2O). Both tools remain outside the chamber until use.
2. Wait 10–20 min for the single pancreatic cells to settle at the bottom of the chamber. Then, identify a solitary single acinar cell.
3. Perfuse the cell with a hypotonic solution through the glass pipette given access to the cell by a manipulator.
4. Allow the cell to swell more than double its normal size and then make a small cut on the cell surface with the glass needle (Fig. 18.2). The swollen nucleus immediately pops out and recovers to normal size soon after the glass pipette is withdrawn. It is now time to apply the standard patch-clamp techniques to the nuclear preparation.

18.2.3. Outer and Inner Nucleic Membrane

Inositol 1,4,5- triphosphate (IP_3)-induced Ca^{2+}-releasing channels (6, 8) and several K^+ (7, 9) and Cl^- channels (9) have been reported in the outer nucleic membrane using patch-clamp techniques.

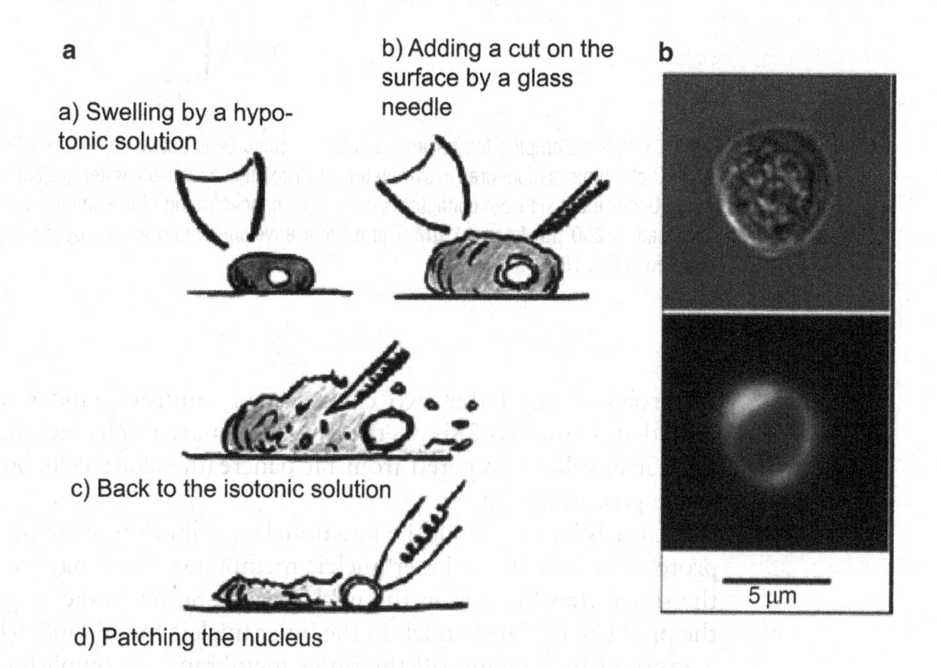

Fig. 18.2. Collecting a nucleus from a single cell by rupturing the cell membrane. (**a**) Enucleation procedure. (**b**) Nuclear envelope prepared from a single pancreatic acinar cell. (BODIPY-thapsigargin stain).

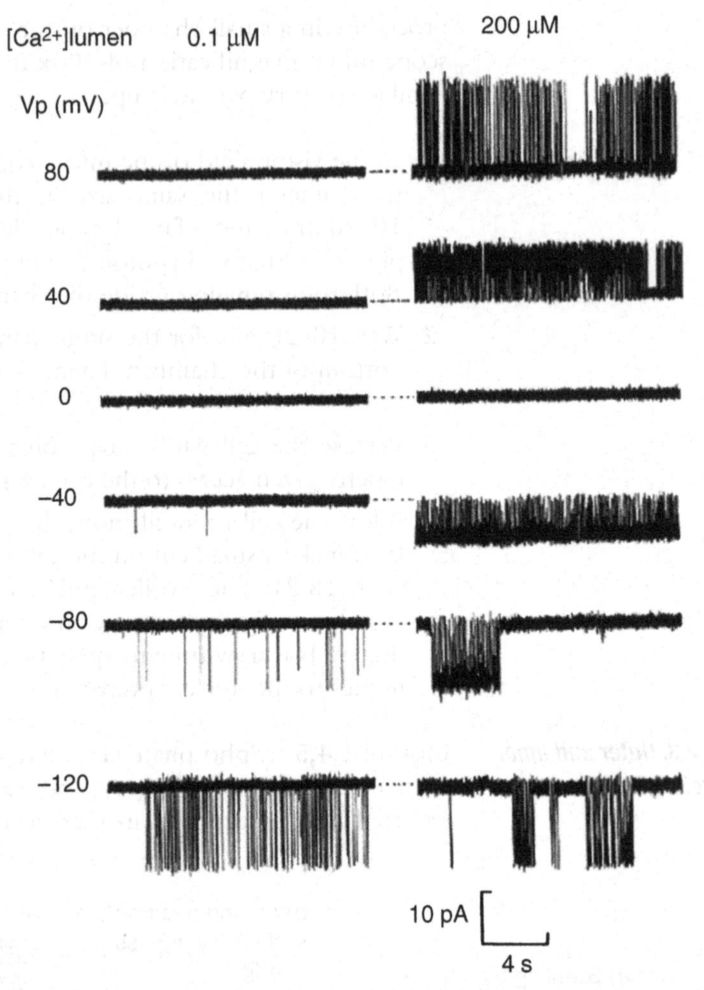

Fig. 18.3. Patch-clamping the nuclear envelope in pancreatic acinar cells. The activity of maxi-K^+ channels is monitored in symmetrical KCl solution across the outer nuclear membrane. Under the inside-out configuration, the Ca^{2+} concentration of the bathing side was increased to 200 µM from 0.1 µM. The nuclear envelope was prepared by the method depicted in Fig. 18.2.

The roles of the latter two channels are multitude and remain unfolded. Figure 18.3 shows the activity of maxi-K^+ channels in the nuclear envelope prepared from rat pancreatic acinar cells by the above procedure (7).

Little is known about the functional contribution of membrane proteins located in the inner nucleic membrane. They may control the signal transduction in the nucleoplasm. Some studies suggest the presence of ion channels in the inner nucleic membrane, which is exposed by scraping off the outer membrane, accomplished by shaking the nuclear envelopes in a Na citrate solution (10, 11).

18.3. Comments

Incorporation of channel proteins into artificial lipid bilayer membranes is other choice for monitoring organelle channel activities. Such attempts were reported in lysosomes (12) and secretory granules (13). However, to avoid contamination of the artificial lipids, disturbance of the channel orientation, and loss of the channel's interaction with other organelle proteins, giga-seal patch-clamping must be tried in the organelles to maintain their native morphology. We find it sensible to use the patch-clamp method in every field where the membranes exist. Also, at the present time it is the only way to monitor ion channel activities precisely in native organelles.

References

1. Wieckowski M, Giorgi C, Lebiedzinska M, Duszynski J, Pinton P (2009) Isolation of mitochondria-associated membranes and mitochondria from animal tissues and cells. Nature Protocol 4:1582–1590
2. Borecky J, Jezek P, Siemen D (1997) 108-pS channels in brown fat mitochondria might be identical to the inner membrane anion channel. J Biol Chem 272:19282–19289
3. Keller B, Hedrich R (1992) Patch-clamp techniques to study ion channels from organelles. Methods Enzymol 207:673–681, 146
4. Zoratti M, De Marchi U, Gulbins E, Szabo I (2009) Novel channels of the inner mitochondrial membrane. Biochim Biophys Acta 1789:351–363
5. Franco-Obregon A, Wang H, Clapham D (2000) Distinct ion channel classes are expressed on the outer nuclear envelope on T- and B-lymphocyte cell lines. Biophys J 79:202–214
6. Ionescu L, Cheung K-H, Vais H, Mak D-O, White C, Foskett K (2006) Graded recruitment and inactivation of single InsP3 receptor Ca^{2+}-release channels: implications for quantal Ca^{2+} release. J Physiol 573:645–662
7. Maruyama Y, Shimada H, Taniguchi J (1995) Ca^{2+}-activated K^{+}-channels in the nuclear envelope isolated from single pancreatic acinar cells. Pflugers Arch 430:148–150
8. Mak D, Foskette K (1994) Single-channel inositol 2,4,5-trisphosphate receptor currents revealed by patch clamp of isolated Xenopus oocyte nuclei. J Biol Chem 269:29375–29378
9. Matzke A, Weiger T, Matzke M (2010) Ion channels at the nucleus: electrophysiology meets the genome. Mol Plant 4:642–652
10. Humbert J-P, Matter N, Artault J-C, Koppler P, Malviya A (1996) Inositol 1,4,5-trisphosphate receptor is located to the inner nuclear membrane vindicating regulation of nuclear calcium signaling by inositol 1,4,5-trisphosphate. J Biol Chem 271:475–478
11. Marchenko S, Yarotskyy V, Kovalenko T, Kostyuk P, Thomas R (2005) Spontaneously active and InsP3-activated ion channels in cell nuclei from rat cerebellar Purkinje and granule neurons. J Physiol 565:897–910
12. Schieder M, Rotzer K, Bruggemann A, Biel M, Wahl-Schott C (2010) Characterization of two-pore channel 2 (TPCN2)-mediated Ca^{2+} currents in isolated lysosomes. J Biol Chem 285:21219–21222
13. Lee W, Torchalski B, Roussa E, Thvenod F (2008) Evidence for KCNQ1 K^{+} channel expression in rat zymogen granule membranes and involvement in cholecystokinin-induced pancreatic acinar secretion. Am J Physiol 294:C879–C892

Chapter 19

Role of Ion Channels in Plants

Rainer Hedrich, Dirk Becker, Dietmar Geiger, Irene Marten, and M. Rob G. Roelfsema

Abstract

When the second patch-clamp book of Sakmann and Neher appeared in 1995 (Sakmann and Neher, Single-channel recording, 2nd edn. Plenum Press, New York, 1995), the molecular nature of plant ion channels was still in its infancy. Since 1995, various members of the Shaker-, Two-Pore-, and KCO-type potassium channels have been identified; and their cellular and subcellular localizations have been resolved. The function of major K^+ channels has been characterized in its natural environment of plant cells and after heterologous expression. Just a few years ago, the first genes encoding plant plasma membrane anion channels were identified and shown to encode channels mediating Slow/SLAC-type and Rapid/QUAC-type currents. Distinct members of the potassium and anion channel families are involved in volume regulation, nutrient sensing, and uptake. Among them the K^+ channel AKT1 and anion channel SLAC1 are addressed in a calcium-dependent manner. Thereby, protein kinase–channel interaction and transphosphorylation are the keys to channel opening. In contrast to animal cells, plant cells are equipped with a large central vacuole. This acidic internal organelle provides for dynamic storage of ions and nutrients. Using isolated vacuoles from the model plant *Arabidopsis thaliana* in combination with transient overexpression approaches, major and low abundant ion channels and transporters could be characterized. This chapter provides insights into the current state of the plant ion channel field and introduces new approaches with patch-clamping plant cells and vacuoles.

19.1. Plant Potassium Channels

Plant potassium channels were first identified by patch-clamp studies on guard cell protoplasts and pulvinus motor cells (2–6). Follow-up studies identified potassium channels in the plasma and vacuolar membranes from various plant cells (2, 7–9). In plant membranes, K^+ selective channels often are the dominant means of conductance. In preparations with interfering anion channel activity, anion currents can be eliminated using gluconate or glutamate – organic anions impermeable to anion channels – as counter-ions for potassium in patch-clamp solutions.

Yasunobu Okada (ed.), *Patch Clamp Techniques: From Beginning to Advanced Protocols*, Springer Protocols Handbooks, DOI 10.1007/978-4-431-53993-3_19, © Springer 2012

In 1992, the molecular identity of the first K^+ channels in *Arabidopsis thaliana* 1 (KAT1) were identified, and the *Arabidopsis* K^+ transporter 1 (AKT1) was cloned (10, 11). Both K^+ channels belong to a gene family sharing structural homology to *Drosophila* Shaker channels (12). The *A. thaliana* genome comprises 15 genes encoding potassium channels. Among them, nine members are structurally related to voltage-dependent Shaker-like potassium channels (K_V), five share structural homology with voltage-independent, two-pore K^+ channels (TPK/K2P), and a single member, KCO3, encodes for a K^+ channel similar to the *Streptomyces lividans* potassium channel KcsA (13). The TPKs and KCO3 are addressed in the section on vacuole channels, below.

19.1.1. Shaker-Like K^+ Channels

19.1.1.1. Inward Rectifiers

Patch-clamp studies by Brüggemann et al. (14) have shown that the inward rectifying K^+ channel KAT1, active in *Arabidopsis* guard cells, is capable of mediating potassium uptake from media containing as little as 10 μM of external K^+. Thus, this K^+ uptake channel is working in the concentration range of high-affinity K^+ uptake systems whenever the membrane potential is both negative to the Nernst equilibrium potential of K^+ and to the activation threshold of the channel. Voltage-dependent inward rectifying K^+ channels usually exhibit a fixed voltage threshold for activation. Thus, starting from a holding potential (V_H)- K_{in}^+ were channels are closed at around −20 mV-channel activity is triggered by hyperpolarizing pulses to, for example, −180 mV. Depending on activation kinetics and thus time to reach steady-state currents (I_{SS}), the duration of the activation pulse may range from 0.5 s (e.g., KAT1) up to 5 s (Solanum KST1). Based on their biophysical properties, the heterologous expression system, and the potassium concentration in the bath solution, inward rectifying currents are elicited by hyperpolarizing voltages. Most of the biophysical studies with inward rectifying K^+ channels on plant cells have been performed in the whole-cell mode of the patch-clamp technique because single-channel conductances are in the order of 5–15 pS only and thus detailed single-channel recordings render a challenging task.

Inward rectifying K^+ channels are expressed in growing and dividing cells and those undergoing reversible volume changes (e.g., stomatal guard cells). For stomata to open, guard cells accumulate potassium, a process associated by hyperpolarization-activated, pH-sensitive K^+ currents. H^+ pumps hyperpolarize the plasma membrane, thereby generating the voltage gradient that drives K^+ inflow. Guard cell K^+ channels – e.g., KST1 and KAT1 from *Solanum tuberosum* and *A. thaliana*, respectively (15, 16) – appear acid-activated. Extracellular and cytosolic protons act on discrete His residues within the guard cell K^+ channel protein. Protonation increases the channels' voltage sensitivity (15, 17–20). Thus, during stomatal opening, when pumping activity of the H^+-ATPase is high, an increased number of channels open. Guard cells are characterized by a high density of K_{in} channels encoded by *KAT1*, *KAT2*, *AKT1*, *AKT2*, and *AtKC1* Shaker subunits (21). Brought about by heteromeric assembly, the biophysical and pharmacological properties of the "guard cell inward rectifier" thus resemble a combination of individual α-subunits. Specific subunit combinations expressed in guard cells or other cell types provide for cell-type specific voltage dependency and calcium or pH sensitivity (22–24). Mutants lacking only the *KAT1* gene show little stoma phenotype (25). The *Arabidopsis* mutant ***kincless***, expressing a dominant negative ***kat2*** construct, completely lacks inward potassium currents. The fact that in ***kincless*** the stomatal opening is strongly impaired underpins the general importance of plant Shaker-type channels for the regulation of volume and turgor (26).

19.1.1.2. Outward Rectifiers and Weakly Voltage-Dependent K^+ Channels

In contrast to inward rectifiers, K_{out} channels (e.g., *Arabidopsis* GORK and SKOR) activate at depolarized membrane potentials; and gating strongly depends on the extracellular K^+ concentration. This allows channel opening at membrane potentials more positive than the equilibrium potential of potasium (E_K). Based on this K^+ sensitivity, even in the presence of a steep driving force for potassium – micromolar external K^+ concentrations and about 100 mM in the cytoplasm – K^+ release channels are kept in the closed state. This prevents potassium loss from cells under unfavorable conditions of K^+ supply (27, 28). AKT2-like channels seem to represent the plant *Shaker* channel type that operates in two gating modes. Depending on its phosphorylation state, the *Arabidopsis* AKT2 channel shifts from a voltage-independent mode to a K_{in}-like, voltage-dependent mode (29).

Voltage Regulation

Structure–function studies based on site-directed mutagenesis and domain swapping between K_{in}-, K_{out}-, and AKT2-type channels have identified key residues involved in voltage gating (18, 30–37). The finding that several regions in Shaker-type plant K^+ channels – including the cytosolic N- and C-termini, S4 domain, P region, and TM6 – contribute to gating control suggests that distributed residues interact in concert to control the biophysical properties of plant potassium channels (for review see (38, 39)).

Phosphorylation Regulation

Apart from guard cells, root hairs represent a cellular model for studying K^+ transport in the context of potassium nutrition and sensing. They predominantly express the Shaker-type α-subunits AKT1, AtKC1, and GORK (40, 41). AKT1 appears essential for the overall root K^+ uptake because *akt1-1* mutant exhibits severe growth retardation under K^+ starvation (42–45). Despite the presence of AtKC1 transcripts, patch-clamp studies on root hair protoplasts revealed a complete absence of K_{in} currents in the *akt1-1* mutant (46), indicating that the AtKC1 Shaker-like protein represents a modulatory subunit active with heteromeric K^+ uptake channel complexes only. Recent work has shown that during K^+

starvation the AKT1 channel protein is subject to activation via a Ca^{2+}-dependent signaling network. Limiting K^+ in the soil supply is suggested to feed back on cytosolic calcium signals sensed via distinct calcineurin-B-like (CBL) proteins, which contain four calcium-binding EF hands. Upon assembly with their interacting protein kinase (CIPK) the CBL–CIPK complex is targeted to the plasma membrane to activate the AKT1 channel in a phosphorylation-dependent manner (47–50). Whereas AKT1 phosphorylation leads to channel activation, channel dephosphorylation by the PP2C phosphatase AIP1 inactivates AKT1 (50). Heterologous co-expression of Ca^{2+} sensors (CBLs), CBL-interacting protein kinases (CIPKs) and AKT1 in *Xenopus* oocytes allowed functional characterization of AKT1 channels and elucidating the role of the root hair-expressed AtKC1 channel (Fig. 19.1a,b). Well in agreement with patch-clamp studies on root hair protoplast from wild-type and AKT1 and ATKC1 mutants, AKT1 can be classified as a

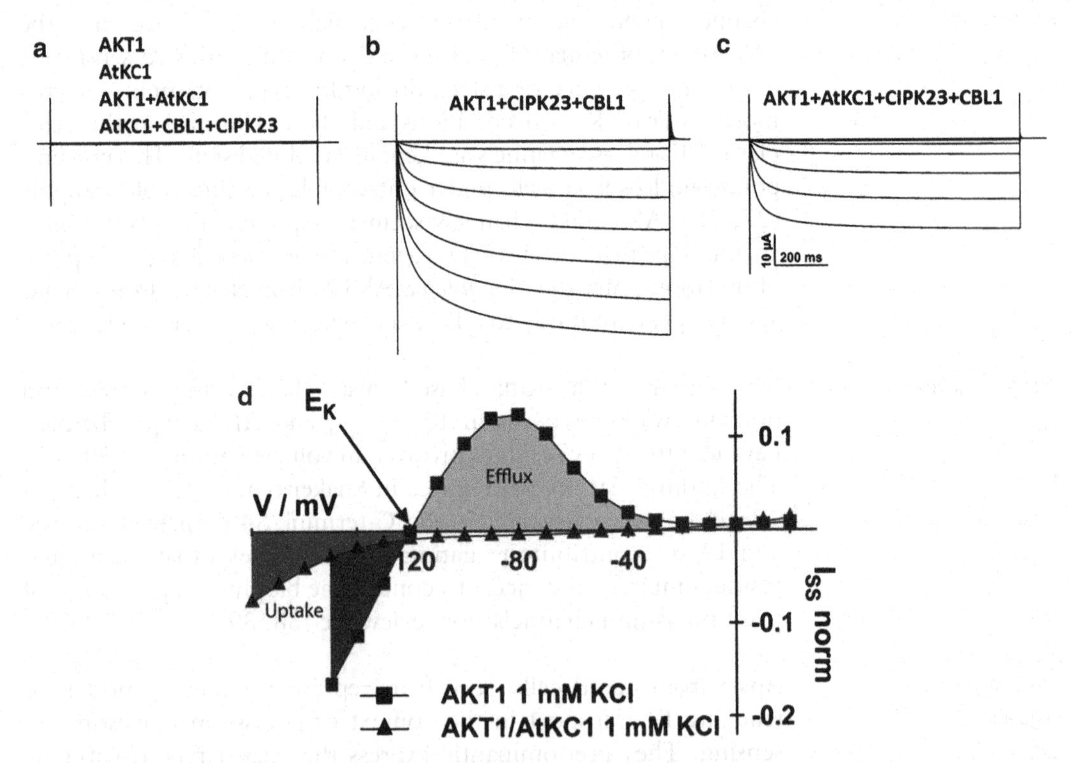

Fig. 19.1. Activation of root K^+ uptake channels requires calcium-dependent phosphorylation. (**a**) *Arabidopsis* root Shaker K^+ channel α-subunits AKT1, AtKC1, or combinations of the two do not give rise to voltage activation of potassium currents upon heterologous expression in *Xenopus* oocytes. (**b**) Voltage-dependent activation of time-dependent, inward potassium currents upon co-expression of the calcium sensor CBL1 and its interacting kinase CIKP23 with AKT1. (**c**) Additional expression of AtKC1 reduces inward K^+ currents by decreasing the open probability of the K^+ uptake channel. (**d**) Normalized current–voltage relation depicting the effect of the modulatory AtKC1 subunit on AKT1-mediated K^+ currents (see text for explanation). Currents in a, b, and c were recorded by means of the two-electrode voltage-clamp method using pulse protocols ranging from +60 to −160 mV in 10-mV steps. Holding potential was −30 mV. Bath solution in (**a**) and (**b**) contained (in mM): 30 K^+, 1 $MgCl_2$, 2 $CaCl_2$, 10 MES/Tris pH 5.6. (Data modified after Geiger et al. (51), © American Society for Biochemistry and Molecular Biology).

voltage-dependent inward rectifier and AtKC1 as its corresponding regulatory α-subunit (51). As shown for KAT1, voltage-dependent activation of AKT1 sets in at a rather fixed voltage threshold (ca. −80 mV) and is not dependent on the K^+ gradient. Under potassium starvation, the equilibrium potential of potassium (E_K) is far negative of AKT1's activation threshold (e.g., at 100 μM K_{soil}^+ and 100 mM K_{cyt}^+, $E_K = -180$ mV). Under these experimental conditions, voltage activation of AKT1 elicits pronounced K^+ efflux (Fig. 19.1c). In plants, however, heteromeric assembly of AKT1/AtKC1 shifts the activation threshold of the root K^+ uptake channel complex toward very negative membrane potentials (negative of E_K). Thus, potassium loss under unfavorable low soil K^+ level is prevented (46, 51, 52).

19.2. Plant Anion Channels

Plant anion channels represent key players of a broad spectrum of pivotal physiological functions. They are involved in volume and turgor regulation, ion homeostasis, and signaling processes with phytohormones and microbes. The first notion of plant anion conductance dates back to the 1980s (53–56). To differentiate currents through anion channels from the large potassium background conductance, patch-clamp studies are usually carried out with salts of cations that are impermeable for K^+ channels (e.g., TEA^+ or Cs^+). Based on their preference to conduct certain anions (e.g., chloride, sulfate, nitrate, malate), different types of anion channels could be resolved and analyzed when the respective anion represents the major solute of the patch-clamp buffers (Table 19.1).

The molecular representatives of the anion currents recorded in plants have been identified only recently (57–59). This breakthrough initiated research in various fields, including functional genomics, anion channel-linked signaling pathways, and structure–function analysis of channel protein and interacting regulators. This section illustrates the anion gene families dominating anion conductance of the plant plasma membrane and elaborates on their electrical properties and regulation.

19.2.1. SLAC-Type Anion Channels at the Plant Plasma Membrane

Due to their autonomous character and their electrical isolation from other plant tissues, guard cells represent a perfect model system to study ion channels and abscisic acid (ABA) and calcium signaling pathways that are maintained even in single isolated cells. Guard cells in the epidermis of plants balance the uptake of CO_2 from the atmosphere and the concomitant loss of water from leaves. When water supply is limited, the drought hormone ABA triggers release of anions and K^+ from guard cells and thus stomatal closure (55, 60). The initial steps in ABA

Table 19.1
Patch-clamp studies with protoplasts from *Arabidopsis thaliana*

Tissue/cell type	Major ion	Channel type	Gene(s)	References
Guard cell	Potassium	Inward K^+ rectifier	*KAT1*	(14, 21, 25, 170–172)
	Potassium	Outward K^+ rectifier	*GORK*	(27, 171, 173)
	Chloride	S-type anion	*SLAC1* family	(59, 63, 64, 70, 170, 171, 174–176)
	Nitrate	S-type anion	*SLAH3*	(77)
	Sulfate, malate	R-type anion	*ALMT12/QUAC1*	(57)
	Chloride	R-type anion	*ALMT12*	(177)
Mesophyll	Potassium	Outward K^+ rectifier		(178)
	Chloride	Mechanosensitive anion		(179)
Root	Chloride	R-type anion	*QUAC* family	(180)
Root hairs	Potassium	Inward, outward K^+ rectifier	*AKT1, GORK*	(41, 46, 48)
Rhizodermis	Potassium	Outward K^+ rectifier	*GORK*	(181, 182)
Hypocotyl	Potassium	Inward K^+ rectifier	*KAT1, KAT2*	(183)
	Chloride	S-type anion	*SLAC* family	(184, 185)
	Sulfate, nitrate	R/S-type anion	*SLAC/QUAC* family	(186)
	Chloride	R-type anion	*QUAC* family	(187)
Pollen	Potassium	Inward K^+ rectifier	*SPIK*	(187, 188)
Tumor	Potassium	Inward K^+ rectifier	*AKT1, AKT2/3*	(189)
Cell suspension	Chloride, glutamate	S-type anion	*SLAC* family	(190)
Callus	Chloride	Anion		(191)

The major ion in the patch-clamp solutions for studying a respective channel type in the cited reference(s) is given. If known, the corresponding ion channel gene/family is indicated.

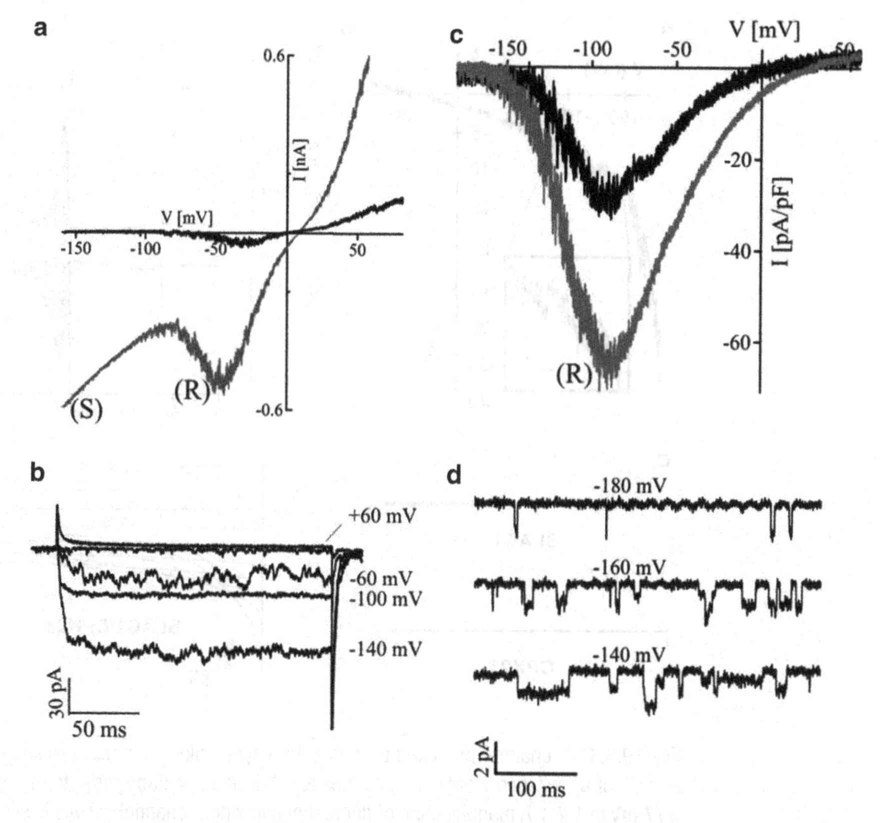

Fig. 19.2. Macroscopic abscisic acid (ABA)-induced anion currents in guard cells. (**a**) Whole-cell currents were elicited from *Vicia faba* guard cells with (*gray*) and without 10 μM cytosolic ABA (*black*) in response to voltage ramps from −158 to −82 mV in 1500 ms. In the absence of ABA, only R-type currents were recorded. A rise in R-type (*R*) and S-type (*S*) currents were observed with cytosolic ABA. (Data modified after Levchenko et al. (61), © National Academy of Sciences USA). (**b**) Fast activation and deactivation of macroscopic R-type currents in *Arabidopsis thaliana* guard cells. Voltage pulses were applied from a holding voltage of −180 mV. (**c**) Stimulation of R-type currents from *A. thaliana* guard cells in response to cytosolic ABA (50 μM) under sulfate- and malate-based experimental conditions. *Black trace*, control, without ABA; *gray trace*, with ABA. Voltage ramps were applied from −180 to +60 mV in 13 s. (**d**) Single-channel fluctuations monitored from outside-out membrane patches of *A. thaliana* guard cells under sulfate- and malate-based experimental conditions at membrane voltages as indicated. (**b**–**d** Courtesy of Mumm et al., unpublished).

signal transduction have been shown to address guard cell anion channels (Fig. 19.2) in a calcium-dependent and calcium-independent manner (61, 62). An ABA- and CO_2/O_3-insensitive mutant was shown to lack a gene encoding a putative guard cell anion transporter named SLAC1 (slow anion channel associated 1) (58, 59).

Upon expression of SLAC1 in *Xenopus* oocytes in the presence of certain protein kinases (see 63, 64 and following sections in paragraph 19.2.1) macroscopic anion currents could be elicited (Fig. 19.3c). Prolonged voltage pulses to hyperpolarized membrane potentials trigger the SLAC1-mediated anion currents with rapid activation/slow deactivation kinetics, similar to the slow anion channel in guard cells (65, 66). The half-maximal activation

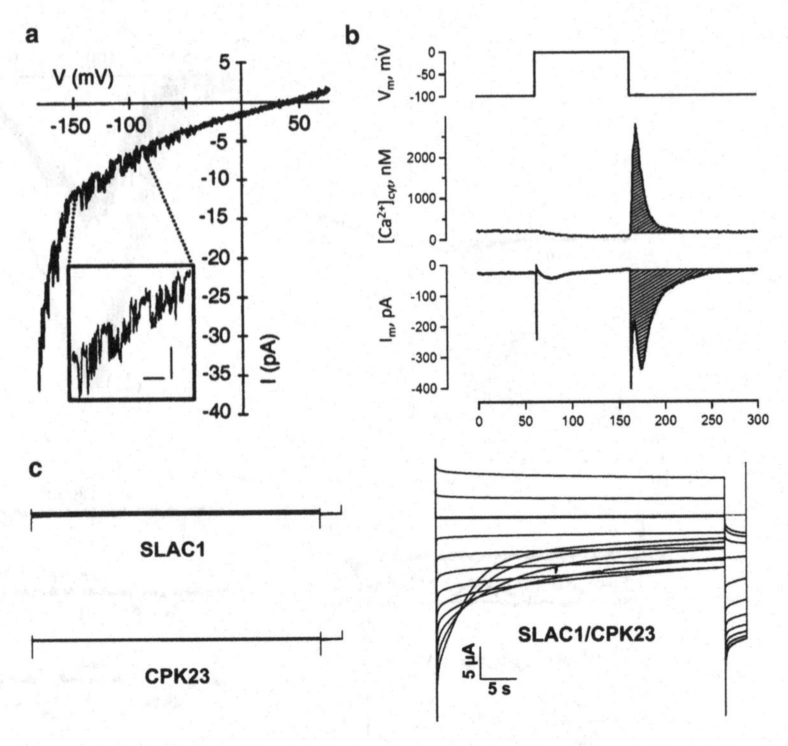

Fig. 19.3. Ca^{2+} channel-mediated activation of S-type anion channels. (**a**) Single-channel activity of a Ca^{2+}-permeable channel measured during a voltage ramp from −183 mV to +77 mV in 1.2 s. A magnification of fluctuations in single-channel activity is shown in the *box*; *vertical bar* 2 pA, *horizontal bar* 10 mV. (**b**) Voltage-dependent cytosolic free Ca^{2+} concentration changes in a tobacco guard cell loaded with the Ca^{2+}-sensitive fluorescent dye FURA2. *Upper trace*. The guard cell was clamped from a holding potential of −100 mV to 0 mV for 2 min. *Middle trace*. The cytosolic free Ca^{2+} concentration first decreases after depolarizing the clamp potential from −100 mV to 0 mV, followed by a transient overshoot after returning to −100 mV. *Lower trace*. A transient increase of inward current mirrors the voltage-induced cytosolic Ca^{2+} concentration overshoot. (**c**) Voltage-clamp measurement in *Xenopus laevis* oocytes injected with cRNA from slow anion channel associated 1 (SLAC1), calcium-dependent protein kinase 23 (CPK23), and both. Note that currents carried by slow activating anion channels are observed only after co-injection of SLAC1 and CPK23. [Data are from (**a**) Stoelzle et al. (111), © National Academy of Sciences USA; (**b**) Stange et al. (107), © John Wiley and Sons; (**c**) Geiger et al. (63), © National Academy of Sciences USA].

voltage ($V_{1/2}$) of SLAC1 shifts by about 50 mV upon increasing bath Cl^- concentration from 10 to 100 mM, indicating that the voltage dependence of SLAC1 is modulated by external chloride. In agreement with anion-permeable conductance, variation of the external chloride and nitrate concentration results in Nernstian-like behavior of the reversal potential. Note that SLAC1 is annotated as a bacteria-like dicarboxylate carrier sharing homology with yeast MAE1 malate transporter (67), although SLAC1 does not conduct malate (64, 68).

Elements in ABA signaling have been identified by genetic screens, which have revealed ABA-insensitive open-stomata plant mutants with deregulated guard cell volume control. A Snf1-related protein kinase 2 (SnRK2.6 named OST1 (69)) and Ca^{2+}-dependent kinases (CDPKs (70)) and protein phosphatases of the PP2C family (ABI1, ABI2 (71, 72)), exhibited the strongest phenotypes. Split YFP-based protein–protein interaction assays, using SLAC1 as the bait, identified guard cell-expressed OST1 from the SnRK2 family (69) and protein kinases CPK21 and 23 from the CDPK family (73, 74) as physically interacting SLAC1 partners (63, 64). The activation of the protein kinase OST1 in response to ABA is suppressed in the dominant *abi1-1* mutant, indicating that the PP2C phosphatase ABI1 negatively regulates ABA signal transduction upstream of OST1 (69).

When SLAC1 is co-expressed with both a SLAC1-activating kinase and ABI1, the guard cell anion channel activity is suppressed. Biochemical analysis involving the cytosolic ABA receptor RCAR1 (75, 76) provided unequivocal evidence of a bifurcated fast ABA signaling pathway leading to the activation of SLAC1 (77). SLAC1 belongs to a family of five related potential anion channel proteins. In guard cells, SLAC1 is located side by side with the SLAC1 homolog SLAH3 in the plasma membrane. SLAH3 transports nitrate rather than chloride (77) and is the major component of the large nitrate conductance of the SLAC-type channel in guard cells (78). SLAH3 is under the control of RCAR1, ABI1, and CPK21 but not the OST1 branch of the ABA signaling pathway.

The structure of SLAC1 was recently modeled to crystal coordinates obtained with the bacterial SLAC1 homolog TehA (68). It was shown that three SLAC1 subunits likely assemble into a trimeric complex. Each subunit, consisting of ten transmembrane α-helical domains, forms a central five-helix transmembrane pore that is gated by an extremely conserved phenylalanine residue. This structural model of SLAC1 will guide future structure–function studies with SLAC1 and SLAC1 homologs (SLAHs).

19.2.2. R-Type Anion Channels at the Plant Plasma Membrane

Anion efflux from guard cells involves anion channels of the SLAC type and the rapid/quick (QUAC) type (Fig. 19.2). These two anion channels are characterized by different voltage dependence, kinetics, and susceptibility to blockers (79–84). QUAC-type anion channels mediate strongly voltage-dependent plasma membrane anion fluxes (53–56, 85). Upon depolarization QUAC activates with fast kinetics, and hyperpolarization causes rapid deactivation. (53, 86, 87). Using kinetic data for fitting QUAC-type currents could successfully be described in terms of the Hodgkin–Huxley equations (86, 87). The Cl^- dependence of inward currents (Cl^- release) is characterized by a maximum single-channel conductance of 89 pS, half-saturating at 87 mM cytosolic chloride. In addition to this substrate saturation, anion release is dependent on external anion activity (88–91). Interestingly, extracellular rather than cytosolic anions affect the gating process of QUAC-type channels.

Thus, the anion concentration and species determine permeation and gating of this plant anion channel type.

The nature of the first QUAC-type channel QUAC1 has been associated with a member of the ALMT family (57). Upon patch clamp studies with root cells some other ALMT channels (aluminum-activated malate transporter) have been characterized as plasma membrane, Al^{3+}-activated malate channels (92–94). Interestingly, one member of the ALMT protein family, ZmALMT1, activates in an Al^{3+}-independent manner and transports inorganic anions such as Cl^-, NO_3^-, and SO_4^{2-} rather than malate (95). AtALMT12 represents a guard cell QUAC-type anion channel (57). Plants lacking AtALMT12 are impaired in dark- and CO_2-induced stomatal closure and in response to the drought-stress hormone ABA. Patch-clamp studies on guard cell protoplasts isolated from *atalmt12* mutants revealed reduced QUAC-type currents compared to wild-type plants when malate was present in the bath media in patch-clamp studies. Following expression of AtALMT12 in *Xenopus* oocytes, voltage-dependent anion currents reminiscent of QUAC channels could be activated and, consistent with SLAC1, ALMT12 was named QUAC1 (quick activating anion channel 1 (82)). Because the *almt12* loss of function mutant was only reduced in QUAC-type currents but did not completly lack R-type activity, other members of the ALMT family likely encode QUAC2 or QUAC3. Former studies showed that in addition to Cl^-, NO_3^-, and SO_4^{2-} QUAC-type channels exhibit permeability for malate (55, 85, 89). Extracellular malate has been postulated to play an important role in activating anion release from guard cells by inducing a shift in the voltage dependence of QUAC (82, 89, 96). Similar to the situation in its natural environment of the guard cell, in *Xenopus* oocytes QUAC1 appears regulated by extracellular malate. Hence, this signaling anion shifts QUAC1's voltage-dependent relative open probability toward more negative (physiological) membrane potentials (57).

19.3. Plant Ca^{2+} Channels

As in animals, plant cells maintain high Ca^{2+} concentration gradients between the cytosol and intracellular compartments. Whereas the Ca^{2+} concentration in the cytosol is generally as low as 10^{-7} M, it approximates 10^{-4} M in the cell wall and vacuole (see (97) and the references therein). Because of the large concentration gradient across the plasma membrane and intracellular membranes, opening of Ca^{2+} channels leads to elevation of the cytosolic free Ca^{2+} concentration. As with animal cells, plant cells use intracellular changes in the cytosolic free Ca^{2+} concentration as a signal to induce a variety of responses (98, 99). Ca^{2+} signals were shown to be involved in the growth of root hairs, pollen tubes, and closure of stomatal pores. For this reason, the mechanism through which Ca^{2+} signals trigger specific responses has become a central question in plant biology (97, 100).

19.3.1. Classes of Ca^{2+} Channels

Most studies on Ca^{2+} channels in plants have focused on the plasma membrane. Within this membrane these Ca^{2+} channels do not only evoke Ca^{2+} signals, they also play an essential role in uptake of Ca^{2+} from the soil and its distribution throughout the plant (101). Most channels implied for plant Ca^{2+} uptake do not show high selectivity for Ca^{2+} but, instead, seem to function as nonselective cation channels (102). Exceptions to this rule are the stretch-activated channels in guard cells (103) and onion epidermal cells (104). The nonselective cation channels can be grouped according to their voltage dependence into hyperpolarization-activated and voltage-independent channels (97, 105).

In contrast to animals, hyperpolarization-activated Ca^{2+}-permeable channels seem to be omnipresent in the plasma membrane of plant cells (Fig. 19.3a,b). As a result, hyperpolarization normally provokes an increase in the cytosolic free Ca^{2+} concentration (106, 107), whereas depolarization causes the opposite effect (107). This is different from that in animal cells, in which depolarization leads to activation of Ca^{2+} channels (108). Hyperpolarization-activated Ca^{2+} channels appear stimulated by reactive oxygen species (ROS) (109), fungus-derived elicitors (110), and blue light irradiation (111), suggesting that the latter stimuli provoke Ca^{2+} signals in combination with a hyperpolarized membrane potential.

The second class of nonselective cation channels does not require hyperpolarization as channels open in a voltage-independent manner. This property implies that these channels are deactivated under most conditions because a membrane potential-independent influx of Ca^{2+} would be lethal under most conditions. Like hyperpolarization-activated channels, voltage-independent Ca^{2+} channels were found to be activated by pathogen-derived elicitors (112). Furthermore, these channels are involved in cold signaling (113) and uptake of Ca^{2+} from the soil by root cells (114).

19.3.2. How to Resolve Calcium Channels?

As with plant plasma and vacuolar membranes, K^+ channels generally represent the dominant means of cation conductance. Thus, patch-clamp settings to measure Ca^{2+} channels are largely based on buffers containing K^+ channel blockers (109–111). Often Ba^{2+} has been included in the bath and pipette media. This divalent cation blocks most K^+ channels but is conducted by nonselective (Ca^{2+}-permeable) channels (102). Apart of K^+ channels, anion channels may obscure studying Ca^{2+} channels; and for this reason the bath and pipette solutions normally contain anions that do not permeate plant anion channels, such as gluconate (see K^+ channel section, above). Unfortunately, the pharmacology of Ca^{2+}-permeable channels in plants has not received much attention. Ca^{2+} channels are blocked by phenylalkylamines (e.g., verapamil) at high concentrations (102), which are well known inhibitors of L-type Ca^{2+} channels in animal cells (108). At high concentrations, however, these blockers also inhibit K^+-selective channels (115, 116). For this reason, these Ca^{2+} channel blockers do not represent selective Ca^{2+} channel inhibitors. In addition to the phenylalkylamides, trivalent cations such as La^{3+}, Gd^{3+}, and Al^{3+} often are used to block Ca^{2+}-permeable channels. Again, however, these blockers are rather unspecific and affect a large range of cation channels. Furthermore, Al^{3+} is well known for its ability to stimulate several members of the aluminium-activated malate transporter family (ALMT).

19.3.3. Genes Encoding Plant Ca^{2+} Channels

Even though plant genomes harbor several gene families with homology to animal Ca^{2+} channels, so far no plant gene has been shown to encode a Ca^{2+} channel unequivocally. One of these gene families is the group of cyclic nucleotide-gated channels (CNGCs). Several mutants with nonfunctional CNGCs were found to display constitutive pathogen defense responses (117, 118), suggesting that these genes encode channels that are involved in Ca^{2+} responses upon microorganisms attack. However, only equivocal data about the nature of plant-specific CNG channels is available as the expression of plant CNGC genes in heterologous systems seldom resulted in functional channels (105). A role of plant CNGC genes in Ca^{2+} signaling also is supported by indirect evidence gained with pollen tubes (119). Loss of the *At*CNGC18 gene disables apical outgrowth of the tubular pollen cell. For pollen tubes, the establishment of a Ca^{2+} gradient from the growing tip toward the base of the cell directs polar growth.

In addition to the CNGC genes, glutamate receptor-like channel genes may encode Ca^{2+} channels in plants (120, 121). Application of glutamate to plant tissues provokes elevation of the cytosolic free Ca^{2+} level, which is associated with strong depolarization of the plasma membrane (122, 123). Further evidence for the function of glutamate receptor-like genes as Ca^{2+} channels was obtained by expression of a chimerical construct based on an animal ionotropic glutamate receptor with a pore domain of a homologous plant gene (124).

The difficulty of activating the products of plant CNGC- and glutamate receptor-like genes in patch-clamp studies with isolated protoplast or after heterologous expression in animal cells indicates that these channels differ from their homologs in animals. Possibly, plant channels are equipped with an additional emergency mechanism, not known from their animal counterparts, that prevents their activity in heterologous systems. Future studies may uncover ligands, or proteins that interact with these channels and shift them toward the active state. In addition, new gene families encoding Ca^{2+} channels may be uncovered through genome studies and/or with screens for signaling mutants. The search for mutants defective in nodule formation for instance, has led to the identification of cation channels that are involved in the generation of nuclear Ca^{2+} signals (125–127).

19.4. Vacuolar Ion Channels and Transporters

Soon after patch-clamp studies with the plasma membrane of plant cells were established (2, 128), ion channels were identified in plant-specific organelles such as the chloroplast (129, 130) and the large central vacuole membrane (131, 132). The central vacuole represents the major storage compartment of the plant cell and can take up as much as 90% of its volume. Consequently, most of the osmotic active solutes appear sequestered in the vacuole. Apart of the important role of the vacuole in volume and turgor control, it serves as a dynamic pool for ion and metabolite homeostasis in the cytoplasm (133–135). Patch-clamp studies have been the key to discover the nature of predominant vacuolar channels and pumps and to dissect the molecular mechanism of proton-coupled sugar and polyol transporters, which are active in this intracellular membrane.

19.4.1. Protoplast and Vacuole Isolation

For patch-clamp measurements with vacuoles, protoplasts first have to be isolated from the respective cell type. For this, cell walls are enzymatically removed via treatment with a fungal cellulase and pectinase cocktail. Release of viable protoplasts has to take place with osmotic potential of the enzyme solution slightly lower than that of the tissue (Fig. 19.4a). For experiments with the model plant *Arabidopsis* and most other plants, selective osmotic shock lysis of a selected single protoplast or an entire population has been the method of choice to liberate fresh vacuoles rapidly (136). Accordingly, protoplasts are ruptured by bath perfusion or local perfusion with application pipettes (137) using a solution of lower osmotic potential than the cell wall–free cell. The protoplasts initially swell before the plasma membrane ruptures and the central vacuole becomes accessible for patch-clamp studies. So far, patch-clamp studies with *Arabidopsis* vacuoles have been addressed to mesophyll cells, guard cells, and suspension-cultured cells (Table 19.2).

19.4.2. Sign Convention

Unlike protoplast, isolated vacuoles are approached with the patch pipette from the cytosolic side. Thus, after establishing the whole-vacuole configuration, the pipette solution is in contact with the vacuolar lumen, instead of the cytosol as with protoplasts. According to the sign convention of 1992 (138), cation currents flowing into the cytosol are denoted as negative values, irrespective of the membrane that is studied. Based on this convention, the vacuolar lumen

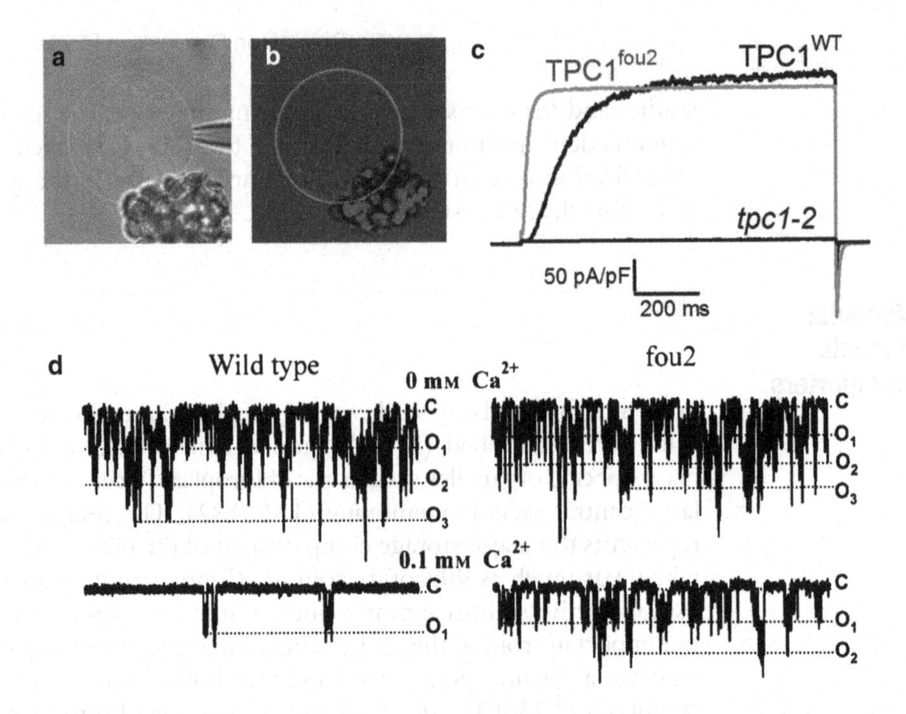

Fig. 19.4. Whole-vacuolar patch clamp recordings of SV/TPC1 channels from *A. thaliana* mesophyll vacuoles. (**a**) After the patch pipette was attached to the vacuolar membrane and a tight giga-ohm seal between pipette and membrane was formed, the membrane patch in the pipette tip was ruptured by simultaneously applying a short voltage pulse in the range of 600–800 mV and sudden negative pressure via the pipette. (**b**) Mesophyll protoplasts isolated from *tpc1-2* loss-of-function mutants were subjected to transient transformation with green fluorescent protein (GFP)-labeled SV channel constructs. Vacuoles with a GFP-fluorescent membrane were used for subsequent patch-clamp analysis. (**c**) Whole vacuolar current recordings from vacuoles of nontransformed and transformed *tpc1-2* mesophyll protoplasts were performed during a voltage pulse of +110 mV from a holding voltage of −60 mV under symmetrical K^+ and pH conditions 2 days after transient transformation. SV currents were activated only from *tpc1-2* vacuoles with transiently expressed *fou2* and wild-type (WT) *TPC1* channel genes. Note the faster activation of the $TPC1^{fou2}$ channels compared to the $TPC1^{WT}$ channels. (Courtesy of Dadacs-Narloch and Hedrich, unpublished). (**d**) Single-channel fluctuations were recorded from vacuoles of wild-type and *fou2* plants in the absence or presence of luminal Ca^{2+}. The cytoplasmic side of the excised vacuolar membrane was facing the bath medium. In contrast to wild-type SV channels, *fou2* single-SV-channel fluctuations were still observed at an increased luminal Ca^{2+} level, pointing to reduced sensitivity of fou2 SV channels toward blocking luminal Ca^{2+}. (Modified after Beyhl et al. (147). © John Wiley and Sons).

Table 19.2
Patch-clamp studies with vacuoles from *Arabidopsis thaliana*

Tissue/cell type	Major ion	Transporter type	Gene	References
Cell suspension	Potassium	SV channel	*TPC1*	(140)
Guard cell, mesophyll	Potassium	SV channel	*TPC1*	(192)
Mesophyll	Potassium, calcium	SV channel	*TPC1*	(23, 147, 193, 194)
	Nitrate	Nitrate/proton antiporter	*CLCa/b*	(195, 196)
	Malate	Malate channel	*ALMT9*	(197)
	Malate	Malate transporter	*AttDT*	(198)

The major ion in the patch-clamp solutions for studying the vacuolar channel or carrier type in the cited reference and the corresponding transporter gene are given.

is thus equivalent to the extracellular space, which in many respects is in line with its composition. Note that due to the sign convention, the voltage-dependent, slow-activating vacuolar cation channel of large conductance, which was discovered and classified as an inward rectifier in 1986, was reclassified as an outward rectifier (for review, see (139)).

19.4.3. Slow Vacuolar Channels

The initial patch-clamp study on isolated plant vacuoles identified the slow vacuolar (SV) channel as dominating ion conductance at high Ca^{2+} concentrations at the cytosolic side of the membrane (131). Since then, this nonselective cation channel has puzzled the field of plant ion transport. Under experimental conditions with equal K^+ concentrations on both sides of the vacuolar membrane, SV channels open at strongly depolarized potentials (140). Based on this property, one would expect that SV channels assist with the import of cations into the vacuole. However, neither the gradient of most cations nor the vacuolar membrane potential supports this direction of transport. Instead, SV channels more likely act as cation release valves. In line with this function, the activation potential of SV channels is modulated via several cations in the cytosol and in the vacuolar lumen (132, 140–143). Depending on the cation (gradient and/or concentration), the activation potential of SV channels probably shifts toward the membrane potential to favor extrusion of certain cations from the vacuole into the cytosol (140, 141, 144–147).

The SV channels in the model plant *A. thaliana* are encoded by the single-copy gene *AtTPC1* (148). *AtTPC1* encodes a relatively large protein, comprised of two fused Shaker-like units, each containing six transmembrane domains and a pore-forming domain (149). Plant genomes lack genes encoding archetypical voltage-dependent calcium channels. Because of some distant homology of TPC1 to the half of the α-subunit of L-type calcium channels from animals, TPC1 channels were initially treated as the first plant calcium channels of known molecular nature (149–152). Today, it is generally accepted that the plant-specific TPC1 channel is rather homologous with the human TPC2 channel, which is located in lysosomes and is involved in nicotinic acid adenine dinucleotide phosphate (NAADP)-dependent release of Ca^{2+} (153).

19.4.4. Expression System for Vacuolar Channels

The lack of an expression system for ion channels in the vacuolar membrane has long hampered their structure–function analysis. Recently, this limitation has been overcome for TPC1 channels using a transient expression system with mesophyll protoplasts from the *Attpc1-2* loss-of-function mutant (Fig. 19.4b, c). Mesophyll protoplasts were isolated from *tpc1-2* mutants, as basically described by Beyhl et al. (147). They were subjected to polyethylene glycol (PEG)-mediated transfection with the respective DNA constructs of the green fluorescent protein (GFP)-labeled

TPC1 ortholog (154). After 2 days, GFP-labeled vacuoles can be approached with the patch-clamp technique. Using this experimental approach it can be shown that TPC1 channels of higher plants and moss are active in the vacuolar membrane of *A. thaliana* mesophyll cells. For calcium-dependent activation, the cytosolic calcium-binding EF hands of TPC1 are indispensable (Petra Dietrich, personal communication). In contrast to cytosolic calcium ions, luminal ones block the SV channel of wild-type *Arabidopsis* but not that of the hyperactive *fou2* mutant (155). This low-affinity Ca^{2+} block is correlated with a binding site involving an aspartic acid at the fou2 position 454 and adjacent glutamates (156). Both calcium-binding domains in TPC1 are pointing to a central role of the signaling cation in the regulation of SV channel activity (Fig. 19.4d).

In *fou2*, the reduced luminal calcium sensitivity correlates with increased calcium levels in the storage organelle and thus enlarges the calcium gradient across the vacuolar membrane. Unexpectedly, this single point mutation in the SV channel also leads to higher levels of the plant hormone jasmonate. This hormone is well known for its role in responses to wounding, touch, and cold, suggesting that these stimuli regulate jasmonate release through a mechanism involving Ca^{2+} release from the vacuole.

19.4.5. Vacuolar K^+ and Anion Channels

Plant cells can accumulate potassium in the vacuole to concentrations higher than the cytosolic level of approximately 100 mM. However, this store may contain considerably less K^+ if plants are coping with K^+ starvation (157). Depending on the growth conditions, plant cells maintain the potassium balance in the cytoplasm via K^+ transport across the vacuolar membrane. K^+ uptake into vacuoles is most likely facilitated by NHX-type cation transporters, whereas several classes of K^+ channels can mediate K^+ release from the vacuole or uptake at low K^+ levels into the vacuolar lumen.

Vacuolar K^+ channels are encoded by *KCO3* and the tandem pore K^+ channel (*TPK*) genes. KCO3 contains two membrane-spanning domains connected by a pore domain, and this structure is duplicated in TPK channels. Functional KCO3 tetrameric or TPK dimeric channels contain four pore domains, an assembly common for Shaker-like channels. Four of the five TPK channels are localized to the vacuole; only TPK4 is located in the plasma membrane of pollen tubes. Tandem pore channels lack a voltage-sensing domain and respond to voltage changes such as a K^+ leak (158, 159). Channel opening, however, is controlled by cytosolic factors, such as calcium ions and 14-3-3 proteins. These factors address the N- and C-terminal domains, which have Ca^{2+}-binding EF hands and a binding site for 14-3-3 proteins, respectively.

Vacuoles can contain a large number of inorganic and organic anions (135). Apart of their function as counter-ions for vacuolar potassium, organic anions are important in complexing and detoxifying heavy metals. Candidate genes for vacuolar anion channels are found in two gene families: aluminum-activated malate transporter (*ALMT*) and chloride channels (*CLC*). However, so far *CLC* genes were shown to encode only vacuolar carriers.

19.4.6. Vacuolar H^+ Pumps and H^+ Coupled Transport

Transport processes across the tonoplast are energized by two proton pumps: the vacuolar H^+-ATPase (V-ATPase) and H^+-pyrophosphatase (V-PPase). Patch-clamp studies unequivocally demonstrated the presence and functionality of both pumps (131, 160). Upon pumping cytosolic protons into the vacuolar lumen, a proton gradient and membrane voltage are generated, enabling the cell to create and maintain the ionic or metabolite gradient, respectively. The importance of the vacuolar V-ATPase was recently shown with transformed *Arabidopsis* plants in which genes encoding a subunit of this H^+ pump were silenced (161). Such plants lacking V-ATPase activity in the vacuolar membrane are entirely dwarfed and contain lower nitrate levels (161). This phenotype indicates that the H^+-ATPase and H^+-coupled transporters such as CLC work hand in hand.

In addition to the CLC-type anion and NHX-type cation carriers, the vacuolar membrane harbors a large number of membrane proteins that couple the transport of metabolites to that of protons. A function in plant cell sugar homeostasis has been attributed to the vacuolar tonoplast monosaccharide transporter TMT1/2 (Fig. 19.5) and to the sucrose release carrier SUC4 (162, 163). The function of these membrane proteins was studied in the whole-vacuole mode. Sucrose- and glucose-induced proton currents carried by TMT transporters in an antiport mechanism could be provoked in the presence of a pH gradient (163, 164). As documented with TMTs and SUC4, this patch-clamp approach in combination with stable or transient overexpression (see section 19.4.4) (162) enables high-resolution recordings with otherwise low-abundant or weak electrogenic transport systems.

19.5. Conclusion and Perspectives

Patch-clamp techniques with plant cells have gone a tremendous way, from the first characterization of ion channels in 1984 (2, 128) to the identification of genes that encode them and regulatory mechanisms controlling their activity. Nevertheless, several milestones are likely to await us in coming decade. First, we are still lacking the genes that encode Ca^{2+} channels in plants, even though the importance of Ca^{2+} signals for regulation of K^+ and anion transport is evident. It is likely that the identification of these channel genes is of major importance to understand how ion homeostasis in plants is regulated. For instance, regulation of the AKT1 channel by K^+ supply in the soil is likely to involve Ca^{2+} signals (49), but the

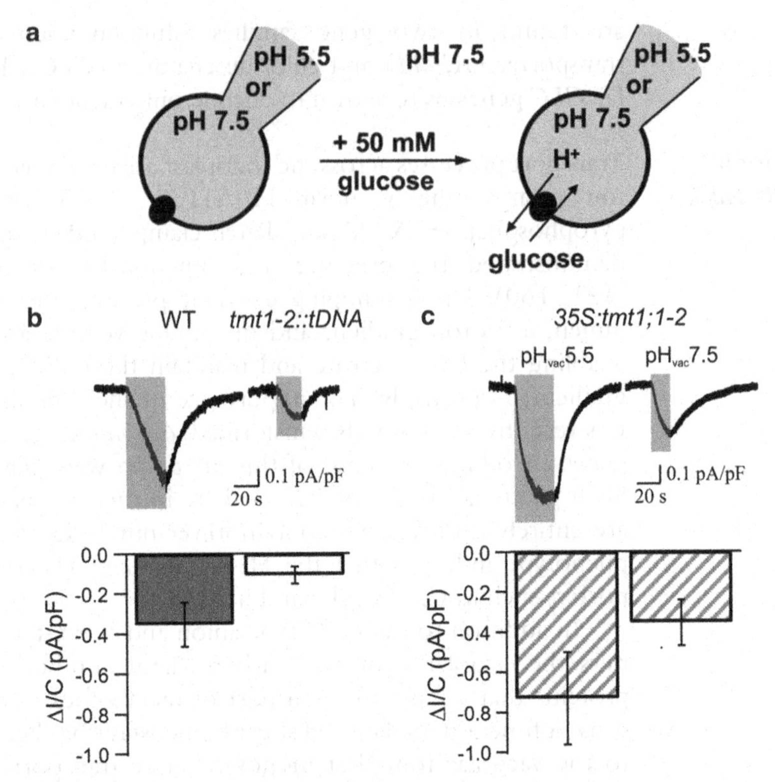

Fig. 19.5. Functional patch-clamp analysis of the vacuolar H^+-coupled glucose transporter TMT1/2. (**a**) Experimental conditions for recording low carrier-mediated current amplitudes. In response to cytosolic glucose application under constant pH across the membrane (e.g., either symmetrical pH 7.5 or physiological pH gradient), currents were monitored in the whole-vacuolar mode. Otherwise symmetrical ionic conditions were chosen. (**b**) Glucose-induced downward deflections of the current trace were much smaller in *tmt1/2* loss-of-function mutants (*tmt1-2:tDNA, white bar*) than in wild-type vacuoles (*WT, dark bar*). Current responses were monitored at 0 mV in the presence of a physiological pH gradient pointing to the function of TMT1/2 as an antiporter. Cytosolic glucose treatment is indicated by *superimposed bars* of the respective current traces. (**c**) Compared to a cytosol-directed pH gradient, reduced glucose-induced currents were recorded from vacuoles of a stable TMT1-overexpressing line (*35S:tmt1;1–2*) under symmetrical pH conditions. Thus, TMT1-mediated glucose transport is driven by both the pH and glucose gradient. (**a–c**) (Data from Wingenter et al. (162). © American Society of Plant Biologists).

following two related questions are still unsolved: What is the sensor for potassium starvation? How does the sensor trigger cytoplasmic calcium signals?

Regulation of ion transporters by Ca^{2+}-dependent protein kinases seems to be a common mechanism in plant cells (47). The root AKT1 K^+ channel and the nitrate transporter/sensor NRT1-1 (CHL1) are stimulated by a CIPK–CBL complex (49, 165) consisting of a Ca^{2+} sensor protein that interacts with a protein kinase. Likewise, the SLAC1 and SLAH3 anion channels are activated by CPKs (63), in which the Ca^{2+} sensor and protein kinase are united within a single protein. Other K^+ channels, anion channels, or calcium transporters may be regulated by homologous complexes assembled by distinct combinations of the 10–34 members from the CBL, CIPK, and CPK gene families (47, 166). A major task for plant ion transport research will be to identify common regulatory protein complexes interacting with ion transport proteins. In addition to the Ca^{2+}-dependent complexes, SLAC1 was found to be activated through a Ca^{2+}-independent mechanism. Here, the drought stress hormone ABA binds to a PYR/PYL/RCAR protein that interacts with a protein phosphatase, causing activation of an SnRK protein kinase (167). The Ca^{2+}-dependent CPK and Ca^{2+}-independent SnRK protein kinases are homologous, suggesting a common evolutionary origin of both regulation mechanisms. Future projects are likely to show to which extent protein complexes that regulate ion channels are similar and where they divert.

Even though plants lack a central nervous system, transport processes within a cell or between cells and tissues need to be coordinated. It is likely that with the genes in hand we will be able to tackle various long-distance signal transduction mechanisms in the near future. Within single cells, plastids and vacuoles are known to influence ion transport at the plasma membrane (168). However, the signaling modules that coordinate transport between these domains have remained elusive. Plant cells are equipped with a large number of receptors that enable them to recognize molecular patterns of other cell types or microorganisms (169). It is well known that these receptors regulate ion transport, but the signaling chain that connects these transporters to an ion channel still needs to be identified. Finally, cells within organs are known to communicate over long distances for transport of ions, but the signaling mechanisms involved are mostly still unknown. Long-distance signaling appears to be associated with hydraulic or electrical signals. Identifying the molecular mechanisms that support plant-specific action potentials would certainly be a milestone for learning about ion transport in plants.

References

1. Sakmann B, Neher E (1995) Single-channel recording, 2nd edn. Plenum Press, New York
2. Schroeder JI, Hedrich R, Fernandez JM (1984) Potassium-selective single channels in guard-cell protoplasts of *Vicia faba*. Nature 312(5992):361–362
3. Hedrich R, Schroeder JI, Fernandez JM (1987) Patch-clamp studies on higher-plant cells: a perspective. Trends Biochem Sci 12(2):49–52
4. Moran N, Ehrenstein G, Iwasa K, Mischke C, Bare C, Satter RL (1988) Potassium channels in motor cells of *Samanea saman*: a patch-clamp study. Plant Physiol 88(3):643–648
5. Satter RL, Moran N (1988) Ionic channels in plant-cell membranes. Physiol Plant 72(4): 816–820
6. Iijima T, Hagiwara S (1987) Voltage-dependent K channels in protoplasts of trap-lobe cells of *Dionaea muscipula*. J Membr Biol 100(1):73–81
7. Schroeder JI, Hedrich R (1989) Involvement of ion channels and active transport in osmoregulation and signaling of higher plant cells. Trends Biochem Sci 14(5):187–192
8. Hedrich R, Schroeder JI (1989) The physiology of ion channels and electrogenic pumps in

higher-plants. Annu Rev Plant Physiol Plant Mol Biol 40:539–569

9. Hedrich R, Becker D (1994) Green circuits–the potential of plant specific ion channels. Plant Mol Biol 26(5):1637–1650
10. Sentenac H, Bonneaud N, Minet M, Lacroute F, Salmon JM, Gaymard F, Grignon C (1992) Cloning and expression in yeast of a plant potassium ion transport system. Science 256(5057):663–665
11. Anderson JA, Huprikar SS, Kochian LV, Lucas WJ, Gaber RF (1992) Functional expression of a probable *Arabidopsis thaliana* potassium channel in *Saccharomyces cerevisiae*. Proc Natl Acad Sci USA 89(9):3736–3740
12. Jan LY, Jan YN (1997) Cloned potassium channels from eukaryotes and prokaryotes. Annu Rev Neurosci 20:91–123
13. Hedrich R, Anschütz U, Becker D (2011) The plant plasma membrane, plant cell monographs. In: Murphy AS, Peer W, Schulz B (eds) Biology of plant potassium channels, vol 19. Springer, Heidelberg, pp 253–274
14. Brüggemann L, Dietrich P, Becker D, Dreyer I, Palme K, Hedrich R (1999) Channel-mediated high-affinity K^+ uptake into guard cells from *Arabidopsis*. Proc Natl Acad Sci USA 96(6):3298–3302
15. Muller-Rober B, Ellenberg J, Provart N, Willmitzer L, Busch H, Becker D, Dietrich P, Hoth S, Hedrich R (1995) Cloning and electrophysiological analysis of KST1, an inward rectifying K^+ channel expressed in potato guard cells. EMBO J 14(11):2409–2416
16. Nakamura RL, McKendree WL Jr, Hirsch RE, Sedbrook JC, Gaber RF, Sussman MR (1995) Expression of an *Arabidopsis* potassium channel gene in guard cells. Plant Physiol 109(2):371–374
17. Hoth S, Dreyer I, Dietrich P, Becker D, Muller-Rober B, Hedrich R (1997) Molecular basis of plant-specific acid activation of K^+ uptake channels. Proc Natl Acad Sci USA 94(9):4806–4810
18. Hoth S, Geiger D, Becker D, Hedrich R (2001) The pore of plant K^+ channels is involved in voltage and pH sensing: domain-swapping between different K^+ channel alpha-subunits. Plant Cell 13(4):943–952
19. Dietrich P, Sanders D, Hedrich R (2001) The role of ion channels in light-dependent stomatal opening. J Exp Bot 52(363): 1959–1967
20. Mouline K, Very AA, Fdr G, Boucherez J, Pilot G, Devic M, Bouchez D, Thibaud J-B, Sentenac H (2002) Pollen tube development and competitive ability are impaired by disruption of a Shaker K^+ channel in *Arabidopsis*. Genes Dev 16(3):339–350
21. Szyroki A, Ivashikina N, Dietrich P, Roelfsema MR, Ache P, Reintanz B, Deeken R, Godde M, Felle H, Steinmeyer R, Palme K, Hedrich R (2001) KAT1 is not essential for stomatal opening. Proc Natl Acad Sci USA 98(5): 2917–2921
22. Ivashikina N, Deeken R, Fischer S, Ache P, Hedrich R (2005) AKT2/3 subunits render guard cell K^+ channels Ca^{2+} sensitive. J Gen Physiol 125(5):483–492
23. Latz A, Ivashikina N, Fischer S, Ache P, Sano T, Becker D, Deeken R, Hedrich R (2007) In planta AKT2 subunits constitute a pH- and Ca^{2+}-sensitive inward rectifying K^+ channel. Planta 225(5):1179–1191
24. Dreyer I, Poree F, Schneider A, Mittelstadt J, Bertl A, Sentenac H, Thibaud JB, Mueller-Roeber B (2004) Assembly of plant Shaker-like K_{out} channels requires two distinct sites of the channel alpha-subunit. Biophys J 87(2): 858–872
25. Kwak JM, Murata Y, Baizabal-Aguirre VM, Merrill J, Wang M, Kemper A, Hawke SD, Tallman G, Schroeder JI (2001) Dominant negative guard cell K^+ channel mutants reduce inward-rectifying K^+ currents and light-induced stomatal opening in *Arabidopsis*. Plant Physiol 127(2):473–485
26. Lebaudy A, Vavasseur A, Hosy E, Dreyer I, Leonhardt N, Thibaud JB, Very AA, Simonneau T, Sentenac H (2008) Plant adaptation to fluctuating environment and biomass production are strongly dependent on guard cell potassium channels. Proc Natl Acad Sci USA 105(13):5271–5276
27. Ache P, Becker D, Ivashikina N, Dietrich P, Roelfsema MRG, Hedrich R (2000) GORK, a delayed outward rectifier expressed in guard cells of *Arabidopsis thaliana*, is a K^+-selective, K^+-sensing ion channel. FEBS Lett 486(2): 93–98
28. Johansson I, Wulfetange K, Poree F, Michard E, Gajdanowicz P, Lacombe B, Sentenac H, Thibaud JB, Mueller-Roeber B, Blatt MR, Dreyer I (2006) External K^+ modulates the activity of the *Arabidopsis* potassium channel SKOR via an unusual mechanism. Plant J 46(2):269–281
29. Michard E, Dreyer I, Lacombe B, Sentenac H, Thibaud JB (2005) Inward rectification of the AKT2 channel abolished by voltage-dependent phosphorylation. Plant J 44(5): 783–797
30. Blatt MR (1992) K^+ channels of stomatal guard-cells: characteristics of the inward

rectifier and its control by pH. J Gen Physiol 99(4):615–644

31. Gajdanowicz P, Garcia-Mata C, Gonzalez W, Morales-Navarro SE, Sharma T, Gonzalez-Nilo FD, Gutowicz J, Mueller-Roeber B, Blatt MR, Dreyer I (2009) Distinct roles of the last transmembrane domain in controlling *Arabidopsis* K^+ channel activity. New Phytol 182(2):380–391
32. Geiger D, Becker D, Lacombe B, Hedrich R (2002) Outer pore residues control the H^+ and K^+ sensitivity of the *Arabidopsis* potassium channel AKT3. Plant Cell 14(8):1859–1868
33. Marten I, Hoshi T (1998) The N-terminus of the K channel KAT1 controls its voltage-dependent gating by altering the membrane electric field. Biophys J 74(6):2953–2962
34. Lai HC, Grabe M, Jan YN, Jan LY (2005) The S4 voltage sensor packs against the pore domain in the KAT1 voltage-gated potassium channel. Neuron 47(3):395–406
35. Li L, Liu K, Hu Y, Li D, Luan S (2008) Single mutations convert an outward K^+ channel into an inward K^+ channel. Proc Natl Acad Sci USA 105(8):2871–2876
36. Poree F, Wulfetange K, Naso A, Carpaneto A, Roller A, Natura G, Bertl A, Sentenac H, Thibaud JB, Dreyer I (2005) Plant K_{in} and K_{out} channels: approaching the trait of opposite rectification by analyzing more than 250 KAT1-SKOR chimeras. Biochem Biophys Res Commun 332(2):465–473
37. Tang XD, Marten I, Dietrich P, Ivashikina N, Hedrich R, Hoshi T (2000) Histidine(118) in the S2-S3 linker specifically controls activation of the KAT1 channel expressed in *Xenopus* oocytes. Biophys J 78(3):1255–1269
38. Dreyer I, Blatt MR (2009) What makes a gate? The ins and outs of K_v-like K^+ channels in plants. Trends Plant Sci 14(7):383–390
39. Latorre R, Munoz F, Gonzalez C, Cosmelli D (2003) Structure and function of potassium channels in plants: some inferences about the molecular origin of inward rectification in KAT1 channels. Mol Membr Biol 20(1): 19–25
40. Daram P, Urbach S, Gaymard F, Sentenac H, Cherel I (1997) Tetramerization of the AKT1 plant potassium channel involves its C-terminal cytoplasmic domain. EMBO J 16(12):3455–3463
41. Ivashikina N, Becker D, Ache P, Meyerhoff O, Felle HH, Hedrich R (2001) K^+ channel profile and electrical properties of *Arabidopsis* root hairs. FEBS Lett 508(3):463–469
42. Desbrosses G, Josefsson C, Rigas S, Hatzopoulos P, Dolan L (2003) AKT1 and TRH1 are required during root hair elongation in *Arabidopsis*. J Exp Bot 54(383): 781–788
43. Spalding EP, Hirsch RE, Lewis DR, Qi Z, Sussman MR, Lewis BD (1999) Potassium uptake supporting plant growth in the absence of AKT1 channel activity: inhibition by ammonium and stimulation by sodium. J Gen Physiol 113(6):909–918
44. Dennison KL, Robertson WR, Lewis BD, Hirsch RE, Sussman MR, Spalding EP (2001) Functions of AKT1 and AKT2 potassium channels determined by studies of single and double mutants of *Arabidopsis*. Plant Physiol 127(3):1012–1019
45. Rubio F, Nieves-Cordones M, Aleman F, Martinez V (2008) Relative contribution of *At*HAK5 and *At*AKT1 to K^+ uptake in the high-affinity range of concentrations. Physiol Plant 134(4):598–608
46. Reintanz B, Szyroki A, Ivashikina N, Ache P, Godde M, Becker D, Palme K, Hedrich R (2002) AtKC1, a silent *Arabidopsis* potassium channel alpha-subunit modulates root hair K^+ influx. Proc Natl Acad Sci USA 99(6): 4079–4084
47. Hedrich R, Kudla J (2006) Calcium signaling networks channel plant K^+ uptake. Cell 125(7):1221–1223
48. Li L, Kim BG, Cheong YH, Pandey GK, Luan S (2006) A Ca^{2+} signaling pathway regulates a K^+ channel for low-K response in *Arabidopsis*. Proc Natl Acad Sci USA 103(33):12625–12630
49. Xu J, Li HD, Chen LQ, Wang Y, Liu LL, He L, Wu WH (2006) A protein kinase, interacting with two calcineurin B-like proteins, regulates K^+ transporter AKT1 in *Arabidopsis*. Cell 125(7):1347–1360
50. Lee SC, Lan WZ, Kim BG, Li L, Cheong YH, Pandey GK, Lu G, Buchanan BB, Luan S (2007) A protein phosphorylation/dephosphorylation network regulates a plant potassium channel. Proc Natl Acad Sci USA 104(40):15959–15964
51. Geiger D, Becker D, Vosloh D, Gambale F, Palme K, Rehers M, Anschuetz U, Dreyer I, Kudla J, Hedrich R (2009) Heteromeric AtKC1/AKT1 channels in *Arabidopsis* roots facilitate growth under K^+-limiting conditions. J Biol Chem 284(32):21288–21295
52. Wang Y, He L, Li HD, Xu J, Wu WH (2010) Potassium channel alpha-subunit AtKC1 negatively regulates AKT1-mediated K^+ uptake in *Arabidopsis* roots under low-K^+ stress. Cell Res 20(7):826–837
53. Hedrich R, Busch H, Raschke K (1990) Ca^{2+} and nucleotide dependent regulation of voltage

dependent anion channels in the plasma membrane of guard cells. EMBO J 9(12): 3889–3892
54. Hedrich R, Jeromin A (1992) A new scheme of symbiosis: ligand- and voltage-gated anion channels in plants and animals. Philos Trans R Soc Lond B Biol Sci 338(1283):31–38
55. Keller BU, Hedrich R, Raschke K (1989) Voltage-dependent anion channels in the plasma membrane of guard cells. Nature 341(6241):450–453
56. Marten I, Lohse G, Hedrich R (1991) Plant growth hormones control voltage-dependent activity of anion channels in plasma membrane of guard cells. Nature 353(6346):758–762
57. Meyer S, Mumm P, Imes D, Endler A, Weder B, Al-Rasheid KAS, Geiger D, Marten I, Martinoia E, Hedrich R (2010) *At*ALMT12 represents an R-type anion channel required for stomatal movement in *Arabidopsis* guard cells. Plant J 63(6):1054–1062
58. Negi J, Matsuda O, Nagasawa T, Oba Y, Takahashi H, Kawai-Yamada M, Uchimiya H, Hashimoto M, Iba K (2008) CO_2 regulator SLAC1 and its homologues are essential for anion homeostasis in plant cells. Nature 452(7186):483–486
59. Vahisalu T, Kollist H, Wang YF, Nishimura N, Chan WY, Valerio G, Lamminmaki A, Brosche M, Moldau H, Desikan R, Schroeder JI, Kangasjarvi J (2008) SLAC1 is required for plant guard cell S-type anion channel function in stomatal signalling. Nature 452(7186): 487–491
60. Lebaudy A, Very AA, Sentenac H (2007) K^+ channel activity in plants: genes, regulations and functions. FEBS Lett 581(12):2357–2366
61. Levchenko V, Konrad KR, Dietrich P, Roelfsema MR, Hedrich R (2005) Cytosolic abscisic acid activates guard cell anion channels without preceding Ca^{2+} signals. Proc Natl Acad Sci USA 102(11):4203–4208
62. Schroeder JI, Hagiwara S (1989) Cytosolic calcium regulates ion channels in the plasma membrane of *Vicia faba* guard cells. Nature 338:427–430
63. Geiger D, Scherzer S, Mumm P, Marten I, Ache P, Matschi S, Liese A, Wellmann C, Al-Rasheid KA, Grill E, Romeis T, Hedrich R (2010) Guard cell anion channel SLAC1 is regulated by CDPK protein kinases with distinct Ca^{2+} affinities. Proc Natl Acad Sci USA 107(17):8023–8028
64. Geiger D, Scherzer S, Mumm P, Stange A, Marten I, Bauer H, Ache P, Matschi S, Liese A, Al-Rasheid KA, Romeis T, Hedrich R (2009) Activity of guard cell anion channel SLAC1 is controlled by drought-stress signaling kinase-phosphatase pair. Proc Natl Acad Sci USA 106(50):21425–21430
65. Linder B, Raschke K (1992) A slow anion channel in guard cells, activating at large hyperpolarization, may be principal for stomatal closing. FEBS Lett 313(1):27–30
66. Schroeder JI, Keller BU (1992) Two types of anion channel currents in guard cells with distinct voltage regulation. Proc Natl Acad Sci USA 89(11):5025–5029
67. Camarasa C, Bidard F, Bony M, Barre P, Dequin S (2001) Characterization of *Schizosaccharomyces pombe* malate permease by expression in *Saccharomyces cerevisiae*. Appl Environ Microbiol 67(9):4144–4151
68. Chen YH, Hu L, Punta M, Bruni R, Hillerich B, Kloss B, Rost B, Love J, Siegelbaum SA, Hendrickson WA (2010) Homologue structure of the SLAC1 anion channel for closing stomata in leaves. Nature 467(7319): 1074–1080
69. Mustilli A-C, Merlot S, Vavasseur A, Fenzi F, Giraudat J (2002) *Arabidopsis* OST1 protein kinase mediates the regulation of stomatal aperture by abscisic acid and acts upstream of reactive oxygen species production. Plant Cell 14(12):3089–3099
70. Mori IC, Murata Y, Yang Y, Munemasa S, Wang YF, Andreoli S, Tiriac H, Alonso JM, Harper JF, Ecker JR, Kwak JM, Schroeder JI (2006) CDPKs CPK6 and CPK3 function in ABA regulation of guard cell S-type anion- and Ca^{2+}-permeable channels and stomatal closure. PLoS Biol 4(10):e327
71. Koornneef M, Reuling G, Karssen CM (1984) The isolation and characterization of abscisic acid insensitive mutants of *Arabidopsis thaliana*. Physiol Plant 61(3):377–383
72. Leung J, Bouvier-Durand M, Morris PC, Guerrier D, Chefdor F, Giraudat J (1994) *Arabidopsis* ABA response gene ABI1: features of a calcium-modulated protein phosphatase. Science 264(5164):1448–1452
73. Harmon AC, Yoo BC, McCaffery C (1994) Pseudosubstrate inhibition of CDPK, a protein kinase with a calmodulin-like domain. Biochem 33(23):7278–7287
74. Harper JF, Huang JF, Lloyd SJ (1994) Genetic identification of an autoinhibitor in CDPK, a protein kinase with a calmodulin-like domain. Biochem 33(23):7267–7277
75. Ma Y, Szostkiewicz I, Korte A, Moes D, Yang Y, Christmann A, Grill E (2009) Regulators of PP2C phosphatase activity function as abscisic acid sensors. Science 324(5930):1064–1068
76. Park SY, Fung P, Nishimura N, Jensen DR, Fujii H, Zhao Y, Lumba S, Santiago J,

Rodrigues A, Chow TF, Alfred SE, Bonetta D, Finkelstein R, Provart NJ, Desveaux D, Rodriguez PL, McCourt P, Zhu JK, Schroeder JI, Volkman BF, Cutler SR (2009) Abscisic acid inhibits type 2C protein phosphatases via the PYR/PYL family of START proteins. Science 324(5930):1068–1071

77. Geiger D, Maierhofer T, Al-Rasheid KA, Scherzer S, Mumm P, Ache P, Grill E, Marten I, Hedrich R (2011) Fast abscisic acid signalling of stomatal closure via guard cell anion channel SLAH3 and ABA-receptor RCAR1. Science Signalling 17;4(173):ra32
78. Schmidt C, Schroeder JI (1994) Anion selectivity of slow anion channels in the plasma membrane of guard cells: large nitrate permeability. Plant Physiol 106(1):383–391
79. Dietrich P, Hedrich R (1994) Interconversion of fast and slow gating modes of GCAC1, a guard cell anion channel. Planta 195:301–304
80. Marten I, Busch H, Raschke K, Hedrich R (1993) Modulation and block of the plasma membrane anion channel of guard cells by stilbene derivatives. Eur Biophys J 21:7
81. Marten I, Zeilinger C, Redhead C, Landry DW, al-Awqati Q, Hedrich R (1992) Identification and modulation of a voltage-dependent anion channel in the plasma membrane of guard cells by high-affinity ligands. EMBO J 11(10):3569–3575
82. Raschke K (2003) Alternation of the slow with the quick anion conductance in whole guard cells effected by external malate. Planta 217(4):651–657
83. Schroeder JI, Schmidt C, Sheaffer J (1993) Identification of high-affinity slow anion channel blockers and evidence for stomatal regulation by slow anion channels in guard cells. Plant Cell 5(12):1831–1841
84. Schwartz A, Ilan N, Schwarz M, Scheaffer J, Assmann SM, Schroeder JI (1995) Anion channel blockers inhibit S-type anion channels and abscisic acid responses in guard cells. Plant Physiol 109(2):651–658
85. Roberts SK (2006) Plasma membrane anion channels in higher plants and their putative functions in roots. New Phytol 169(4):647–666
86. Kolb HA, Marten I, Hedrich R (1995) Hodgkin-Huxley analysis of a GCAC1 anion channel in the plasma membrane of guard cells. J Membr Biol 146(3):273–282
87. Schulz-Lessdorf B, Lohse G, Hedrich R (1996) GCAC1 recognizes the pH gradient across the plasma membrane: a pH-sensitive and ATP-dependent anion channel links guard cell membrane potential to acid and energy metabolism. Plant J 10(6):993–1004
88. Dietrich P, Hedrich R (1998) Anions permeate and gate GCAC1, a voltage-dependent guard cell anion channel. Plant J 15(4): 479–487
89. Hedrich R, Marten I (1993) Malate-induced feedback regulation of plasma membrane anion channels could provide a CO_2 sensor to guard cells. EMBO J 12(3):897–901
90. Hedrich R, Marten I, Lohse G, Dietrich P, Winter H, Lohaus G, Heldt HW (1994) Malate-sensitive anion channels enable guard-cells to sense changes in the ambient CO_2 concentration. Plant J 6(5):741–748
91. Lohse G, Hedrich R (1995) Anions modify the response of guard-cell anion channels to auxin. Planta 197(3):546–552
92. Kollmeier M, Dietrich P, Bauer CS, Horst WJ, Hedrich R (2001) Aluminum activates a citrate-permeable anion channel in the aluminum-sensitive zone of the maize root apex: a comparison between an aluminum-sensitive and an aluminum-resistant cultivar. Plant Physiol 126(1):397–410
93. Pineros MA, Kochian LV (2001) A patch-clamp study on the physiology of aluminum toxicity and aluminum tolerance in maize: identification and characterization of Al^{3+}-induced anion channels. Plant Physiol 125(1):292–305
94. Ryan PR, Skerrett M, Findlay GP, Delhaize E, Tyerman SD (1997) Aluminum activates an anion channel in the apical cells of wheat roots. Proc Natl Acad Sci USA 94(12): 6547–6552
95. Pineros MA, Cancado GM, Maron LG, Lyi SM, Menossi M, Kochian LV (2008) Not all ALMT1-type transporters mediate aluminum-activated organic acid responses: the case of *Zm*ALMT1—an anion-selective transporter. Plant J 53(2):352–367
96. Konrad KR, Hedrich R (2008) The use of voltage-sensitive dyes to monitor signal-induced changes in membrane potential-ABA triggered membrane depolarization in guard cells. Plant J 55(1):161–173
97. Roelfsema MRG, Hedrich R (2010) Making sense out of Ca^{2+} signals: their role in regulating stomatal movements. Plant Cell Environ 33(3):305–321
98. Dodd AN, Kudla J, Sanders D (2010) The language of calcium signaling. Annu Rev Plant Biol 61:593–620
99. Hetherington AM, Brownlee C (2004) The generation of Ca^{2+} signals in plants. Annu Rev Plant Biol 55:401–427

100. McAinsh MR, Pittman JK (2009) Shaping the calcium signature. New Phytol 181(2): 275–294
101. White PJ, Broadley MR (2003) Calcium in plants. Ann Bot 92(4):487–511
102. Demidchik V, Davenport RJ, Tester M (2002) Nonselective cation channels in plants. Annu Rev Plant Biol 53:67–107
103. Cosgrove DJ, Hedrich R (1991) Stretch-activated chloride, potassium, and calcium channels coexisting in plasma-membranes of guard cells of *Vicia faba* L. Planta 186(1): 143–153
104. Ding JP, Pickard BG (1993) Mechanosensory calcium-selective cation channels in epidermal cells. Plant J 3(1):83–110
105. Demidchik V, Maathuis FJM (2007) Physiological roles of nonselective cation channels in plants: from salt stress to signalling and development. New Phytol 175(3):387–404
106. Grabov A, Blatt MR (1998) Membrane voltage initiates Ca^{2+} waves and potentiates Ca^{2+} increases with abscisic acid in stomatal guard cells. Proc Natl Acad Sci USA 95(8): 4778–4783
107. Stange A, Hedrich R, Roelfsema MRG (2010) Ca^{2+}-dependent activation of guard cell anion channels, triggered by hyperpolarization, is promoted by prolonged depolarization. Plant J 62(2):265–276
108. Dolphin AC (2006) A short history of voltage-gated calcium channels. Br J Pharmacol 147:S56–S62
109. Pei ZM, Murata Y, Benning G, Thomine S, Klusener B, Allen GJ, Grill E, Schroeder JI (2000) Calcium channels activated by hydrogen peroxide mediate abscisic acid signalling in guard cells. Nature 406(6797):731–734
110. Gelli A, Higgins VJ, Blumwald E (1997) Activation of plant plasma membrane Ca^{2+}-permeable channels by race-specific fungal elicitors. Plant Physiol 113(1):269–279
111. Stoelzle S, Kagawa T, Wada M, Hedrich R, Dietrich P (2003) Blue light activates calcium-permeable channels in *Arabidopsis* mesophyll cells via the phototropin signaling pathway. Proc Natl Acad Sci USA 100(3):1456–1461
112. Zimmermann S, Nurnberger T, Frachisse JM, Wirtz W, Guern J, Hedrich R, Scheel D (1997) Receptor-mediated activation of a plant Ca^{2+}-permeable ion channel involved in pathogen defense. Proc Natl Acad Sci USA 94(6):2751–2755
113. Carpaneto A, Ivashikina N, Levchenko V, Krol E, Jeworutzki E, Zhu JK, Hedrich R (2007) Cold transiently activates calcium-permeable channels in *Arabidopsis* mesophyll cells. Plant Physiol 143(1):487–494
114. White PJ, Davenport RJ (2002) The voltage-independent cation channel in the plasma membrane of wheat roots is permeable to divalent cations and may be involved in cytosolic Ca^{2+} homeostasis. Plant Physiol 130(3): 1386–1395
115. Fairley K, Laver D, Walker NA (1991) Whole-cell and single-channel currents across the plasmalemma of corn shoot suspension cells. J Membr Biol 121(1):11–22
116. Roelfsema MRG & Prins HBA (1997) Ion channels in guard cells of *Arabidopsis thaliana* (L) Heynh. Planta 202(1):18–27
117. Clough SJ, Fengler KA, Yu IC, Lippok B, Smith RK, Bent AF (2000) The *Arabidopsis* dnd1 "defense, no death" gene encodes a mutated cyclic nucleotide-gated ion channel. Proc Natl Acad Sci USA 97(16):9323–9328
118. Jurkowski GI, Smith RK, Yu IC, Ham JH, Sharma SB, Klessig DF, Fengler KA, Bent AF (2004) *Arabidopsis* DND2, a second cyclic nucleotide-gated ion channel gene for which mutation causes the "defense, no death" phenotype. Mol Plant Microbe Interact 17(5): 511–520
119. Frietsch S, Wang YF, Sladek C, Poulsen LR, Romanowsky SM, Schroeder JI, Harper JF (2007) A cyclic nucleotide-gated channel is essential for polarized tip growth of pollen. Proc Natl Acad Sci USA 104(36): 14531–14536
120. Lacombe B, Becker D, Hedrich R, DeSalle R, Hollmann M, Kwak JM, Schroeder JI, Le Novere N, Nam HG, Spalding EP, Tester M, Turano FJ, Chiu J, Coruzzi G (2001) The identity of plant glutamate receptors. Science 292(5521):1486–1487
121. Dietrich P, Anschutz U, Kugler A, Becker D (2010) Physiology and biophysics of plant ligand-gated ion channels. Plant Biol 12:80–93
122. Dennison KL, Spalding EP (2000) Glutamate-gated calcium fluxes in *Arabidopsis*. Plant Physiol 124(4):1511–1514
123. Meyerhoff O, Muller K, Roelfsema MR, Latz A, Lacombe B, Hedrich R, Dietrich P, Becker D (2005) AtGLR3.4, a glutamate receptor channel-like gene is sensitive to touch and cold. Planta 222(3):418–427
124. Tapken D, Hollmann M (2008) *Arabidopsis thaliana* glutamate receptor ion channel function demonstrated by ion pore transplantation. J Mol Biol 383(1):36–48
125. Ane JM, Kiss GB, Riely BK, Penmetsa RV, Oldroyd GED, Ayax C, Levy J, Debelle F, Baek JM, Kalo P, Rosenberg C, Roe BA, Long SR, Denarie J, Cook DR (2004) *Medicago truncatula* DMI1 required for bacterial and

fungal symbioses in legumes. Science 303(5662):1364–1367

126. Imaizumi-Anraku H, Takeda N, Charpentier M, Perry J, Miwa H, Umehara Y, Kouchi H, Murakami Y, Mulder L, Vickers K, Pike J, Downie JA, Wang T, Sato S, Asamizu E, Tabata S, Yoshikawa M, Murooka Y, Wu GJ, Kawaguchi M, Kawasaki S, Parniske M, Hayashi M (2005) Plastid proteins crucial for symbiotic fungal and bacterial entry into plant roots. Nature 433(7025):527–531
127. Parniske M (2008) Arbuscular mycorrhiza: the mother of plant root endosymbioses. Nat Rev Microbiol 6(10):763–775
128. Moran N, Ehrenstein G, Iwasa K, Bare C, Mischke C (1984) Ion channels in plasmalemma of wheat protoplasts. Science 226(4676):835–838
129. Schönknecht G, Hedrich R, Junge W, Raschke K (1988) A voltage-dependent chloride channel in the photosynthetic membrane of a higher plant. Nature 336:589–592
130. Pottosin II, Schonknecht G (1996) Ion channel permeable for divalent and monovalent cations in native spinach thylakoid membranes. J Membr Biol 152(3):223–233
131. Hedrich R, Flügge U-I, Fernandez JM (1986) Patch-clamp studies of ion transport in isolated plant vacuoles. FEBS Lett 204:228–232
132. Hedrich R, Neher E (1987) Cytoplasmic calcium regulates voltage-dependent ion channels in plant vacuoles. Nature 329:833–835
133. Conn S, Gilliham M (2010) Comparative physiology of elemental distributions in plants. Ann Bot 105(7):1081–1102
134. Meyer S, De Angeli A, Fernie AR, Martinoia E (2010) Intra- and extra-cellular excretion of carboxylates. Trends Plant Sci 15(1):40–47
135. Martinoia E, Maeshima M, Neuhaus HE (2007) Vacuolar transporters and their essential role in plant metabolism. J Exp Bot 58(1):83–102
136. Hedrich R, Barbier-Brygoo H, Felle H, Flügge UI, Lüttge U, Maathuis FJM, Marx S, Prins HBA, Raschke K, Schnabl H, Schroeder JI, Struve I, Taiz L, Ziegler P (1988) General mechanisms for solute transport across the tonoplast of plant vacuoles: a patch-clamp survey of ion channels and proton pumps. Bot Acta 101:7–13
137. Schulz-Lessdorf B, Hedrich R (1995) Protons and calcium modulate SV-type channels in the vacuolar-lysosomal compartment: channel interaction with calmodulin inhibitors. Planta 197:655–671
138. Bertl A, Blumwald E, Coronado R, Eisenberg R, Findlay G, Gradmann D, Hille B, Kohler K, Kolb HA, MacRobbie E et al (1992) Electrical measurements on endomembranes. Science 258(5084):873–874
139. Hedrich R, Marten I (2011) TPC1–SV channels gain shape. Mol Plant 4(3):428–441
140. Ivashikina N, Hedrich R (2005) K^+ currents through SV-type vacuolar channels are sensitive to elevated luminal sodium levels. Plant J 41(4):606–614
141. Pottosin II, Tikhonova LI, Hedrich R, Schönknecht G (1997) Slowly activating vacuolar channels cannot mediate Ca^{2+}-induced Ca^{2+} release. Plant J 12(6):1387–1398
142. Pei ZM, Ward JM, Schroeder JI (1999) Magnesium sensitizes slow vacuolar channels to physiological cytosolic calcium and inhibits fast vacuolar channels in fava bean guard cell vacuoles. Plant Physiol 121(3):977–986
143. Carpaneto A, Cantu AM, Gambale F (2001) Effects of cytoplasmic Mg^{2+} on slowly activating channels in isolated vacuoles of *Beta vulgaris*. Planta 213(3):457–468
144. Pottosin II, Martinez-Estevez M, Dobrovinskaya OR, Muniz J, Schonknecht G (2004) Mechanism of luminal Ca^{2+} and Mg^{2+} action on the vacuolar slowly activating channels. Planta 219(6):1057–1070
145. Pottosin II, Martinez-Estevez M, Dobrovinskaya OR, Muniz J (2005) Regulation of the slow vacuolar channel by luminal potassium: role of surface charge. J Membr Biol 205(2):103–111
146. Perez V, Wherrett T, Shabala S, Muniz J, Dobrovinskaya O, Pottosin I (2008) Homeostatic control of slow vacuolar channels by luminal cations and evaluation of the channel-mediated tonoplast Ca^{2+} fluxes in situ. J Exp Bot 59(14):3845–3855
147. Beyhl D, Hortensteiner S, Martinoia E, Farmer EE, Fromm J, Marten I, Hedrich R (2009) The *fou2* mutation in the major vacuolar cation channel TPC1 confers tolerance to inhibitory luminal calcium. Plant J 58(5): 715–723
148. Peiter E, Maathuis FJ, Mills LN, Knight H, Pelloux J, Hetherington AM, Sanders D (2005) The vacuolar Ca^{2+}-activated channel TPC1 regulates germination and stomatal movement. Nature 434(7031):404–408
149. Furuichi T, Cunningham KW, Muto S (2001) A putative two pore channel AtTPC1 mediates Ca^{2+} flux in *Arabidopsis* leaf cells. Plant Cell Physiol 42(9):900–905
150. Hashimoto K, Saito M, Matsuoka H, Iida K, Iida H (2004) Functional analysis of a rice putative voltage-dependent Ca^{2+} channel, *Os*TPC1, expressed in yeast cells lacking its

homologous gene CCH1. Plant Cell Physiol 45(4):496–500

151. Kadota Y, Furuichi T, Ogasawara Y, Goh T, Higashi K, Muto S, Kuchitsu K (2004) Identification of putative voltage-dependent Ca^{2+}-permeable channels involved in cryptogein-induced Ca^{2+} transients and defense responses in tobacco BY-2 cells. Biochem Biophys Res Commun 317(3):823–830
152. Wang YJ, Yu JN, Chen T, Zhang ZG, Hao YJ, Zhang JS, Chen SY (2005) Functional analysis of a putative Ca^{2+} channel gene *Ta*TPC1 from wheat. J Exp Bot 56(422):3051–3060
153. Ruas M, Rietdorf K, Arredouani A, Davis LC, Lloyd-Evans E, Koegel H, Funnell TM, Morgan AJ, Ward JA, Watanabe K, Cheng X, Churchill GC, Zhu MX, Platt FM, Wessel GM, Parrington J, Galione A (2010) Purified TPC isoforms form NAADP receptors with distinct roles for Ca^{2+} signaling and endolysosomal trafficking. Curr Biol 20(8):703–709
154. Yoo SD, Cho YH, Sheen J (2007) *Arabidopsis* mesophyll protoplasts: a versatile cell system for transient gene expression analysis. Nat Protoc 2(7):1565–1572
155. Bonaventure G, Gfeller A, Rodriguez VM, Armand F, Farmer EE (2007) The *fou2* gain-of-function allele and the wild-type allele of two pore channel 1 contribute to different extents or by different mechanisms to defense gene expression in *Arabidopsis*. Plant Cell Physiol 48(12):1775–1789
156. Dadacz-Narloch B, Beyhl D, Larisch C, López-Sanjurjo E, Reski R, Kuchitsu K, Müller T, Becker D, Schoenknecht G, Hedrich R (2011) A novel calcium binding site in the slow vacuolar cation channel TPC1 senses luminal calcium levels Plant Cell 23(7):2696–2707
157. Walker DJ, Leigh RA, Miller AJ (1996) Potassium homeostasis in vacuolate plant cells. Proc Natl Acad Sci USA 93(19): 10510–10514
158. Becker D, Geiger D, Dunkel M, Roller A, Bertl A, Latz A, Carpaneto A, Dietrich P, Roelfsema MR, Voelker C, Schmidt D, Mueller-Roeber B, Czempinski K, Hedrich R (2004) *At*TPK4, an *Arabidopsis* tandem-pore K^+ channel, poised to control the pollen membrane voltage in a pH- and Ca^{2+}-dependent manner. Proc Natl Acad Sci USA 101(44):15621–15626
159. Latz A, Becker D, Hekman M, Muller T, Beyhl D, Marten I, Eing C, Fischer A, Dunkel M, Bertl A, Rapp UR, Hedrich R (2007) TPK1, a Ca^{2+}-regulated *Arabidopsis* vacuole two-pore K^+ channel is activated by 14-3-3 proteins. Plant J 52(3):449–459
160. Hedrich R, Kurkdjian A (1988) Characterization of an anion-permeable channel from sugar beet vacuoles: effect of inhibitors. EMBO J 7(12): 3661–3666
161. Krebs M, Beyhl D, Gorlich E, Al-Rasheid KA, Marten I, Stierhof YD, Hedrich R, Schumacher K (2010) *Arabidopsis* V-ATPase activity at the tonoplast is required for efficient nutrient storage but not for sodium accumulation. Proc Natl Acad Sci USA 107(7):3251–3256
162. Wingenter K, Schulz A, Wormit A, Wic S, Trentmann O, Hoermiller II, Heyer AG, Marten I, Hedrich R, Neuhaus HE (2010) Increased activity of the vacuolar monosaccharide transporter TMT1 alters cellular sugar partitioning, sugar signaling, and seed yield in *Arabidopsis*. Plant Physiol 154(2): 665–677
163. Wormit A, Trentmann O, Feifer I, Lohr C, Tjaden J, Meyer S, Schmidt U, Martinoia E, Neuhaus HE (2006) Molecular identification and physiological characterization of a novel monosaccharide transporter from *Arabidopsis* involved in vacuolar sugar transport. Plant Cell 18(12):3476–3490
164. Schulz A, Beyhl D, Marten I, Wormit A, Neuhaus E, Poschet G, Büttner M, Schneider S, Sauer N, Hedrich R. Plant J. 2011 Oct;68(1):129–136. doi: 10.1111/j.1365-313X.2011.04672.x. Epub 2011 Jul 27.
165. Ho CH, Lin SH, Hu HC, Tsay YF (2009) CHL1 functions as a nitrate sensor in plants. Cell 138(6):1184–1194
166. Hrabak EM, Chan CWM, Gribskov M, Harper JF, Choi JH, Halford N, Kudla J, Luan S, Nimmo HG, Sussman MR, Thomas M, Walker-Simmons K, Zhu JK, Harmon AC (2003) The *Arabidopsis* CDPK-SnRK superfamily of protein kinases. Plant Physiol 132(2):666–680
167. Cutler SR, Rodriguez PL, Finkelstein RR, Abrams SR (2010) Abscisic acid: emergence of a core signaling network. Annu Rev Plant Biol 61:651–679
168. Marten I, Deeken R, Hedrich R, Roelfsema MRG (2010) Light-induced modification of plant plasma membrane ion transport. Plant Biology 12:64–79
169. Boller T, Felix G (2009) A renaissance of elicitors: perception of microbe-associated molecular patterns and danger signals by pattern-recognition receptors. Annu Rev Plant Biol 60:379–406
170. Marten H, Konrad KR, Dietrich P, Roelfsema MR, Hedrich R (2007) Ca^{2+}-dependent and -independent abscisic acid activation of plasma membrane anion channels in guard cells of

Nicotiana tabacum. Plant Physiol 143(1): 28–37

171. Wang XQ, Ullah H, Jones AM, Assmann SM (2001) G protein regulation of ion channels and abscisic acid signaling in *Arabidopsis* guard cells. Science 292(5524):2070–2072
172. Weschke W, Panitz R, Sauer N, Wang Q, Neubohn B, Weber H, Wobus U (2000) Sucrose transport into barley seeds: molecular characterization of two transporters and implications for seed development and starch accumulation. Plant J 21(5):455–467
173. Hosy E, Vavasseur A, Mouline K, Dreyer I, Gaymard F, Poree F, Boucherez J, Lebaudy A, Bouchez D, Very AA, Simonneau T, Thibaud JB, Sentenac H (2003) The *Arabidopsis* outward K^+ channel GORK is involved in regulation of stomatal movements and plant transpiration. Proc Natl Acad Sci USA 100(9): 5549–5554
174. Munemasa S, Hossain MA, Nakamura Y, Mori IC, Murata Y (2011) The *Arabidopsis* calcium-dependent protein kinase, CPK6, functions as a positive regulator of methyl jasmonate signaling in guard cells. Plant Physiol 155(1): 553–561
175. Pei ZM, Kuchitsu K, Ward JM, Schwarz M, Schroeder JI (1997) Differential abscisic acid regulation of guard cell slow anion channels in *Arabidopsis* wild-type and *abi1* and *abi2* mutants. Plant Cell 9(3):409–423
176. Pei ZM, Ghassemian M, Kwak CM, McCourt P, Schroeder JI (1998) Role of farnesyltransferase in ABA regulation of guard cell anion channels and plant water loss. Science 282(5387):287–290
177. Sasaki T, Mori IC, Furuichi T, Munemasa S, Toyooka K, Matsuoka K, Murata Y, Yamamoto Y (2010) Closing plant stomata requires a homolog of an aluminum-activated malate transporter. Plant Cell Physiol 51(3):354–365
178. Romano LA, Miedema H, Assmann SM (1998) Ca^{2+}-permeable, outwardly-rectifying K^+ channels in mesophyll cells of *Arabidopsis thaliana*. Plant Cell Physiol 39(11):1133–1144
179. Qi Z, Kishigami A, Nakagawa Y, Iida H, Sokabe M (2004) A mechanosensitive anion channel in *Arabidopsis thaliana* mesophyll cells. Plant Cell Physiol 45(11):1704–1708
180. Diatloff E, Roberts M, Sanders D, Roberts SK (2004) Characterization of anion channels in the plasma membrane of *Arabidopsis* epidermal root cells and the identification of a citrate-permeable channel induced by phosphate starvation. Plant Physiol 136(4):4136–4149
181. Shabala S, Demidchik V, Shabala L, Cuin TA, Smith SJ, Miller AJ, Davies JM, Newman IA (2006) Extracellular Ca^{2+} ameliorates NaCl-induced K^+ loss from *Arabidopsis* root and leaf cells by controlling plasma membrane K^+-permeable channels. Plant Physiol 141(4): 1653–1665
182. Demidchik V, Cuin TA, Svistunenko D, Smith SJ, Miller AJ, Shabala S, Sokolik A, Yurin V (2010) *Arabidopsis* root K^+-efflux conductance activated by hydroxyl radicals: single-channel properties, genetic basis and involvement in stress-induced cell death. J Cell Sci 123(9):1468–1479
183. Philippar K, Ivashikina N, Ache P, Christian M, Luthen H, Palme K, Hedrich R (2004) Auxin activates KAT1 and KAT2, two K^+-channel genes expressed in seedlings of *Arabidopsis thaliana*. Plant J 37(6):815–827
184. Cho MH, Spalding EP (1996) An anion channel in *Arabidopsis* hypocotyls activated by blue light. Proc Natl Acad Sci USA 93(15): 8134–8138
185. Lewis BD, KarlinNeumann G, Davis RW, Spalding EP (1997) Ca^{2+}-activated anion channels and membrane depolarizations induced by blue light and cold in *Arabidopsis* seedlings. Plant Physiol 114(4):1327–1334
186. Colcombet J, Lelievre F, Thomine S, Barbier-Brygoo H, Frachisse JM (2005) Distinct pH regulation of slow and rapid anion channels at the plasma membrane of *Arabidopsis thaliana* hypocotyl cells. J Exp Bot 56(417):1897–1903
187. Thomine S, Zimmermann S, Guern J, BarbierBrygoo H (1995) ATP-dependent regulation of an anion channel at the plasma membrane of protoplasts from epidermal cells of *Arabidopsis* hypocotyls. Plant Cell 7(12):2091–2100
188. Frachisse JM, Thomine S, Colcombet J, Guern J, Barbier-Brygoo H (1999) Sulfate is both a substrate and an activator of the voltage-dependent anion channel of *Arabidopsis* hypocotyl cells. Plant Physiol 121(1):253–261
189. Deeken R, Ivashikina N, Czirjak T, Philippar K, Becker D, Ache P, Hedrich R (2003) Tumour development in *Arabidopsis thaliana* involves the Shaker-like K^+ channels AKT1 and AKT2/3. Plant J 34(6):778–787
190. Ghelis T, Dellis O, Jeannette E, Bardat F, Cornel D, Miginiac E, Rona JP, Sotta B (2000) Abscisic acid specific expression of RAB18 involves activation of anion channels in *Arabidopsis thaliana* suspension cells. FEBS Lett 474(1):43–47
191. Lew RR (1991) Substrate regulation of single potassium and chloride-ion channels in *Arabidopsis* plasma-membrane. Plant Physiol 95(2):642–647

192. Rienmuller F, Beyhl D, Lautner S, Fromm J, Al-Rasheid KA, Ache P, Farmer EE, Marten I, Hedrich R (2010) Guard cell-specific calcium sensitivity of high density and activity SV/TPC1 channels. Plant Cell Physiol 51(9):1548–1554
193. Ranf S, Wunnenberg P, Lee J, Becker D, Dunkel M, Hedrich R, Scheel D, Dietrich P (2008) Loss of the vacuolar cation channel, AtTPC1, does not impair Ca^{2+} signals induced by abiotic and biotic stresses. Plant J 53(2): 287–299
194. Scholz-Starke J, Carpaneto A, Gambale F (2006) On the interaction of neomycin with the slow vacuolar channel of *Arabidopsis thaliana*. J Gen Physiol 127(3):329–340
195. De Angeli A, Monachello D, Ephritikhine G, Frachisse JM, Thomine S, Gambale F, Barbier-Brygoo H (2006) The nitrate/proton antiporter AtCLCa mediates nitrate accumulation in plant vacuoles. Nature 442(7105):939–942
196. von der Fecht-Bartenbach J, Bogner M, Dynowski M, Ludewig U (2010) CLC-b-mediated NO_3^-/H^+ exchange across the tonoplast of *Arabidopsis* vacuoles. Plant Cell Physiol 51(6):960–968
197. Kovermann P, Meyer S, Hortensteiner S, Picco C, Scholz-Starke J, Ravera S, Lee Y, Martinoia E (2007) The *Arabidopsis* vacuolar malate channel is a member of the ALMT family. Plant J 52(6):1169–1180
198. Hurth MA, Suh SJ, Kretzschmar T, Geis T, Bregante M, Gambale F, Martinoia E, Neuhaus HE (2005) Impaired pH homeostasis in *Arabidopsis* lacking the vacuolar dicarboxylate transporter and analysis of carboxylic acid transport across the tonoplast. Plant Physiol 137(3):901–910

Chapter 20

Automated Patch Clamping

Kohei Sawada and Takashi Yoshinaga

Abstract

Patch-clamping is an essential technique in the field of ion channel drug discovery and safety pharmacology. Conventional patch-clamp methods provide high data quality and can be applied to a wide variety of cells, but the methods are laborious and time-consuming. Automated patch-clamp (APC) systems in which multi-well planar patch plates are used have been developed to overcome these disadvantages, and they allow data from hundreds to thousands of cells to be obtained in a single day. The recording success rate highly depends on cell types and cell conditions because cells are positioned on the patch holes in a blind manner and only cells that fit to patch holes can be used for recording. The first APC system that was developed, the IonWorks HT system, has been used widely, but it has limitations in regard to success rates and high-throughput ability. The next system, IonWorks Quattro system, uses a population patch-clamp method, and it has provided a high success rate and higher throughput ability. Sophion's QPatch system yielded data whose quality is similar to that of conventional patch-clamp methods, although precise adjustment of the experimental conditions is necessary to obtain high success rates. This chapter describes the principle of APC systems and their characteristics.

20.1. Introduction

One of the initial hurdles for novices in ion channel research is mastering the conventional patch-clamp techniques. A variety of skills such as those required for cell preparation, fabrication of patch electrodes, sophisticated operation of micromanipulators to make giga-ohm seal formation between electrodes and cell membranes under a microscope, and for adjustment of negative pressure or Zap pulse levels to rupture the patch membrane must be learned to become a good patch-clamper. Even after acquiring these skills during the course of a long training period, patch-clamp experiments remain laborious work.

Investigators engaged in ion channel research thought that the work involved in conducting ion channel research, especially research related to drug discovery, would be greatly facilitated if all

Yasunobu Okada (ed.), *Patch Clamp Techniques: From Beginning to Advanced Protocols*, Springer Protocols Handbooks, DOI 10.1007/978-4-431-53993-3_20, © Springer 2012

of these experimental operations were automated and the technical hurdles were lowered. Although various patch-clamp automation methods have been explored during the past decade, the major automated methods currently used are planar patch-clamp methods (1, 2). The procedure used to perform planar patch-clamp methods can be outlined as follows. Cells are poured into a chamber such as a bathtub with a drain hole in the bottom. After plugging the hole with a cell by applying suction below the hole, a whole-cell patch configuration is established by rupturing the patch membrane. The solution in the bathtub is the extracellular solution, and the solution underneath the drain hole is the intracellular solution.

Automated patch-clamp (APC) methods have more technical limitations than conventional patch-clamp methods. Only cells of suitable size and shape to fit the hole can be used, and it is often difficult to obtain and maintain high giga-ohm seals. The current clamp mode is available in only a few APC systems. Therefore, although the APC methods cannot completely replace conventional patch-clamp methods, the high throughput ability of an APC system – as many as several thousand data points a day – greatly facilitates ion channel drug discovery.

20.2. Principle

The recording chambers have very small volumes, ranging from 10 to 100 μl, and a hole approximately 1 μm in diameter at the bottom (Fig. 20.1). Several microliters of cell suspension solution containing dispersed cells at a concentration of $1–4 \times 10^6$ cells/ml is added to the chamber, and a cell that comes onto the hole is tightly locked there by mild suction, forming a cell-attached clamp configuration. A whole-cell clamp configuration can be established either by rupturing the patch membrane by applying suction or Zap pulses or by applying an ionophore such as amphotericin B (perforated-patch configuration). Once the cell suspension, compound solutions, and microplates have been set up in the machine, experiments can be carried out automatically under the control of computer programs.

20.3. Cells

Cell lines used for heterologous expression of ion channels are generally used in APC experiments, although they are not limited to the established cell lines. HEK293, CHO, CHL, and BHK cells are the main adhesive cells that are used; Jurkat cells, which are nonadhesive, are also used. The rates of success in patch-clamp

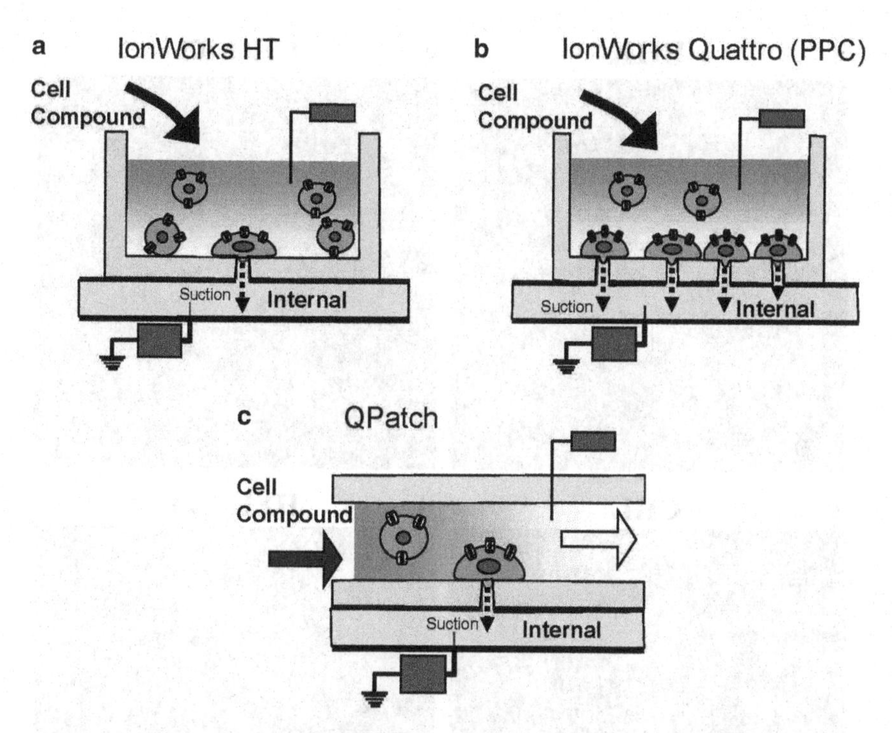

Fig. 20.1. Recoding chambers of automated patch-clamp systems. (**a**) IonWorks HT. Each well has a single hole at the bottom. Compounds are added to the recording chamber from above. (**b**) IonWorks Quattro. Each well has 64 holes at its bottom, and the average current from 64 cells is recorded. This is the population patch-clamp (PPC) method. (**c**) QPatch. Extracellular solutions can be replaced by perfusion.

recording vary with the cell lines. A difference in cell size between cell lines, size variation within the same cell line, adhesive properties, and the nature of the cell surface are likely to affect the success rate.

Differences in the shapes of BHK, CHO-K1, CHL, and HEK-293 cells are shown in Fig. 20.2. The cells shown in Fig. 20.2 were isolated by application of ethylenediamine tetraacetic acid (EDTA) or trypsin/EDTA for 3–10 min at 37°C. CHO-K1 cells and CHL cells are similar morphologically, although CHL cells are the smallest of all of the cell lines examined. Their shape is globular and uniform. Many BHK cells are also globular, but they are larger than the CHL and CHO-K1 cells. On the other hand, many HEK293 cells are ellipsoidal, and they vary in size. The order of these cell lines in terms of the current recording success rate is CHL, CHO > BHK > HEK293. The current recording success rate of HEK293 cells increased greatly when the cells were incubated at 37°C for 30 min in the culture medium after treatment with trypsin/EDTA.

Other factors affecting the success rates range from the assay solution to the cell clone. Osmolarity is an important factor in regulating the cell shape, and it should be checked when the assay solutions are prepared. Uniformity of cell size and shape should be carefully controlled because the cells for current recording in APC experiments are selected in a random manner. Because the compatibility between the cells and APC system is a key factor, selection of

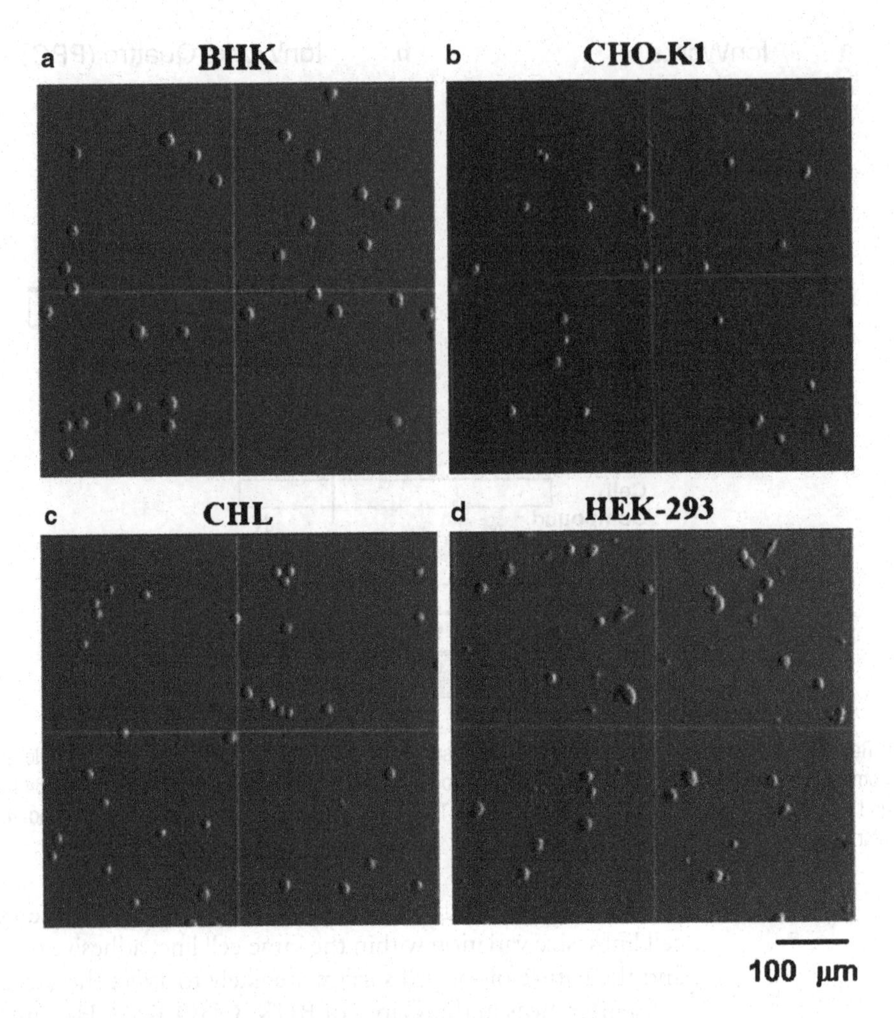

Fig. 20.2. Morphology of freshly isolated cells. Cultured cells were isolated by treatment with ethylenediamine tetraacetic acid (EDTA) or trypsin/EDTA, and their morphology was examined under a microscope.

the most suitable cell clone for the APC system used in the experiment is desirable. The nature of cell clones (e.g., stable expression of channels after repeated freezing and thawing or passage) and a moderate proliferation rate are also important because ion channel studies performed by APC methods often require many cells over a long period of time.

20.4. Characteristics of APC Systems

20.4.1. IonWorks HT System

IonWorks HT system was the first commercially available APC machine. The patch plate has 384 wells, and each well has a hole at its bottom (Fig. 20.1a). The intracellular solution chamber is located under the plate, and a cell is positioned over the hole by negative pressure in the intracellular chamber. After a tight seal has been formed, the intracellular solution is replaced with a solution containing an ionophore to establish a whole-cell patch configuration.

Because the seal resistance does not usually reach the giga-ohm level in this system, leak current subtraction is required for current recording and analysis. The leak current subtraction procedure is performed as follows: A small step pulse that does not activate target channels is applied from a holding potential. The leak current–voltage relation is linearly extrapolated, and the calculated leak current is subtracted from the current elicited by test pulses. There is no series resistance or capacitive compensation function. Because only two additions of compound solutions are available in the same well, it is difficult to obtain full concentration–response relation curves using the same cell. Despite these limitations, this system has been widely used to examine the effects of compounds on a variety of ion channels including hERG and Nav1.5. Moreover, it has yielded results that are essentially the same as the results obtained by conventional patch-clamp methods and by other APC systems.

Figure 20.3a shows the effects of flecainide on the Nav1.5 current. The recorded peak inward current was not completely accurate because no series resistance or capacitance compensation was performed, and the maximum sampling frequency was 10 kHz. However, the concentration–response curve for flecainide was similar to the curve obtained with the QPatch system, which has better recording quality (Fig. 20.3d).

The leak subtraction method cannot be applied to non-voltage-gated channels such as Kir channels, which require that a test compound solution be added first and a specific blocker solution be applied in the second addition without a leak subtraction function. The effects of the test compound can be assessed by comparing the current amplitudes before adding drugs and in the presence of a test compound after subtraction of the residual current in the presence of a specific blocker.

The failure in current recoding and inter-well variation in current amplitude are annoying problems in the 384-well measurement. Figure 20.4 shows current traces and amplitude variation of the hERG current recorded with the IonWorks HT system and the Quattro system. The level of hERG channel expression in the cell line shown in Fig. 20.4 is very high, but the success of recording with the IonWorks HT system was about 80%, and there was wide variation in the current amplitude. The ultimate success rate was lower, because the wells of small current levels were excluded from the analysis to maintain the reliability of the results. Because four wells are used to examine the same compound solution to attenuate the risk for data loss, the maximum number of compounds that can be assessed in a single plate is 96. This implies that the high throughput ability is not fully utilized in the IonWorks HT system. Population patch-clamp (PPC) methods have been developed to overcome these problems (3), and a system that uses this method, the IonWorks Quattro system, is now commercially available.

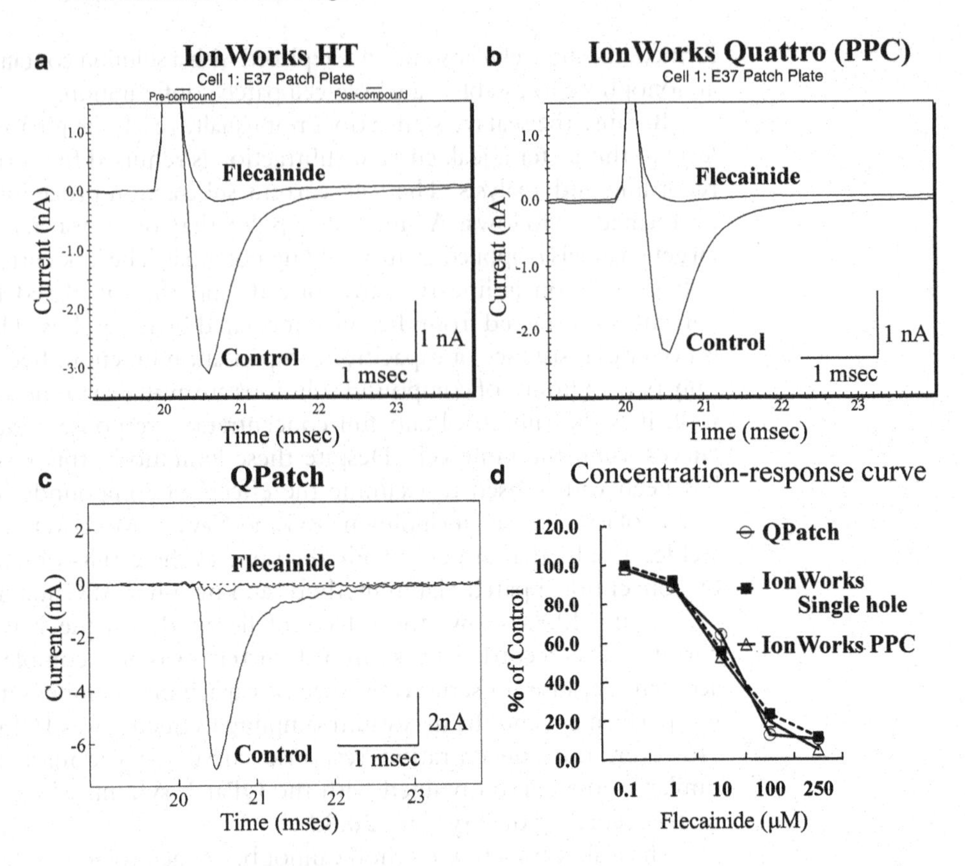

Fig. 20.3. Inhibitory effects of flecainide on the Nav1.5 current. Nav1.5 currents were elicited with a step pulse to −10 mV from a holding potential of −100 mV in CHL cells stably expressing Nav1.5 channels. (**a–c**) Superimposition of the baseline current trace and the current trace recorded after complete inhibition by 500 μM of flecainide. (**d**) Concentration–response curves for the inhibitory activity of flecainide at concentrations ranging from 0.1 to 250 μM.

20.4.2. IonWorks Quattro System

The basic principle of the IonWorks Quattro system is identical to that of the IonWorks HT system. The most striking difference between the systems is that each well of patch plates for PPC methods has 64 holes in its bottom (Fig. 20.1b). Even if 64 cells are correctly positioned over 64 holes, the series resistance value is lowered to several tens of mega-ohms because the inverse of the resistance across the well and internal solution is equal to the sum of the inverse of the resistance of the 64 individual holes. The current trace is almost the same as that recorded with the IonWorks HT system because the total current recorded from the 64 cells is divided by 64 and displayed. However, the inter-well variation in current amplitude is much lower because the current represents the average of the 64 cells (Fig. 20.4c).

Furthermore, the failure rate is significantly lower because the effects of nonexpressing cells and poorly sealed holes are minimized. The recording success rate for hERG-expressing cells was 97%, and about 370 compounds could be tested in a single experiment. It takes about 90 min to perform a single experiment. Therefore,

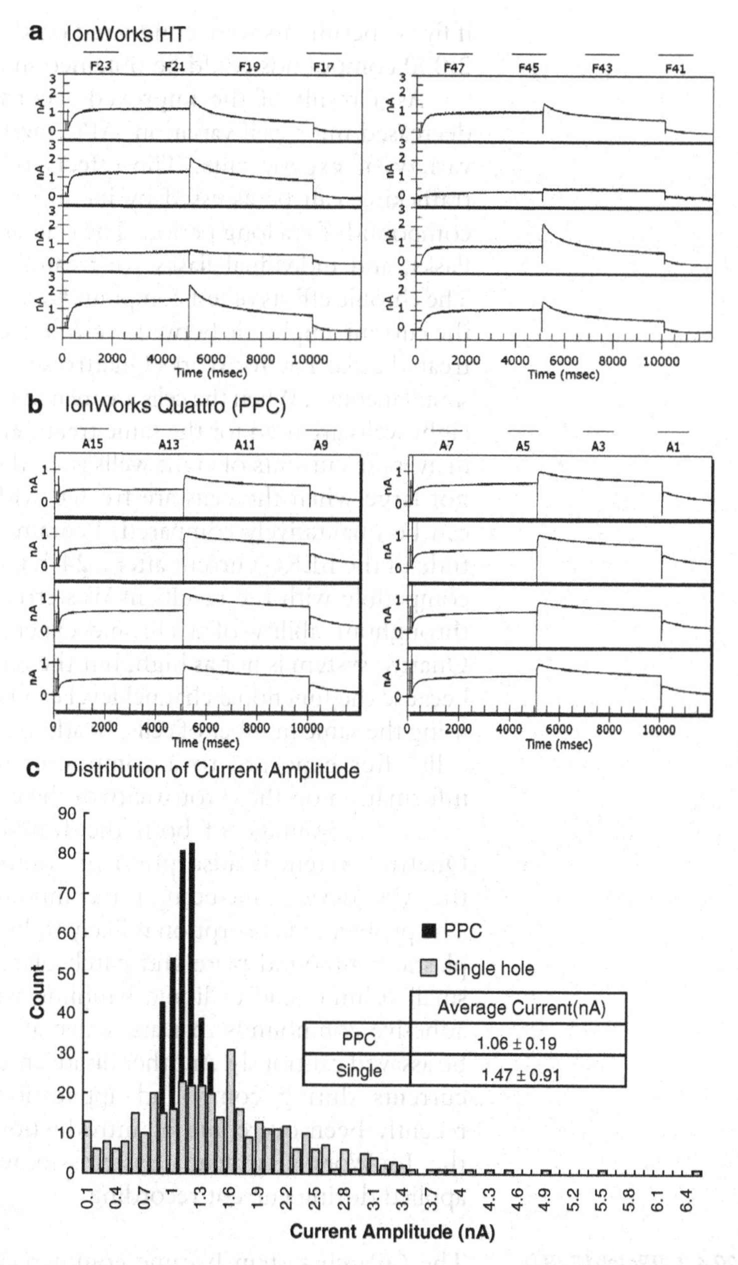

Fig. 20.4. Variation in the hERG current amplitude. The hERG current was elicited by a step pulse to +20 mV for 5 s from a holding potential of −90 mV. The amplitude of the hERG current was measured as the tail current on repolarization to −50 mV from a step pulse to +20 mV. (**a**) Current traces obtained from eight single-well holes. A large variation in current amplitude was observed. (**b**) Current traces of eight wells recorded by the population patch clamp (PPC) mode. The current amplitudes recorded from the well were almost the same. (**c**) Comparison of current amplitude distribution recorded by the single-hole mode and the PPC mode. The variation in current amplitude recorded by the PPC mode was smaller than when recorded by the single-hole mode.

if five experiments were conducted on the same day, data for about 2,000 compounds could be obtained in a single day.

As a result of the improved recording success rate and the decreased inter-well variation, APC methods are now applied to a variety of experiments. The effects of compounds on channel trafficking can be assessed by incubating cells in the presence of compounds for a long period. The cells are cultured in small culture flasks, and individual flasks are treated with various compounds. The chronic effects of test compounds can be evaluated by comparing the current amplitude between vehicle-treated cells and compound-treated cells. The IonWorks Quattro system can test 48 kinds of cell simultaneously. When the cells are treated with 48 kinds of compound, eight wells are used for the same treatment. Because the differences in average currents of eight wells (i.e., the average of 512 cells) are not large when the cells are treated with vehicle, 48 compounds can be qualitatively compared. Pentamidine decreased the amplitude of the hERG current after a 24-h incubation, a result that was compatible with the results in Western blot experiments (4). The throughput ability of a chronic experiment with the IonWorks Quattro system is not as high, but the results are extremely reliable because the functional channel level can be estimated and compared using the same number of cells. In addition, the number of surviving cells after exposure to a compound for 24 h provides useful information on the cytotoxicity of the compound.

A disadvantage of both the IonWorks HT system and the Quattro system is adsorption of compounds on the surface of the APC devices, including the compound plate and patch plates. The problem of adsorption is likely to be mainly attributable to the plastic compound plate and patch plate, which has wells of very small volume, and to liquid handling with plastic tips. Especially, adhesive compounds that are active at low concentrations should be assayed cautiously. Another disadvantage is an inability to record currents during compound application, but this problem has recently been overcome by introduction of a novel APC system, the IonWorks Barracuda system, in which compounds can be applied during current recording.

20.4.3. QPatch16 (HT) System

The QPatch system became commercially available at nearly the same time as the other APC systems (5). Only the whole-cell clamp mode is available with this system, but the procedure for voltage-clamp recording is not different from conventional patch-clamp methods in that multiple application of compound solutions and continuous current recording with programmed voltage protocols can be carried out. Full concentration–response curves and inactivation and activation curves can be constructed from the recordings made in a single cell. Because compounds can be applied during voltage control, effects of compounds on ligand-gated channels can be studied. This function broadened the application of APC systems in ion channel drug discovery.

The patch plate used in the QPatch16 system has 16 wells (48 wells in the QPatch HT system), a fluid path for the intracellular solution, a fluid path for the extracellular solution, and a small chamber with a hole at the intersection between the two paths (Fig. 20.1c). After the integrity of the hole has been confirmed by applying the test voltage, a small volume of cell suspension solution is allowed to flow into the chamber. A cell is drawn over the hole by negative pressure from the internal solution compartment, and after formation of the giga-ohm seal the whole-cell clamp mode is established by breaking the patch membrane (by applying additional negative pressure or Zap). All of these procedures and fine-tuning the negative pressure or Zap levels are under software control and are implemented automatically.

Series resistance or capacitance compensation functions and leak-subtraction functions are available, and the maximum sampling frequency is 50 kHz. Thus, as shown in Fig. 20.3c, sodium channel currents, which have rapid channel opening and closing kinetics, can be recorded with the same quality as is obtained with conventional patch-clamp methods.

When six concentrations of a compound are applied sequentially to a single well and recording from every well is successful, a 16-well plate can provide 96 data points in a single experiment. The QPatch system is equipped with a cell suspension solution reservoir, and more than ten patch plates can be set. When the experiment with one plate has been completed, the next plate is mounted, and cells are dispensed automatically. If a large volume of cell suspension has been prepared, experiments can be continued until all of the cell suspensions or plates are used. However, because the condition of the cells deteriorates over time, experiments for five plates in one cycle is generally conducted, assuming that one plate experiment takes 1 h. If every experiment is successful, a total of 480 data points are obtained with the QPatch16 system and 1,440 data points with the QPatch HT system. Because the success rate depends on the type of host cells, clones of expressing cells, and the protocol, these conditions must be precisely adjusted to achieve a high success rate.

The inner surfaces of the flow paths and the chamber are glass-coated, ensuring that less of the compound is adsorbed than is found on plastic surfaces. If glass compound plates are used to assay adhesive compounds, the effect of adsorption is minimized. When repeated application of the same concentration of adhesive compound is conducted until a steady effect is attained, the results obtained are almost the same as those obtained with conventional patch-clamp techniques.

20.5. Conclusion

Although APC techniques have achieved an important place in studies of ion channel drugs, improvements in terms of reliability, success rate, and expansion of the application have been pursued. The cells used in APC systems have been expanded to include acute isolated primary cells and cardiomyocytes derived from human ES/iPS cells (stem cells). However, many problems remain to be solved, including how to acquire large numbers of cells, how to establish tight seals for cells that are not globular in shape, and how to improve the overall success rate.

References

1. Wang X, Li M (2003) Automated electrophysiology: high throughput of art. Assay Drug Dev Technol 1:695–708
2. Dunlop J, Bowlby M, Peri R, Vasilyev D, Arias R (2008) High-throughput electrophysiology: an emerging paradigm for ion-channel screening and physiology. Nat Rev Drug Discov 7:358–368
3. Finkel A, Wittel A, Yang N, Handran S, Hughes J, Costantin J (2006) Population patch clamp improves data consistency and success rates in the measurement of ionic currents. J Biomol Screen 11:488–496
4. Kuryshev YA, Ficker E, Wang L, Hawryluk P, Dennis AT, Wible BA, Brown AM, Kang J, Chen XL, Sawamura K, Reynolds W, Rampe D (2005) Pentamidine-induced long QT syndrome and block of hERG trafficking. J Pharmacol Exp Ther 312:316–323
5. Mathes C (2006) QPatch: the past, present and future of automated patch clamp. Expert Opin Ther Targets 10:319–327

Chapter 21

Patch-Clamp Biosensor Method

Seiji Hayashi and Makoto Tominaga

Abstract

Cells or membranes with ligand-gated ion channels endogenously expressed in cultured cells or heterologously expressed in HEK293 cells can be used as biosensors with a patch-clamp technique to detect small amounts of locally released bioactive substances that are known to activate ionotropic receptors with high affinity. We describe a detailed procedure for a patch-clamp biosensor method by focusing on the detection of ATP using undifferentiated PC12 and HEK293 cells expressing $P2X_2$ ion channels.

21.1. Introduction

A biosensor is generally considered to be an analytical device that incorporates biological "sensing" elements to detect selectively small amounts of physiologically relevant molecules. It can be used for broad applications, including clinical diagnosis, food engineering, and environmental pollution monitoring. As a biological "sensing" elements, biological materials such as microorganisms, enzymes, and antibodies have been employed as microbial sensors, enzyme electrodes, and immune sensors, respectively. As specific receptors for various physiologically reactive substances are present on the plasma membranes of most living cells, measurable responses can be detected as a result of their activation when a specific ligand binds to its receptor.

Most plasma membrane receptors can be classified into two large categories, metabotropic and ionotropic. Metabotropic receptors coupled to guanine nucleotide-binding proteins (G proteins) share a common architecture consisting of seven transmembrane domains. They can activate target effectors directly or indirectly through various signal transduction pathways. Responses of ion channels activated by metabotropic receptors are usually slow. In contrast, because ionotropic receptors are themselves ion

Yasunobu Okada (ed.), *Patch Clamp Techniques: From Beginning to Advanced Protocols*, Springer Protocols Handbooks, DOI 10.1007/978-4-431-53993-3_21, © Springer 2012

channels, their resulting responses are fast. Thus, ligand-gated ion channels in isolated patch membranes in an outside-out configuration or in entire cell plasma membranes under whole-cell configuration can serve as biosensors that can detect neurotransmitters released from nerve endings or bioactive substances released from endocrine or exocrine cells in real time. In this chapter, we describe a cell-based biosensor method for real-time detection of adenosine 5′-triphosphate (ATP) locally released from target cells.

21.2. Outline of the Patch-Clamp Biosensor Method

This method is based on the patch-clamp technique using outside-out or whole-cell configuration to measure ionic currents caused by activation of ligand-gated cation channels. There are many kinds of receptor on the membrane surface of the cell, including receptors for nicotinic acetylcholine, γ-aminobutyric acid (GABA), glycine, serotonin (5-HT_3), glutamate [*N*-methyl-D-aspartic acid (NMDA), α-amino-3-hydroxy-5-methylisoxazole-4-propionic acid (AMPA), kainic acid], and ATP. These receptors are activated by their selective bioactive substances. To measure local concentrations of bioactive substances in the vicinity of the cell surface in real time, a sensor cell or patch membrane is placed in close proximity to the target cell, and the ligand-gated currents induced by bioactive substances released from the target cell surface are measured under outside-out or whole-cell configuration with a patch-clamp technique.

21.2.1. Patch-Clamp Biosensor by Patch Membrane

In 1983, two reports describing the biosensor method using a patch-clamp technique appeared simultaneously. In these reports, patch membranes containing nicotinic acetylcholine (ACh) receptors excised from skeletal muscle cells of chick (1) or *Xenopus* (2) were used as a biosensor that could detect ACh released from the growth cones of embryonic chick ganglion neurons or embryonic *Xenopus* neurites, respectively (Fig. 21.1). Notably, the establishment of a stable outside-out configuration with the patch-clamp technique was more difficult than that of a stable whole-cell configuration (see Chap. 2).

21.2.2. Patch-Clamp Biosensor in Whole-Cell Mode

Similar to the case of patch membrane biosensors, cells in which ligand-gated cation channels are expressed or stably transfected with recombinant cDNAs can be used as biosensors in whole-cell mode, as originally reported in 1987 by Schwartz (3). In that article, a cell isolated from retinas was used as a sensor to monitor the release of GABA from retinal neurons after establishing a whole-cell configuration. Briefly, under conditions of horizontal and bipolar cell co-culture, a whole-cell configuration was established

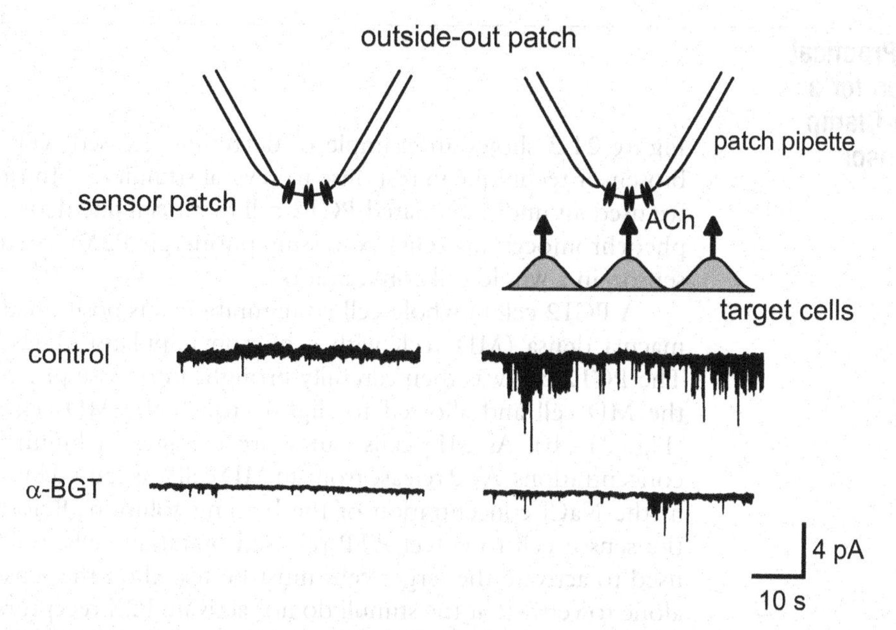

Fig. 21.1. Detection of acetylcholine (*ACh*) released from growth cones using a patch membrane biosensor method. After excising a patch membrane containing the nicotinic ACh receptors from skeletal muscle cells in an outside-out configuration, the patch membrane is used as a biosensor to detect spontaneous ACh release from the growth cone. In the absence of growth cones, there is no signal detected in the patch membrane (*left*). When the patch membrane is brought into close proximity to the growth cone, spontaneously released ACh binds to nicotinic ACh receptors, which can cause ACh-dependent opening of ligand-gated cation channels of nicotinic ACh receptors expressed in skeletal muscle cells (*right*). It is readily apparent that the activated currents result from released ACh as pretreatment with α-bungarotoxin (*α-BGT*), an irreversible nicotinic ACh receptor antagonist, prominently reduces the current amplitude. (Modified from Young and Poo (2)).

in each cell. A bipolar cell was lifted from the bottom of the chamber with a micromanipulator and carefully brought into close proximity to a horizontal cell. A depolarizing pulse was then applied to the horizontal cell to induce GABA release, which is sensed by a biosensor in the bipolar cell. The whole-cell biosensor method was also used to detect glutamate released from chick cochlea hair cells following activation of NMDA glutamate receptors expressed in granule cells (4). Similarly, we have applied our biosensor technique to various cell types to detect released ATP (5–10) or glutamate (10). For measurement of ATP in particular, it is desirable to determine the released ATP concentrations in the vicinity of the cell surface instead of bulk ATP concentrations because of the rapid ATP degradation by ecto-ATPases existing on the cell surface and delayed diffusion due to unstirred layer effects. Although new methods have been developed to avoid these problems (11, 12), the requirement for specially designed cell lines and sophisticated detection apparatus limits their practical application for routine measurements. In contrast, our cell-based biosensor method allows simple real-time measurement of local ATP concentrations at a given site on the cell surface.

21.3. Practical Method for a Patch-Clamp Biosensor

Figure 21.2 shows an example of detecting released ATP by the biosensor technique in response to several stimuli (5). In this case, we used an undifferentiated PC12 cell (a rat cell line derived from pheochromocytoma cells) expressing purinergic P2X receptors as a sensor, in a whole-cell configuration.

A PC12 cell in whole-cell configuration was positioned near a macula densa (MD) cell with a micromanipulator (Fig. 21.2a). The PC12 cell was then carefully brought into close proximity to the MD cell and allowed to slightly touch the MD cell surface (Fig. 21.2b). As MD cells can sense changes in luminal NaCl concentrations, ATP release from an MD cell was caused by changes in the NaCl concentration of the bathing solution. Before using the sensor cell to detect ATP released by target cells, the stimuli used to activate the target cells must be tested on the sensor cells alone to verify that the stimuli do not activate P2X receptors in the absence of target cells. For MD cells, therefore, it should be verified that a PC12 cell alone does not cause any currents in response to changes in the NaCl concentration of the bathing solution.

Similar to glomeruli with MD plaques, the release of ATP (5–10) or glutamate from various cell types was measured using the following procedure.

21.3.1. Sensor Cells

Appropriate sensor cells were prepared for the bioactive substances to be measured. For example, HEK293 cells expressing $P2X_2$ receptor subunits were prepared for measuring ATP release (8, 10), and HEK293 cells expressing glutamate receptor subunits NR1/NR2B

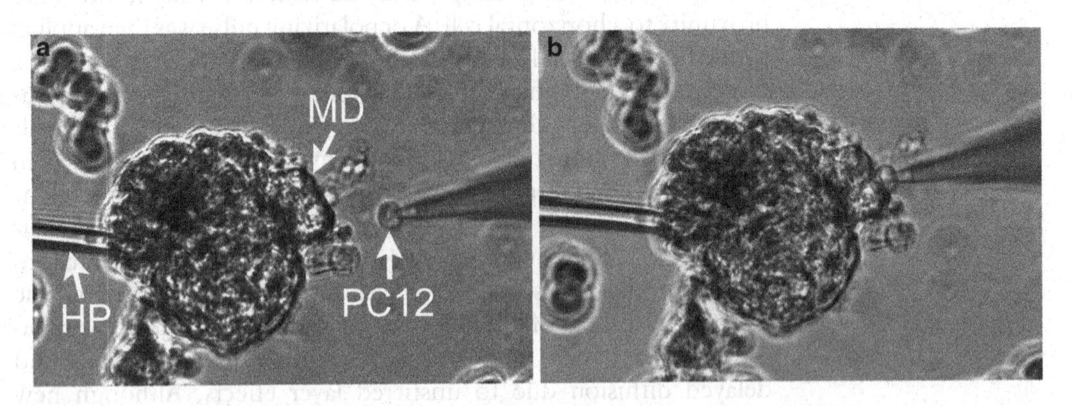

Fig. 21.2. Adenosine triphosphate (ATP) release from macula densa cells detected by a bioassay technique using an undifferentiated PC12 cell (*PC12*). Glomeruli containing the macula densa (*MD*) plaque were isolated from rabbit kidneys. The glomerulus was mechanically fixed to the bottom of an experimental chamber with a holding pipette (*HP*) to expose the basolateral membrane of the MD plaque. (**a**) After establishing a whole-cell configuration, a PC12 cell was lifted from the bottom of the chamber with a micromanipulator and was carefully brought into close proximity to an MD cell. (**b**) The patch pipette with the PC12 cell in whole-cell configuration is positioned to touch the MD cell surface. ATP released from the MD cell was detected in response to an increase in NaCl concentration of the bathing solution. (Modified from Bell et al. (5)).

were used for measuring glutamate release (10). In the case of ATP release measurement, HEK293 cells expressing $P2X_2$ receptor subunits are preferable because $P2X_2$ receptor-mediated currents are known to exhibit less desensitization compared to other types of P2X receptors (10). Alternatively, cells endogenously expressing certain receptors can also be used. Nerve growth factor (NGF)-differentiated PC12 cells have been widely used for studies on apoptotic death or regulation of neural transmitter release. On the other hand, undifferentiated PC12 cells express several P2X receptor subtypes (except $P2X_6$) on the cell surface (13) and are easy to culture, suggesting that undifferentiated PC12 cells are preferable for use as sensor cells.

21.3.2. Experimental Procedure for ATP Release Detection by a Cell-Based Biosensor

Figure 21.3 diagrams the experimental procedure for detecting ATP release with a cell-based biosensor. Sensor cells are grown in culture dishes with suitable medium in a CO_2 incubator. Immediately prior to use, the cells are detached from the culture dishes by trituration and transferred to a sterile glass vial containing a sterile magnetic stirrer bar on the bottom. The sensor cells are then cultured in suspension in a CO_2 incubator with gentle stirring so as to be well isolated from each other. A drop of the culture medium containing sensor cells is placed on the bottom of the experimental chamber filled with bath solution.

A patch pipette is attached to a sensor cell to establish a whole-cell configuration. A sensor cell is lifted from the bottom of the chamber using a micromanipulator and brought into the corner of the experimental chamber. On occasion, a sensor cell may adhere too tightly to the bottom of the experimental chamber to permit lifting of the cell during whole-cell recording. In this case, the cell should be pulled slightly upward with a micromanipulator after establishing a whole-cell configuration, and the cell should be gradually and gently shaken back and forth with the micromanipulator. Once the cell can be slightly detached from the bottom of the experimental chamber, upward pulling and gentle shaking should be repeated until the sensor cell is completely detached from the surface. The success rate of this operation is lower if performed in the cell-attached configuration.

The poly-L-lysine-coated glass coverslip (3×10 mm) with target cells is placed very carefully into the experimental chamber to minimize disruption of the surface of the bathing solution without disturbing the sensor cell in a whole-cell configuration. The sensor cell is then carefully brought into close proximity to the target cell or allowed to slightly touch the target cell. Whether the sensor cell has already touched the target cell can be judged by visually monitoring the cell shape under a microscope. The target cell is then stimulated with the desired treatment to induce the release of bioactive substances, and the evoked current is recorded in the sensor cell.

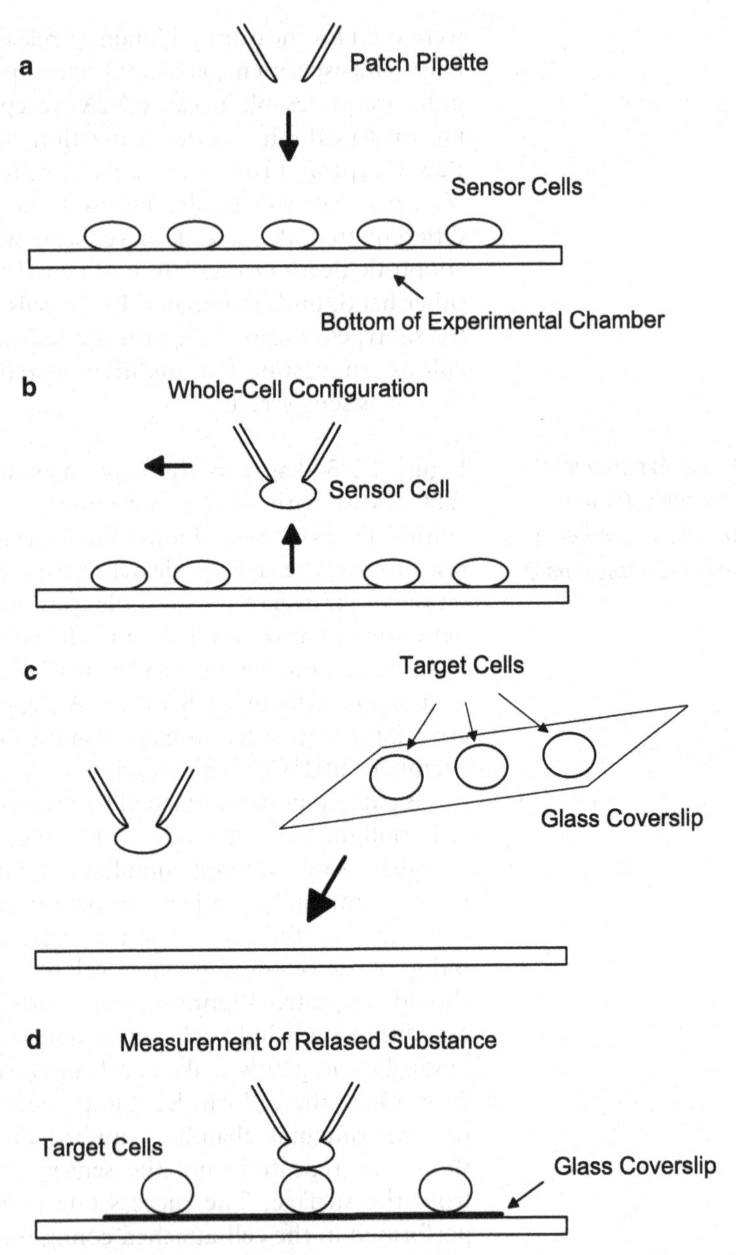

Fig. 21.3. Schema of the experimental procedure for detecting ATP release by a cell-based biosensor. (**a**) A drop of culture medium containing singly isolated sensor cells is placed in the experimental chamber. (**b**) After establishing whole-cell configuration, a sensor cell is lifted from the bottom of the chamber with a micromanipulator. (**c**) The glass coverslip with target cells is placed carefully into the experimental chamber. (**d**) After the sensor cell is brought into close proximity to the target cell, the desired stimulation is applied to the target cell. (Modified from Hayashi et al. (8)).

21.3.3. Calibration Curve

To quantify the stimulation-evoked signals recorded from target cells in terms of the local concentration of a bioactive substance, a calibration curve is constructed. The procedure to construct a concentration-response curve for ATP is as follows. The sensor cell (an undifferentiated PC12 cell or a $P2X_2$-expressing HEK293 cell) in the whole-cell configuration is lifted with a micromanipulator and positioned near a local microperfusion device containing several inlet tubes (Fig. 21.4a). After each inlet tube is filled with

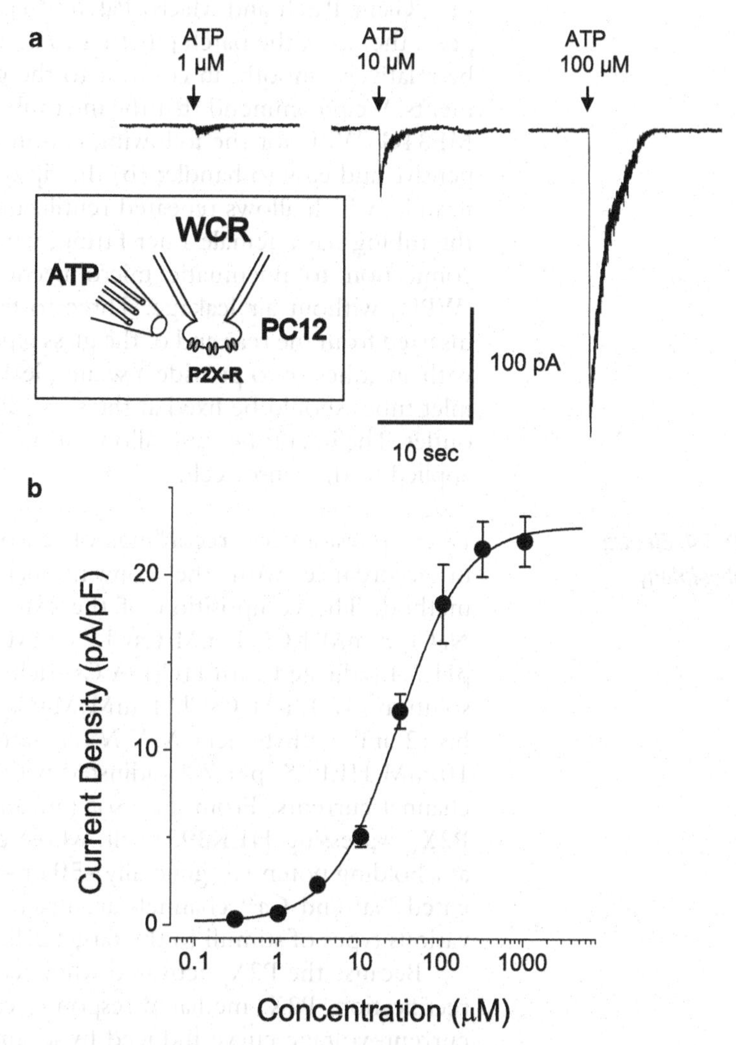

Fig. 21.4. Construction of a concentration–response curve for calibration of ATP responses. (**a**) ATP-induced current responses in a PC12 cell expressing P2X receptors (*P2X-R*). The sensor, an undifferentiated PC12 cell, in a whole-cell recording configuration (*WCR*) is positioned near a glass pipette containing several inlet tubes filled with known ATP concentrations. ATP responses are obtained by puff applications from each inlet tube. Because purinergic P2X receptors expressed in undifferentiated PC12 cells are nonselective cation channels, an inward current is observed in response to ATP at a holding potential of −50 mV. (**b**) Concentration–response curve for ATP-induced currents in undifferentiated PC12 cells. (Modified from Hayashi et al. (8)).

various known concentrations of ATP, ATP-evoked currents are recorded upon puff applications of ATP from each inlet tube (Fig. 21.4a), and the current amplitudes are plotted against the log of ATP concentrations to construct a concentration-response curve (Fig. 21.4b). To minimize the desensitization effects, the sensor cell's exposure to ATP should be brief and number less than five times in total.

A patch pipette with a large diameter used for the local microperfusion device is made using a procedure similar to that shown for the "Giant Patch and Macro Patch" (see Chap. 14). For this purpose, the tip of the patch pipette should have a large diameter and be relatively smooth, in contrast to the giant patch-clamp experiments. We recommend that the inlet tubes be made from MicroFil MF34G (WPI) for the following reasons: (a) this material is inexpensive and easy to handle; (b) the tip of the needle is sturdy and flexible, which allows repeated reutilization of the device; and (c) the tubing has a female Luer fitting, which allows easy and direct connection to pneumatic microinjection systems [e.g., PV820 (WPI)] without air leakage. Three to five inlet tubes are usually inserted from the rear end of the glass pipette and bonded together with an adhesive to provide a secure, leak-free seal. The tips of the inlet tubes should be fixed at the same distance from the common outlet. The inlet tubes thus allow various ATP concentrations to be applied to the sensor cell.

21.3.4. Electrophysiology

Electrophysiological recordings of sensor currents are performed in accordance with the conventional whole-cell patch-clamp method. The composition of the external solution is 145 mM NaCl, 5 mM KCl, 1 mM $CaCl_2$, 1 mM $MgCl_2$, 10 mM HEPES, pH 7.4 (adjusted with HCl). A Cs^+-rich solution is used as pipette solution (150 mM CsCl, 1 mM $MgCl_2$, 10 mM ethyleneglycol-bis-(2-aminoethylether)-N,N,N',N'-tetraacetic acid (EGTA), 10 mM HEPES, pH 7.4 (adjusted with CsOH)) to eliminate K^+ channel currents. From a sensor (an undifferentiated PC12 or a $P2X_2$-expressing HEK293) cell, whole-cell currents are recorded at a holding potential (generally −50 or −60 mV) at which voltage-gated Na^+ and Ca^{2+} channels are inactivated upon application of various types of stimuli to the target cell.

Because the $P2X_2$-activated whole-cell currents exhibit inward rectification, $P2X_2$-mediated responses can be confirmed with the current–voltage curve induced by a ramp pulse (from −100 mV to +100 mV for 500 ms) applied during a 5-s interval (10), as shown in Fig. 21.5. A quantitative estimate of the amount of released ATP near the target cell extracellular surface can be provided to extrapolate the amplitude of $P2X_2$-activated current from the ATP calibration curve. It should be noted that the bathing solution (external solution) should not be perfused in the experimental chamber to avoid the loss of released ATP.

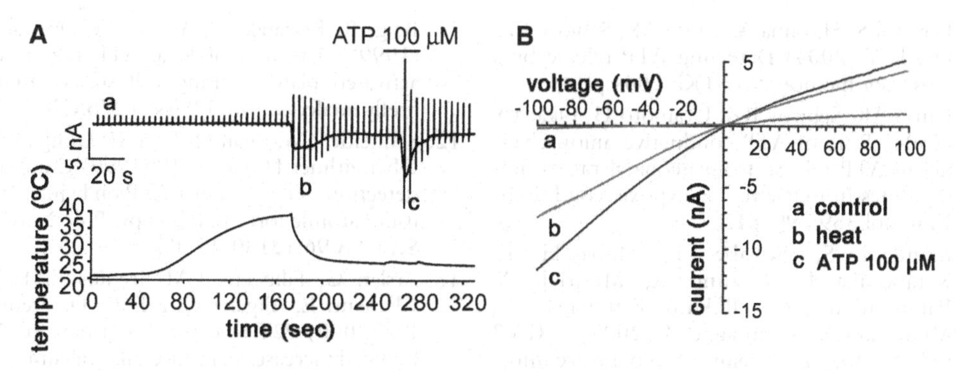

Fig. 21.5. Measurement of ATP release from keratinocytes using HEK293 cells transfected with $P2X_2$ cDNA as a biosensor. (**A**) After establishing a whole-cell configuration of $P2X_2$-expressing HEK293 cells that were co-cultured with mouse keratinocytes, the patch-clamped cell was lifted from the bottom of the chamber. The cell was placed very close to the keratinocytes, to which a heat stimulus (from 25°C to 40°C) was applied using a microscope stage heater, leading to $P2X_2$ receptor activation in the biosensor HEK293 cell at a holding potential of −60 mV. (**B**) Current–voltage curves induced by ramp-pulses (from −100 mV to +100 mV for 500 ms) with 5-s intervals for a control (*a*), current upon heat application (*b*) and ATP-evoked current (*c*) in the current trace (**A**). Both *b* and *c* show inward rectification, indicating activation of $P2X_2$ currents. (Modified from Mandadi et al. (10)).

21.4. Conclusion

We describe a biosensor system and outline a cell-based biosensor method to detect local ATP released from a target cell. The biosensor method with optimal ligand-gated ion channels allows effective evaluation of the local release of desired neurotransmitters from neurons or bioactive substances from secretory cells in real time with quantitative assessment. Mutants of some ligand-gated ion channels may also be used to generate new models, which could make release that is undetectable by conventional methods detectable. Future development and advances in the application of this cell-based biosensor method are expected.

References

1. Hume RI, Role LW, Fischbach GD (1983) Acetylcholine release from growth cones detected with patches of acetylcholine receptor-rich membranes. Nature 305:632–634
2. Young SH, Poo MM (1983) Spontaneous release of transmitter from growth cones of embryonic neurones. Nature 305:634–637
3. Schwartz EA (1987) Depolarization without calcium can release gamma-aminobutyric acid from a retinal neuron. Science 238:350–355
4. Kataoka Y, Ohmori H (1994) Activation of glutamate receptors in response to membrane depolarization of hair cells isolated from chick cochlea. J Physiol 477:403–414
5. Bell PD, Lapointe J-Y, Sabirov R, Hayashi S, Peti-Peterdi J, Manabe K, Kovacs G, Okada Y (2003) Macula densa cell signaling involves ATP release through a maxi anion channel. Proc Natl Acad Sci USA 100:4322–4327
6. Hazama A, Hayashi S, Okada Y (1998) Cell surface measurements of ATP release from single pancreatic beta cells using a novel biosensor technique. Pflugers Arch 437:31–35
7. Hazama A, Shimizu T, Ando-Akatsuka Y, Hayashi S, Tanaka S, Maeno E, Okada Y (1999) Swelling-induced, CFTR-independent ATP release from a human epithelial cell line: lack of correlation with volume-sensitive Cl^- channels. J Gen Physiol 114:525–533

8. Hayashi S, Hazama A, Dutta AK, Sabirov RZ, Okada Y (2004) Detecting ATP release by a biosensor method. Sci STKE 2004:pl14
9. Dutta AK, Sabirov RZ, Uramoto H, Okada Y (2004) Role of ATP-conductive anion channel in ATP release from neonatal rat cardiomyocytes in ischaemic or hypoxic conditions. J Physiol 559:799–812
10. Mandadi S, Sokabe T, Shibasaki K, Katanosaka K, Mizuno A, Moqrich A, Patapoutian A, Fukumi-Tominaga T, Mizumura K, Tominaga M (2009) TRPV3 in keratinocytes transmits temperature information to sensory neurons via ATP. Pflugers Arch 458:1093–1102
11. Beigi R, Kobatake E, Aizawa M, Dubyak GR (1999) Detection of local ATP release from activated platelets using cell surface-attached firefly luciferase. Am J Physiol 276:C267–C278
12. Schneider SW, Egan M, Jena BP, Guggino WB, Oberleithner H, Geibel JP (1999) Continuous detection of extracellular ATP on living cells by using atomic force microscopy. Proc Natl Acad Sci USA 96:12180–12185
13. Arslan G, Filipeanu CM, Irenius E, Kull B, Clementi E, Allgaier C, Erlinge D, Fredholm BB (2000) P2Y receptors contribute to ATP-induced increases in intracellular calcium in differentiated but not undifferentiated PC12 cells. Neuropharmacology 39:482–496

Chapter 22

Temperature-Evoked Channel Activation: Simultaneous Detection of Ionic Currents and Temperature

Makoto Tominaga and Kunitoshi Uchida

Abstract

Since the cloning of the first thermosensitive TRP channel, a capsaicin receptor TRPV1, in 1997, nine thermosensitive TRP channels have been identified. Thermosensitive TRP channels can be used to detect local temperatures by expression in HEK293 cells as a kind of biosensor, as described in Chap. 21, with a patch-clamp method. Because analyses of the thermosensitive ion channels have not yet been widely performed, the thermosensitive ion channels and detailed methods for their analysis are introduced here.

22.1. Introduction

It has become clear that cells utilize ion channels to detect ambient temperature, and a physical stimulus, cloned thermosensitive channel proteins, can be used as temperature detectors in heterologous expression systems. However, analyses of the thermosensitive ion channels have not yet been widely performed. Accordingly, we introduce thermosensitive ion channels here and describe in detail methods for their analysis.

22.2. Thermosensitive TRP Channels as Thermosensors

In 1997, it was proposed that ion channels activated by heat might be present in dorsal ganglion neurons. That same year, the gene encoding TRPV1 protein was isolated from a rodent sensory neuron cDNA library by a group in the United States (1). TRPV1 was the first discovered mammalian molecule to be activated by temperature. Since then, a large group of TRP channels, the

Yasunobu Okada (ed.), *Patch Clamp Techniques: From Beginning to Advanced Protocols*, Springer Protocols Handbooks, DOI 10.1007/978-4-431-53993-3_22, © Springer 2012

so-called thermosensitive TRP channels have been identified in mammals. The name TRP (transient receptor potential) comes from the prototypical member, which was found to be deficient in a *Drosophila* mutant exhibiting abnormal transient responsiveness to continuous light. TRP channels are now divided into seven subfamilies (TRPC, TRPV, TRPM, TRPML, TRPN, TRPP, TRPA), and there are six subfamilies (except TRPN) with 27 channels in humans. One subunit of the TRP channel is composed of six transmembrane domains and a putative pore region with both amino and carboxyl termini on the cytosolic side. It is thought that the subunits form functional channels as homo- or hetero-tetramers. TRP channels are important for detecting various physical and chemical stimuli in vision, taste, olfaction, hearing, touch, and thermosensation (2) .

Nine thermosensitive TRP channels have been reported in mammals (Table 22.1, Fig. 22.1) (3–5). They belong to the TRPV, TRPM, and TRPA subfamilies; and their temperature thresholds for activation are in the range of the physiological temperatures that we can discriminate. Thermosensitive TRP channels work as "multimodal receptors," which respond to various chemical and physical stimuli. TRPV1, the first identified thermosensitive TRP channel, was found to be a receptor for capsaicin and later was found to have thermosensitivity. The temperature threshold for activation of rat TRPV1 expressed in HEK293 cells is around 42°C, a temperature known to evoke pain sensation in humans. The temperature thresholds are decreased by repeated exposure to heat stimuli. TRPV2 is activated by a much higher temperature stimulus (>52°C). TRPV3 is activated by warm temperatures, and its activity is enhanced as the temperature increases. TRPV3 also exhibits sensitization (increased activity) by repeated heat stimuli. The temperature-evoked activity of TRPV4 is increased in the range of 27–41°C, and its temperature threshold for activation is increased (desensitized) by repeated heat stimuli. TRPM2, TRPM4, and TRPM5 are also activated by warm temperatures. TRPM8 is a cold receptor activated by temperatures lower than 26°C. Menthol, the main ingredient of mint, which gives a cool sensation, functions as a ligand for TRPM8, and the temperature threshold is increased (sensitized) by concomitant application of menthol. TRPA1 was initially reported to be activated by cold temperatures (<17°C).

Temperature-mediated channel activities were observed in excised patch membranes expressing TRPV1, TRPV3, TRPM2, TRPM5, TRPM8, and TRPA1 among the nine thermosensitive TRP channels, suggesting direct activation of the channels by temperatures without involvement of any cytosolic components. Many thermosensitive TRP channels are activated by stimuli in addition

Table 22.1
Nine thermosensitive TRP channels reported in mammals

TRP channel	Temperature threshold (°C)	Tissue distribution	Other stimuli
TRPV1	<42	Sensory neuron, brain, skin	Capsaicin, proton, shanshool, allicin, camphor, resiniferatoxin, vanillotoxin, 2-APB, propofol, anandamide, arachidonic acid metabolic products (by lipoxygenases), NO, extracellular cation
TRPV2	<52	Sensory neuron, brain, spinal cord, lung, liver, spleen, colon, heart, immunocyte	Probenecid, 2-APB, cannabidiol, mechanical stimulation
TRPV3	<32	Skin, sensory neuron, brain, spinal cord, stomach, colon	Camphor, carvacrol, menthol, eugenol, thymol, 2-APB
TRPV4	<27–41	Skin, sensory neuron, brain, kidney, lung, inner ear, bladder	4α-PDD, bisandrographolide, citric acid, arachidonic acid metabolic products (by epoxygenases), anandamide, hypoosmolality, mechanical stimulation
TRPM2	<36	Brain, pancreas	(Cyclic) ADP ribose, β-NAD, H_2O_2
TRPM4, TRPM5	Warm	Heart, liver, taste cell, pancreas	Intracellular Ca^{2+}
TRPM8	<27	Sensory neuron	Menthol, icilin, eucalyptol
TRPA1	<17	Sensory neuron, inner cell	Allyl isothiocyanate, carvacrol, cinnamaldehyde, allicin, acrolein, icilin, tetrahydrocannabinol, menthol (10–100 μM), formalin, H_2O_2, alkalization, intracellular Ca^{2+}, NSAIDs, propofol/ isoflurane/ desflurane/etomidate/octanol/hexanol

to temperature, and it is known that temperature thresholds for activation of TRPV1 and TRPM8 shift (decrease and increase, respectively) in the presence of minimal concentrations of their chemical agonists (capsaicin and menthol, respectively). Furthermore, the temperature threshold for TRPV1 can be decreased to around 30°C upon phosphorylation, which indicates that TRPV1 is activated at physiological body temperature and could explain the spontaneous pain in acute inflammation. The properties of the thermosensitive TRP channels are shown in Table 22.2.

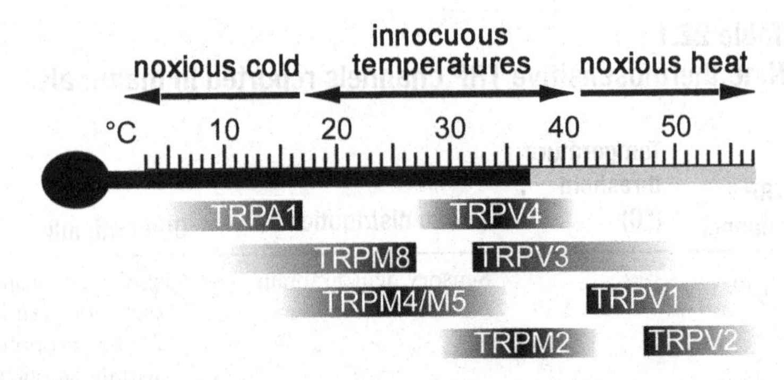

Fig. 22.1. Thermosensitive TRP channels in mammals. Nine thermosensitive TRP channels with their specific temperature thresholds are shown. They belong to the TRPV, TRPM, and TRPA subfamilies. Other TRP channels in TRPV and TRPM subfamilies do not have thermosensitivity. TRPA1, TRPV1-4, and TRPM8 expressed in primary sensory neurons or skin are believed to be involved in ambient temperature detection. TRPM2, TRPM4, and TRPM5 are thought to be involved in temperature-dependent physiological functions in other tissues.

22.3. Analysis of Temperature Detection with a Patch-Clamp Biosensor Method

22.3.1. Electrophysiology

The bath solution contains 140 mM NaCl, 5 mM KCl, 2 mM $MgCl_2$, 2 mM $CaCl_2$, and 5 mM HEPES (pH 7.4 adjusted with NaOH), and the pipette solution contains 140 mM KCl, 10 mM EGTA, and 5 mM HEPES (pH 7.4 adjusted with KOH). A cover glass (diameter 12 mm) on which HEK293 cells expressing thermosensitive TRP channels are seeded is placed in a 12 mm wide chamber so the cover glass does not move even under the solution flow. The patch-clamped cell is lifted up after making a whole-cell configuration as described in Chap. 21, and moved to the center of the chamber (Fig. 22.2). Lifting up the patch-clamped cell is an essential procedure because small bubbles under the cover glass become bigger upon heating, moving the cover glass, which leads to breaking the giga-seal. Because all of the known thermosensitive TRP channels function as nonselective cation channels, it is best to observe inward currents at a holding potential of −60 mV.

The heat stimulus can be applied in three ways: (1) perfusing preheated solution (up to 50–65°C), (2) heating the solution with an inline heater (one-line heater SH-27B or six-line heater SHM-6; Warner Instrument Corporation, Hamden, CT, USA) before reaching the chamber, or (3) heating a metal chamber holder (platform PH-1; Warner Instrument Corporation). Although it takes a long time (<1°C/s increase) to heat the cells with the third method (heating the metal chamber holder), it could be preferable when detailed activation kinetics are under analysis. Generally, the heating speeds do not matter because it is accepted that many of the thermosensitive TRP channels sense the absolute temperature, not the heating speed. As described in Chap. 21, the heating chamber itself should be heated when measuring the molecules released from the cells upon heating to avoid loss of the released molecules by the solution flow. An automatic temperature controller (TC-334; Warner Instrument Corporation) is recommended with the above materials when using an inline heater or heating the chamber holder. A chilled (to 0°C) solution can be used when applying cold stimulus to the cells. It would be better to expose the cells to a warm solution (around 30°C) before the cold stimulus to activate TRPM8 because naive TRPM8 is activated at room temperature. It should be noted that simultaneous monitoring of the precise temperature around the patch-clamped cell is very important. Therefore, the temperature probe (TA-30; Warner Instrument Corporation) should be located in close proximity to the cell (<100 μm distance), and both the temperature information and current signals should be recorded simultaneously through the AD converter (Digidata 1440A; Warner Instrument Corporation) with specific software such as pClamp (Molucular Devises, Sunnyvale, CA, USA). A manipulator with rough moving can be used to place the temperature probe in the correct position. It is recommended that a patch pipette be applied with the fine manipulator from the right side and the temperature probe with the rough manipulator from the left side (Fig. 22.3). It is also important to place the patch-clamped cell in the center of the chamber because the temperature at the periphery fluctuates owing to the turbulent flow.

22.3.2. Data Analysis

Temperature and current data recorded simultaneously can be plotted as a function of time, as shown in Fig. 22.4. Temperature-dependent current activation of TRPV1 is more easily recognized when plotting the currents as a function of temperature (Fig. 22.5). Furthermore, an Arrhenius plot in which log values of current (picoamperes) are plotted against reciprocal values of absolute temperatures provides the precise temperature threshold as a flex point of the curve. More practically, the flex point can be determined as the intersection of the tangential lines as shown in Fig. 22.6.

Although ion movement is generally accelerated by temperature increases, Q_{10} values for ion channels are not increased upon heating (<3.0) (6). On the other hand, Q_{10} values for the thermosensitive TRP channels are very high, often >10.0, suggesting huge effects of temperature changes on the channel functions.

The temperature threshold for TRPM8 activation by the cold stimulus is similarly analyzed by a temperature-response profile, as shown in Fig. 22.7.

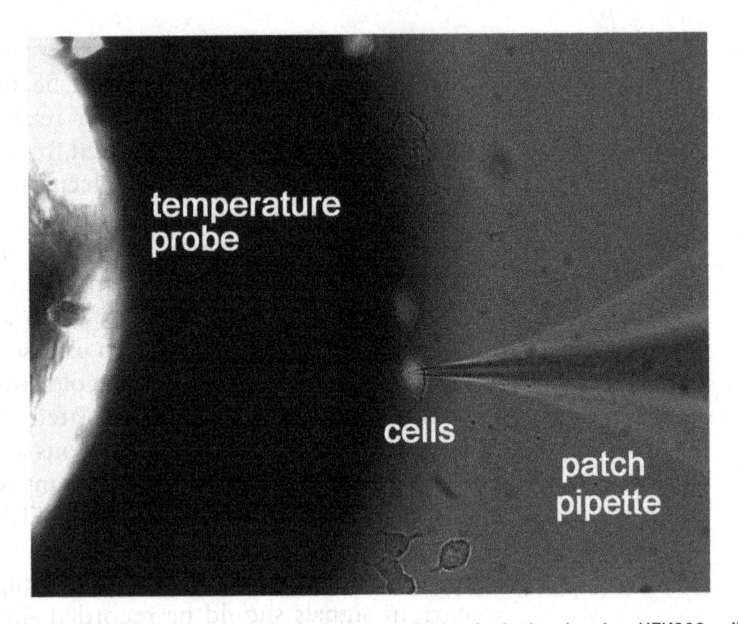

Fig. 22.2. Cells, a patch pipette, and a thermosensing probe in the chamber. HEK293 cells expressing thermosensitive TRP channels in the chamber are chosen with green fluorescent protein (GFP) signals as markers under the fluorescence microscope. The patch-clamped cell is moved to the center of the chamber after being lifted up, and the temperature probe is placed in close proximity to the cell (<100 μm distance).

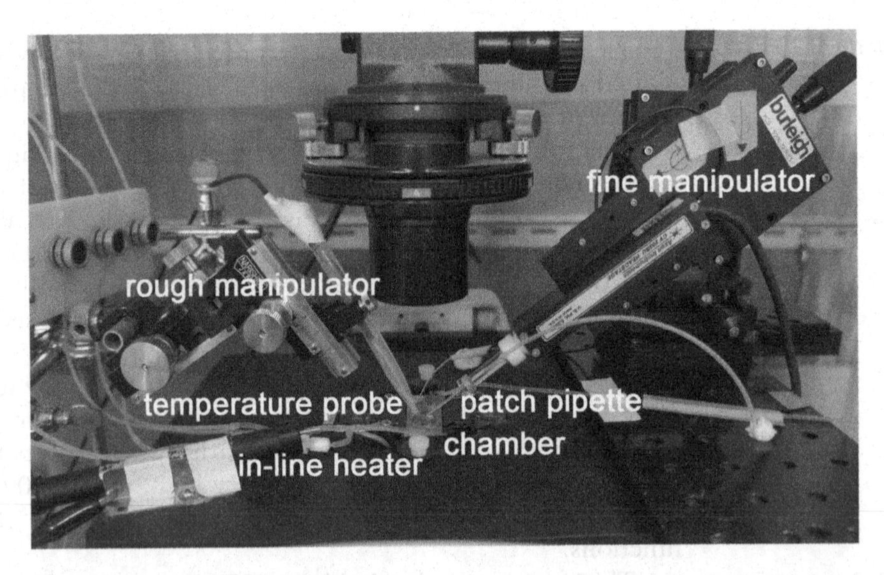

Fig. 22.3. Patch-clamp experimental setup for detecting the current and temperature. The fine manipulator for the patch pipette is on the right, and the rough manipulator for the temperature probe is on the left. The chamber is fixed with two metal chamber holders that can be heated. The solution can also be heated with the in-line heater.

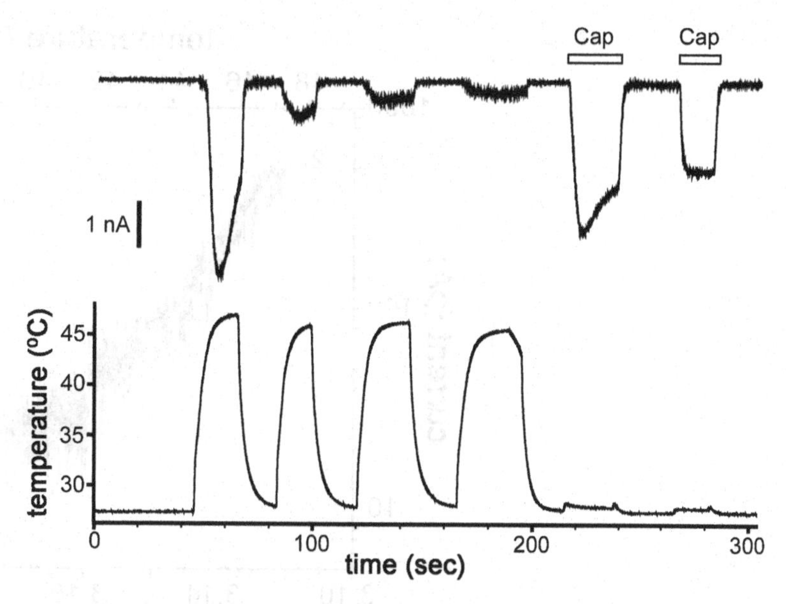

Fig. 22.4. TRPV1-mediated current responses evoked by heat and capsaicin (*Cap*). Heat-evoked and capsaicin (1 μM)-activated inward currents are observed at −60 mV. Both responses are desensitized (decreased) in the presence of extracellular Ca^{2+}. Because Cap-activated currents are still observed even after almost complete desensitization upon repeated heat stimuli, it is thought that their activation mechanisms are different or that heat up to 45°C is not a full stimulus for TRPV1 activation. The *lower trace* indicates temperature. *White bars* indicate capsaicin application.

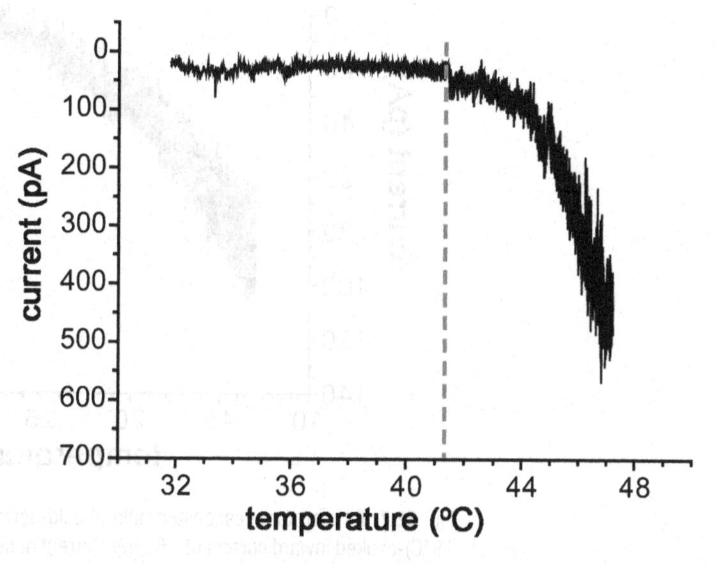

Fig. 22.5. Temperature-response profile of a heat-activated TRPV1 current. Note the temperature dependence of the heat-evoked inward current response at −60 mV. Current activation can be observed at or above 42°C, with the current increasing as the temperature rises.

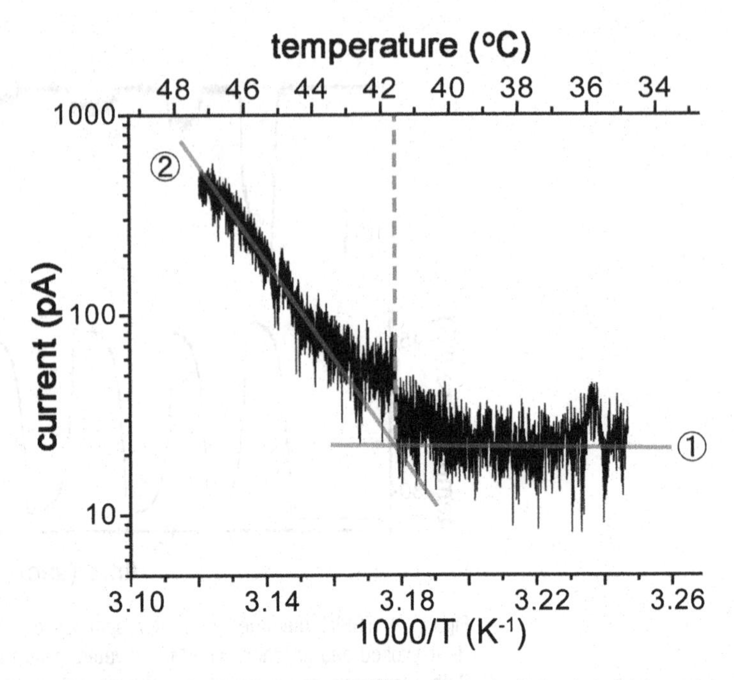

Fig. 22.6. Arrhenius plot of heat-activated TRPV1 currents. An Arrhenius plot (log values of current (picoamperes) versus reciprocal values of absolute temperatures) from the data in Fig. 22.5. There are two clear phases: a phase showing no current changes regardless of temperature change (*1*) and a phase showing a clear current increase as the temperature elevates (*2*). The temperature threshold is determined as the temperature at the intersection of the tangential lines (41.7°C in the particular cell shown by the *dashed line*).

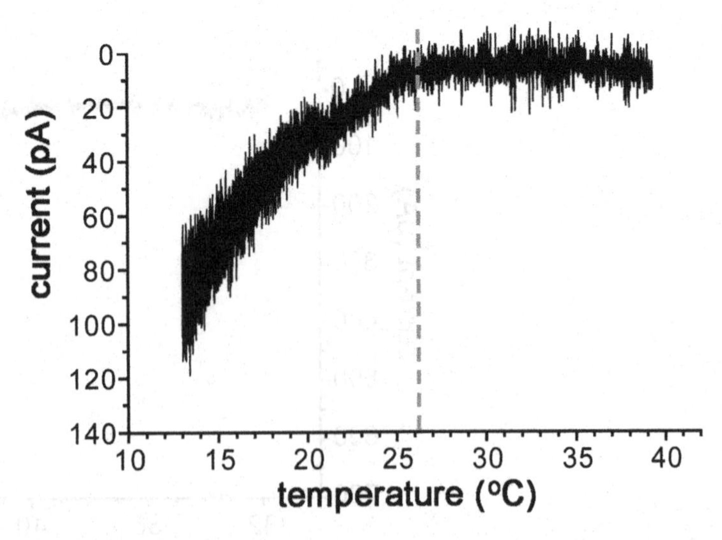

Fig. 22.7. Temperature-response profile of cold-activated TRPM8 current. Cold (down to 15°C)-evoked inward current at –60 mV. Current activation is observed at <26°C, and the current increases as the temperature is reduced.

22.4. Conclusion

We described a method for recording channel activation by temperature and a way to analyze the data. The thermosensitive TRP channels introduced here are expressed not only in sensory neurons but also in many other cell types. Therefore, it is possible to record the channel activities upon temperature stimulation in native cells as well as in HEK293 cells. Moreover, it has become evident that not only mammals but also other living things utilize thermosensitive TRP channels to detect ambient temperatures. Accordingly, analyzing temperature-activated currents in *Xenopus* oocytes or insect cells will become popular in the future.

References

1. Caterina MJ, Schumacher MA, Tominaga M, Rosen TA, Levine JD, Julius D (1997) The capsaicin receptor: a heat-activated ion channel in the pain pathway. Nature 389:816–824
2. Venkatachalam K, Montell C (2007) TRP channels. Annu Rev Biochem 76:387–417
3. Tominaga M, Caterina MJ (2004) Thermosensation and pain. J Neurobiol 61:3–12
4. Dhaka A, Viswanath V, Patapoutian A (2006) TRP ion channels and thermosensation. Annu Rev Neurosci 29:135–161
5. Talavera K, Nilius B, Voets T (2008) Neuronal TRP channels: thermometers, pathfinders and life-savers. Trends Neurosci 31:287–295
6. Hille B (2001) Ion channels of excitable membranes, 3rd edn. Sinauer, Sunderland, MA

Chapter 23

Heterologous Expression Systems and Analyses of Ion Channels

Kazuharu Furutani and Yoshihisa Kurachi

Abstract

This chapter describes practical techniques for electrophysiological experiments in the heterologous expression systems using mammalian cell cultures and *Xenopus laevis* oocytes. These two experimental systems are the most commonly used for expression of ion channels and transporters, contributing to numerous discoveries of their function and molecular properties. When planning an experiment, researchers need to determine whether either system is appropriate for use. It is crucial to understand the advantages and disadvantages of each system. The practical techniques discussed here include preparation of cells, heterologous expression of ion channel and other proteins, and the electrophysiological techniques available for recording from these systems. Also, we show the applications of these systems to analyses of K^+ channels.

23.1. Introduction

For decades, expression techniques in heterologous systems have provided invaluable contributions to the studies of membrane proteins, especially ion channels. These techniques serve as the bridge between their molecular biology and electrophysiology, and they are indispensable for analyses of functionality of the cloned message or gene for ion channels in isolation or in combination with other proteins. They address which specific protein regions, or even single amino acids, are involved in determining or influencing various molecular properties, including ion selectivity, voltage dependence, ligand binding, subunit composition, modulation by signal pathways, and protein transport. In the current era of rapid progress in both structural and computational biology, these techniques have gained importance by providing the functional basis for structural models.

This chapter deals with two heterologous expression systems: mammalian cell lines and *Xenopus laevis* oocytes. Both expression

Yasunobu Okada (ed.), *Patch Clamp Techniques: From Beginning to Advanced Protocols*, Springer Protocols Handbooks, DOI 10.1007/978-4-431-53993-3_23, © Springer 2012

systems are widely used and have proven to be extremely useful, with the more appropriate system chosen according to the desired results. This chapter describes the practical methods for heterologous expression of ion channels in these two systems, electrophysiological techniques available for recording from these systems, and their respective advantages and disadvantages.

23.2. Heterologous Expression Methods Using Mammalian Cells

23.2.1. Principles and Experimental Outline

Mammalian cell lines with low expression of endogenous ion channels are often used as heterologous expression systems. The authors use human embryonic kidney (HEK) 293 cells, although these techniques can be applied to any similarly modified cell line; HeLa, COS-1, and Chinese hamster ovary (CHO) cells are common alternatives. All are easy to culture and be introduced to a foreign gene (or combinations) of interest.

There are a number of techniques to introduce genes into mammalian cells. First, gene transfer can be mediated by viruses such as the adenovirus or retrovirus. However, virus-mediated techniques are not suitable for studies recording multiple recombinant channels as these techniques take a long time to generate virus. In addition, these methods are relatively toxic and require special facilities. Common, alternative methods of nonviral gene transfer include the calcium phosphate method, lipofection, DEAE-dextran, and electroporation. General characteristics of each are described as follows.

23.2.1.1. Calcium Phosphate Method

When plasmid cDNA is mixed in a solution containing calcium chloride and then added to a phosphate-buffered solution, a fine precipitate of the calcium phosphate–DNA complex forms in the solution. When this solution is added to cells in culture medium, the precipitate incorporates into the cell by endocytosis and is believed subsequently to migrate to the nucleus.

- No special equipment
- Inexpensive, simple, and easy technique to perform

23.2.1.2. DEAE-Dextran Method

DEAE-dextran is another chemical that interacts with DNA. It is thought that DEAE-dextran also forms an insoluble precipitate with DNA, which is taken into the host cell via an unknown mechanism.

- Excellent reproducibility with less variation in results
- Lower transfection efficiency compared to the calcium phosphate and lipofection methods
- Cytotoxicity

23.2.1.3. Lipofection Method

Positively charged liposomes, which are lipid bilayer vesicles, are complexed with DNA, binding with the cell membrane, and subsequently introducing DNA into the cytoplasm (1).

- High transfection efficiency
- No special equipment
- Applications to a wide range of cells

- Narrow optimal transfection conditions (cell density, DNA content, liposome quantity, incubation time and medium)
- Expensive reagents

23.2.1.4. Electroporation Method

Electroporation involves application of a high-voltage electrical shock to a mixture of DNA and cells in suspension, destabilizing the cell membrane and opening up pores through which DNA may enter.

- Simple and easy technique to perform
- High transfection efficiency

- Narrow optimal transfection conditions (nature of the electrical pulse, the ionic strength of the buffer, the nature of the cells)
- Expensive equipment

23.2.1.5. FuGENE 6 Transfection Reagent Method

FuGENE 6 transfection reagent is lipid-based with the following characteristics.

- High transfection efficiency for many types of cells
- Low cytotoxicity
- Transfection in the presence of serum

Among these techniques, the lipofection and FuGENE transfection reagent methods can deliver DNA into the cells more efficiently than chemical cDNA precipitation with calcium phosphate or DEAE-dextran. They also offer the advantages of lower cost and less toxicity compared to electroporation.

The authors use transient transfection instead of stable transfection because of its advantages for expeditious analysis of genes, although stable transfection ensures long-term, reproducible, well-defined gene expression.

23.2.2. Instructions

23.2.2.1. Preparation of Cells

Feeder cell cultures should be maintained in large culture flasks and can be kept for ~20 passages. Excessive number of passages may decrease transfection efficacy. Similarly, newly thawed cells are suboptimal for electrophysiological experiments owing to their poor membranes, a condition that lasts for at least 2 weeks.

23.2.2.2. Transfection

Expression Vector

Plasmid vectors are generally used to carry a gene of interest into a target mammalian cell. These plasmid vectors must contain a promoter for high-level expression of cloned DNA inserts in the host cell. Many commercially available vectors may be used for transfection and for molecular biology experiments such as site-direct mutagenesis. Plasmid DNA is dissolved in sterile water or Tris/EDTA (TE) buffer.

Procedure A: Lipofection (Lipofectamine 2000)

1. One day prior to transfection, cells are seeded into a 12-well culture plate so the cells are ~80% confluent on the day of the electrophysiological experiment. During incubation, growth medium should be composed of Dulbecco's modified Eagle's medium (DMEM) + 10% fetal bovine serum (FBS).
2. For each transfection sample, dilute ~0.5 μg of DNA in 50 μl of a serum-free medium (e.g., Opti-MEM; Life Technologies, Carlsbad, CA, USA) with 2 μl PLUS Reagent (Life Technologies). Mix gently.
3. For each transfection sample, dilute 3 μl of Lipofectamine 2000 (Life Technologies) in 50 μl of Opti-MEM gently and incubate for 5 min at room temperature.
4. After 5 min of incubation, combine the diluted DNA (step 2) with diluted Lipofectamine 2000 (step 3) (total volume 100 μl of complex per transfection sample). Mix gently and incubate for 20 min at room temperature.
5. While waiting for this incubation, wash cells with serum-free medium (Opti-MEM) to remove FBS.
6. Add the 100 μl of complex to each well containing cells and serum-free medium (~400 μl). Mix gently by rocking the plate back and forth.
7. Incubate cells at 37°C in a CO_2 incubator for 18–48 h prior to testing for transgene expression and replace the medium (containing transfection reagents) with serum-containing medium (without transfection reagents) after 4–6 h. Expression typically takes 18–24 h after completion of transfection until it is ready for electrophysiological assay.
8. Some researchers transfect cells on a coated coverslip and continue to renew the medium until expression and recording. However, because a large number of cells on a single coverslip is not necessary, preparing as many coverslips as required for additional electrophysiological recordings is preferable. We recommend then replating the appropriate density of cells on these additional coverslips before electrophysiological assay.
9. If necessary, optimizing the ratio of transfection reagents to DNA may further increase efficacy.

Procedure A: FuGENE6

FuGENE 6 transfection regent (Roche, Basel, Switzerland) is commonly chosen as the transfection reagent due to its aforementioned

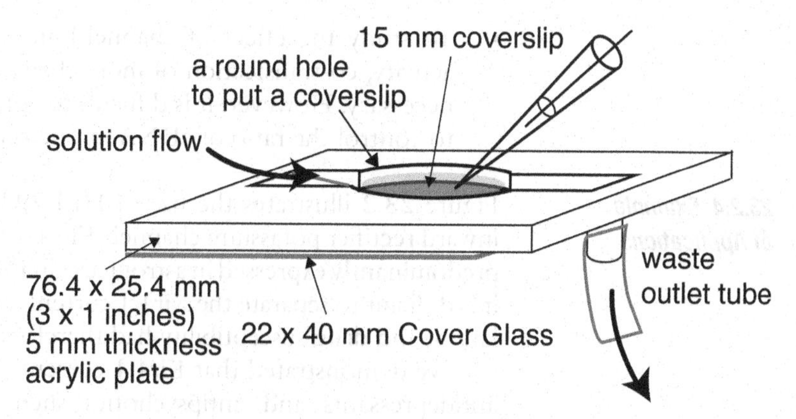

Fig. 23.1. Recording chamber for mammalian cells. Make a round hole in the acrylic plate and attach a cover glass under the plate for putting the coverslip inside. This chamber rests on the platform of an inverted microscope with slide holder.

advantages. The standard transfection protocol with FuGENE is similar to that of lipofection, but there is no need to remove serum during transfection or reagent prior to the assay.

23.2.2.3. Electrophysiological Recordings

The instructions below assume a basic understanding of patch-clamp recording. Detailed patch-clamp recordings and data analysis are described in Chap. 2. A glass coverslip rests on a recording chamber (Fig. 23.1) mounted on an inverted microscope.

23.2.3. Advantages and Limitations of Using Mammalian Cells

- The process of transient expression is simple and relatively easy. Transfection techniques are now commonly used in most laboratories.
- Easier to analyze channel activity at the single-channel level in mammalian cells compared to *Xenopus* oocytes.
- Easy to control intracellular medium in the whole-cell configuration and excised patch.
- Intracellular signal transduction cascades have been well characterized in studies of mammalian cell lines.
- Stable transfected cell lines can be obtained. This is particularly useful when a large number of cells are needed.

- With transient transfection, only a fraction of cells pick up the expression vector and express the gene product. It is often useful to co-transfect cells with a fluorescent reporter gene such as *GFP* or to construct a bicistronic vector that co-expresses both the target gene and a marker.
- It is necessary to separate endogenous and exogenous ion channels. Which endogenous ion channels are present in the host cells must be clearly defined prior to assay.

- To study the effect of channel-binding proteins on channel activity, co-transfection of more than one cDNA into a cell is necessary. However, it is difficult to confirm co-expression and to control the ratio of cDNA in one cell.

23.2.4. Example of Applications

Figure 23.2 illustrates the use of HEK293 cells for expression of inward rectifier potassium channel (Kir)4.1 with examples. Kir4.1 is predominantly expressed in astroglial cells in the brain (2). However, it is difficult to separate the Kir4.1 current; and the function, modulation, and drug susceptibility had therefore remained unclear.

We demonstrated that Kir4.1 currents are inhibited by several antidepressants and antipsychotics such as haloperidol (3–5). Figure 23.2a shows that extracellular application of haloperidol inhibits currents in the whole-cell mode, and Fig. 23.2b shows that intracellular application of haloperidol inhibits channel activity in the inside-out mode. Haloperidol decreases the open probability

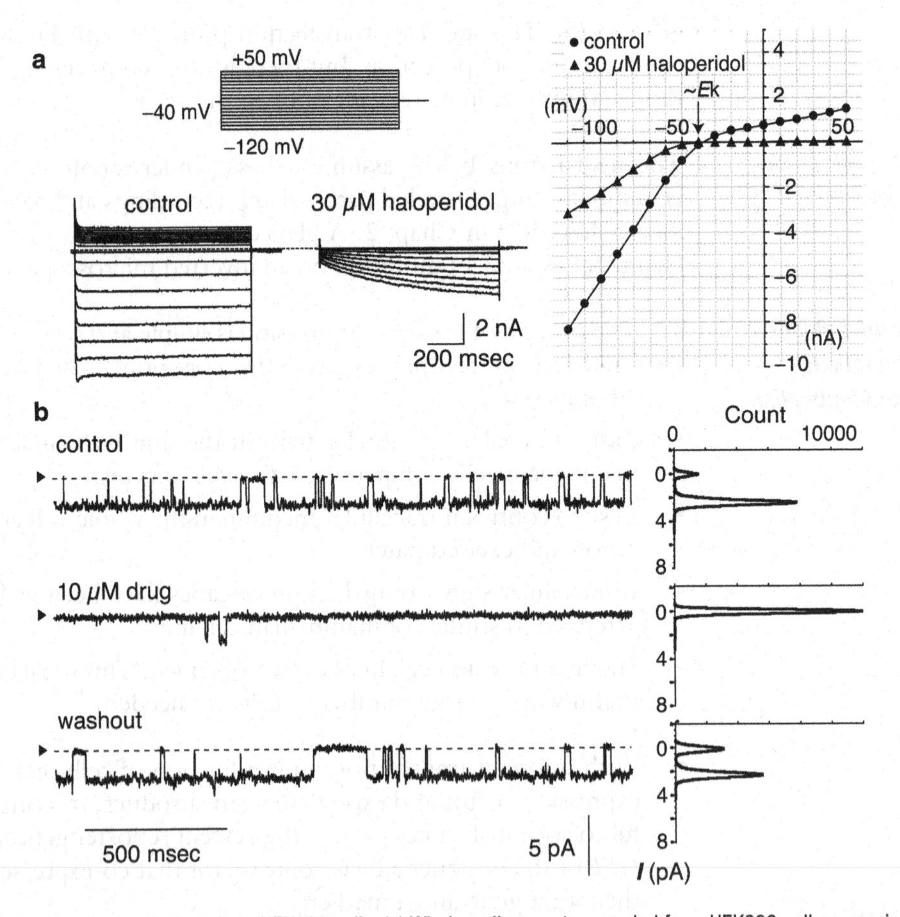

Fig. 23.2. Patch-clamp recording from HEK293 cells. (**a**) Whole-cell currents recorded from HEK293 cell expressing Kir4.1 in 30 mM K+ Tyrode solution. Representative traces of the Kir4.1 channel current response to 30 μM haloperidol at different potentials. (**b**) Single-channel Kir4.1 activity recorded continuously at −100 mV from a HEK293 cell expressing Kir4.1. Potassium (152 mM) is in both the bath and pipette. After inside-out patch recording, the intracellular surface of the patch was exposed to 10 μM haloperidol. *Dashed line* shows the zero current level.

and shortens the dwell time without having an effect on the single-channel conductance. Such currents were not observed in untransfected HEK293 cells.

23.3. Heterologous Expression Methods Using *Xenopus* Oocytes

23.3.1. Introduction

Oocytes of the African clawed frog, *Xenopus laevis*, are widely used for studying ion channels (6–8). These oocytes can be easily manipulated and have the ability to translate foreign genetic information efficiently and faithfully. For most studies, oocytes are microinjected with RNA that has been transcribed in vitro from a cDNA clone (7, 9–11).

23.3.2. Instructions

23.3.2.1. Preparation of Xenopus Oocytes

Development of *Xenopus* oocytes (12) can be divided into six stages (I–VI) based on the morphological appearance and size (13). It is possible to express exogenous protein from foreign RNA in all stages. Stage I oocytes (50–300 μm diameter) are transparent with no animal or vegetal pole but with a clear nucleus. Stage II oocytes are characterized by a diameter of 300–450 μm with opaque white cytoplasm. Stage III oocytes are usually 450–600 μm in diameter and fully pigmented. From stage IV, some polarity is newly evident. In these cells (600–1,000 μm diameter), pigment is primarily concentrated in the animal hemisphere. Stage V oocytes (diameter 1.0–1.2 mm) exhibit clear differentiation of animal and vegetal poles, with the animal pole appearing lighter in color compared to oocytes at stage IV. At stage VI, the oocytes are identified by their unpigmented equatorial band; they have diameters of 1.2–1.3 mm. The grown stage V/VI *Xenopus* oocytes are commonly used for microinjection because they are the largest cells, frequently enabling the development of large membrane currents; they are also able to receive injections of up to 100 nl of solution. We always specify to animal suppliers to deliver the largest, most mature female frogs carrying the most fully grown oocytes.

Xenopus is an aquatic air-breathing animal that can rapidly become dehydrated, eventually perishing if denied access to water. Regarding their environment and housing, the temperature of the tank water is controlled between 18°C and 22°C by thermostatically controlling the room temperature under an artificial lighting cycle of 12 h illumination and 12 h darkness.

A typical electrophysiological study rarely requires more than 50–100 oocytes. Because female *Xenopus* lives long and contains great quantities (~30,000) of oocytes when fully mature, a single animal can be reused several times. Therefore, it is important to identify each frog. Individual frogs can be marked by various means, including tattooing, toe clipping, or colored threads tied between the toes (although these can easily fall off). For tattooing, use 5 N HCl to draw a number on the back of the frog.

Procedure B: Isolation of Oocytes from *Xenopus laevis*

1. Anesthetize female *Xenopus* with 0.2% (may use up to 0.35% safely) w/v solution of ethyl-m-aminobenzoate (Tricaine; Sigma, St. Louis, MO, USA), ensuring that the frog is completely immersed. Anesthesia is usually obtained 10 min after immersion in 0.2% tricaine.
2. Transfer the *Xenopus* and place dorsal side down on a flat bed of ice in a shallow tray to maintain anesthesia.
3. Make a small incision about 1 cm long with a surgical knife in the right or left lower abdominal skin from medial to anterolateral.
4. Another incision should be made through the muscle sheet in the same manner, after which the ovary wall and the dark oocytes should be exposed. While making this second incision, damage to the frog's internal organs must be avoided by lifting up the exposed muscle sheet using forceps.
5. Isolate as many oocytes as required by pulling out the lobes of the ovary with a fine pair of forceps, separating them from the remaining ovarian tissue using small scissors, and transfer to a 35-mm tissue culture dish containing Ca^{2+}-free oocyte Ringer's solution 2 (OR2) (see "Recipes," below).
6. Put the remaining ovary back into the abdomen.
7. Suture the skin and muscle layer separately (approximately three or four stitches each).

The entire operation should be done under clean conditions, although sterile surgical procedure is not necessary because magainin antimicrobial peptide secreted from the animal's skin allows the frog to be returned to the water immediately after surgery with low likelihood of subsequent infection.

It is possible to isolate oocytes multiple times from the same animal. A 6- to 8-week recovery period should follow each harvesting.

Procedure C: Preparation of Oocytes

1. Wash the ovary tissue (or clumps of oocytes) thoroughly in Ca^{2+}-free OR2 several times shortly after isolation from the animal. This removes any debris and dilutes enzymes from damaged cells that may damage the integrity of oocytes.
2. At this point, each oocyte is contained in a small ovarian sac composed of epithelial cells and enveloped in a tough follicle cells layer (Fig. 23.3a). Because these cell layers are difficult to pierce with a needle, strip off individual oocytes from the clump and then remove the follicle layer before microinjection *(description of the procedure continues with step 3 after the paragraphs on defolliculation)*.

   Manual Defolliculation

   Manual defolliculation of oocyte can be achieved at the same time the oocyte is separated from the clump of ovarian tissue.

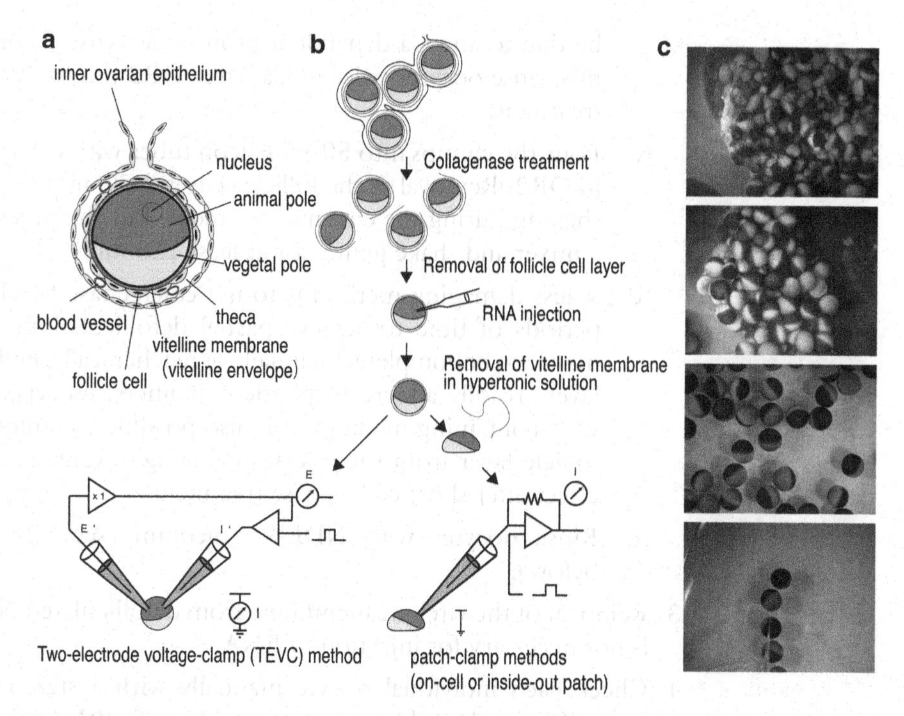

Fig. 23.3. Preparation of oocytes. (**a**) *Xenopus* oocyte illustrating the various anatomical features including the cell and membrane layers surrounding the oocyte plasma membrane. (**b**) Steps involved in preparation of oocytes for two electrophysiological recordings of heterologous ion channels. (**c**) *From top to bottom*: a portion of the ovary from a female *Xenopus*, clump of oocytes, the selected oocytes after defolliculation, and impalement of an oocyte for RNA injection.

Watchmaker's forceps (no. 5) are used carefully and gently to remove the epithelial and follicular layers from around the oocyte while avoiding touching or damaging the oocyte. This method may not always result in complete removal of the follicular cell layer.

Enzymatic Defolliculation

Defolliculated oocytes can also be obtained by enzymatic treatment with collagenase l–2 mg/ml (approximately 0.2–0.5 U/ml) (type lA, Sigma; or type I, Life Technologies) in OR2 at room temperature (18–20°C). We recommend testing each lot of collagenase prior to large-scale use, given its variations in effectiveness and toxicity. The use of collagenase has an advantage over manual defolliculation in that many oocytes can be treated at once, although the procedure may be detrimental to long-term oocyte survival.

(a) Tear apart the large clumps of oocytes with forceps and divide the tissue with forceps into smaller sections; this step facilitates the action of collagenase.

(b) Collagenase is particularly damaging to oocytes if the incubation is carried out in the presence of calcium, which may be due to any Ca-dependent protease activity. To prevent this, rinse oocytes twice in Ca^{2+}-free OR2 prior to enzyme treatment.

(c) Pour the clumps into 50-ml Falcon tubes with collagenase in OR2. Removal of the follicle cells is facilitated by gentle shaking during the enzyme treatment. Put the tubes onto a mixer and shake gently for at least 120 min.

(d) A less damaging method is to use collagenase for shorter periods of time to achieve partial defolliculation, which may then be completed manually or mechanically. Follicular layers readily adhere to plastic containers, especially in a Ca^{2+}-containing medium; it is also possible to remove the follicle layer from the oocyte by taking advantage of this effect after short collagenase treatment.

(e) Rinse oocytes with NDE-96 medium (see "Recipes," below).

3. Removal of the vitelline membrane from defolliculated oocytes is not necessary for injection of RNA.
4. Check each individual oocyte manually with a stereomicroscope. If stage V and VI oocytes are chosen for RNA injection, oocytes with diameters >1 mm that have a distinct boundary between hemispheres should be selected. The remaining, less dense, immature stage or damaged oocytes may be discarded.
5. After selection, oocytes should be rested for several hours before injection to allow for any additional oocyte death to occur. There is no disadvantage to incubating oocytes overnight prior to RNA injection. In addition, NDE-96 medium (see "Recipes," below) typically contains gentamicin or penicillin/streptomycin (unless otherwise stated) to prevent bacterial contamination.

Recipes

- NDE-96 medium: 96 mM NaCl, 2 mM KCl, 1 mM $MgCl_2$, 1.8 mM $CaCl_2$, 5 mM HEPES, pH 7.6 (adjusted by NaOH)
- OR2: 82.5 mM NaCl, 2 mM KCl, 1 mM $MgCl_2$, 5 mM HEPES, pH 7.5 (adjusted by NaOH)

23.3.2.2. RNA Preparation

The RNA can be extracted from tissues or synthesized in vitro from cloned cDNA (14). The injection of tissue-extracted mRNA serves to archive the expression of specific genes not yet cloned or to supply unknown accessory protein factors that might influence the function of a known channel by co-injection with a synthetic RNA.

Higher levels of expression of the coding proteins can be achieved by injection of RNA. Cloned DNA may be subcloned into a transcription vector from which in vitro production of synthetic complementary RNA (cRNA) may be directed by the RNA polymerase II-binding site of a bacteriophage such as SP6, T7, or T3. If the message does not translate well in the oocyte because of poor stability or poor ribosome binding, the coding region can be cloned in a plasmid designed for expression in *Xenopus* oocytes, such as pSP64T or pBSTA. These plasmids contain the *Xenopus* β-globin 5′ and 3′ untranslated mRNA regions, with a poly(A) tail on the 3′ end. This enhances both stability and translatability of some messages and can result in significant increases in the level of expression.

Procedure D: In Vitro Transcription

1. The plasmid DNA should be cut with a restriction enzyme that linearizes the plasmids past the 3′ end of the coding region. For linearization, it is best to use a restriction enzyme that leaves either a 5′ overhang or blunt end because a 3′ overhang can function as a primer for synthesis in the wrong direction, making antisense RNA with the potential to interfere with translation of RNA.
2. After linearization, DNA should be extracted with phenol–chloroform–isoamyl alcohol, precipitated with isopropanol, and resuspended in RNase-free water for use as a transcription template.
3. In our laboratory, in vitro transcription is performed from the linearized plasmid using the mMESSAGEmMASHINE high-yield capped RNA transcription kit (Life Technologies), according to the manufacturer's instructions.
4. RNA for injection should be precipitated with ethanol or LiCl to remove excess salt and detergents, which can have a markedly deleterious effect on the health of oocytes. After precipitation, the RNA should be resuspended at an appropriate concentration in RNase-free water. It can be stored at −70°C.

23.3.2.3. RNA Injection and Expression

RNA Injection Setup

RNA injection (15) into *Xenopus* oocytes can be performed using simple experimental equipment that includes the following: a stereomicroscope with variable magnification and a good depth of field; a coarse micromanipulator (e.g., MM3; Narishige, Tokyo, Japan) permitting smooth vertical and horizontal movement needed to impale oocytes; a microinjector (e.g., Picoliter Pressure Injector PLI-100; Warner Instruments, Hamden, CT, USA); sterile glass tubing for fabricating microneedles; and a fiberoptic light source.

Procedure E: RNA Injection

1. Thaw the RNA solution at room temperature and spin down samples to collect the RNA at the bottom of the Eppendorf tube. Store the sample on ice until it is required for use.
2. Make glass injection micropipettes by using a pipette puller. After pulling pipettes with one step, break the glass tips off with a sterile needle so the tip diameter is 20–40 μm.
3. Fill the injection micropipette tip with 1–3 μl of RNA solution.
4. After mounting the micropipette onto the microinjector, adjust the injection setup to expel the desired injection volume of RNA solution.

5. Place the selected oocytes in NDE in a 35-mm tissue culture dish under a stereomicroscope. Polypropylene mesh can be fixed to the bottom of the dish to keep the oocytes from moving.
6. Impale the oocyte using the manipulator. Successful impalement is identified by sudden disappearance of the dimpled membrane surface around the micropipette tip. Inject up to 100 nl of RNA (typically 40–50 nl is preferred) into each oocyte.
7. Retract the injection micropipette from the oocyte. Expel a small volume of RNA solution between each injection to ensure that the micropipette tip does not become blocked and to avoid dilution of RNA with intracellular contents from previously injected oocytes.
8. Repeat the procedure with six or seven additional oocytes.
9. After microinjection, transfer oocytes into sterile dishes containing NDE at 18–20°C. Any oocytes acutely damaged by the injection procedure quickly become evident and should be removed.
10. NDE should be replaced every 24 h with fresh sterile NDE and any damaged or dead oocytes removed. To permit adequate ion channel expression, oocytes injected with RNA usually require incubation for up to 2–3 days.
11. Once the oocytes express the receptor/ion-channel proteins of interest, they should be stored at ~10°C and the NDE replaced every 3–4 days (any damaged oocytes should continue to be removed daily). The low temperature prolongs survival of the oocytes. How long oocytes can be used for electrophysiology depends on the type of ion channels being expressed and the peak current amplitudes.

Note: The amount of RNA required for injection depends on the type of RNA. Usually, 50 nl/oocyte of 0.1–1.0 μg cRNA/μl (i.e., 5–50 ng/oocyte) is used for expression studies. It is important to optimize the amount of cRNA injection for each message.

23.3.2.4. Electrophysiological Recordings

Two-Electrode Voltage-Clamp

Ion channels expressed in *Xenopus* oocytes can be studied using the two-electrode voltage-clamp (TEVC) technique (10, 16). Multielectrode voltage-clamp amplifiers such as GeneClamp 500B, Axoclamp 900A (Molecular Devices, Sunnyvale, CA, USA), Oocyte Clamp OC-725C (Warner Instruments), and iTEV 90 (HEKA, Lambrecht, Germany) are designed to have high-voltage output compliance to clamp the large currents generated by ion channel-expressing oocytes. The oocyte is impaled with two microelectrodes. One electrode is used to record the intracellular potential, and the second is used to pass the current necessary to clamp the membrane potential to the command voltage in a negative feedback loop.

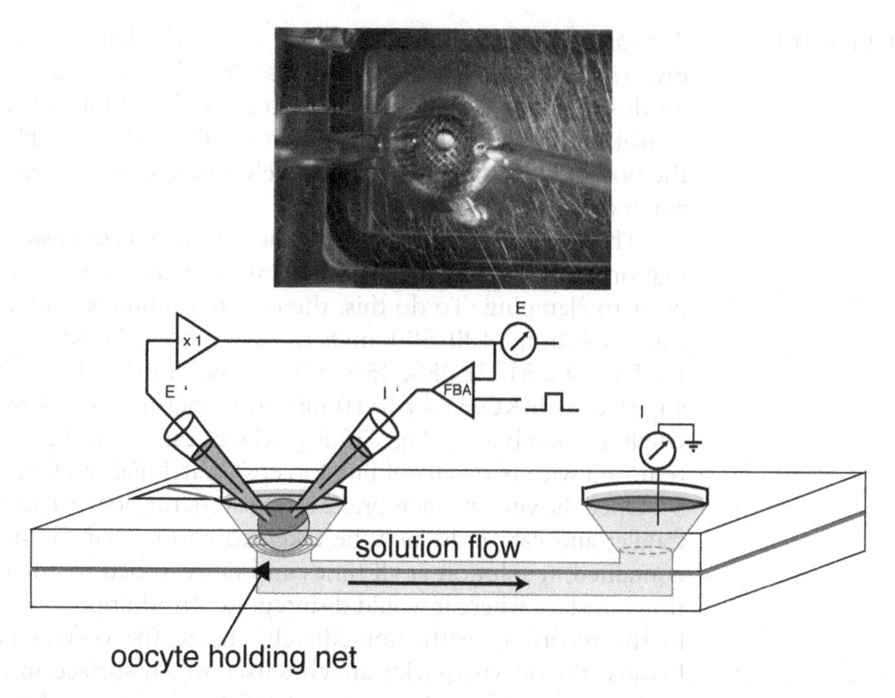

Fig. 23.4. Recording chamber for *Xenopus* oocytes. This perfusion chamber is designed for oocyte two-electrode voltage-clamping. An oocyte is confined on a polypropylene mesh fixed to the chamber. Solution can be introduced via gravity-feed and removed via suction. It features a narrow well for applications requiring rapid perfusion exchange.

The large size of the oocyte makes this feasible. The follicular cell envelope can be removed or left in situ for this recording method. The oocyte needs to be held firmly in position after impalement with the recording electrodes. Immobilization of the oocyte in the recording chamber can be achieved by positioning within a trench cut in the base of the bath previously coated with a layer of Sylgard resin (Dow Corning, Seneffe, Belgium). Alternatively, the oocyte can be placed on a polypropylene mesh fixed to the chamber. The authors utilize a custom-built chamber, as shown in Fig. 23.4.

Standard intracellular electrodes are made from capillary glass containing thin filaments giving a resistance of 0.4–1.0 MΩ. The low resistance is required to allow the passage of sufficient current to voltage-clamp the oocyte. These glass electrodes can be back-filled with 3 M KCl or K acetate solution and placed into each of the two holders, making sure that the Ag/AgCl wire makes contact with the pipette solution. The silver chloride coat is typically renewed before each experiment.

Impalement of an oocyte causes dimpling of the cell surface. Successful impalement of the electrodes is registered by the appearance of an electrotonic potential recorded by the voltage electrode. Gradual disimpalement of the electrodes to reduce membrane surface dimpling, followed by sealing of the electrodes, results in a viable oocyte, allowing continuous, stable recording. These recordings are similar to other voltage-clamp experiments in small cells.

Excised Patch Method

Xenopus oocytes can also be used for patch-clamp experiments involving single-channel recordings (9, 17, 18). Cell-attached, inside-out, and outside-out patches can be formed using methods similar to those employed for much smaller cells (19). However, the oocyte is too large to be effectively whole-cell clamped using a patch electrode.

The vitelline membrane surrounding the oocyte obstructs formation of the seal on the oocyte membrane and must be removed prior to clamping. To do this, the oocyte is immersed in a hypertonic solution (400–500 mOsm) (e.g., 60 mM KCl, 10 mM EGTA, 40 mM HEPES, 250 mM sucrose, 8 mM $MgCl_2$, pH 7.0 adjusted with KOH) for 5–10 min to retract the oocyte from the vitelline membrane (Fig. 23.5), which can then be manually removed with two pairs of fine forceps (Watchmaker's forceps, no. 5). Once the vitelline membrane is removed, the oocyte is extremely fragile; and care must now be taken to ensure that the oocyte is contained in solution at all times and not exposed to an air–solution interface where it would disintegrate. In addition, once placed in the recording bath, immediately ensure the correct position because the oocyte quickly attaches itself to any surface and subsequent movement results in rupture of the plasma membrane with leakage of yolk platelets.

The method of obtaining a high-resistance seal on an oocyte is generally similar to that used for smaller cells, except it is not usually possible to see when the patch pipette is about to touch the surface of the oocyte (i.e., blind approach). Seal resistance increases upon touching the oocyte surface, and gentle suction often results in the development of a high-resistance seal. The formation of inside-out patches can be achieved by slow withdrawal of the micropipette from the oocyte. Outside-out patches require formation of the

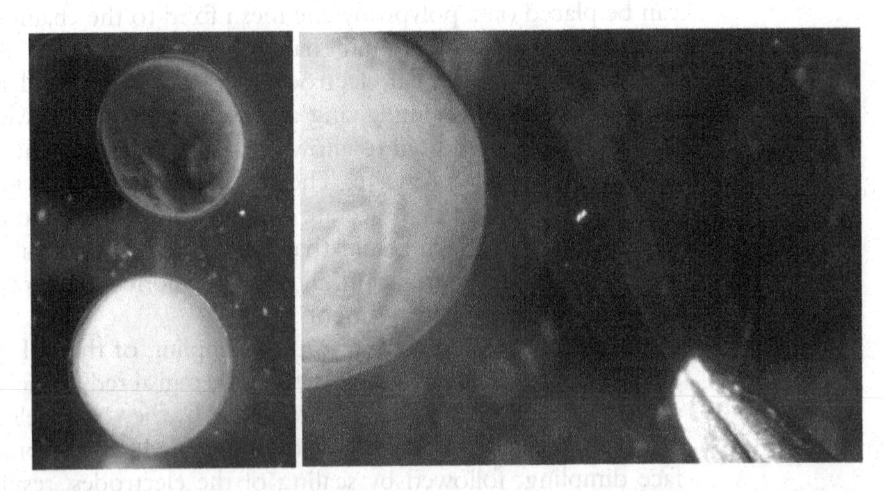

Fig. 23.5. Removal of the vitelline membrane surrounding an oocyte. *Left.* Hypertonicity-induced separation between vitelline and plasma membranes in oocytes. *Right.* Manual removal of vitelline membrane with fine forceps.

whole-cell recording mode before withdrawing the micropipette, as described previously (19).

Formation of giant or macro-patches requires the use of micropipettes with large tip diameters (2–8 μm) and low resistance (0.6–2.0 MΩ) (16). Detailed methods are described in Chap. 14.

There are a few advantages of using the macro-patch method over conventional TEVC. First, it enables current flowing through many ion channels to be recorded with better time resolution (50–200 μs for macro-patch compared to 0.5–2.0 ms for TEVC) because of the reduced membrane capacitance. In addition, when using the macro-patch method in an inside-out configuration, access is allowed to the cytoplasmic side of the membrane.

23.3.3. Advantages and Limitations of TEVC

- The *Xenopus* oocyte contains a highly efficient and faithful expression system that accepts a variety of RNAs from different species and produces functional receptor/ion-channel proteins.
- TEVC experiments are relatively easy to perform and can be used as a tool for first-line screening for a variety of wild-type and mutant RNAs. Moreover, TEVC is stable for a prolonged period, and extracellular solution changes are well tolerated.
- Currents from the RNA-injected oocyte are usually large.
- It is possible to express multiple proteins in the same cell in controlling the concentration of RNAs encoding each protein and thus examine channels with relatively well-controlled compositions.
- Oocytes express only a few endogenous channels that contribute but a small fraction of the total current in injected oocytes.
- The large size of the oocyte allows direct injection of RNA and membrane-impermeable compounds.
- *Xenopus* is easy maintained with minimal care and facilities.

- The major disadvantage of recording from the entire oocyte is that the large size and extensive membrane invaginations result in an extremely large membrane capacitance (approximately 100–250 nF). The large capacitance causes a slow clamp settling time when the membrane potential changes. Sometimes, small oocytes of earlier stages (II–III) may be used, providing faster decay to the transient capacitative current following brief voltage steps under voltage-clamp. If the speed of the clamp is insufficient, it may be more appropriate to use an alternative expression system.
- There is a certain degree of variability in the level of expression between oocyte batches. Oocyte quality is subject to wide variation due to seasonal and other factors, being worse particularly in summer.
- The presence of endogenous channels, including stretch-activated channels (18), Ca^{2+}-activated Cl^- channels (20–22),

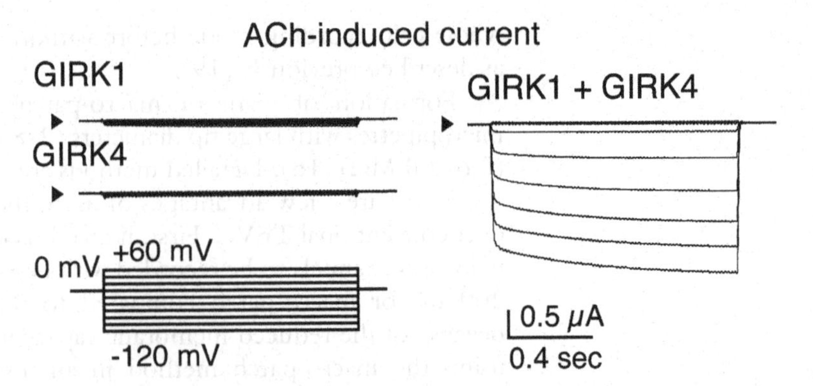

Fig. 23.6. Two-electrode voltage-clamp recording from *Xenopus* oocytes. Whole-cell currents recorded from *Xenopus* oocytes expressing GIRK1, GIRK4, and GIRK1 plus GIRK4, indicated at the top of each trace. All oocytes were also injected with cRNA encoding m2-muscarinic receptor. TEVC recording was done with 90 mM KCl in the bath solution. Representative traces show currents recorded in the presence of 1 μM acetylcholine (*ACh*).

and delayed rectifier K^+ channels (23), may interfere with the recording of currents, especially small currents.

- Channel distribution is not uniform in the cell membrane but tends to cluster in regions (24).

23.3.4. Example of Applications

Figure 23.6 illustrates the current from an oocyte co-expressing G-protein-gated potassium (GIRK) channel with m2 muscarinic receptor. GIRK channels are also a type of inwardly rectifying K^+ channels. There are four mammalian GIRK subunits: GIRK1–4 (Kir3.1–Kir3.4). GIRK1 combines with GIRK2 in the central nervous system and with GIRK4 in cardiomyocytes to form heterotetramers (2, 25). When the GIRK channel was co-expressed with the m2-muscarinic receptor in oocytes, the GRIK channel was activated by an agonist (acetylcholine) and demonstrated inward-rectifying currents.

23.4. Conclusion

This chapter describes practical techniques for electrophysiological experiments in heterologous expression systems using mammalian cell cultures and *Xenopus* oocytes. These two experimental systems are the most commonly used for expression of ion channels and transporters, contributing to numerous discoveries about their function and molecular properties. When planning an experiment, researchers need to determine whether either system is appropriate for use. It is crucial to understand the advantages and disadvantages of each system. Single-channel analysis can be carried out relatively easily in mammalian cells compared with *Xenopus* oocytes, and channel kinetics can be analyzed in detail. On the other hand, the *Xenopus* oocyte expression system ensures simultaneous introduction of multiple cRNAs into a cell. For this reason, this system is considered to have the advantage when performing experiments to verify the interaction of two or more proteins, such as receptor-induced modulation of ion channels or the formation of ion channel complexes. In addition, TEVC of oocytes is easily applied to rapid screening of ion channel function, particularly in pharmacological experiments.

References

1. Felgner PL, Gadek TR, Holm M et al (1987) Lipofection: a highly efficient, lipid-mediated DNA-transfection procedure. Proc Natl Acad Sci USA 84:7413–7417
2. Hibino H, Inanobe A, Furutani K et al (2010) Inwardly rectifying potassium channels: their structure, function, and physiological roles. Physiol Rev 90:291–366
3. Furutani K, Ohno Y, Inanobe A et al (2009) Mutational and in silico analyses for antidepressant block of astroglial inward-rectifier Kir4.1 channel. Mol Pharmacol 75:1287–1295
4. Ohno Y, Hibino H, Lossin C et al (2007) Inhibition of astroglial Kir4.1 channels by selective serotonin reuptake inhibitors. Brain Res 1178:44–51
5. Su S, Ohno Y, Lossin C et al (2007) Inhibition of astroglial inwardly rectifying Kir4.1 channels by a tricyclic antidepressant, nortriptyline. J Pharmacol Exp Ther 320:573–580
6. Gundersen CB, Miledi R, Parker I (1983) Voltage-operated channels induced by foreign messenger RNA in *Xenopus* oocytes. Proc R Soc Lond B Biol Sci 220:131–140
7. Gurdon JB, Lane CD, Woodland HR et al (1971) Use of frog eggs and oocytes for the study of messenger RNA and its translation in living cells. Nature 233:177–182
8. Sumikawa K, Houghton M, Emtage JS et al (1981) Active multi-subunit ACh receptor assembled by translation of heterologous mRNA in *Xenopus* oocytes. Nature 292:862–864
9. Conforti L, Sperelakis N (2001) The patch-clamp technique for measurement of K^+ channels in *Xenopus* oocytes and mammalian expression systems. In: Archer SL, Rusch NJ (eds) Potassium channels in cardiovascular biology. Springer, New York
10. Dascal N (2001) Voltage clamp recordings from *Xenopus* oocytes. Curr Protoc Neurosci Chapter 6: Unit 6.12
11. Smart TG, Krishek BJ (1995) *Xenopus* oocyte microinjection and ion-channel expression. In: Boulton AA, Baker GB, Walz W (eds) Patch-clamp applications and protocols. Humana Press, New York
12. Goldin AL (1992) Maintenance of *Xenopus laevis* and oocyte injection. Methods Enzymol 207:266–279
13. Dumont JN (1972) Oogenesis in *Xenopus laevis* (Daudin) I. Stages of oocyte development in laboratory maintained animals. J Morphol 136:153–179
14. Goldin AL, Sumikawa K (1992) Preparation of RNA for injection into *Xenopus* oocytes. Methods Enzymol 207:279–297
15. Soreq H, Seidman S (1992) *Xenopus* oocyte microinjection: from gene to protein. Methods Enzymol 207:225–265
16. Stuhmer W (1992) Electrophysiological recording from *Xenopus* oocytes. Methods Enzymol 207:319–339
17. Choe H, Sackin H (1997) Improved preparation of *Xenopus* oocytes for patch-clamp recording. Pflugers Arch 433:648–652
18. Methfessel C, Witzemann V, Takahashi T et al (1986) Patch clamp measurements on *Xenopus laevis* oocytes: currents through endogenous channels and implanted acetylcholine receptor and sodium channels. Pflugers Arch 407: 577–588
19. Hamill OP, Marty A, Neher E et al (1981) Improved patch-clamp techniques for high-resolution current recording from cells and cell-free membrane patches. Pflugers Arch 391:85–100
20. Barish ME (1983) A transient calcium-dependent chloride current in the immature *Xenopus* oocyte. J Physiol 342:309–325
21. Miledi R (1982) A calcium-dependent transient outward current in *Xenopus laevis* oocytes. Proc R Soc Lond B Biol Sci 215:491–497

22. Schroeder BC, Cheng T, Jan YN et al (2008) Expression cloning of TMEM16A as a calcium-activated chloride channel subunit. Cell 134:1019–1029
23. Lu L, Montrose-Rafizadeh C, Hwang TC et al (1990) A delayed rectifier potassium current in *Xenopus* oocytes. Biophys J 57:1117–1123
24. Lopatin AN, Makhina EN, Nichols CG (1998) A novel crystallization method for visualizing the membrane localization of potassium channels. Biophys J 74:2159–2170
25. Yamada M, Inanobe A, Kurachi Y (1998) G protein regulation of potassium ion channels. Pharmacol Rev 50:723–760

Chapter 24

Patch-Clamp and Single-Cell Reverse Transcription–Polymerase Chain Reaction/Microarray Analysis

Akihiro Yamanaka

Abstract

If gene expression could be analyzed after electrophysiological recording of a single cell, it would enhance our molecular understanding of neuronal physiology and help to interpret the results obtained from patch-clamp recordings. For instance, after completing a recording from a neuron in a brain slice preparation, analysis of the cell's mRNA could reveal the kind of neurotransmitter the cell used. Detection of an ion channel's mRNA would provide molecular evidence of the presence of the ion channel as predicted by the electrophysiological experiment. However, the amount of mRNA in single (neuronal) cell is very small. Therefore, analysis of mRNA from a single cell requires special isolation techniques for the mRNA and an accurate amplification step. Recent remarkable advances in molecular biology now allow us to apply reverse transcription–polymerase chain reaction methods to the analysis of specific gene expression as well as microarray analysis of the entire transcriptome from a single cell.

24.1. Patch-Clamp Reverse Transcription–Polymerase Chain Reaction Method

This chapter introduces a single-cell reverse transcription–polymerase chain reaction (single-cell RT-PCR) method that can be used after slice patch-clamp recording from a neuron. This method can also be applied to the analysis of gene expression in a single cell from a cultured cell line. The procedure is the same in each case. After electrophysiological recordings, mRNA is collected from the cytoplasm by use of a fine glass pipette and negative pressure in the pipette. Next, cDNA is synthesized from mRNA by adding reverse transcriptase. The target gene is then detected using the PCR.

Yasunobu Okada (ed.), *Patch Clamp Techniques: From Beginning to Advanced Protocols*, Springer Protocols Handbooks, DOI 10.1007/978-4-431-53993-3_24, © Springer 2012

24.2. Preparation of the Glass Pipette and Solutions

The glass pipette is washed with 100% ethanol and rinsed several times with distilled water (MilliQ water). The glass pipette is then boiled in MilliQ water in a microwave oven and sterilized by dry heating. After this step, the glass pipette should be handled with sterilized tweezers or sterile gloves. Note that the center of the glass pipette, which contains a part to be the tip of an electrode after pulling, must not be touched. The glass pipette holder and the silver wire are cleaned by treatment with RNasesZAP (Ambion AM9780), rinsed with MilliQ water, and dried well.

24.3. Constitution of Pipette Solution

The K-gluconate pipette solution consists of the following: 145 mM K-gluconate, 3 mM $MgCl_2$, 10 mM HEPES, and 0.2 mM EGTA, pH 7.2, in nuclease-free water. Sterilized glassware should be used to prepare the pipette solution. The pipette solution is filter-sterilized by passage through a 0.2-μm filter before use. Samples are aliquoted into sterilized tubes with O-rings and stored in a deep freezer (−80°C). On the day of the experiment, the pipette solution is taken up in a new sterile syringe (1 ml), which is attached to a new sterilizing filter (NALGENE, 0.2 μm, 176–0020). To fill the glass pipette with the pipette solution, a disposable fine loading tip (Eppendorf, Microloader, 5242956. 003) is attached to the filter. Connection between the loading tip and the filter is adjusted via an adapter, such as a yellow tip, which is cut to an appropriate length.

24.4. Whole-Cell Recording

Please refer to other chapters for the preparation of brain slices. The tip of the glass pipette should be >1 μm. Electrical resistance of about 1–3 MΩ is preferable. Light positive pressure is applied to the solution when approaching the cell. For the usual slice patch-clamp, after making a giga-seal the patch membrane is ruptured by applying further negative pressure or zapping to make a whole cell. Under this condition electrophysiological recordings are performed. Note, however, that this cell condition is not good for extended recording.

24.5. Retrieving the Cytoplasm

After checking the state of the giga-seal with a seal test, the cytoplasm is slowly withdrawn into the glass pipette by applying negative pressure. The giga-seal should be maintained during retrieval. The cell gradually shrinks during the retrieval, which can be seen with infrared differential interference contrast (IR-DIC) imaging (Fig. 24.1). At this point, if the nucleus (which is very firm) plugs the tip of the glass pipette, collection of cytoplasm stops. Ideally, more than 90% of the cytoplasm is collected into the pipette, but 50–60% is sufficient in some cases.

24.6. Reverse Transcription

After collecting a sufficient quantity of cytoplasm, the glass pipette is slowly detached from the cell. At this point, if the negative pressure is too high the pipette can aspirate unwanted material. Therefore, strong negative pressure is not recommended (around 1 atmosphere of pressure is recommended). Adherence of cellular material to the glass pipette might also cause contamination. The pipette is slowly raised and carefully pulled out from the extracellular fluid. *The glass pipette should not be pulled out of the pipette holder.* The tip of the pipette is carefully inserted into a 1.5- or 0.2-ml tube. The tip of the pipette is broken by gently touching it to the bottom of the tube, which collects the cytoplasm and broken glass. Gentle contact is enough to transfer the cytoplasm into the tube. Note that the cytoplasm from several cells can be collected into one tube. The tube is then placed on finely crushed Dry Ice, and the preceding steps can be repeated. By repeating this step, the cytoplasm from several cells or dozens of cells can be collected in a single tube. After collection is completed, 10 μl of the following solution is immediately added to the tube and stirred well by vortexing or tapping.

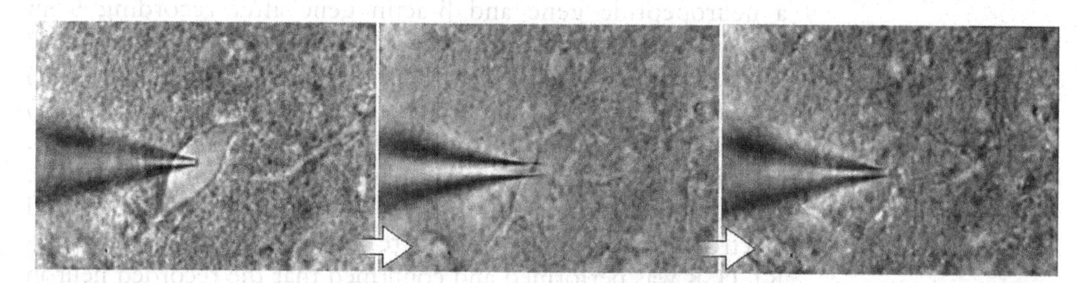

Fig. 24.1. Cytoplasm retrieval after slice patch-clamp recording. The retrieval into a glass pipette was performed by applying negative pressure (*from left to right*). During retrieval the cell volume decreases, and the edge of the cell becomes unclear (*middle panel*), with only the nucleus remaining (*right panel*).

Solution composition

Nuclease-free water (Promega)	7.5 μl
RNase Inhibitor (40 U/μl; Promega)	0.5 μl
$(dT)_{15}$ Primer or Random 6 Primer (50 μM)	1 μl
dNTP mix (2.5 mM each; TaKaRa)	1 μl
Total (volume of pipette solution is ignored)	10 μl

The tube is then incubated for 5 min at 65°C followed by chilling on ice.

Reverse transcriptase (PrimeScript Reverse Transcriptase, Cat. #2680A; TaKaRa) is added and stirred gently. The tube is then incubated for 10 min at 30°C and for 60 min at 42°C. During this step, reverse transcriptase synthesizes the complementary first strand cDNA from mRNA.

Above-described template RNA/primer mixture	10 μl
5× PrimeScript Buffer (TaKaRa)	4 μl
PrimeScript RT (TaKaRa)	1 μl
Nuclease-free water	5 μl
Total	20 μl

The tube is incubated for 15 min at 70°C and kept at 4°C. Sterile tubes and tips with filters are used until the end of the reverse transcription step. It is also recommended that the worker wear a mask and that the surface of the bench is wiped with RNase Zap or overlayed with new aluminum foil or new wrap during the procedure.

24.7. Detection of Gene Expression

Using this method, the author successfully detected expression of a neuropeptide gene and β-actin gene after recording from hypothalamic neurons in a mouse brain slice preparation. Transgenic mice in which orexin-producing neurons (orexin neuron) specifically expressed enhanced green fluorescent protein (EGFP) were used. The brains were isolated from these transgenic mice, and slices were prepared at a thickness of 350 μm using a vibratome. Orexin neurons were identified by EGFP fluorescence, and electrophysiological analyses were performed. After recording, RT-PCR was performed and confirmed that the recorded neuron contained prepro-orexin mRNA (1). All primers used for RT-PCR are designed in the exon to sandwich the intron.

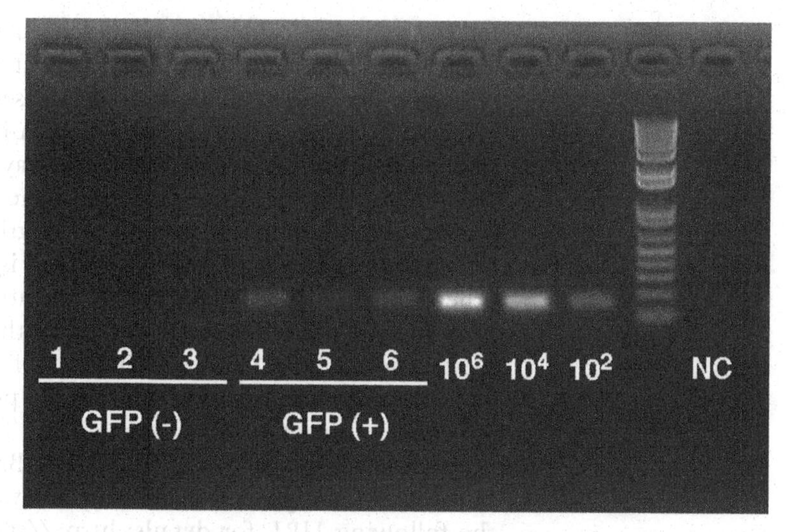

Fig. 24.2. Detection of gene expression. *GFP*, green fluorescent protein.

2× Go Taq buffer (Promega, M5112)	5 μl
cDNA	1 μl
Nuclease-free water (Promega)	3.92 μl
Forward primer (100 μM)	0.04 μl
Reverse primer (100 μM)	0.04 μl
Total	10 μl

PCR is performed using the following protocol. The tube is incubated at 95°C for 3 min. Next, the tube is incubated at 95°C for 15 sec, 55°C for 15 sec and 72°C for 15 sec. This step is repeated for 40 times. Then the tube is incubated at 72°C for 3 min. This should be adjusted in each primer pair. 95°C, 3 min; (95°C for 15 s; 55°C for 15 s; 72°C for 15 s) × 40, 72°C 3 min.

Using brain slice preparations from transgenic mice in which orexin neurons specifically expressed EGFP, single neuronal cell RT-PCR was performed using primers for the prepro-orexin gene (Fig. 24.2). Lanes 1, 2, and 3 in Fig. 24.2 are from a neuron not expressing GFP (non-orexin neuron); and lanes 4, 5, and 6 are from a neuron expressing GFP (orexin neuron). Lanes 7, 8, and 9 are 1×10^6 to 1×10^2 copy numbers of plasmid; and NC is a negative control (pipette solution only).

24.8. Microarray Analysis from a Single Cell

Microarray gene expression analysis is a powerful tool for analyzing all the genes expressed in a single cell – that is, the transcriptome.

Recent remarkable advances in molecular biology, which include improvements in reagents and equipment for analyses, enable the researcher to analyze all the genes expressed in a single cell (2–4). However, the amount of isolated mRNA from a single cell is too small to analyze by conventional microarray techniques. Therefore, preliminary amplification using PCR is necessary. Because of bias during amplification (sequence or length of genes), however, quantitative analysis of the microarray is problematic. On the other hand, qualitative analysis is not affected by amplification bias. Included here is a brief outline of the methods that can be used. They consist of a combination of commercially available gene amplification kits and microarray products available from Affymetrix Corporation (Tokyo, Japan). The method introduced here is a modification of the Integrative Brain Research Program, "Protocol for single-cell microarray analysis" (2). Please refer to the following URL for details: http://srv02.medic.kumamoto-u.ac.jp/dept/morneuro/r_micro.html/.

- The cytoplasm is collected and transferred to a tube using the same method for single-cell RT-PCR. cDNA is synthesized using a commercially available reagent kit (TaKaRa Bio, Shiga, Japan, SMARTer Pico PCR cDNA Synthesis kit, #634928) that permits synthesis of cDNA from a very small amount of mRNA. Subsequently, cDNA is amplified by PCR.
- The quality and amplification of genes is verified using real-time PCR for quantitative determination of a housekeeping gene or a target gene.
- Biotin-labeled cDNA synthesis (in vitro transcription, IVT): Biotin-labeled cDNA is synthesized using the IVT Labeling Kit (#900449; Affymetrix) (37°C, 16 h). The product is then purified through the use of the Sample Cleanup Module (#900371; Affymetrix).
- cRNA quantitation: Synthesized cRNA is quantitated with the Bioanalyzer 2100, RN A6000 nano kit (Agilent, CA, USA).
- cRNA fragmentation: Synthesized cRNA is fragmented through the use of fragmentation buffer in the sample cleanup module incubation at 94°C for 35 min. Fragmentation of cRNA is checked with the Bioanalyzer.
- Preparation of the hybridization cocktail: The hybridization cocktail is prepared by adding fragmented cRNA and incubating the mixture at 99°C for 5 min. The cocktail is then incubated at 45°C for 5 min, after which it is centrifuged at 15,000 rpm for 5 min. The supernatant is used for hybridization.
- Hybridization for microarray: Various types of GeneChips are available. We use the Mouse Genome 430A 2.0 Array GeneChip, which permits analysis of 14,000 genes. Hybridization buffer is applied. Then the GeneChip is

prehybridized at 45°C at 60 rpm for 10 min in the Hybridization Oven 645. Hybridization buffer (1×) is replaced by the hybridization cocktail. Hybridization is done at 45°C at 60 rpm for 16–18 h.

- Washing, fluorescent dye labeling, and scanning: The GeneChip is washed and labeled with a fluorescent dye using the Wash and Stain kit (#900720; Affymetrix). Fluidics Station 450 automates this step. After the washing step, the GeneChip is analyzed by a scanner (Scanner 3000 7G).
- Data analysis: First, hybridization of control RNA or expression of a housekeeping gene is examined to determine whether hybridization and labeling were performed correctly. The gene expression is then analyzed. Genespring GX (Agilent Technologies) analysis software is useful for various statistical analyses. Excel can be used for simple analyses of present/absent calls. The experiment is repeated several times (use several arrays). Gene expression is determined by checking the "present" calls in more than one array.

References

1. Yamanaka A, Beuckmann CT, Willie JT, Hara J, Tsujino N, Mieda M et al (2003) Hypothalamic orexin neurons regulate arousal according to energy balance in mice. Neuron 38(5):701–713
2. Esumi S, Wu SX, Yanagawa Y, Obata K, Sugimoto Y, Tamamaki N (2008) Method for single-cell microarray analysis and application to gene-expression profiling of GABAergic neuron progenitors. Neurosci Res 60(4):439–451
3. Kurimoto K, Saitou M (2010) Single-cell cDNA microarray profiling of complex biological processes of differentiation. Curr Opin Genet Dev 20(5):470–477
4. Tsuzuki K (1998) Analysis of molecular basis of neuronal properties using the patch-clamp RT-PCR method. Nippon Rinsho 56(7):1681–1687

Chapter 25

Smart-Patch Technique

Ravshan Z. Sabirov, Yuri E. Korchev, and Yasunobu Okada

Abstract

When a patch pipette approaches a cellular surface at distances comparable to the size of the tip opening, the monitored pipette current drops before the pipette touches the membrane. Fine nano-tipped glass electrodes can effectively function as "spherical sensors" that can "roll" over surface irregularities in the specimen without damaging it. By keeping the pipette current constant (so the distance between the pipette tip and the sample surface remains constant) and scanning the sample, it is possible to obtain an image of its surface without physical contact between the pipette and the surface. Thus, one can first image the cellular surface using fine-tipped patch pipettes with scanning ion conductance microscopy (SICM) methodology and then position the same pipette precisely at locations of interest (e.g., scallop crests, T-tubules in cardiac myocytes, neuronal dendrites); finally, a giga-ohm seal can be formed according to the conventional patch-clamp procedure. This smart combination of patch-clamp and SICM methods is called the smart-patch technique. It allows precise mapping of ion channels over the cellular surface.

25.1. Introduction

When studying cells by the conventional patch-clamp technique, the precise position of the pipette tip over the cell surface is often uncertain. Researchers tend to place the micropipette electrode at the cell membrane somewhere above the nucleus because this is the thickest region of the cell body, and there is therefore less danger of colliding with the bottom of the experimental chamber. With the conventional patch-clamp technique, it is still possible to patch different areas of the cell by using higher magnification and good optics. However, even using a phase-contrast or differential interference contrast (DIC) microscope, fine details of the cellular surface (e.g., grooves, protrusions, microvilli) are not clearly visible. The recently developed scanning ion conductance microscopy (SICM) methodology allows imaging of the cellular surface with submicron or even nanometer resolution using very fine patch

Yasunobu Okada (ed.), *Patch Clamp Techniques: From Beginning to Advanced Protocols*,
Springer Protocols Handbooks, DOI 10.1007/978-4-431-53993-3_25, © Springer 2012

pipettes. Following surface imaging, the same pipette can be positioned precisely at the area of interest and used to form giga-ohm contacts and electrically isolate membrane patches. This smart combination of patch-clamp and SICM methods is called the smart-patch technique.

25.2. Principle of SICM

Scanning probe microscopy (SPM) is a versatile tool that generates images of physical surfaces using a probe that interacts with the surface of interest and scans the specimen. Raster scans result in a line-by-line record of the probe–surface interaction as a function of probe position, which is then converted to a pseudo-colored image by a computer program. The probe used in SPM can vary in nature. The most widely used SPM, atomic force microscopy (AFM), employs a sharp tip at the end of a flexible cantilever as a mechanical probe. The information is gathered by "feeling" the surface with this probe. However, imaging the surface of living cells has been rather difficult because the mechanical probe often damages the soft cellular surface.

In 1989, Hansma and colleagues (1) suggested employing a glass microelectrode filled with an electrolyte solution as a probe for SPM. The ionic current flowing through the electrode tip decreases when the micropipette approaches the surface of a non-conducting substrate. Thus, by using the ionic current through the pipette tip as a feedback signal for SPM, it was possible to obtain images of the surface of polymeric films immersed in conductive fluids (2). The first high-resolution images of living cells were obtained by Korchev and colleagues (3–7) using fine glass microelectrodes pulled with standard pipette pullers; the ionic currents were measured with standard patch-clamp electronics.

The basic principles of surface sensing by a microelectrode are illustrated in Fig. 25.1. The overall resistance of the tip is a combination of the resistance of the micropipette itself (R_P) and the access resistance (R_{AC}) of the micropipette opening with radius r. The pipette resistance R_P can be considered the resistance of the electrolyte inside the pipette tip, which can be represented as a frustum of a right circular cone. The total current (I) flowing through the pipette can be expressed as follows:

$$I = \frac{V}{(R_P + R_{AC})} \tag{25.1}$$

where V is the voltage applied to the electrode.

The access resistance is defined as the resistance along the convergent paths from the bath to the micropipette opening. It is the

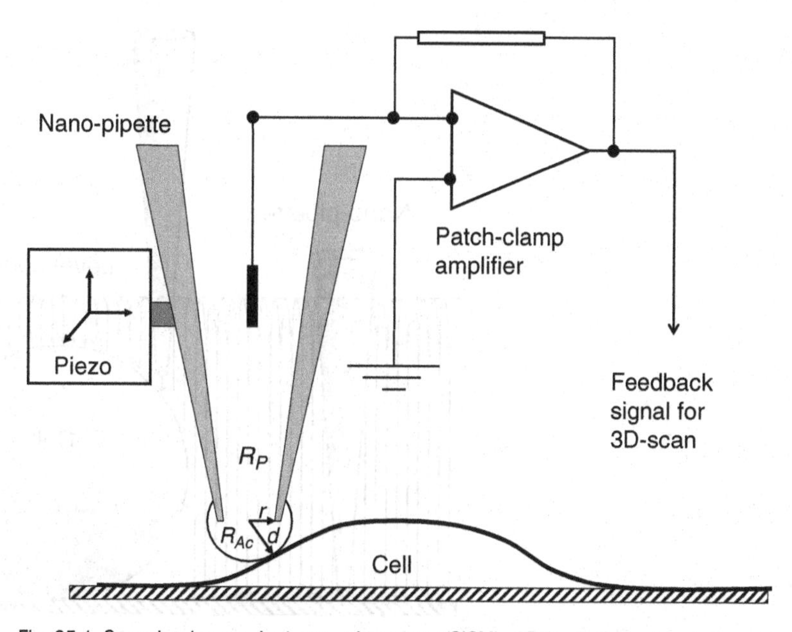

Fig. 25.1. Scanning ion conductance microscopy (SICM). *3D* three-dimensional, R_P resistance of the micropipette, R_{AC} access resistance of the micropipette, *r* radius of the micropipette, *d* distance between the tip and the sample.

access resistance that changes depending on the distance from the tip to the cellular surface. R_{AC} is a complex function of the distance (*d*) between the tip and the sample and the geometry and electrochemical properties of the sample surface. A fine nano-tipped glass electrode effectively functions as a "spherical sensor" of diameter *d* that can "roll" over surface irregularities in the specimen without damaging it, as depicted in Fig. 25.1. By keeping the pipette current constant (and hence a constant distance *d* between the pipette tip and the sample surface) and scanning the sample, it is possible to obtain an image of its surface without physical contact between the pipette and the surface. Instead of measuring the direct current (DC) at a constant voltage, one can measure the pipette resistance by applying either voltage steps in the voltage-clamp mode or current steps in the current-clamp mode (termed pulse-mode SICM (8, 9). To diminish the effects of pipette DC current (I_{DC}) drifting, one can use vertical distance modulation (10) with an amplitude of about one tip diameter and a frequency of 200–1,000 Hz. This modulation produces a modulated current (I_{MOD}), which is then used as a feedback signal instead of I_{DC}. The modulated current is minimal when the probe is far from the surface but increases faster than I_{DC} upon approach of the sample.

In its classic form, SICM using the pipette current I_{DC}, the modulated current I_{MOD}, or the pipette resistance as feedback signals can produce good images of relatively flat cells (e.g., epithelial cells, fibroblasts, cultured astrocytes). However, when the pipette

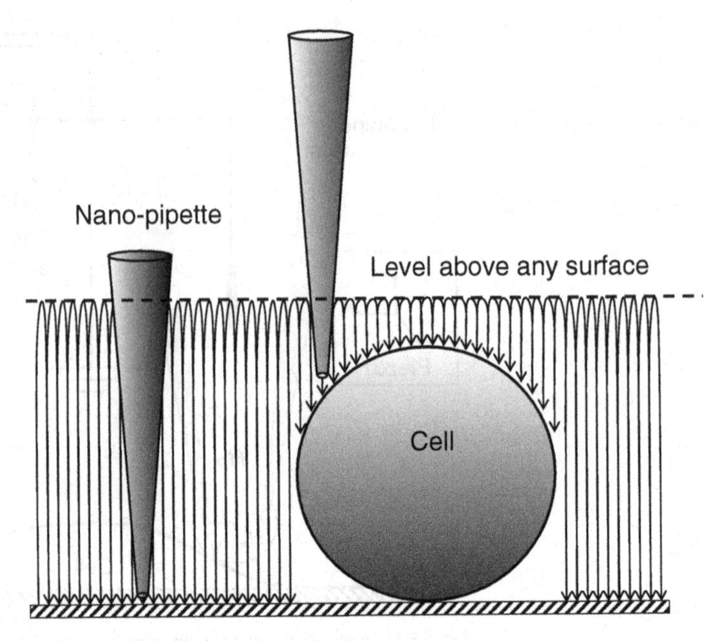

Fig. 25.2. Principle for hopping probe ion conductance microscopy (HPICM).

encounters a steep vertical surface such as that of the spherical cell illustrated in Fig. 25.2, it may collide before the pipette tip "senses" the presence of the surface. The recently developed hopping probe ion conductance microscopy (HPICM) (11) circumvents this problem and can obtain clear images of samples with very steep surfaces and complex convoluted structures. The principle of HPICM is illustrated in Fig. 25.2. Here, the pipette is kept at a level well above any surface, and the reference current is measured. The pipette is then lowered until it approaches the surface, and the current is reduced by a predefined value, which is usually approximately 0.25–1.0% of its initial value. The *Z*-value at which this is achieved is recorded as the height of the sample surface at this specific imaging point. The pipette is then lifted back to the starting level, above any surface. The sample is moved laterally to a new imaging point, and the procedure is repeated. Because lateral movement occurs while the pipette is raised, the pipette never collides with the sample. One advantage of HPICM is that the imaging points can be selected in any order. Thus, the sample can be imaged first at low resolution by dividing the overall surface into relatively large squares and assessing the heights at the corners of the squares. Then, the areas of high "roughness" can be selectively imaged at higher resolution. This approach, called an "adaptive scan," allows the imaging process to be sped up greatly. A similar method, named the "back-step mode" of SICM (8, 12), has been used for long-term imaging and cell volume measurements of single neuronal cells.

25.3. Experimental Design for SICM

The basic arrangement and operation of the scanning ion conductance microscope have been described in detail elsewhere (3–6, 13–15). With the SICM method, a fine-tipped pipette, called a nano-pipette, is used as a probe. The pipettes can be pulled from borosilicate glass capillaries (0.5–1.0 mm outer diameter with inner filaments) using a standard puller and a two-step pulling protocol. When filled with standard extracellular solutions, pipettes with a tip diameter of 50–100 nm give a resistance of approximately 80–100 MΩ. For smart-patch application, one may need to employ wider-tipped pipettes if the channel density is relatively low. Generally, cells can be imaged even with pipettes of 2–3 MΩ and a tip diameter of about 1 μm. However, the resolution of the images obtained is substantially lower than those of images obtained with nano-sized pipettes. For current measurements and patch-clamp, the nano-pipette is connected to a normal patch-clamp electronic circuit, such as that of an Axopatch 200A patch-clamp amplifier coupled to a DigiData 1322A interface (Axon Instruments, Sunnyvale, CA, USA). Certainly, any other currently available amplifiers can be used successfully.

Precise control of the tip position is achieved by mounting the probe on a piezo translation stage (e.g., Tritor; Piezosystem, Jena, Germany). The ionic currents flowing into the pipette are measured at an applied DC voltage of 50–200 mV. After amplification, it is used in the feedback loop that controls the probe's vertical position. Scanning can be performed in two ways. With one method, the sample is kept in a fixed position, and the nano-pipette probe is driven in the *x*-, *y*- and *z*-directions by the same piezo translator. Alternatively, the specimen can be moved in the *x*- and *y*-directions by a two-dimensional piezo translator while the probe is kept in a fixed *x*-, *y*-position and only its height (*z*-axis) controlled by an electronically driven one-dimensional actuator. Each method has its pros and cons and should be selected depending on the experimental design. The information on both the lateral and vertical positions is recorded and used to generate three-dimensional topographical images. Two examples of such images are shown in Fig. 25.3. These images were obtained using I_{MOD} as a feedback signal with modulation at 200 Hz. Images obtained in the hopping mode of SICM are shown in Fig. 25.4.

25.4. Smart-Patch Technique

Coordinates from the SICM images can be used subsequently for precise positioning of the pipette on a defined area of interest on the cell surface. This method is known as high-resolution scanning

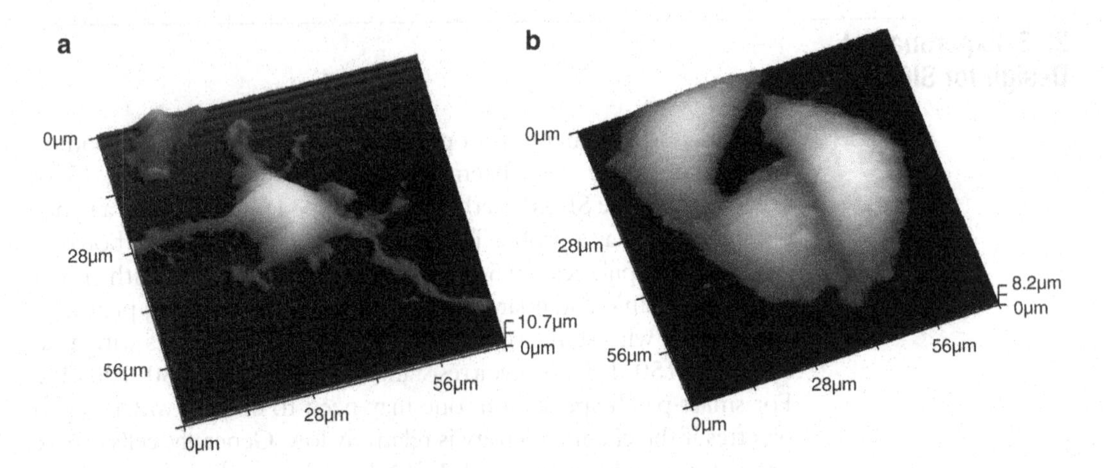

Fig. 25.3. SICM images of a single mouse mammary C127 cell (**a**) and a group of human epithelial intestine 407 cells (**b**). (Courtesy of Sabirov, Korchev, Shevchuk, and Okada, unpublished data).

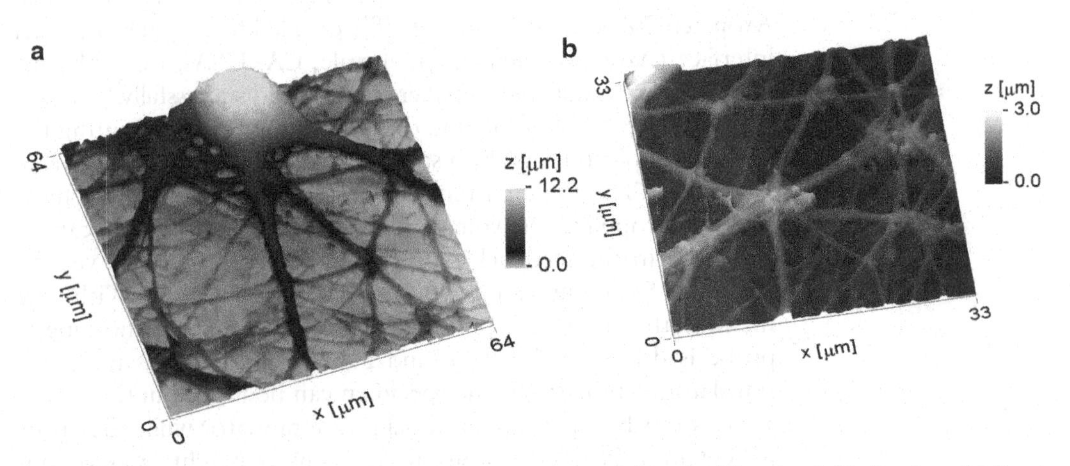

Fig. 25.4. HPICM images of a single hippocampal neuron (**a**) and part of the dendrite network (**b**). (Courtesy of P. Novak, unpublished data).

patch-clamp, or the "smart" patch-clamp technique (6, 13, 14). After imaging, the patch pipette is vertically placed and lowered by piezo control until it contacts the cell surface, usually after moving it down by one tip radius. Light suction is often necessary to form a giga-ohm seal. Tight seals with a resistance of around 5–10 GΩ can routinely be maintained in cell-attached recordings, and patches can be excised for inside-out experiments. An example of a smart-patch study is shown in Fig. 25.5. Whole-cell recording is difficult to carry out with nano-pipettes owing to the difficulty of rupturing the membrane patches and series resistance problems.

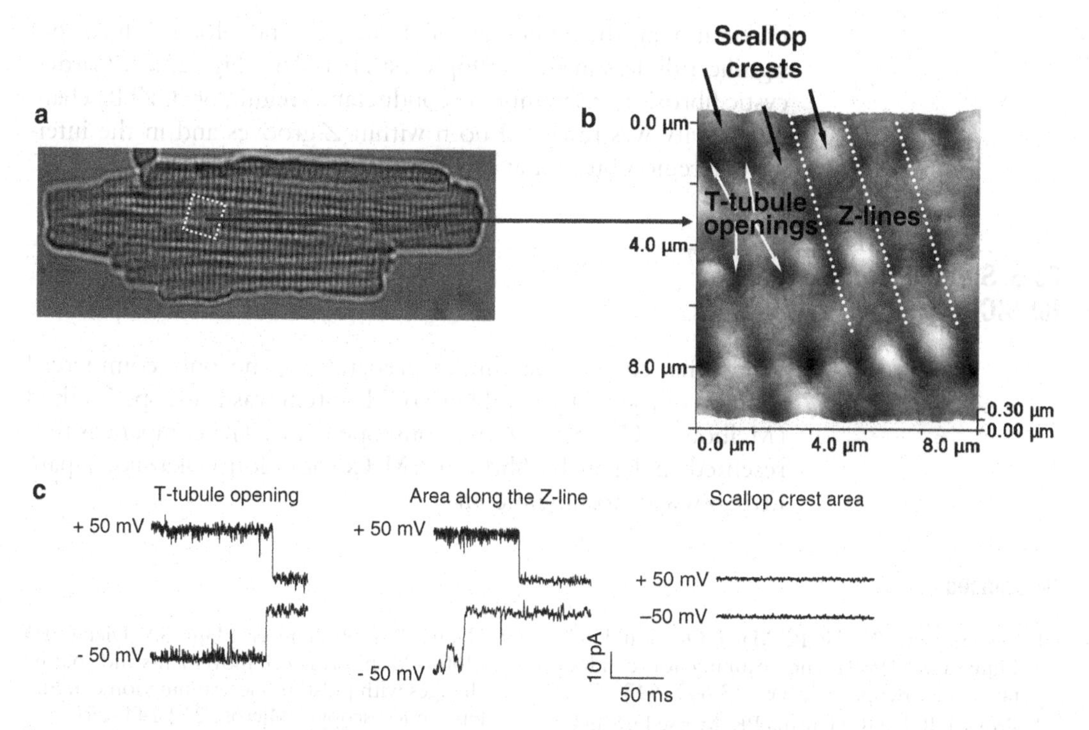

Fig. 25.5. Example of a smart-patch application. (**a**) Experiments were performed in a singly isolated rat ventricular myocyte observed under phase-contrast microscopy. The SICM image was generated for a small area (shown in **a** as a *white dotted square*) using a nano-pipette. (**b**) Single-channel patch-clamp recordings were then performed in this area. (**c**) Maxi-anion channel activity was detected in the inside-out mode at the openings of T-tubules and along the *Z*-lines but not on the scallop crests. (Modified from (16)).

25.5. Applications of the Smart-Patch Technique

The SICM method is highly effective for resolving fine structures on the cellular surface, which may have complex topography, such as the T-tubular system in muscle, microvilli in epithelia, or complex networks of axons and dendrites in the nervous system (Fig. 25.4). We may expect that ion-transporting pathways are distributed unevenly over the cell surface, and this distribution is likely to reflect the physiological function of these pathways. It should be noted that the spatial arrangement of ion channels and transporters is poorly investigated at present, and the smart-patch technique offers a unique methodology for this kind of study. Several successful applications of the smart-patch technique have proved its effectiveness. Direct smart-patch experiments have demonstrated, for instance, that K_{ATP} potassium channels (6), L-type Ca^{2+} channels, and three chloride channels with intermediate unitary amplitudes (13) are concentrated at the openings of T-tubules of cardiomyocytes. The activity of the maxi-anion channel was also found to be maximum at the openings of T-tubules and along *Z*-lines but significantly less in the scallop crest area (16) (Fig. 25.5). Cardiac cystic fibrosis transmembrane conductance regulator (CFTR) channel activity was recorded both within *Z*-grooves and in the intergroove region but not at the mouths of T-tubules (17).

25.6. Supplier for SICM

When this chapter was under preparation, the only commercial supplier of the fully assembled SICM system was Ionscope Limited (Melbourn, UK; http://www.ionscope.com). The company is represented in Japan by Shoshin EM Corporation (Okazaki, Japan; http://www.shoshinem.com).

References

1. Hansma PK, Drake B, Marti O, Gould SA, Prater CB (1989) The scanning ion-conductance microscope. Science 243:641–643
2. Proksch R, Lal R, Hansma PK, Morse D, Stucky G (1996) Imaging the internal and external pore structure of membranes in fluid: TappingMode scanning ion conductance microscopy. Biophys J 71:2155–2157
3. Korchev YE, Milovanovic M, Bashford CL, Bennett DC, Sviderskaya EV, Vodyanoy I, Lab MJ (1997) Specialized scanning ion-conductance microscope for imaging of living cells. J Microsc 188:17–23
4. Korchev YE, Bashford CL, Milovanovic M, Vodyanoy I, Lab MJ (1997) Scanning ion conductance microscopy of living cells. Biophys J 73:653–658
5. Korchev YE, Gorelik J, Lab MJ, Sviderskaya EV, Johnston CL, Coombes CR, Vodyanoy I, Edwards CR (2000) Cell volume measurement using scanning ion conductance microscopy. Biophys J 78:451–457
6. Korchev YE, Negulyaev YA, Edwards CR, Vodyanoy I, Lab MJ (2000) Functional localization of single active ion channels on the surface of a living cell. Nat Cell Biol 2:616–619
7. Korchev YE, Raval M, Lab MJ, Gorelik J, Edwards CR, Rayment T, Klenerman D (2000) Hybrid scanning ion conductance and scanning near-field optical microscopy for the study of living cells. Biophys J 78:2675–2679
8. Mann SA, Hoffmann G, Hengstenberg A, Schuhmann W, Dietzel ID (2002) Pulse-mode scanning ion conductance microscopy: a method to investigate cultured hippocampal cells. J Neurosci Methods 116:113–117
9. Happel P, Hoffmann G, Mann SA, Dietzel ID (2003) Monitoring cell movements and volume changes with pulse-mode scanning ion conductance microscopy. J Microsc 212:144–151
10. Shevchuk AI, Gorelik J, Harding SE, Lab MJ, Klenerman D, Korchev YE (2001) Simultaneous measurement of Ca^{2+} and cellular dynamics: combined scanning ion conductance and optical microscopy to study contracting cardiac myocytes. Biophys J 81:1759–1764
11. Novak P, Li C, Shevchuk AI, Stepanyan R, Caldwell M, Hughes S, Smart TG, Gorelik J, Ostanin VP, Lab MJ, Moss GW, Frolenkov GI, Klenerman D, Korchev YE (2009) Nanoscale live-cell imaging using hopping probe ion conductance microscopy. Nat Methods 6:279–281
12. Happel P, Dietzel ID (2009) Backstep scanning ion conductance microscopy as a tool for long term investigation of single living cells. J Nanobiotechnol 7:7
13. Gu Y, Gorelik J, Spohr HA, Shevchuk A, Lab MJ, Harding SE, Vodyanoy I, Klenerman D, Korchev YE (2002) High-resolution scanning patch-clamp: new insights into cell function. FASEB J 16:748–750
14. Gorelik J, Gu Y, Spohr HA, Shevchuk AI, Lab MJ, Harding SE, Edwards CR, Whitaker M, Moss GW, Benton DC, Sanchez D, Darszon A, Vodyanoy I, Klenerman D, Korchev YE (2002) Ion channels in small cells and subcellular structures can be studied with a smart patch-clamp system. Biophys J 83:3296–3303
15. Gorelik J, Zhang Y, Shevchuk AI, Frolenkov GI, Sanchez D, Lab MJ, Vodyanoy I, Edwards CR, Klenerman D, Korchev YE (2004) The use of scanning ion conductance microscopy to image A6 cells. Mol Cell Endocrinol 217:101–108

16. Dutta AK, Korchev YE, Shevchuk AI, Hayashi S, Okada Y, Sabirov RZ (2008) Spatial distribution of maxi-anion channel on cardiomyocytes detected by smart-patch technique. Biophys J 94:1646–1655
17. James AF, Sabirov RZ, Okada Y (2010) Clustering of protein kinase A-dependent CFTR chloride channels in the sarcolemma of guinea-pig ventricular myocytes. Biochem Biophys Res Commun 391:841–845

Chapter 26

Ion Channel Pore Sizing in Patch-Clamp Experiments

Ravshan Z. Sabirov and Yasunobu Okada

Abstract

The physical dimensions of an ion channel pore can be estimated in patch-clamp experiments using various charged and noncharged molecular probes. Measuring the permeability to organic ions of different size and shape yields an estimate of the size of the narrowest part of the selectivity filter. Open-channel blockers represent another frequently used tool to estimate the size of an ion channel pore near the binding pocket. The special case of permeable blockage yields the size of the narrowest portion of the ion-transporting pathway. Noncharged molecular probes may also affect ionic currents through channels either by blocking them or by decreasing the effective ionic mobility within the pore. Size-dependent suppression of single-channel amplitudes by neutral polymers, such as polyethylene glycols, can be interpreted in terms of molecular partitioning between the bulk solution and pore interior. This allows us to gauge the size of channel vestibules based on standard patch-clamp data.

26.1. Introduction

The shape and physical dimensions of an ion channel pore can be estimated using various probing methods in patch-clamp experiments. The size of the narrowest part of the channel lumen should correspond to the size of the largest molecule, charged or uncharged, that is able to cross the pore. Away from the narrowest portion, which is often called a selectivity filter, a channel may have wider extracellular and intracellular vestibules. In this chapter we consider the commonly used approaches for pore size determination in patch-clamp experiments.

Yasunobu Okada (ed.), *Patch Clamp Techniques: From Beginning to Advanced Protocols*, Springer Protocols Handbooks, DOI 10.1007/978-4-431-53993-3_26, © Springer 2012

26.2. Pore Sizing Using Permeable Ions

Ion channels vary greatly in their ability to pass ions and other charged molecules selectively. For instance, voltage-gated sodium, potassium, and calcium channels in excitable membranes are highly selective for Na^+, K^+, and Ca^{2+} ions (1). However, even these highly selective pores can conduct other ions. This makes it possible to estimate the size of the pore by observing different ions of different size pass or be conducted through it. Among inorganic ions, Li^+, Rb^+, Cs^+, Tl^+, and NH_4^+ can pass through cation-selective channels with different efficiencies. However, this list of inorganic cations is too short to yield a reliable pore size estimate. It can be extended by including alkyl derivatives of ammonium in which one to four N-linked hydrogens of NH_4^+ are substituted with methyl, ethyl, or propyl radicals. For channels with very narrow pores, this would be sufficient. However, wider pores (e.g., ligand-gated ion channels, or LGIC) may pass even larger cations, such as choline, triethanolamine, trimethylsulfonium, trimethylsulfoxonium, piperazine, glucosamine, guanidine and its derivatives, formamidine, imidazole, tris(hydroxymethyl)aminomethane (Tris), histidine, lysine, and arginine. (For a list of 40 positively charged molecular probes used for pore sizing of the skeletal muscle end-plate acetylcholine receptor channel, see (2).)

Chloride channels also have highly variable patterns of ionic selectivity. Thus, ligand-gated γ-aminobutyric acid (GABA)-ergic and glycine-ergic receptors pass all halides, bicarbonate, and formate but are poorly permeable to acetate (3). The cystic fibrosis transmembrane conductance regulator (CFTR) Cl^- channel permits passage of large organic anions, such as propanoate, pyruvate, glutamate, glucuronate, 2-(*N*-morpholino)ethanesulfonic acid (MES), HEPES, and 2-{[tris(hydroxymethyl)methyl] amino}ethanesulfonic acid (TES), only when they are present on the intracellular, but not extracellular, side of the membrane (4). The volume-sensitive outwardly rectifying (VSOR) Cl^- channel is permeable to the amino acids glutamate and aspartate (5), whereas the maxi-anion channel passes not only these amino acids but also more bulky anions such as glucoheptonate and lactobionate, and even some nucleotides such as ADP^{3-}, ATP^{4-}, and UTP^{4-} (see for reviews (6–8)).

In electrophysiological experiments, the permeability of ions is usually expressed as a permeability ratio relative to a title ion: Na^+, K^+, Ca^{2+}, and Cl^- for sodium, potassium, calcium, and chloride channels, respectively. The permeability ratio is calculated based on reversal potential (E_{rev}) measurements and a suitable form of the Goldman–Hodgkin–Katz equation. Because complicated pipette and bath solutions make unequivocal and adequate analysis difficult, it is a good idea to simplify the solutions as much as possible and use either the absolute values of E_{rev} in experimental conditions that are close to being bionic (i.e., when one side of the channel contains only the title ion and the other side contains only the test ion) or the shift of E_{rev} after equimolar replacement of part or all of the title ion by the test ion. As ionic strength does not change much in these experiments, researchers often assume that the activity coefficients for the title ion and the test ions are equal. This assumption might not be always true. However, standard activity coefficient data are not available for most organic ions, and the assumption is therefore thought to be reasonable.

After measuring the permeability ratios for a series of ions, one would plot the permeability ratio as a function of the ionic radius. For ionic radii, it is simplest to use crystal radii, which are available in the chemical literature for most inorganic cations and anions (e.g., see (9)). In very narrow channels (e.g., potassium-selective channels), ions pass through the pore in a dehydrated state, so using the crystal radius is appropriate. However, even the voltage-gated sodium channel can accommodate one Na^+ ion together with one or two water molecules. Using the crystal radius in such a case may therefore not be correct. The hydrodynamic radius supposedly represents the size of the hydrated ion, and it is calculated from limiting ionic mobility based on the Stokes–Einstein relation for diffusion coefficients (values for widely used ions are given in (1)). However, using the Stokes radius is somewhat controversial because it is always uncertain how many water molecules are retained while ions move through the channel pore. Organic ions normally are poorly hydrated and easily lose their weakly bound water shell when passing through channels. Therefore, most researchers use either the crystal radius for organic ions if they are available in the literature or simply calculate the equivalent size as the geometric mean of the three dimensions derived from space-filling models. Such models, called Corey–Pauling–Koltun (CPK) models, can be made from a ready-to-assemble kit or constructed using molecular drawing and modeling software (e.g., Molecular Modeling Pro; Norgwyn Montgomery Software, North Wales, PA, USA).

The size and shape of the ion exhibiting the lowest measurable permeability gives an estimate of the size and shape of the channel pore. The cutoff size for permeable ions is usually obvious from the plot of the permeability ratio against the ionic radius. The minimum pore diameter can also be calculated using the excluded area theory (2), which is based on the idea that the permeability is proportional to the cross-sectional area of the selectivity filter (the most constricted part of the pore):

$$\frac{P_X}{P_{Title}} = k\left[1 - \frac{R_X}{R_p}\right]^2 \tag{26.1}$$

where $\frac{P_X}{P_{Title}}$ is the permeability ratio of the test ion to the title ion, R_X is the radius of the test ion, R_p is the radius of the pore, and k is a proportionality constant. Taking into account the force of friction, the following equation is obtained (2):

$$\frac{P_X}{P_{Title}} = (k / R_X)\left[1 - \frac{R_X}{R_p}\right]^2 \tag{26.2}$$

The pore size obtained from (26.2) is usually somewhat larger than that calculated from (26.1).

26.3. Pore Sizing Using Blockers

Most ion channels are blocked by charged molecules in a voltage- and side-dependent manner. If the mechanism of blockage has been identified as an open-channel block caused by occlusion of the permeation pathway, systematically varying the size and shape of the blockers may yield an estimate of the pore dimensions near the selectivity filter. It should be noted, however, that the binding site (or sites) for blockers may be located in very different part(s) of the channel pore. Precisely speaking, the cutoff size of the blocker therefore reflects the size of the binding pocket, rather than the true pore size. In cases in which the blocker binds to the same site as the permeant ion, one could make arguments about the size of the pore in the vicinity of the selectivity filter.

Permeable block is a special case of voltage-dependent open-channel blockage under conditions in which the blocker occludes the permeation path at moderate potentials but can “punch through” the pore at larger transmembrane voltages. In this case, the blocker passes through the pore, and its dimensions therefore give an estimate of the narrowest pore region. For instance, calixarenes, basket-shaped compounds bearing several negatively charged sulfonic groups, were found to cause a voltage-dependent permeable block of the VSOR Cl^- channel (10), giving a minimum pore radius estimate of 0.57 nm.

26.4. Pore Sizing Using Nonelectrolytes

When charged particles (permeant ions and charged blockers) are used for pore sizing, the results are greatly affected by the electrostatic forces (between permeant ions and/or between a permeant ion and the charged wall of the channel pore) that govern the movement of ions in channel pores. Nonelectrolytes have an obvious advantage as probes for pore sizing, as they are relatively inert with respect to the electrostatics. Macroscopic nonelectrolyte fluxes through ion channels have been successfully measured for ion channels inserted into planar lipid bilayers (see (11)); there have been no reports, however, for ion channels in cellular plasma membranes under patch-clamp. This is likely due to experimental difficulties in measuring the net nonelectrolyte fluxes with radiolabeled tracers. Also, it is difficult to prove that the electrically silent flux of noncharged probe molecules occurs only through the specific channel under consideration but not via other routes.

26.4.1. Principle of Nonelectrolyte Partitioning Method

An alternative way to detect the presence of noncharged molecules inside the channel pore was suggested based on studies of ionic conductivity and viscosity of electrolyte solutions containing polyethylene glycols (PEGs) of different molecular mass and the effects of these polymers on the single-channel amplitudes of toxin-formed ion channels in planar lipid bilayer membranes (12–14). In these studies, the authors found that PEGs obey the viscosity law for random coils in aqueous solutions; the electrical conductivity of PEG-containing solutions, however, did not follow the macroscopic viscosity. For instance, solutions containing equal mass concentrations of PEG 300 and PEG 4000 had practically equivalent electrical conductivity, but the viscosity of the PEG 4000 solution was about four times greater than that of the PEG 300 solution. This suggests that the mobility of ions in polymer solutions depends mainly on the hydration of the monomeric unit of PEG (which is ethylene glycol) and thus is not affected by the size of the polymer. The authors pictured PEGs as sponge-like spheres, inside of which ions can move as freely as in a solution of the monomer dissolved at the same mass (percent) concentration. If the polymeric sphere (or more precisely, stochastic random coil) is small enough to access the pore interior freely, it changes the mobility of ions inside the channel lumen in the same way it does in the bulk of the bathing solution. In contrast, if the polymeric sphere is too large to get into the channel pore, it is effectively "filtered out" by the channel. As a result, the channel conductance is close to the control value measured in the absence of polymer despite the fact that the mobility of ions in the surrounding bathing solution is greatly decreased. If we vary the molecular weight of the polymer, the accessibility of the channel lumen for polymers of different molecular weight (and thus different size) changes from full partitioning (for small molecules) to complete exclusion (for large polymers). This can be detected as a gradual PEG size-dependent change in the single-channel amplitude without a change in the bulk electrical conductivity.

The greatest benefit of the electrical method of detection of nonelectrolyte access to the ion channel lumen is that it permits probing of the pore size with nonelectrolytes at the single-channel level. Therefore, one may neglect other pathways for the passage of nonelectrolytes, such as transporters or electrically silent water channels. The method has been widely used to estimate the pore size of channels in lipid bilayer experiments (15, 16) and has yielded pore dimensions consistent with X-ray crystallographic data for staphylococcal α-hemolysin (17) and bacterial porin OmpF (18). PEG exclusion estimates of pore dimensions for the mitochondrial porin VDAC were consistent with electron microscopic data (19–21). However, despite its simplicity and proven consistency, employing high concentrations of nonelectrolytes was thought inevitably to destroy membrane patches in patch-clamp experiments owing to large osmotic gradients. Therefore, until recently the method was used exclusively for channels incorporated into artificial lipid bilayers.

26.4.2. Use of Nonelectrolyte Partitioning in Patch-Clamp

We have demonstrated that, with some care, the access of polymeric probes to the ion channel interior can be successfully detected in patch-clamp experiments as well (22, 23). The principal behavior of PEG molecules in solutions suitable for patch-clamp experiments was found to be roughly the same as that observed in the rather nonphysiological 100–1,000 mM KCl solutions often used in bilayer studies. The slightly different values for the hydrodynamic radii found in a standard Ringer's solution could be due to the presence of divalent cations and the different buffers used. As anticipated, the stability of membrane patches was greatly reduced by the application of PEGs either from the intracellular side of membrane patches (infusion of PEG-containing solutions into the bath in the inside-out mode) or from the extracellular side (PEG addition to the pipette solution in the cell-attached or inside-out mode). However, some patches (particularly smaller ones formed on 3- to 5-MΩ pipettes) survived for a longer time (up to 10–20 min), permitting accurate determination of single-channel current amplitudes. It should be noted, however, that outside-out membrane patches were immediately destroyed upon perfusion of polymer-containing solutions and that the polymers could therefore not be applied from the outside in the outside-out mode. Symmetrical two-sided PEG application was possible through infusion of PEG-containing solutions into the bath after establishing the inside-out configuration with pipettes filled with PEG-containing solutions.

Single maxi-anion channels observed in membrane patches excised from C127 mouse mammary tumor-derived cells responded with a decrease in unitary conductance when a small PEG (PEG 200) (R_h = 0.455 nm) was present in either the pipette or bath solution; however, a large PEG (PEG 4000) (R_h = 1.91 nm) did not produce such an effect on the single-channel current amplitude (22). Examples of such recordings are shown in Fig. 26.1.

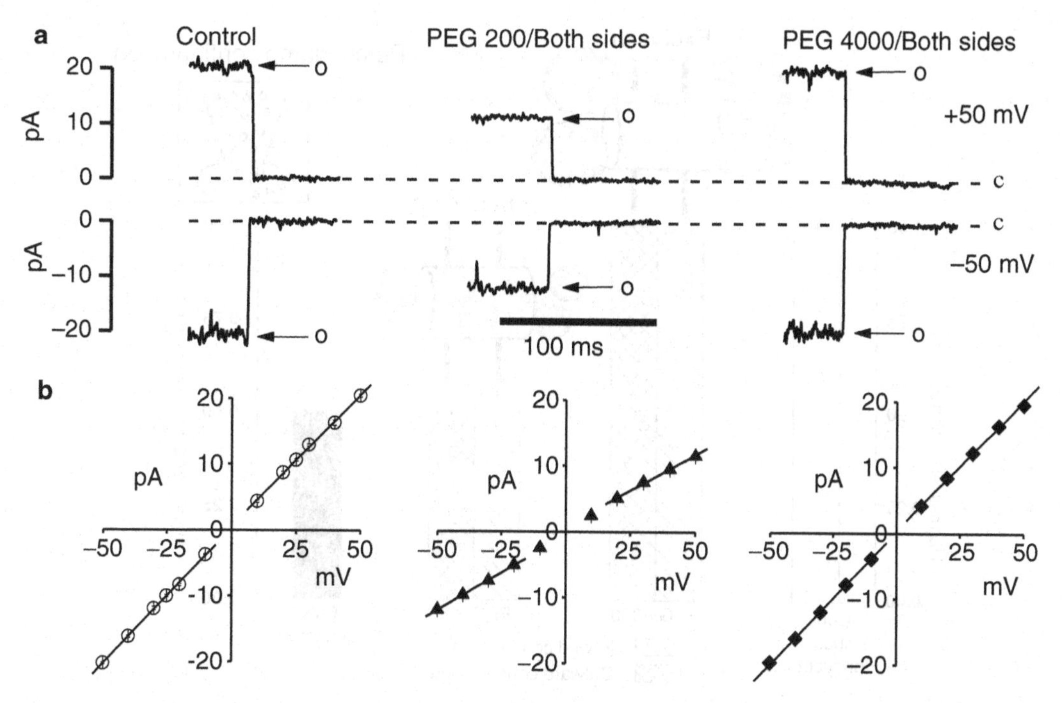

Fig. 26.1 Small polyethylene glycol (*PEG*) 200, but not large PEG 4000 decreases the unitary current amplitude of the maxi-anion channel. The polymers were added to both sides of the membrane patch. (**a**) Representative single-channel current traces. (**b**) Unitary current-to-voltage relations. (Modified from (22))

When PEG 200 was applied from both sides of the membrane patch, the ratio of the channel conductance in the presence of PEG to that in its absence was similar to the ratio of the bulk aqueous conductivity with polymer to that without polymer (Fig. 26.2). This suggests that the ionic environment in the pore interior is almost identical to that in the bulk solution and is affected by small polymeric molecules in the same way. PEG 4000 also decreased the bulk conductivity as much as PEG 200 did. However, it was excluded from the pore of the maxi-anion channel and thus was unable to affect the mobility of ions inside the pore. The effects of PEG 400 to PEG 1540 on the single-channel current amplitude declined in a differential, size-dependent manner, corresponding to the varying degrees of partial access of each polymer to the channel interior. Beginning from a certain molecular weight (e.g., from intracellularly applied PEG 1540 or 2000 in the case of the maxi-anion channel, as shown in Fig. 26.3), the ion channel becomes insensitive to polymers of large size, suggesting complete exclusion of these large polymers from the channel lumen. It should be emphasized that the bulk solution conductivity was almost insensitive to the polymer molecular weight and, hence, the polymer size (open triangles in Fig. 26.3). Symmetrical two-sided application of PEGs yields effects that are intermediate to those observed in the one-sided experiments (Fig. 26.4a).

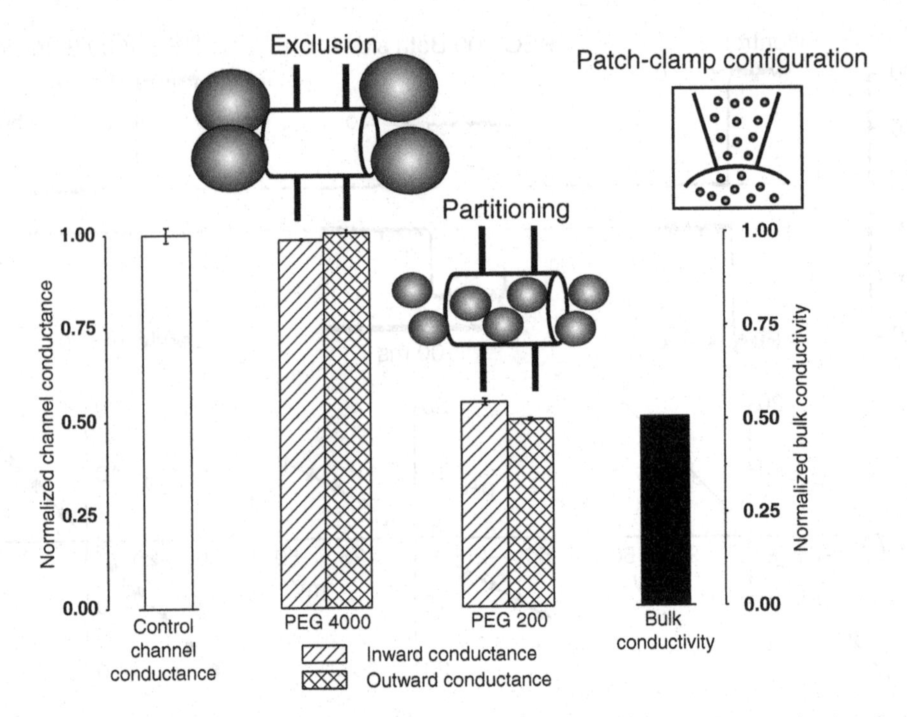

Fig. 26.2. How polymer partitioning is related to the relative single-channel amplitude of the maxi-anion channel and relative bulk conductivity. The large PEG 4000 cannot access the pore and does not affect the channel amplitude, whereas the small PEG 200 is freely distributed between the bulk solution and the pore lumen and decreases the channel conductance to the same extent as it does the bulk conductivity. (Modified from (22)).

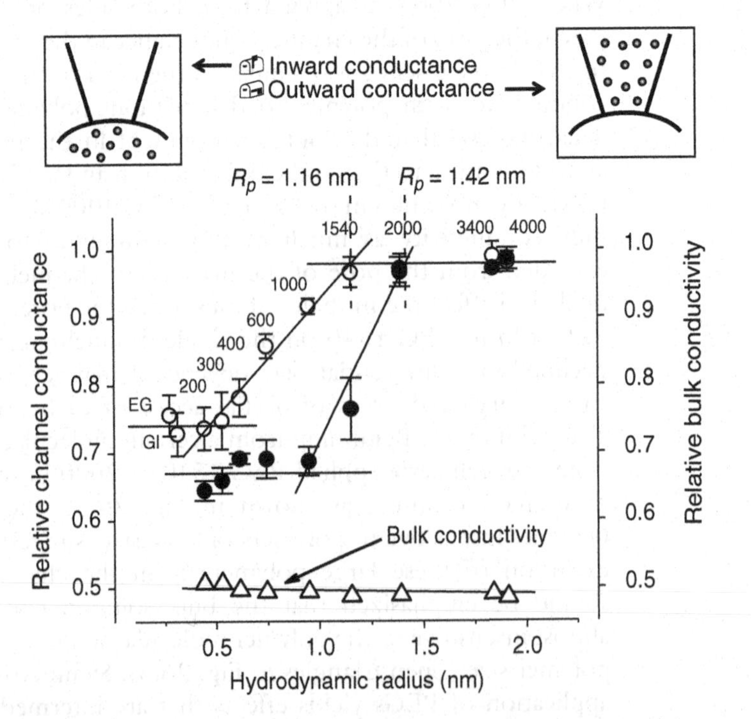

Fig. 26.3 One-sided application of polymeric probes permits separate estimation of the size of the outer and inner vestibules of the maxi-anion channel from relative single-channel conductances. (Modified from (22))

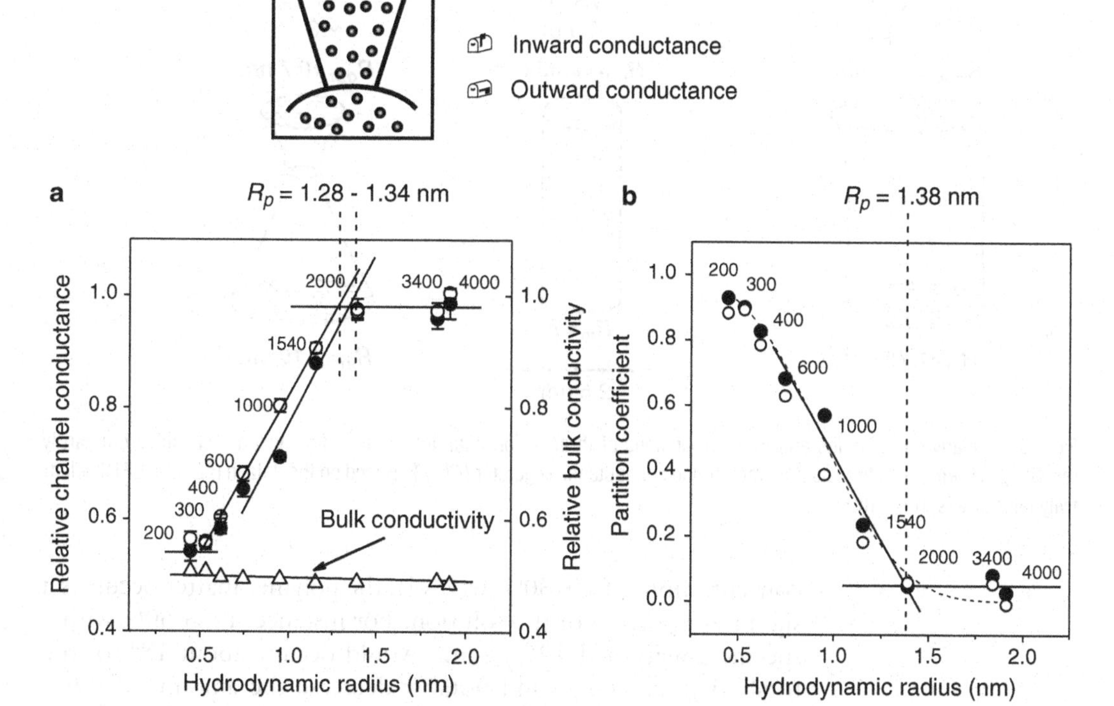

Fig. 26.4 Relative single-channel conductance of the maxi-anion channel in the presence of PEG applied symmetrically from both sides of the membrane patch. (**a**) Outward and inward conductances yield slightly different cutoff sizes due to the asymmetrical structure of the maxi-anion channel pore. (**b**) Partition coefficient is calculated for inward conductance using (26.3) and is plotted against the polymer size. The cutoff size is close to that obtained from the relative conductances in (**a**). Dashed sigmoidal line is a fit to the scaling law (26.4) with $R_p = 0.97$ nm. (Modified from (22))

Volume-sensitive VSOR Cl^- channels (23) and cAMP-activated CFTR Cl^- channels (24, 25) gave qualitatively similar but quantitatively different responses in the presence of polymeric probes. Thus, dependence of the single-channel conductance on the hydrodynamic radius of PEGs had a unique shape for each channel, reflecting individual differences in the size and shape of each channel pore (Fig. 26.5).

26.4.3. Solutions for Nonelectrolyte Partitioning

Relatively high concentrations (expressed as percent weight/volume) of nonelectrolytes are employed in nonelectrolyte partitioning (NPM) experiments. Using solutions of equal percent concentration is more appropriate for NPM experiments because bulk conductivity is constant in solutions of PEGs that are of equal percentage, not of equal molarity. There are two ways to prepare nonelectrolyte-containing solutions for NPM. With the first method, the concentrations of salt components are calculated for the final volume of the nonelectrolyte-containing solution. All components including the nonelectrolyte are then weighed, and water is added to adjust to the final volume. This is an apparently correct method of preparation. However, when PEGs are used at

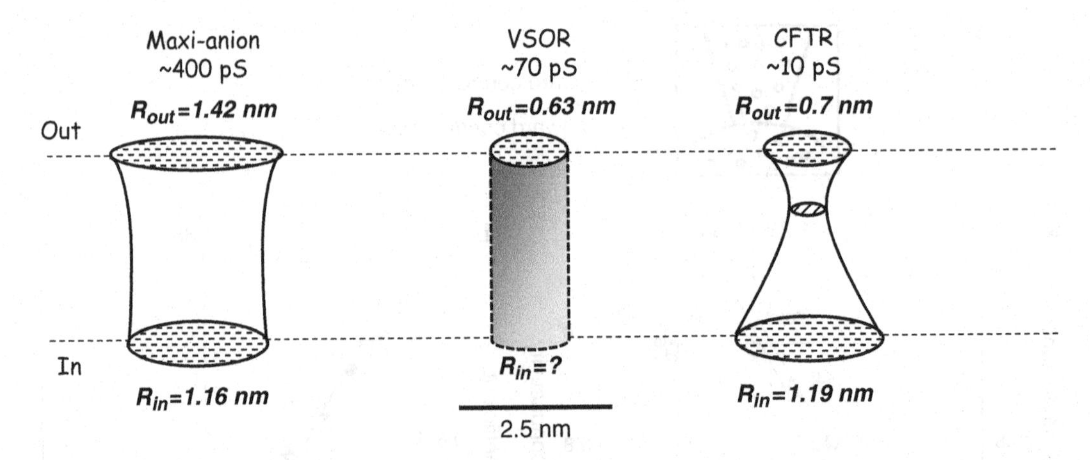

Fig. 26.5. Relative amplitudes and pore dimensions of three anion channels: maxi-anion, volume-sensitive outwardly rectifying (*VSOR*), cystic fibrosis transmembrane conductance regulator (*CFTR*). (Based on the data in (22–25). CFTR selectivity filter size is from (4)).

concentrations of 10–30% (wt/vol), the polymer matter occupies a significant portion of the solution. For instance, PEG 600, with a specific gravity of 1.128 g/cm^3, would occupy about 18% of the total volume. This would notably decrease the amount of water available for salt components and thus increase their effective concentrations (activities), leading to an increase in the channel conductance in the presence of completely excluded large polymers. With the second method, the necessary amount of PEG (powder or liquid) is weighed; and then a premade bath or pipette solution is added to the final desired volume. For instance, to make a 20% solution, 20 g of PEG is placed in a beaker and a premade solution is added to the final volume of 100 ml. This way, effective ionic concentrations (ionic activities) would remain unchanged because the volume accessible for the ions has not changed. Certainly, this notion is not precisely true because PEG molecules do interact with water and change its activity (and hence the ionic activities). However, the effect is small enough that it can be neglected in NPM experiments.

Every nonelectrolyte solution should be tested for its electrical conductivity. This can be done with any type of commercially available conductivity meter. In our hands, an inexpensive B-173 conductivity meter from Horiba (Kyoto, Japan), which requires only 100–150 μl of a test solution, gave reliable and reproducible results. Using more sophisticated meters did not significantly improve the measurements, although they required larger volumes and sometimes a complicated calibration procedure.

26.4.4. Data Analysis in Nonelectrolyte Partitioning

The easiest way to analyze NPM data is to plot relative single-channel conductance (γ) as a function of the polymer hydrodynamic radius (R_h). PEG solutions often produce a shift in the reversal potential. Therefore, care should be taken to avoid misinterpretation of single-channel current amplitude measurements. The single-channel chord conductance can be used only when the reversal potential shift has been evaluated by some other means (e.g., from macroscopic current measurements with and without polymers). Using the slope conductance calculated from the single-channel current-to-voltage relation is preferable, although the contribution of amplitude variations at low driving forces could be substantial and should be carefully evaluated – and excluded if necessary.

A graph of the relative single-channel conductance γ plotted against R_h has an ascending portion (partial partitioning of polymer between the bathing solution and channel), as seen in Figs. 26.3 and 26.4a, which then turns into a plateau (complete exclusion of polymers from the pore). The ascending part can be fitted with a linear function, and the intersection of this line with the plateau line gives an estimate of the pore size. Another way to analyze the NPM results is to normalize the change in single-channel conductance by the change in the bulk solution conductivity. This procedure yields a parameter (ν) that was initially interpreted as a permeability parameter (12, 14, 26) but was later reinterpreted as a partition coefficient (18, 27):

$$\nu = \frac{(\gamma - \gamma_0)/\gamma_0}{(\chi - \chi)/\chi_0} \qquad (26.3)$$

where γ and χ are the channel conductance and bulk solution conductivity in the presence of nonelectrolytes, respectively; and γ_0 and χ_0 are the same parameters in the control solution that does not contain nonelectrolytes. The partition coefficient has a value of $\nu=1$ for full penetration and $\nu=0$ for complete exclusion (Fig. 26.4b). The pore cutoff size is obtained by determining the point of intersection of straight lines fitted to the data points in the descending part of the curve (partial access zone) and the lower plateau level (no access zone).

Instead of simply finding a cutoff size, the pore size can be evaluated by fitting the ν data to a scaling law of the following form (17, 18, 27):

$$\nu = \exp\left[-\left(\frac{R_h}{R_p}\right)^{\alpha}\right] \qquad (26.4)$$

where R_h and R_p are the radii of the nonelectrolyte (PEG) and the pore, respectively; and α characterizes the sharpness of the transition between regimens of exclusion and penetration. An example of such a fit is shown in Fig. 26.4b (dotted curve).

Another useful parameter for NPM data analysis is the filling coefficient (F), defined as follows (26):

$$F = \frac{(\gamma_0 - \gamma)/\gamma}{(\chi_0 - \chi)/\chi} \quad (26.5)$$

where γ and χ are the channel conductance and the bulk solution conductivity in the presence of nonelectrolytes, respectively; and γ_0 and χ_0 are the same parameters in the control solution without nonelectrolytes. Formula (26.5) is derived based on the assumption that the pore electrical resistance can be considered a sum of two resistances in series: that of a polymer-filled part (with a resistance of $F/(A\chi)$) and that of an unfilled part (with a resistance of $F/(A\chi_0)$), where $A = \pi R_p^2/L$, R_p is the pore radius, and L is the pore length. The filling coefficient is thought to represent the fraction of the pore filled with the polymer. The plot of the filling coefficient as a function of the polymer hydrodynamic radius is similar to the plot for the partition coefficient, and further analysis can be performed in the same way as for v.

26.4.5. Interpretation of Nonelectrolyte Partitioning Data

The most straightforward interpretation to make is that of the cutoff size, which is determined as a point of intersection between the zones of partial access and complete exclusion. This value gives an estimate of the size of the channel vestibule, extracellular or intracellular, depending on the side of polymer application. The two-sided application of polymers yields an intermediate value, which is more like an average pore size.

Theoretically, the absolute values of the partition coefficient v and filling coefficient F and their differences depending on the side of polymer application should provide important information about the geometry of the channel pore. It is particularly useful to interpret the NPM results in combination with ion permeation and open-channel blockage data. For instance, significantly smaller pore dimensions obtained from ion permeation experiments compared to those from NPM suggest the existence of a constriction inside the pore. As the filling coefficient is derived considering the pore electrical resistance as a sum of the resistances of two parts in series – one filled with polymer and another unfilled – one may relate F values obtained in a one-sided polymer application to the location of the selectivity filter, which would effectively restrict the passage of even small PEGs to the opposite side. For example, studying the CFTR pore, we observed the value $F = 0.1$ for extracellular application of PEG 300 ($R_h = 0.53$ nm) and $F = 0.9$ for intracellular application of the same molecule (25). This means that only chloride ions moving one-tenth of the total length of the channel, where the voltage drop occurs, sense the presence of PEG added from the outside. Thus, the narrowest constriction (selectivity filter) that serves as a barrier for PEG 300 is located not closer than 10% of the total electrical distance for chloride ions from the outside. It was consistently found that chloride ions moving up to approximately 90% from the intracellular mouth sense the presence of PEG 300 added from the intracellular side. Therefore, although the single CFTR Cl^- channel has a current-to-voltage relation that is quite symmetrical, the channel pore is highly asymmetrical. Figure 26.5 represents the overall geometries of pores for three anion channels tested with NPM.

Although a simple interpretation of the NPM data in terms of molecular size exclusion seems reasonable, the mechanism of polymer partitioning into a nano-scale pore is far from being understood in detail. Theoretical approaches suggested so far include one that incorporates the hard sphere partitioning theory (in which polymers are represented as hard spheres) and a scaling approach based on the difference in entropy of a flexible polymer chain partitioned between the bulk solution and a cylindrical pore (15, 17, 18, 27). The main criterion for evaluating the applicability of either theoretical approach is the steepness of the transition zone between partial partitioning and complete exclusion on the plots of v or F versus R_h. It appears that the hard sphere theory better explains the results obtained at relatively high PEG concentrations, whereas the scaling theory might be more applicable at very low PEG concentrations, when tested on the pore of staphylococcal α-hemolysin (15, 17, 18, 27–29). We believe that molecular dynamics or brownian dynamics simulations would provide a clearer understanding of the mechanism by which polymers affect the ionic fluxes through ion channels.

References

1. Hille B (2001) Ion channels of excitable membranes. Sinauer Associates, Inc, Sunderland, MA
2. Dwyer TM, Adams DJ, Hille B (1980) The permeability of the endplate channel to organic cations in frog muscle. J Gen Physiol 75: 469–492
3. Bormann J, Hamill OP, Sakmann B (1987) Mechanism of anion permeation through channels gated by glycine and gamma-aminobutyric acid in mouse cultured spinal neurones. J Physiol 385:243–286
4. Linsdell P, Hanrahan JW (1998) Adenosine triphosphate-dependent asymmetry of anion permeation in the cystic fibrosis transmembrane conductance regulator chloride channel. J Gen Physiol 111:601–614
5. Okada Y (1997) Volume expansion-sensing outward rectifier Cl channel: a fresh start to the molecular identity and volume sensor. Am J Physiol 273:C755–C789
6. Okada Y, Sato K, Toychiev AH, Suzuki M, Dutta AK, Inoue H, Sabirov R (2009) The puzzles of volume-activated anion channels. In: Alvarez-Leefmans FJ, Delpire E (eds) Physiology and Pathology of Chloride Transporters and Channels in the Nervous System. From Molecules to Diseases, Elsevier, San Diego, pp 283–306
7. Sabirov RZ, Okada Y (2005) ATP release via anion channels. Purinergic Signal 1:311–328
8. Sabirov RZ, Okada Y (2009) The maxi-anion channel: a classical channel playing novel roles through an unidentified molecular entity. J Physiol Sci 59:3–21
9. Robinson RA, Stokes RH (1959) Electrolyte solutions. Butterworths, London
10. Droogmans G, Maertens C, Prenen J, Nilius B (1999) Sulphonic acid derivatives as probes of pore properties of volume- regulated anion channels in endothelial cells. Br J Pharmacol 128:35–40
11. Finkelstein A (1987) Water movement through lipid bilayers, pores, and plasma membranes. Theory and reality. John Willew & Sons, New York
12. Sabirov R, Krasilnikov OV, Ternovsky VI, Merzliak PG, Muratkhodjaev JN (1991) Influence of some nonelectrolytes on conductivity of

bulk solution and conductance of ion channels: determination of pore radius from electric measurements. Biologicheskie Membrany 8: 280–291

13. Sabirov RZ, Krasilnikov OV, Ternovsky VI, Merzliak PG (1993) Relation between ionic channel conductance and conductivity of media containing different nonelectrolytes: a novel method of pore size determination. Gen Physiol Biophys 12:95–111
14. Krasilnikov OV, Sabirov RZ, Ternovsky VI, Merzliak PG, Muratkhodjaev JN (1992) A simple method for the determination of the pore radius of ion channels in planar lipid bilayer membranes. FEMS Microbiol Immunol 5:93–100
15. Bezrukov S, Kasianowicz JJ (2002) Dynamic partitioning of neutral polymers into a single ion channel. In: Kasianowicz JJ, Kellernayer MSZ, Deamer DW (eds) Structure and Dynamics of Confined Polymers. Kluwer Publisher, Dordrecht, The Neatherlads, pp 93–106
16. Krasilnikov OV (2002) Sizing channel with polymers. In: Kasianowicz JJ, Kellernayer MSZ, Deamer DW (eds) Structure and Dynamics of Confined Polymers. Kluwer Publisher, Dordrecht, The Netherlands, pp 73–91
17. Merzlyak PG, Yuldasheva LN, Rodrigues CG, Carneiro CM, Krasilnikov OV, Bezrukov SM (1999) Polymeric nonelectrolytes to probe pore geometry: application to the alpha-toxin transmembrane channel. Biophys J 77:3023–3033
18. Rostovtseva TK, Nestorovich EM, Bezrukov SM (2002) Partitioning of differently sized poly(ethylene glycol)s into OmpF porin. Biophys J 82:160–169
19. Krasilnikov OV, Carneiro CM, Yuldasheva LN, Campos-de-Carvalho AC, Nogueira RA (1996) Diameter of the mammalian porin channel in open and "closed" states: direct measurement at the single channel level in planar lipid bilayer. Braz J Med Biol Res 29:1691–1697
20. Carneiro CM, Krasilnikov OV, Yuldasheva LN, Campos de Carvalho AC, Nogueira RA (1997) Is the mammalian porin channel, VDAC, a perfect cylinder in the high conductance state? FEBS Lett 416:187–189
21. Carneiro CM, Merzlyak PG, Yuldasheva LN, Silva LG, Thinnes FP, Krasilnikov OV (2003) Probing the volume changes during voltage gating of Porin 31BM channel with nonelectrolyte polymers. Biochim Biophys Acta 1612:144–153
22. Sabirov RZ, Okada Y (2004) Wide nanoscopic pore of maxi-anion channel suits its function as an ATP-conductive pathway. Biophys J 87: 1672–1685
23. Ternovsky VI, Okada Y, Sabirov RZ (2004) Sizing the pore of the volume-sensitive anion channel by differential polymer partitioning. FEBS Lett 576:433–436
24. Sabirov RZ, Ternovsky VI, Krasilnikov OV, Okada Y (2009) Gauging the pore size of three putative ATP releasing pathways by polymer partitioning. J Physiol Sci 59 (Suppl 1):392 (Abstract)
25. Krasilnikov OV, Sabirov RZ, Okada Y (2011) ATP hydrolysis-dependent asymmetry of the conformation of CFTR channel pore. J Physiol Sci 61:267–378
26. Krasilnikov OV, Da Cruz JB, Yuldasheva LN, Varanda WA, Nogueira RA (1998) A novel approach to study the geometry of the water lumen of ion channels: colicin Ia channels in planar lipid bilayers. J Membr Biol 161:83–92
27. Bezrukov S, Vodyanoy I, Brutyan R, Kasianowicz JJ (1996) Dynamics and free energy of polymers partitioning into a nanoscale pore. Macromolecules 29:8517–8522
28. Movileanu L, Bayley H (2001) Partitioning of a polymer into a nanoscopic protein pore obeys a simple scaling law. Proc Natl Acad Sci USA 98:10137–10141
29. Movileanu L, Cheley S, Bayley H (2003) Partitioning of individual flexible polymers into a nanoscopic protein pore. Biophys J 85: 897–910

Chapter 27

Digital Recording of Patch-Clamp Data

Shigeru Morishima and Andrew F. James

Abstract

Patch-clamp data, including channel currents and membrane voltage, are usually recorded to a computer via a process known as analog-to-digital conversion. This chapter gives essential information on how to avoid pitfalls and obtain error-free data with optimal resolution.

27.1. Analog-to-Digital Conversion

Analog signals or data can be expressed by a set of real numbers, which are "measurable." Ideally (without any noise and with perfect resolution), analog data can be read as finely as we want; for example, if the current signals were recorded for 4 s, we could read the current signal at 3.0 s, at 3.14 s, at 3.14159 s, or at any other time during the recording period; and we would be able to resolve infinitesimally small increments in the data between each of these time points. Thus, the number of analog data is as infinite as the number of a set of all real numbers, and data values are continuous. On the other hand, "digital" means "that which is countable" on a one-to-one correspondence with our fingers. Digital values are "countable" and discrete and, theoretically, not as accurate as analog values. However, despite the theoretically superior accuracy of analog data, we usually prefer digital recordings, using a personal computer (PC) or other compatible digital device, because digital data can be analyzed, reproduced, and manipulated far more easily than can analog data. Most data in patch-clamp recordings, including current and voltage signals, are derived from the amplifier as analog data. If we want to record the signals to a PC hard drive or other digital data storage device the data must undergo a process known

Yasunobu Okada (ed.), *Patch Clamp Techniques: From Beginning to Advanced Protocols*, Springer Protocols Handbooks, DOI 10.1007/978-4-431-53993-3_27, © Springer 2012

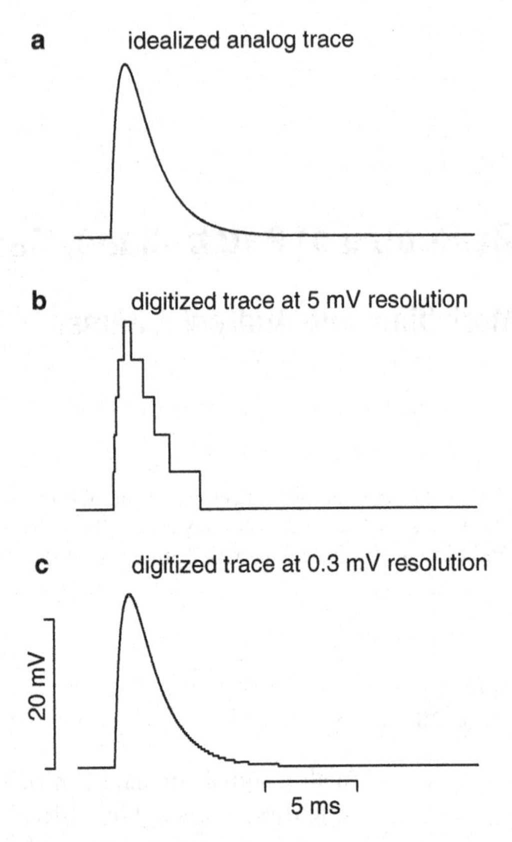

Fig. 27.1. Inaccuracies due to limited resolution following analog-to-digital (A/D) conversion. An idealized analog voltage signal (**a**) is converted to a digital signal with a resolution of 5 mV, as might be achieved with a +10 to −10 V input to a 12-bit A/D converter (**b**) and of 0.3 mV, as might be achieved with a +10 to −10 V input to a 16-bit A/D converter (**c**). Data in **a** were synthesized mathematically to represent an idealized voltage signal as might be recorded from the motor end-plate of rat diaphragm skeletal muscle, assuming no amplification of the voltage signal before A/D conversion. In practice, the output signal of most amplifiers corresponding to membrane potential has been multiplied tenfold so a 20-mV change in membrane potential would be represented by a 200-mV change in signal with the consequence that, in reality, the resolution would be correspondingly tenfold better than what is suggested in this figure.

as analog-to-digital (A/D) conversion, usually via a piece of equipment known eponymously as an A/D converter (1).

Figure 27.1 illustrates a possible consequence of the conversion of an analog voltage signal to a digital signal. Panel A represents an idealized analog voltage signal. In panel B, the same signal has been digitized in 5-mV increments such that the voltages have been converted to the nearest 5-mV value, thus –5, 0, 5, 10, 15, and so on, in millivolts. In consequence, voltages of 0.6 and 1.3 mV would both become 0 mV after A/D conversion, and the errors would be sizable, as is evident by comparing panel B with panel A. If you are looking closely at voltage changes in the order of several tens of millivolts, such errors would be unacceptably large.

However, this does not mean that A/D conversion necessarily produces such an unacceptable error. Depending on the A/D converter, the amplitude interval between each discrete value can be smaller; for example, with an interval of 0.3 mV, which is one-sixteenth of 5 mV, a much more accurate representation of the original analog signal can be obtained (cf. Fig. 27.1c with Fig. 27.1a). The smallest interval in the A/D conversion process is termed the "resolution."

Certainly, however good the resolution may be, the value expressed by a real number would be expected to be even more accurate. Our data, however, always contain some inevitable errors. In just the same way that we would not be able to detect a 1-ml difference with a 1 l graduated cylinder, so an accuracy of ±0.01 pA is not required when the current signal has a root-mean-square noise of ±1 pA. If the noise is much larger than the resolution for A/D conversion, the errors generated by A/D conversion are not practically significant. The resolution should be small in comparison to the noise present within the signals.

27.2. Dynamic Range

In the last section, we considered a potential error generated in the process of A/D conversion. Another important consideration is that no A/D converter can perform an ideal A/D conversion. First, A/D converters can accept only voltage signals within a limited range. Second, the digital value must be expressed within a certain number of figures. As we shall see, these considerations impose a limit on the resolution that can be achieved.

The analog output from the patch-clamp amplifier is a voltage signal. Membrane potentials recorded by the patch-clamp amplifier are amplified, and membrane currents are converted to voltage signals and then amplified. It is these processed signals that are subjected to A/D conversion. Output from amplifiers is usually within a range of around −10 to +10 V, and one should keep in mind that the output range matches the input range of the A/D converter. In many A/D converters, the input ranges that the converter can accept are chosen from predefined ranges (e.g., −10 to +10 V). Once selected, the range is set constant and applied during a series of experiments. Input data outside this range are, in the best case, simply neglected but may damage the converters. (For this reason, A/D converters are usually equipped with a protection circuit.) Hence, the input range into the A/D converter must be determined to accept the output from the amplifier, but the measure by the A/D converter does not directly designate the values of membrane potentials or currents. We will address how this conversion affects data recordings.

The second problem must be considered more carefully. It is also strongly related to the way data are stored on personal computers. Data are stored as a sequence of zeroes and ones, and the amount of information and stored memory can be expressed by bits (2). One bit is a very small amount, and computers usually can process 8 bits (or more) at a time. From the definition, in 8 bits a number of 8 digits in a binary system – that is, 0–255 (i.e., 2^8-1), can be expressed. Similarly, 0–65,535 can be expressed in 16 bits. More than three decades ago, desktop computers could handle only 8-bit digits at a time, but recent PCs can handle 32–64 bits at a time. The larger the number of bits that is handled, the longer the time it takes to process them and the more storage space is required. A/D converters are classified by the number of bits they can handle, which is an important specification of the performance of the A/D converter. A one-bit A/D converter (if ever one should exist) would convert analog signals into either a 0 or a 1. If the input range were between −10 and +10 V, signals between −10 and 0 V would be converted to zero and the rest would become one; and the resolution would be 10 V. Certainly, such a machine would be of no use. In the main, for patch-clamp recordings, 12- to 16-bit A/D converters have been used (of late, 16-bit A/D converters have been most common). Using a 12-bit A/D converter, the range of input signals is divided into 4,096 ($=2^{12}$) steps and recorded. If the range was −10 – +10 V, the resolution would be 20 V/4,096 = 5 mV. On the other hand, with a 16-bit converter, the resolution is estimated to be $20\ \mathrm{V}/2^{16} = 0.3$ mV. Representations of the results of using such a resolution are shown in Fig. 27.1. On the other hand, it is important to note that, in reality, the last 2 to 3 bits would be obscured because of the electric noise and nonlinearity of the converter, so the real resolution would be between four to eight times worse than what is estimated here. The ratio of maximum acceptable signal range to the minimum distinguishable difference of signals is called the "dynamic range." The dynamic range of 8-bit A/D converters is 2^8 (=256), and that of 16-bit converters is 2^{16} (=65,536). Dynamic range is often measured as a logarithmic value. In digital systems, a two-base logarithm is used, and bit is used as a unit of dynamic range. Therefore, we would simply say that the dynamic range of 8-bit A/D converters is 8 bits.

As discussed, any electrical signal is transferred between electrical devices as a voltage signal because voltages can be much more easily handled than can current signals. Thus, in patch-clamp recordings, even when one is recording a current signal the current is converted to a voltage signal proportional to the original current amplitude and read by the A/D converter. In fact, this current-to-voltage conversion is performed as the first step in the patch-clamp amplifier, followed by amplification of voltage. The proportionality constant is often called the "gain" and is usually controlled by a switch on the amplifier panel.

Let us see how the gain affects the resolution during A/D conversion. Suppose that the dynamic range of the A/D is 12 bits, the maximum input range is −10 to +10 V, and the gain is set to 1 mV/pA. When the current input is 10 nA (=10,000 pA), the current is converted to 10 V. Because the maximum input of the A/D converter is 10 V, this is the maximum current that can be measured by this A/D converter system. Currents larger than 10 nA or less than −10 nA would exceed the limit of the A/D converter and would not be measured correctly. From the dynamic range of the A/D converter, the resolution is 5 mV (=20 V/2^{12}), corresponding to a current amplitude of 5 pA. As already mentioned, because of the electrical noise from the A/D converter itself the real resolution is, in fact, more than 20 pA. In other words, currents of <5 pA (or, practically speaking, 20 pA) cannot be measured under this condition. Thus, it would be difficult to measure single-channel currents in this setting. However, if one knows that the maximum current in the experiment does not exceed 100 pA, the gain can be increased to 100 mV/pA. At this gain, the measurable range of the current signal would be −100 to +100 pA, and the resolution becomes 0.05 pA, which is good enough to record most single-channel currents. Using the highest possible gain may also help reduce the noise because the use of a feedback resistor of higher resistance (or a feedback capacitor of higher capacitance) for higher gain would result in less noise than at lower gains.

As this example shows, in digital patch clamp recordings, (1) the dynamic range of the A/D converter, (2) the acceptable input range, and (3) the expected current amplitude must be evaluated, and the gain of the amplifier selected to be as large as possible without exceeding the input range of the A/D converter. In this way, we can optimize the resolution, leading to the most accurate digital recordings of your data.

It should be noted that because a 16-bit A/D converter has a 16-fold better resolution than does a 12-bit A/D converter it can be expected to produce a correspondingly better quality recording. Thus, recording using a 16-bit converter instead of a 12-bit A/D converter achieves the same resolution as can be obtained through use of a 16 times larger gain. Because the acceptable input range is undiminished, a larger current can be simultaneously recorded. This characteristic is especially useful for noise analysis (variance analysis), where noise of limited power is present on a large background current. Although A/D converters with a larger dynamic range perform A/D conversion more slowly, recent technological progress in digital devices can almost compensate for this. However, if multichannel recordings are required, the speed of A/D conversion must be considered carefully.

Recently, 18-bit A/D converters have been gaining popularity. As any A/D converter can be expected to have 2- to 3-bit quantization errors, the two lowest significant bits are usually removed; and 16-bit data are recorded when using an 18-bit converter. As data are usually recorded as a sequence of 2 bytes (i.e., 16 bits), using an 18-bit converter is beneficial in that the data size remains unexpanded and the data thereby become more accurate.

27.3. Sampling Theorem: Digitization of Time

In the last section, we saw how analog data, such as voltage or current signals, are converted to digital data. The input signals are sampled at discrete time points such that the time also becomes digitized in A/D conversion. The digitization of time is another important factor that must be kept in mind. Voltage signals are digitized at very short time intervals. The time interval between the present and the next sampling point is called the "sampling interval," and its inverse is the "sampling frequency." For example, if the data were sampled every 1 ms, the sampling frequency would be 1 kHz. The digitization of time differs from the digitization of signals in that it is not necessary to consider the maximum input range and dynamic range because the total sampling duration is limited only by the space where the data are saved – usually the hard disk capacity (see Sect. 27.4).

As data are sampled during a "short" interval, each value is assumed to be very close, and we can interpolate the signal values between the neighboring sampling points without direct measurement by connecting two adjacent data by a short line segment or a little curved segment. Instinctively, this is true when the sampling interval is sufficiently short compared with the rate of signal change. However, if the signal changes very quickly even within the short sampling interval applied, this assumption does not hold. Then, how short should the sampling interval be to detect "rapidly" changing data correctly? To answer this question, Nyquist predicted, and Shannon and Someya independently proved, the following theory, known as Nyquist–Shannon's sampling theorem (3, 4). Where f_N denotes the maximum frequency contained within the original signal, and f_S is the sampling frequency, the following inequality

$$f_S \geq 2 \cdot f_N \tag{27.1}$$

must hold to reproduce the original signal by the sampled data. Double of f_N is called the Nyquist frequency. For example, if the sampling frequency is 2 kHz (sampling interval = 0.5 ms), the sampling can correctly record data with a maximum frequency of 1 kHz.

We had better clarify the meaning of the maximum frequency contained within the signal of interest. By Fourier's theorem, any periodic function can be expressed as the summation of sine and cosine waves with certain frequencies as follows:

$$f(t) = \frac{a_0}{2} + \sum_{n=1}^{\infty}\left(a_n \cos\frac{2n\pi t}{T} + b_n \sin\frac{2n\pi t}{T}\right) \quad (27.2)$$

where T denotes the cycle of the function, and a_n and b_n are coefficients that can be derived by $f(t)$, as follows:

$$\begin{aligned} a_n &= \frac{2}{T}\int_{-T/2}^{T/2} f(t)\cos\frac{2n\pi t}{T}\,\mathrm{d}t \\ b_n &= \frac{2}{T}\int_{-T/2}^{T/2} f(t)\sin\frac{2n\pi t}{T}\,\mathrm{d}t \end{aligned} \quad (27.3)$$

Each term shows a sine (or cosine) wave with a defined frequency, termed a "frequency component." Simple periodic functions are composed of finite numbers of sine wave components, and thus the function $f(t)$ can be expressed by the summation of the finite terms of a Fourier series. The largest frequency component among the frequency components can be taken as the maximum frequency that the original signal contains (5, 6).

Generally, any signal in our experiments is not usually transformed into a finite number of terms of Fourier series. For example, even a square pulse signal that repeats ±1 V alternatively every 0.5 s is expanded into infinite numbers of Fourier components, although it may be looked upon as a simple periodic function of 1 Hz frequency. This means that most of the given signals do not generally have the highest frequency component, and therefore we stray again back into the unsolved problem of what sampling frequency is adequate to satisfy the Nyquist–Shannon sampling theorem.

In real electrical circuits, noises and errors are inevitable. As we often experience, one of the most problematic noises to remove is high-frequency noise, which often originates from the switching on and off of the circuit. To remove it, most electronic devices, including patch-clamp amplifiers, are equipped with low-pass filters. The low-pass filter cuts off the frequency components higher than a certain limit, the "cutoff frequency." Therefore, the cutoff frequency of the patch-clamp amplifier can be practically considered to be the maximum frequency contained within the signals from the amplifier.

Next, let us see the consequences of not obeying the Nyquist–Shannon criterion and sampling our data at a frequency less than twice the "maximum" frequency component. Figure 27.2 shows a sine wave signal with a maximum and single signal of 10 Hz in the left column and the amplitude of each term of Fourier series plotted against the frequency components in the middle column. When the sampling frequency is larger than 20 Hz, which is double the

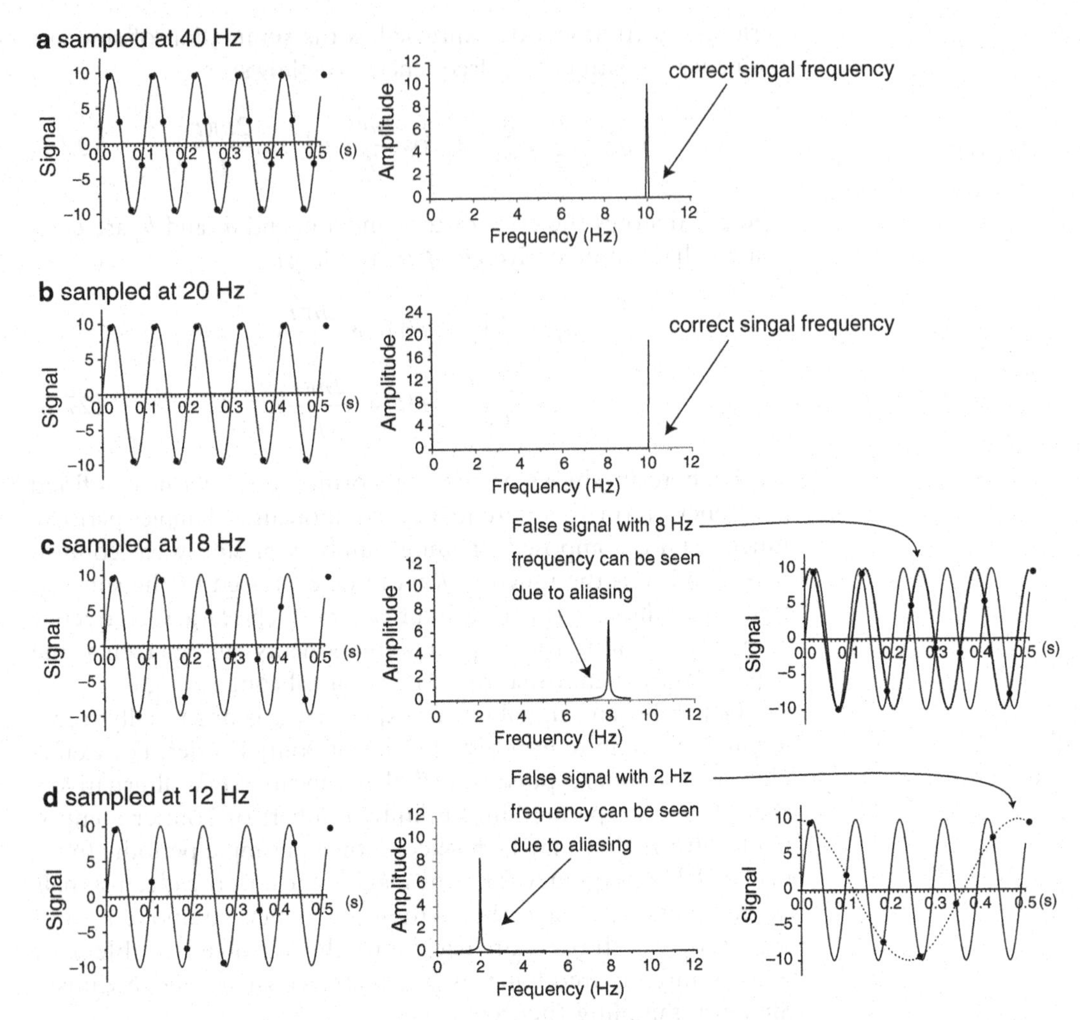

Fig. 27.2. Examples of aliasing. Sine wave signal of 10 Hz frequency was sampled with a sampling frequency of (**a**) 40 Hz, (**b**) 20 Hz, (**c**) 18 Hz, and (**d**) 12 Hz. In *the left panels*, graphs show the original signal and the sampled points (*closed circles*). In *the middle column*, graphs show the result of fast Fourier transformation (FFT) of discrete sampled data. The amplitude of each term of the Fourier series is plotted against frequency components. In this example, the Nyquist frequency is 20 Hz as the frequency of the original signal was 10 Hz. When sampled by a frequency larger than the Nyquist frequency (in **a** and **b**), the original frequency is reconstituted by FFT, as shown in *the middle column*. However in **c** and **d**, the false signals (8 Hz in **c**, 2 Hz in **d**) are detected by FFT. As shown in *the right column*, the false data (*dotted lines*) can be also seen when sampled data points are connected. The phenomenon is called aliasing.

original signal frequency, the sampling successfully reproduces the original signal. However, when the sampling frequency is less than the Nyquist frequency, it seems as if another, virtual signal has appeared with a frequency that was not present in the original signal. This happens because the signal values are favored to be interpolated by a line segment connecting the two neighboring sampled data or, more precisely, by a shortest sine wave fragment of a phase. This phenomenon is called "aliasing." Aliasing happens not only in mathematical theory but also in our perceptions. If you see the wheels of a wagon start rotating in a movie, at first, the wheels appear to move in the appropriate direction and gradually rotate faster. After a while it seems as if they stop rotating, and at the next moment the wheels rotate in the reverse direction. In this example, the frame speed (e.g., 24 frames per second) corresponds to the sampling frequency, and the rotation speed of the wheels corresponds to the frequency of the original signal. At the beginning, as the wagon pulls away, the rotation speed is less than two times the frame speed, and the film correctly projects the rotation; but when the rotation becomes faster, aliasing signals are perceived (7).

In a real system, it is better to set the sampling frequency at more than 2.5–3.0 times, rather than 2.0 times, the cutoff frequency due to the existence of two important artifacts. First, no low-pass filter behaves in an ideal fashion and passes 100% of the signal below the cutoff frequency. Also, it does not completely cut the signal above the cutoff frequency. Second, our current and voltage signals are not periodic. For signal processing, the signals are handled as virtual periodic functions, and the signals are distorted by passing such filters. In short, the Nyquist–Shannon theorem can be stated in the practice of electrophysiology as follows: The sampling frequency used during A/D conversion should be set at least 2.5–3.0 times higher than the cutoff frequency of the amplifier.

27.4. Other Important Factors in A/D Conversion

The accuracy of A/D conversion depends on the specifications of the A/D converter, which include the resolution, dynamic range, maximum input range, maximum sampling frequency, number of input channels, and the data transfer speed. The maximum sampling speed of most A/D converters is 100–500 kHz. However, this speed is attained only when used with a single input from one channel. When four channels are recorded simultaneously (e.g., current and voltage from each of two electrodes) the maximum sampling speed should be divided by the number of input channels. With multiple channel recordings, it may be necessary to limit the sampling frequency to obtain adequate data transfer. To reduce noise, the ground signal should be subtracted before A/D conversion. However, many A/D converters accept only the signal input of BNC cables.

After A/D conversion, the data are transferred to a PC. These days, the transfer speed from an A/D converter to a computer is quite fast and does not usually present a problem, considering that the maximum transfer rate of USB 2.0 reaches up to 60 Mbytes/s. With an A/D converter connected via a slow bus, however, continuous sampling for a long time at a high sampling frequency might be difficult because of the lack of sufficient data transfer speed, and special care must be taken for data not to be dropped from the sampling.

After data are transferred to a computer, they are stored in a file in a sequence. Each sampled datum requires two bytes for storage. For example, if you sample a trace of 100 ms at 2 kHz sampling frequency, the data occupy 400 bytes on your disk. On the other hand, if you obtain data of 20 s duration at 10 kHz with 10 V steps, the data size is 4 Mbytes. Such data file sizes are not a problem unless and until you wish to keep a great many of them on your hard disk.

What should we do if we have recorded data at a certain sampling rate and subsequently notice that the events of interest are faster than the sampling rate? If we have relied on the digitized data as our sole record of the experiment, there is little that we can do other than to repeat the experiment. For such eventualities and to avoid repeating experiments unnecessarily, it is useful to have kept an alternative record of the data. Analog recorders and digital recorders such as DAT have been used in the past. These digital recorders are equipped with A/D converters that perform the A/D conversion at a constant and higher sampling frequency than is used in most patch-clamp recordings (e.g., 44 or 48 kHz). Having an analog or high sampling frequency digital back-up recording enables the well-prepared researcher to resample the data at a sampling frequency appropriate to the events of interest. Sadly, these data recorders are no longer widely available. An alternative strategy might be to record the data simultaneously via another A/D converter and PC. Bear in mind that if data are continuously sampled at 100 kHz over 24 h the data storage capacity consumed is 17.3 Gbytes. In the past, such an option would have been unthinkable because of the limited size of hard disks, but recently hard disk capacity has been enlarged and so becomes a practical choice. In any case, whether you choose simultaneously to record your data to two distinct devices at different sampling frequencies, strategies for data storage and backup using networked storage devices should be considered.

It is important to note that when data are resampled from the storage media, the final cutoff frequency, f, is not the cutoff frequency of the low-pass filter set upon the resampling (f_2); it also depends on the cutoff frequency f_1 applied at the time when data were initially sampled, as follows:

$$\frac{1}{f} = \frac{1}{f_1} + \frac{1}{f_2} \tag{27.4}$$

27.5. Digital-to-Analog Conversion

In patch-clamp experiments, voltage and/or current commands are given to stimulate whole-cell or patches of membrane. The commands are often composed and given from a digital stimulator. In contrast to data sampling, command shapes are composed and sent from the computer to the stimulator, where digital-to-analog (D/A) conversion is performed. This type of stimulator is attached to most patch-clamp systems; and even complicated multistep, sinusoidal, or ramp pulse commands can easily be applied while watching and confirming the command shapes on the computer monitor. The same concepts as for A/D converters – e.g., converting frequency, output range, dynamic range – apply to D/A converters. However, the level of precision required for D/A conversion of the command pulses is less that required for the A/D conversion of recording signals. Nevertheless, an 8-bit D/A converter may cause problems when ramp or sinusoidal pulses should be applied because the voltage step in those pulses are expected to be smooth and continuous, and the limited resolution may result in appreciable steps in the voltage signal.

Another problem in D/A conversion may occur when the given input data are changed. Data are sent from the computer as a sequence of bits. Suppose that binary digital expressions $(1{,}000)_2$ and $(0000)_2$, each consisting of 4 bits, are converted to 10 V and 0 V in a D/A converter. Now, a command of $(1{,}000)_2$ is given, and the stimulator generates a command at 10 V. Suppose a command of $(0111)_2$ was given at the next moment. Maybe $(0111)_2$ means 9.4 V, and the difference between the intended command and the previous command is not so big – but the bits are sent one by one. Thus, when $(0111)_2$ is sent, the first (most significant) bit, "0" arrives at the D/A converter; it may change the highest bit as 0, and the data which the D/A converter has in its digital memory may suddenly become $(0000)_2$ from $(1{,}000)_2$ because the information on the other 3 bits has not arrived. In this case, the D/A converter would suddenly drop the command to 0 V in a short interval before the other bits arrive. This phenomenon, caused by an asynchronous change of digital expression, is called a "glitch." Glitches happen only instantaneously but may cause a big artifact. Well-designed D/A converters do not show glitches (1).

27.6. Data Files and Software

The analysis of patch-clamp data is difficult, time-consuming work. Most common data analyses can be done by popular commercial software, but there are still some analyses that can only be performed with self-made software or macros. Luckily, the data formats used in popular patch-clamp software are open to the public (e.g., (8, 9)). We hope that researchers make their self-made software open to all for free access and continue to help each other in this collegiate fashion.

References

1. Sakmann B, Neher E (1995) Single-channel recording, 2nd edn. Plenum, New York
2. Feynman RP, Hey A (2000) Feynman lectures on computation. Westview Press, New York
3. Nyquist H (1928) Certain topics in telegraph transmission theory. Trans AIEE 47:617–644 (Reprint as classic paper in: Proc. IEEE 90: 280–305, 2002)
4. Shannon CE (1949) Communication in the presence of noise. Proc Inst Radio Eng 37:10–21 (Reprint as classic paper in: Proc. IEEE 86: 447–457, 1998)
5. Hino M (1977) Spectrum analysis (in Japanese). Asakura, Tokyo
6. Bendat JS, Piersol AG (1986) Random data, 2nd edn. Wiley, New York
7. Hamming RW (1983) Digital filters. Prentice-Hall, Englewood Cliffs
8. Sigworth FJ, Affolter H, Neher E (1994) Design of EPC-9, a computer-controlled patch-clamp amplifier. 1. Hardware. J Neurosci Methods 56:195–202
9. Sigworth FJ (1994) Design of EPC-9, a computer-controlled patch-clamp amplifier. 2. Software. J Neurosci Methods 56:203–215
10. Ogawa H (2006) Sampling theory and Isao Someya: A historical note. Inst Electron Inform Comm Eng Sampl Theor Signal Image Process 5:247–256

Chapter 28

Solutions for Patch-Clamp Experiments

Ravshan Z. Sabirov and Shigeru Morishima

Abstract

Solutions used in patch-clamp can be divided into two general categories by the side of the patch membrane with which the solution is in contact. The compositions of extracellular solutions are usually similar to those of the natural extracellular fluids, whereas intracellular solutions resemble cytosol with respect to pH, free Ca^{2+}, and Mg^{2+} concentrations and ATP content. Designing solutions for patch-clamp experiments requires careful consideration of the physicochemical processes controlling the concentrations of protons and divalent cations, ionic strength, and osmotic pressure. Liquid junction potentials arising at the interfaces between different experimental solutions may contribute to the membrane potential and should be measured or calculated and adequately compensated. This chapter describes general strategies and methods for practical calculations of these parameters, which are critically important for successful patch-clamp experiments.

28.1. Introduction

Patch-clamp is compatible with a wide variety of solutions. Both sides of the membrane can be bathed by a defined solution that can be selected depending on the purpose of the experiment. Solutions used in patch-clamp can be divided into two general categories based on the side of the patch membrane with which the solution is in contact. The extracellular side is normally bathed by an "extracellular" solution, which usually but not always has a composition similar to that of the natural extracellular fluid; often it consists mainly of NaCl and contains Ca^{2+} and Mg^{2+} ions in the millimolar range. Intracellular solution is composed mainly of the K^+ ion (or Cs^+ as a substitute for K^+), organic anions, and Cl^-. In many cases, it contains adenosine triphosphate (ATP) and/or guanosine triphosphate (GTP) to maintain intact cellular functions. The free Ca^{2+} concentration in intracellular solutions is usually kept at the

Yasunobu Okada (ed.), *Patch Clamp Techniques: From Beginning to Advanced Protocols*, Springer Protocols Handbooks, DOI 10.1007/978-4-431-53993-3_28, © Springer 2012

Table 28.1
Relation between patch-clamp mode and solutions

Patch-clamp mode	Bath solution	Pipette solution
Whole-cell, outside-out	Extracellular	Intracellular
Inside-out	Intracellular	Extracellular
Cell-attached	Extracellular	Extracellular

submicromolar level, and free Mg^{2+} can be in the millimolar range or its fraction.

Both bath and pipette may contain either the extracellular or the intracellular solution depending on the experimental configuration. The relation of patch-clamp modes and solutions is shown in Table 28.1.

A general principle when designing solutions for electrophysiological studies is to mimic the natural composition of extracellular and intracellular fluids of the living creatures under study. As these fluids vary widely, it is impossible to suggest a standard solution useful for any one case. As a guide, one may use a summary of biological fluids described in several monographs (e.g., (1)). We recommend here only general rules for designing particular solutions for specific experiments. The parameters discussed in the next sections should be considered when designing solutions.

28.2. pH ([H+] Concentration)

The solutions should have an appropriate pH. Normal extracellular solution has a pH of 7.3–7.5, and intracellular solution may have either same or slightly more acidic pH because normally cytosol has a pH of around 7.2.

Biologically, extracellular and intracellular fluids have a characteristic and nearly constant pH. This pH is maintained in a number of ways, one of the most important of which is through buffering systems. Similarly, in experiments if the solution does not have sufficient buffering capacity the pH of a solution may change because of cell metabolism or addition of some acidic or basic drugs. As many ion channels can be affected by pH, it is essential to use buffer solutions for both bath and pipette solutions.

The Henderson–Hasselbalch equation gives a good measure of the acidity of a desired buffer. When an acid (HA) is dissolved in water, the equilibrium can be described as following:

$$HA \rightleftharpoons H^+ + A^- \qquad (28.1)$$

If HA is a strong acid, all of the acid is dissociated, and the equilibrium shifts to the right completely. When HA is a weak acid (e.g., acetic acid), the following equilibrium equation stands:

$$\frac{[H^+][A^-]}{[HA]} = K_a \tag{28.2}$$

where K_a is an equilibrium (dissociation) constant of the (weak) acid. The values of K_a for weak acids range from 10^{-1} to 10^{-7}. From (28.2), we can derive the following equation:

$$\text{pH} = \text{p}K_a + \log\frac{[A^-]}{[HA]} \tag{28.3}$$

Buffer is usually made as a mixture of nearly equimolar weak acid and a salt of its conjugate base (e.g., an aqueous solution of acetic acid and sodium acetate). Suppose that α mole of HA and β mol of its sodium salt (NaA) are dissolved in 1 l of water. Because NaA is usually fully dissociated whereas only a small part of HA is dissociated, $[A^-]$ approximates β unless the solution is much diluted and $\alpha \gg \beta$. Because α is close to β in practical buffering solutions, the pH of the solution can be estimated from the following equation:

$$\text{pH} = \text{p}K_a + \log\frac{\beta}{\alpha} \tag{28.4}$$

This equation is called the Henderson–Hasselbalch equation. More precisely, ionic activities in place of ionic concentrations and corrected K_a (K_a') must be used in the equation.

As it looks similar to (28.3), it is often confused, but one has to remember that the Henderson–Hasselbalch equation estimates, as an approximation, the pH of a buffer – a mixture of a week acid with its conjugate base. If a small amount of acid is added so the final concentration becomes γ mol/l (M) ($\gamma \ll \alpha$), some H^+ binds to A^- to form HA. This buffers the increased H^+, and the pH becomes

$$\text{pH} = \text{p}K_a + \log\frac{\beta-\gamma}{\alpha+\gamma} \tag{28.5}$$

When $\alpha = \beta = 100$ mM (pH equals to pK_a) and a small amount of acid is added ($\gamma = 10$ mM), the pH shifts only by 0.1. Considering that the pH of 10 mM of a strong acid is 2.0, we can realize the strong buffering capacity of the solution. Figure 28.1 shows the relation between pH and the acid/base concentration ratio. The buffering capacity is strongest at $[A^-]/[HA] = 1$, that is, when pH = pK_a.

Important biological buffer systems include the dihydrogen and monohydrogen phosphate system, the carbonic acid and bicarbonate system, the plasma protein buffer system using the characteristics

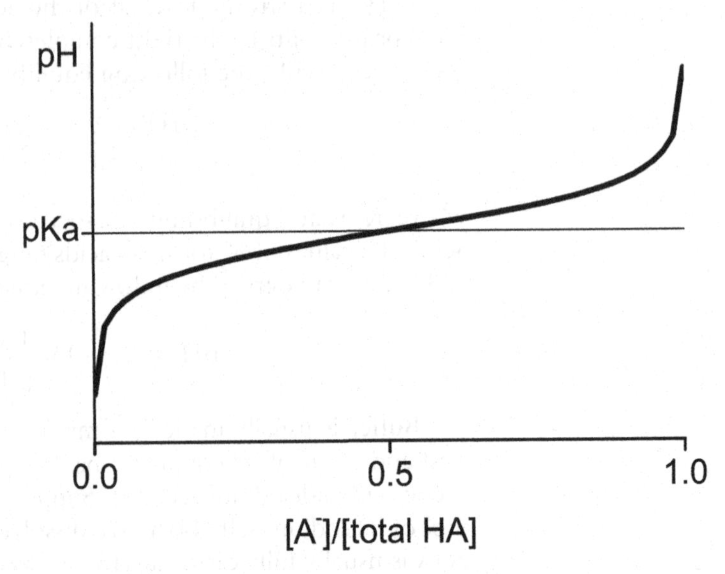

Fig. 28.1. Relation between pH and the ratio of the concentration of an acid HA and its conjugate base A^- (see 28.3) in the buffer solution. The pH value is fairly stable near pK_a despite shifts in the acid concentration.

of such amino acids as zwitter (amphoteric) ions, and the hemoglobin buffer system. In the bicarbonate buffer system, the acid is H_2CO_3, and its conjugate base is HCO_3^- . In most cases, $NaHCO_3$ is used for the salt. In the human body this system cooperates with the respiratory and renal systems, and the gasification and dissolution between H_2CO_3 and CO_2 gas is rapid with the help of carbonic anhydrase. These make the buffering capacity of the bicarbonate system strong and flexible. The pK_a for this buffer is 6.1, and the Henderson–Hasselbalch equation can be written as follows:

$$\mathrm{pH} = \mathrm{p}K_a + \log\frac{[\mathrm{HCO_3^-}]\times 1{,}000}{0.03\times P_{\mathrm{CO_2}}} \tag{28.6}$$

where $[HCO_3^-]$ denotes the molar concentration of bicarbonate salt, and P_{CO_2} (mmHg) denotes the partial pressure of CO_2 gas. For a solution containing 25 mM HCO_3^- under 5% CO_2 gas ($P_{CO_2} = 38$ mmHg), the equation yields a pH of around 7.4.

α-Amino acids contain both a carboxyl group (–COOH) and an amino group (–NH_2). The general formula for an amino acid is given as ^+H_3N–CR–COO^- (R: side chain) when its net charge balances. In an acidic solution, the amino acid acts as a base, and a proton binds to its carboxyl group and becomes: ^+H_3N–CR–COOH. On the contrary, in an alkaline solution, it behaves as an acid, and releases a proton to be H_2N–CR–COO^-. Ions that show both acidic and basic characteristics (e.g., amino acids), are called zwitterions (amphoteric ions), and a solution of a zwitterion can be a buffer by itself. Norman Good developed a series of aminomethanesulfonic acids, a group of zwitterions — 4-(2-hydroxyethyl)piperazine-1-ethanesulfonic acid (HEPES), 2-(*N*-morpholino) ethanesulfonic acid (MES), 1,4-piperazinediethanesulfonic acid (PIPES), 3-morpholino-propanesulfonic acid (MOPS) - and showed that they work as literally "Good" buffers in a physiological range of pH. Every Good's buffer (like other buffers) has an adequate pH range of buffering, and the ionic activity coefficient is close to 1. Therefore, it can be easily chosen for a certain pH and used without complicated calculation. The most widely used is HEPES. Other buffers such as Tris [tris(hydroxymethyl)aminomethane] or glycine can also be used. It should be noted, however, that sometimes they interfere with ion channels under investigation, as some of the buffers may block them. (Tris, or even HEPES, for instance, may inhibit some potassium channels or chloride channels.) Applying inhibitors, one should consider possible interaction with buffers. Lanthanides, for example, form complexes with Tris. The buffer concentration should be sufficient to maintain the desired pH level. Usually, 5–10 mM is a good range. In some special cases (e.g., for proton current measurements) the concentration may be increased up to 50–100 mM, taking care about the total osmolality. The final pH is adjusted with a strong acid or base. It is recommended that a major ion be used as a counterpart. For example, for a NaCl-based solution, NaOH and HCl are the base and acid of choice. For experiments with varying pH, care should be taken to use buffers with sufficient buffering capacity in the desired pH range. The pK_a values and useful pH ranges are usually given in catalogs (see, for example, Sigma-Aldrich webpage http://www.sigmaaldrich.com/life-science/core-bioreagents/biological-buffers/learning-center/buffer-reference-center.html). Usual biological buffers are membrane-impermeable; and thus when applied from extracellular side, they do not affect cytosolic pH. In some cases, however, a weak membrane-permeable acid (e.g., acetate or butyrate) can be included in the extracellular bath solution to control the cytosolic pH (2–4).

28.3. Ionic Strength

Formal ionic strength (I_f) can be calculated according to the formula

$$I_f = \frac{1}{2}\sum_j C_j Z_j^2 \tag{28.7}$$

where Z_j is the charge of ion species j, and C_j is the ionic concentration. For example, 154 mM (0.9%) NaCl solution has an ionic strength of 0.154, and the ionic strength of 10 mM $CaCl_2$ is [(10 $\times 2^2 + 20 \times 1)/2/1{,}000 = 0.030$]. [$I_f$=0.10–0.15] is considered normal, although stable recordings can be obtained even when all monovalents are replaced with mannitol (low ionic strength) or the divalent ion concentration is increased up to 50–100 mM (high ionic strength). Because ionic strength can be an important regulator of channel function, one has to inspect its level with care.

Ionic strength is essential for the estimation of ionic activity. Up to this section, we postulated that all solutions act ideally (like an ideal gas in thermodynamics), and ionic molar concentrations (millimoles or moles) have been used for the calculations of pH or equilibrium equations. Practically, ionic solutions are *not* ideal, and "ionic activity" must be used in place of ion concentration. Ionic activity is a product of an ion activity coefficient and the ion concentration. According to the Debye–Hückel theory, ionic activity coefficient γ can be theoretically derived as follows:

$$\log\gamma = -\frac{1.824\times10^6}{(\varepsilon T)^{2/3}}|z_+z_-|\sqrt{I_f} = -0.509|z_+z_-|\sqrt{I_f} \quad \text{(at 25°C aq.)} \quad (28.8)$$

where I_f denotes ionic strength of the solution, ε the dielectric constant, T temperature (K), and z_+ and z_- charges of cations and anions. In 10 mM NaCl solution, γ becomes 0.89. It should be kept in mind that even the Debye–Hückel theory is valid only for diluted solutions of <10 mM. Calculated and experimentally measured values of activity coefficients can be found in Robinson and Stokes (5).

28.4. Divalent Cations

Divalent cations are important regulators of cellular functions. Usually, 1–2 mM of Mg^{2+} and Ca^{2+} is used in extracellular solutions. Giga-seal formation can resist total omission of one or even both of them, although this can be used only when studying the effect of extracellular divalents on ion channel function. High Mg^{2+} and Ca^{2+} levels up to 50–100 mM can be used temporarily in some special experiments, although changes in surface potential must be kept in mind.

The intracellular Mg^{2+} concentration is usually 1–2 mM. If ATP is included in the solution, an Mg–ATP complex with 1:1 stoicheometry is formed with pK 4.22 (at 25°C); and free Mg^{2+} in a solution containing equimolar Mg^{2+} and ATP (1 mM) would be about 0.2 mM. It is pertinent to remind the reader that ATP poorly discriminates between Mg^{2+} and Ca^{2+}. As free Mg^{2+} is an important regulator of many cellular physiological functions, including activity of phosphatases and some other enzymes, one has to consider not only the total Mg^{2+} concentration but its ionized level as well.

Intracellular Ca^{2+} is an important signaling ion, and its intracellular concentration is normally at around 1/10,000 that of the extracellular fluids. Therefore, the free Ca^{2+} level in intracellular solutions should be carefully adjusted. When making nominally Ca-free solutions, one should keep in mind that even double-distilled deionized water contains micromolar levels of Ca^{2+} (4–20 μM (6–8)). A commonly accepted way to control the free Ca^{2+} concentration is to use Ca buffers (i.e., mixtures of a highly specific chelator in combination with a known total Ca^{2+} concentration). Then, the free Ca^{2+} concentration is calculated based on known dissociation constants of Ca–chelator complexes. The most widely used chelator is EGTA (ethyleneglycol bis(β-aminoethyl ether)-*N,N,N′,N′*-tetraacetic acid), which binds preferentially to Ca^{2+} but not to Mg^{2+}. BAPTA [bis-(o-aminophenoxy)-ethane-*N,N,N′,N′*-tetraacetic acid] is another chelator of choice that has higher buffering capacity and less pH dependence. In contrast to EGTA and BAPTA, EDTA (ethylenediamine tetraacetic acid) binds all divalent and other polyvalent cations (e.g., Ba^{2+}, Cd^{2+}, lanthanides).

The apparent association constant for the Ca–ligand complex K'_{Ca} is defined as:

$$K'_{Ca} = \frac{[\mathrm{Ca-ligand}]}{[\mathrm{Ca}^{2+}][\mathrm{ligand}]} \tag{28.9}$$

where [Ca-ligand], [Ca^{2+}], and [ligand] are equilibrium concentrations of the complex, free Ca^{2+}, and ligand. The prime (′) in K'_{Ca} shows that this equilibrium constant is calculated using ionic activities. [Ca^{2+}] can be obtained from (28.9) by solving the following quadratic equation:

$$[\mathrm{Ca}^{2+}]^2 K'_{Ca} + [\mathrm{Ca}^{2+}]\{1 + K'_{Ca}([\mathrm{Ca_T}] - [\mathrm{ligand_T}])\} - [\mathrm{Ca_T}] = 0 \tag{28.10}$$

where [Ca_T] and [$ligand_T$] are total concentrations of Ca^{2+} and ligand, respectively.

For EGTA, the following equation can be used to calculate K'_{Ca} from association constants of the unprotonated and monoprotonated ligand for calcium (K^1_{Ca} and K^2_{Ca}) and the four acid association constants, K_1, K_2, K_3, and K_4 (9):

$$K'_{Ca} = K^1_{Ca} \Big/ \begin{bmatrix} 1+[\mathrm{H}^+]K_1+[\mathrm{H}^+]^2K_1K_2 \\ +[\mathrm{H}^+]^3K_1K_2K_3+[\mathrm{H}^+]^4K_1K_2K_3K_4 \end{bmatrix} + K^2_{Ca} \Big/ \begin{Bmatrix} (1/[\mathrm{H}^+]K_1)+1+[\mathrm{H}^+]K_2 \\ +[\mathrm{H}^+]^2K_2K_3+[\mathrm{H}^+]^3K_2K_3K_4 \end{Bmatrix} \tag{28.11}$$

The main problem in Ca buffer calculations is to use the proper set of constants. The collection of constants in Table 28.2 is reproduced from Harrison and Bers (10)

Association constants are functions of ionic strength and ambient temperature. Smith and Miller (11) found that a satisfactory fit

Table 28.2
Association constants for chelators used for Ca buffers

Ligand	Ionic strength, temperature (°C)	K_1	K_2	K_3	K_4	K_{CA}^1	K_{CA}^2
EGTA	0.1 M, 20°C	9.47	8.85	2.66	2.0	10.97	5.3
EGTA	0.1 M, 20°C	9.625	9.00	2.813	2.117	11.118	5.509
BAPTA	0.1 M, 22°C	6.36	5.47			6.97	

EGTA ethyleneglycol-bis-(2-aminoethylether)-*N,N,N′,N′*-tetraacetic acid, *BAPTA* 1,2-bis(2-aminophenoxy)ethane-*N,N,N′,N′*-tetraacetic acid
Association constants are expressed as $\log_{10}$ (M^{-1})

to their data could only be achieved if ionic strength was expressed not as a formal I_f calculated according to (28.7), but when ionic strength was expressed in terms of ionic equivalents as:

$$I_e = \frac{1}{2}\Sigma C_j \,|\, Z_j \,| \tag{28.12}$$

Therefore, this expression is commonly used in Ca buffer calculations. When I_e is different from 0.1 M, each constant for protons and metal (K') can be adjusted using the semiempirical form of Debye–Huckel limiting low (10):

$$\log_{10}K' = \log_{10}K + 2xy(\log_{10}f_j - \log_{10}f_j') \tag{28.13}$$

where x and y are the absolute values of the valency of the cation and anion in the complex, f_j is an adjustment factor for ion j in initial ionic strength (at which the constants were measured), and f_j' is that at the desired new condition:

$$\log_{10}f_j = A\{[I_e^{1/2}/(1+I_e^{1/2})] - 0.25I_e\} \tag{28.14}$$

where $A = 1.8246\times10^6/(\varepsilon T)^{1.5}$, ε is dielectric constant of the solvent (water), and T is the absolute temperature. When the temperature is different from 20–22°C, a correction can be made using the Van't Hoff isochore:

$$\log_{10}K' = \log_{10}K - [\Delta H(T-T')/2.303RT^2] \tag{28.15}$$

or in its more correct form (10):

$$\log_{10}K' = \log_{10}K + [H(1/T - 1/T')/2.303RT] \tag{28.16}$$

Because not all constants for the enthalpy values are available, using [ΔH=−8.1 kcal/mol] for K_{Ca}^1 of EGTA in combination with [ΔH=−5.8 kcal/mol] for both K_1 and K_2 (according to (10)) gives practically acceptable results. Unfortunately, no enthalpy data for BAPTA are available. Harrison and Bers (9) determined the enthalpy value for K_{Ca}' at I=0.2 to be 16.6 kJ/mol (4.46 kcal/mol) for

EGTA and 13.9 kJ/mol (3.32 kcal/mol) for BAPTA. These values can be used in (28.15) and (28.16) for temperature corrections.

The above consideration is valid when only Ca^{2+} and one ligand (EGTA or BAPTA) are present. If the solutions contain Mg^{2+} in addition (which binds to both ligands to a certain extent) and ATP (which binds both calcium and magnesium ions), a full set of association constants and enthalpies for all possible complexes should be used for Ca buffer calculations. An algorithm and a computer program written in FORTRAN has been suggested by Fabiato (12) and apparently is available from him on request. A website created by Chris Patton of Stanford University (http://www.stanford.edu/~cpatton/) contains free software (MAXCHELATOR) for Ca buffer calculations. The Ca^{2+} calculations can also be performed directly from the Web browser at http://www.stanford.edu/~cpatton/webmaxcS.htm/.

Using the described calculation methods or computer programs, one should be aware that the actual free Ca^{2+} level in the prepared solution might be significantly different from the calculated values due to use of imperfect approximations (e.g., Debye–Huckel limiting low and Van't Hoff isochore). The differences may be due to the absence of reliable enthalpy values for all association constants and unaccounted dependence of association constants not only on ionic strength and temperature but also on monovalent ion species. One way around this problem is to determine K'_{Ca} directly using Ca-selective electrodes that have linear responses down to the micromolar level [Ca^{2+}] (6–8). Table 28.3 gives some data obtained by this method (see also (9)). An overall association constant can be determined under precisely the same conditions as used for the experiments and projected to the desired

Table 28.3
Apparent stability constants for EGTA and BAPTA estimated by the Ca electrode method using double-log optimization at pH 7.30

Chelator and ionic strength (M)	Temperature (°C)	log K' calculated	log K' measured
EGTA			
0.1	25	7.053	7.127 ± 0.059
0.16	25	7.009	6.966 ± 0.029
BAPTA			
0.1	22	6.908	6.872 ± 0.028
0.15	22	6.706	6.671 ± 0.036
0.2	22	6.573	6.497 ± 0.040
0.2	37	NC	6.686 ± 0.052

From (8)
NC not calculated because of the unavailability of enthalpy values

low $[Ca^{2+}]$ using (28.9). This method does not require an accurate knowledge of all association constants, enthalpies, ionic strength, and pH. On the other hand, it is more laborious compared to a quick calculation when the software is available.

Additional problems arise from impurities of commercially available chemicals. The purity of EGTA was found to be only 93.6–98.8% for different brands, whereas BAPTA purity may be as low as 86%, which significantly affects the calculation results (6–8). A way to solve this problem is to bake the chemicals for 3 h at 150°C in a muffle furnace. This procedure was found to bring the purity to almost 100%, implying that most of the contamination was probably bound water (8).

Calcium chloride is highly hygroscopic; it picks up moisture from the air and becomes liquid if left in open containers. One way to avoid this problem is to use $CaCO_3$ instead of $CaCl_2$, bake it at 110°C for 1 h, and dissolve by titration with HCl (8). Alternatively, a fresh batch of $CaCl_2{\cdot}2H_2O$ or $CaCl_2{\cdot}6H_2O$ can be dissolved immediately after opening to make up a large quantity of 1–3 M solution. The concentration of this solution should be controlled by an analytical method (e.g., oxalate or EDTA titration, flame spectroscopy). An inexpensive, reliable method is to measure the specific gravity (density) of concentrated $CaCl_2$ solutions by a hydrometer. 1 M, 2 M, and 3 M $CaCl_2$ at 20°C have specific gravities of 1.0853, 1.1677, and 1.2463 g/cm^3, respectively (calculated from the data in (13)). The densities of other concentrations at 20°C can be calculated from the formula:

$$d_{20} = 0.9992 + 0.08795\ \mathrm{M} - 0.00186\mathrm{M}^2 \qquad (28.17)$$

where M is the molar $CaCl_2$ concentration. A good hydrometer measures the density with a precision of 0.001, which corresponds to an error of about 0.6% in the estimation of the 2 M $CaCl_2$ solution and about 0.4% for the 3 M $CaCl_2$ solution.

Divalent cations may form insoluble precipitates. These precipitates interfere with giga-ohm seal formation as precipitation occurs at the pipette tip. For example, phosphate buffer in a pipette solution may produce calcium phosphate when used with normal Ringer's solution. Another example is sulfate in the pipette, which forms $BaSO_4$ if Ba^{2+} is used in the bath either to block K^+ channels or as a current currier for Ca^{2+} channels.

28.5. Osmotic Pressure

The osmotic pressure of the solutions is another important parameter that should be considered when designing patch-clamp solutions. Van't Hoff's law defines the osmotic pressure of a solution as:

$$\pi = RT\phi\sum C_s \qquad (28.18)$$

where R is the gas constant, T is absolute temperature, ϕ is the osmotic coefficient that accounts for nonideality of real solutions, and C_s is the concentration of osmotically active particles (osmolytes) formed on dissolution of the solute in water. For example, glucose and mannitol yield only one particle, whereas NaCl ideally yields 2. ΣC_s is called "osmolarity" if C_s is expressed as moles per liter and "osmolality" if C_s is expressed as moles per kilogram of H_2O. Osmolality better describes the osmotic pressure in the Van't Hoff's law and is an output of most osmometers. For mammalian cells, the osmolality of about 300 mOsm/kg H_2O is considered normal. Cells accommodate their volumes to the solutions of 280–330 mOsm/kg H_2O owing to volume regulation mechanisms. Therefore, the solutions with osmolality within this range can be used for patch-clamp as normal isotonic extracellular solutions. The osmolality of the intracellular solutions is a more complicated issue. In the cases of cell-attached, inside-out, and outside-out experiments, they can be simply isotonic to the extracellular ones. With the whole-cell configuration, slowly diffusible or nondiffusible components of the cytoplasm generate an oncotic pressure gradient. Therefore, the whole-cell configuration can be considered as a Donnan system, which results in cell swelling unless this imbalance somehow compensated. A frequently used method is to use slightly hypotonic (by about 15–30 mOsm/kg H_2O) intracellular pipette solution because it prevents spontaneous cell swelling after patch rupturing. However, because an oncotic component of the total intracellular osmotic pressure varies among cell types, the intracellular pipette solution osmolality should be adjusted individually using an inert osmolyte, such as mannitol.

There are three types of osmometer: membrane osmometers, freezing point depressing osmometers, and vapor pressure osmometers. When choosing the apparatus, one should keep in mind that for membrane osmometers the result depends on the permeability of the membrane for the osmolytes, whereas for vapor pressure osmometers the result depends on volatility. For the usual physiological solutions, all three types give consistent results. Standard 300 mOsmol/kg H_2O solution can be prepared by dissolving 9.4484 g NaCl in 1 kg of water (161 mM). The osmolality of the solution can be estimated as $\phi \Sigma C_s$ using [$\phi = 0.93$] for physiological concentrations of NaCl and KCl, [$\phi = 0.85$] for $CaCl_2$, and [$\phi = 1.01$] for sucrose (5). The dependence of the osmolality of mannitol and glucose solutions (measured in our laboratory with a freezing-point depression osmometer # OM802; Vogel, Giessen, Germany) on their molar concentration in the range of 0–300 mmol/l was slightly superlinear and could be well fitted with the following polynomial functions:

$$\Pi_{\text{mannitol}} = 1.0067C + 1.94 \times 10^{-4} C^2 \quad (28.19)$$

$$\Pi_{\text{glucose}} = 1.0054C + 1.81 \times 10^{-4} C^2 \quad (28.20)$$

where C is concentration in millimoles per liter, and $\Pi_{mannitol}$ and $\Pi_{glucose}$ are the osmolality in milliosmoles per kilogram of H_2O for mannitol and glucose, respectively. These equations can be used to calculate the contribution of mannitol and glucose to the total osmolality of the solution.

The contribution of buffers, Ca chelators, and ATP is somewhat ambiguous. According to our measurements, the osmolalities of 10 mM solutions of HEPES, EGTA, and ATP (dissolved in pure water and pH was adjusted to 7.4 by NaOH) were 15.2 ± 0.3, 28.8 ± 0.2, and 35.7 ± 0.3 mOsmol/kg H_2O, respectively (average of six parallel determinations ± SEM). In the range of 0–25 mM, the osmolality of these solutions could be well described by following polynomial functions:

$$\Pi_{HEPES} = 1.58C - 4.05 \times 10^{-3} C^2 \quad (28.21)$$

$$\Pi_{EGTA} = 2.95C - 5.95 \times 10^{-3} C^2 \quad (28.22)$$

$$\Pi_{ATP} = 3.91C - 2.35 \times 10^{-2} C^2 \quad (28.23)$$

where C is the concentration in millimoles per liter. Although these equations give rather precise (within the range of 0.5–1.0 mOsm/kg H_2O) values for the osmolalities of pure water solutions, one should keep in mind that the actual osmotic contribution should depend also on total Ca^{2+} and Mg^{2+} levels for EGTA and ATP.

The osmolality may change due to solvents [e.g., dimethylsulfoxide (DMSO)] added along with drugs. For instance, 0.1% (the amount introduced when the drug diluted from a 1,000× concentrated stock solution) of DMSO, ethanol, and methanol increase the osmolality by 13, 22, and 31 mOsm/kg H_2O, respectively. Slight shrinkage (which is transient owing to limited membrane permeability of the solvents) and related inactivation of volume-sensitive currents may be mistakenly thought to be a drug effect.

28.6. Supplements

Energy sources should be included in both extracellular and intracellular solutions to keep the cells alive during the experiment. Normally, 5–10 mM glucose is added to the extracellular solution. For intracellular solutions, 1–5 mM of ATP or MgATP provides enough energy for cellular metabolism. Several millimoles of pyruvate can be added to provide a substrate for oxidative phosphorylation. Some researchers also add phosphocreatine (up to 25 mM) as an alternative energy source and glutathione to maintain intracellular redox potential. More fine-tuning of the intracellular solution composition might be necessary to prevent the occasionally observed run-down of the currents of interest.

28.7. Liquid Junction Potential

Whenever two solutions with different compositions or a solution and a solid material come into contact, a junction potential is developed. The potential between a solid and a liquid is called the "phase boundary potential," whereas that between two liquids is called the "liquid junction potential." Usually, two solutions can rapidly mix, and such potential subsides. In a case that a solution is somewhat isolated from the other, such as a pipette solution, they are not easily mixed and the potential persists. On the boundary between two solutions, ions move from one side to the other according to their concentration gradients. Each ion has a distinct value of ionic "mobility" that is defined electrochemically; a product of ionic mobility and the concentration gradient determines to which direction each ion diffuses at the interface.

Let us do a thought experiment. Two solutions, a pipette solution containing 20 mM KCl and 130 mM potassium aspartate and a bath solution containing 150 mM NaCl are contacting each other at the tip of the pipette. Ionic mobilities of K^+, Cl^-, and Na^+ ions are roughly similar; but that of aspartate is considerably lower than the other ions because aspartate is a much larger ion. This means that aspartate goes out of the pipette less quickly than Cl^- comes in, whereas cations can be exchanged at a similar speed. As a result, at the interface, the intrapipette solution becomes positive, as though there is a battery with its plus side inside the pipette. More precisely, Kohlraush showed that the junction potential V can be expressed, with mobility u_i, and charge z_i of each ion i, as the following formula (14),

$$V = \left(\frac{RT}{F}\right)\left\{\frac{\sum_i [(z_i u_i)(a_i^S - a_i^P)]}{\sum_i [(z_i^2 u_i)(a_i^S - a_i^P)]}\right\} \ln\left\{\frac{\sum_i z_i^2 u_i a_i^P}{\sum_i z_i^2 u_i a_i^S}\right\} \quad (28.24)$$

where S and P denote the bath and pipette solutions, respectively. Note that V is expressed as the potential of the bath side in reference to the pipette side. This direction is compatible with the standard patch-clamp configuration, where the current or voltage from pipette to bath is always considered to be plus.

We must always pay attention to the junction potential in every patch-clamp experiment. When the patch pipette is immersed in the bath solution, a number of junction potentials and phase boundary potentials are developed in the circuit shown in Fig. 28.2. These potentials are as follows: E_{ip} is the junction (boundary) potential between pipette AgCl electrode and pipette solution; E_J is the junction potential between pipette solution and bath solution; E_{bath} is the junction (boundary) potential between bath solution and reference electrode; and E_{amp} is an artifact potential inside

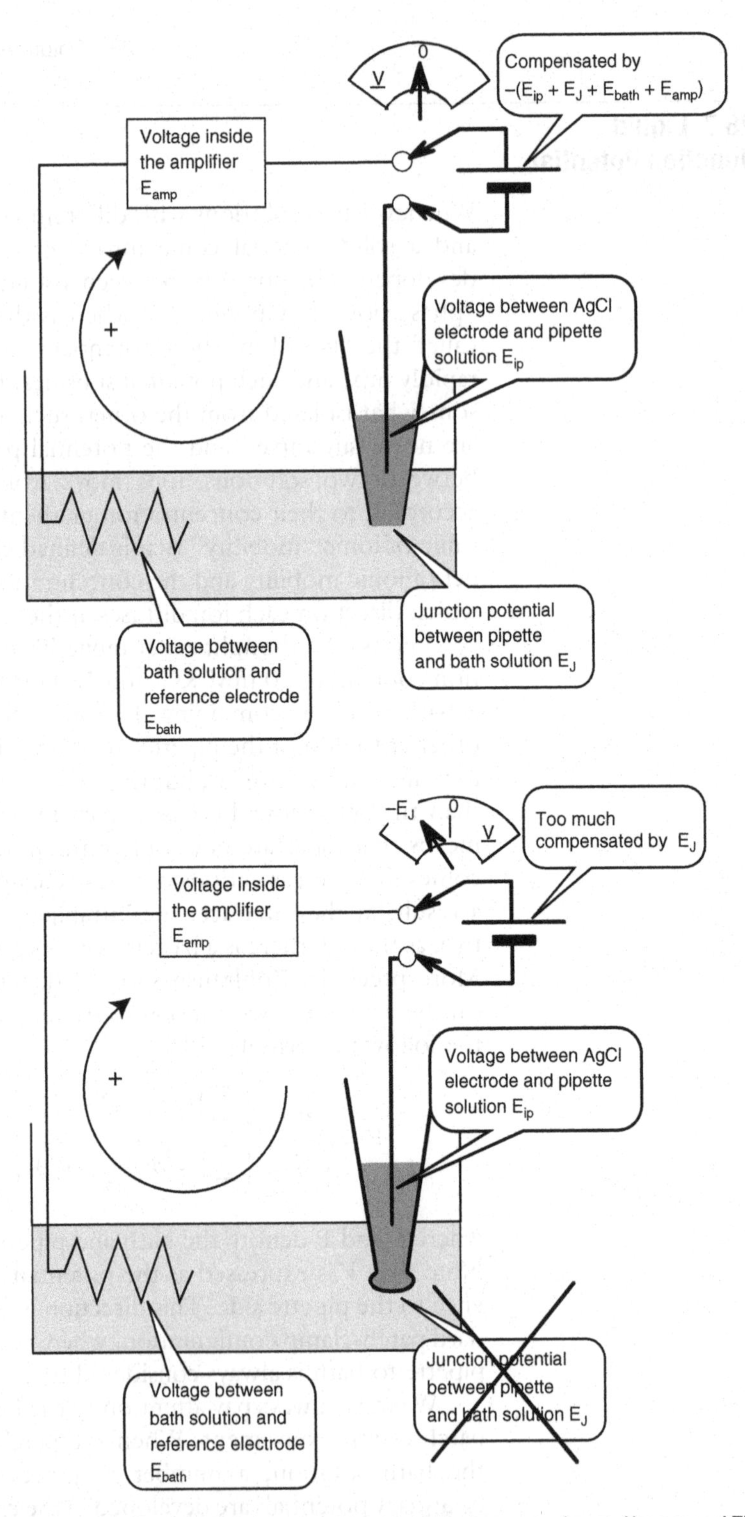

Fig. 28.2. Junction potentials in an outside-out patch-clamp experiment. *Upper panel*. The pipette is immersed in the bath solution, and the pipette offset has been adjusted to zero to cancel the junction potentials. The potential applied to cancel it was $-(E_{ip} + E_J + E_{bath} + E_{amp})$ (see the voltage meter, at the top). *Lower panel*. Outside-out configuration is formed. Because E_J has disappeared immediately after the patch formation, the circuit is over-compensated by the value of E_J.

the amplifier. As a result of a summation of these potentials, the amplifier meter reads certain "pipette offset" in the voltage-clamp mode (or small current in the current-clamp mode). Here, experimenters cancel the pipette offset by adjusting the potentiometer on the amplifier panel to zero. The offset voltage E_{comp} is given in the opposite direction to the pipette offset to correct and cancel it, so E_{comp} can be denoted as

$$E_{ip} + E_J + E_{bath} + E_{amp} - E_{comp} = 0 \quad (28.25)$$

When a patch is made in the outside-out mode, E_{ip}, E_{bath}, and E_{amp} do not change because the correction of the pipette offset. However, after the outside-out patch is formed, the pipette solution and the bath solution no longer are in contact, and thus E_J (the junction potential between the pipette and bath solution) is excluded from the circuit. Therefore, the voltage in the circuit becomes:

$$E_{ip} + E_{bath} + E_{amp} - E_{comp} \quad (28.26)$$

which is equal to $-E_J$ with (28.25). When we give a command voltage Vp, the real voltage applied between both sides of patch membrane, Vm, is:

$$Vm = Vp - E_J \quad (28.27)$$

Amplifiers cannot measure E_J at this step, and Vp can be recorded instead. When the current–voltage (I–V) curve is plotted, Vm (which is smaller than Vp by E_J) must be calculated and used for the values of the membrane (command) voltage.

In the inside-out mode, the direction of the command voltage is reverse, and

$$-Vm = Vp - E_J \quad (28.28)$$

For example, suppose that E_J is +5 mV, and a command voltage (Vp) of −50 mV is given, then +55 mV depolarizing voltage can be expected to be applied to the patch. Note that in some computer-controlled patch-clamp software the direction of the command voltage is automatically inverted when you select the "inside-out mode."

In the whole-cell mode, the same correction as in the outside-out mode should be applied. However, until the pipette solution perfuses the cell interior and replaces the cytoplasm completely (which may take up to 10 min or even longer depending on the cell size and shape), a junction potential should arise between the pipette solution and the intracellular fluid. This potential cannot be estimated precisely, only empirically.

In practice, junction potentials between the pipette and bath solutions can be calculated by (28.24) or directly measured using KCl salt bridges. A program to calculate liquid junction potentials created by Barry and Lynch (14) and modified by the manufacturer is incorporated in Clampex software (Molecular Devices, San Jose, CA, USA).

In most experiments, the liquid junction potential ranges around 10–20 mV. Errors less than 10 mV might be allowed in some studies. However, even a small error is not always negligible. Suppose that reversal potentials are measured to estimate the ion permeability of an anion channel. When an external solution is the same as the internal pipette solution, the reversal potential is zero. Then, suppose the bath is perfused with a solution that replaces 90% of the Cl^- with glutamate. If the measured reversal potential is 47 mV, the permeation ratio P_{Cl}/P_{Glu} is calculated to be 16. However, if the liquid junction potential is compensated, the reversal potential becomes 55 mV, and the ratio is estimated as 70 (15).

When bath solution is changed, not only E_J, but also E_{bath}, should change. Especially, when an AgCl pellet reference electrode is used, the junction (phase boundary) potential is strongly affected by the Cl^- concentration of the bath solution. To avoid this, the use of a salt bridge (usually with 3 M KCl) should be considered.

28.8. Conclusion

Correct solution composition is critical for successful patch-clamp experiments. It is important to consider and, when necessary, to verify the influence of every factor mentioned in this chapter on ion channel function as it may have deep physiological or pathophysiological meaning. Neglecting factors such as the osmotic pressure or junction potentials may result in erroneous results.

References

1. Prosser CL (1973) Inorganic ions. In: Ladd Prosser C (ed) Comparative animal physiology. Saunders, Philadelphia, pp 79–110
2. Sabirov RZ, Okada Y, Oiki S (1997) Two-sided action of protons on an inward rectifier K^+ channel (IRK1). Pflugers Arch 433:428–434
3. Sabirov RZ, Prenen J, Droogmans G, Nilius B (2000) Extra- and intracellular proton-binding sites of volume-regulated anion channels. J Membr Biol 177:13–221
4. Tsai TD, Shuck ME, Thompson DP, Bienkowski MJ, Lee KS (1995) Intracellular H^+ inhibits a cloned rat kidney outer medulla K^+ channel expressed in *Xenopus oocytes.* Am J Physiol 268:C1173–C1178
5. Robinson RA, Stokes RH (1959) Electrolyte solutions. Butterworths, London
6. Miller DJ, Smith GL (1984) EGTA purity and the buffering of calcium ions in physiological solutions. Am J Physiol 246:C160–C166
7. Oiki S, Yamamoto T, Okada Y (1994) Apparent stability constants and purity of Ca-chelating agents evaluated using Ca-selective electrodes by the double-log optimization method. Cell Calcium 15:209–216
8. Oiki S, Yamamoto T, Okada Y (1994) A simultaneous evaluation method of purity and apparent stability constant of Ca-chelating agents and selectivity coefficient of Ca- selective electrodes. Cell Calcium 15:199–208

9. Harrison SM, Bers DM (1987) The effect of temperature and ionic strength on the apparent Ca- affinity of EGTA and the analogous Ca-chelators BAPTA and dibromo-BAPTA. Biochim Biophys Acta 925:133–143
10. Harrison SM, Bers DM (1989) Correction of proton and Ca association constants of EGTA for temperature and ionic strength. Am J Physiol 256:C1250–C1256
11. Smith GL, Miller DJ (1985) Potentiometric measurements of stoichiometric and apparent affinity constants of EGTA for protons and divalent ions including calcium. Biochim Biophys Acta 839:287–299
12. Fabiato A (1988) Computer programs for calculating total from specified free or free from specified total ionic concentrations in aqueous solutions containing multiple metals and ligands. Methods Enzymol 157: 378–417
13. Neher E (1992) Correction for liquid junction potentials in patch clamp experiments. Methods Enzymol 207:123–131
14. D'Ans J, Surawsky H, Synowietz C (1977) In: Schafer Kl (ed) Densities of liquid systems and their heat capacities, Landolt-Bornstein numerical data and functional relationships in science and technology. New Series. Springer, Berlin
15. Barry PH, Lynch JW (1991) Liquid junction potentials and small cell effects in patch-clamp analysis. J Membr Biol 121:101–117

INDEX

Yasunobu Okada (ed.), *Patch Clamp Techniques: From Beginning to Advanced Protocols*, Springer Protocols Handbooks, DOI 10.1007/978-4-431-53993-3,

D

E

Q

R

S

W

X

Z

Printed in the United States
By Bookmasters